MOUNTAIN GEOGRAPHY

Mountain Geography

PHYSICAL AND HUMAN DIMENSIONS

Edited by

MARTIN F. PRICE

ALTON C. BYERS

DONALD A. FRIEND

THOMAS KOHLER

LARRY W. PRICE

UNIVERSITY OF CALIFORNIA PRESS
Berkeley   Los Angeles   London

University of California Press, one of the most distinguished university presses in the United States, enriches lives around the world by advancing scholarship in the humanities, social sciences, and natural sciences. Its activities are supported by the UC Press Foundation and by philanthropic contributions from individuals and institutions. For more information, visit www.ucpress.edu.

University of California Press
Berkeley and Los Angeles, California

University of California Press, Ltd.
London, England

Library of Congress Cataloging-in-Publication Data

Mountain Geography
    Mountain geography : physical and human dimensions / edited by Martin F. Price, Alton C. Byers, Donald A. Friend, Thomas Kohler, Larry W. Price.
        pages   cm
    Revision of: Mountains and man
    Includes bibliographical references and index.
    ISBN 978-0-520-25431-2 (cloth : alk. paper)
    1. Mountains   2. Mountain people.   3. Human geography.
4. Geomorphology.   I. Price, Martin F.   II. Title.
GB501.2.M684   2013
910'.02143—dc23   2013002583

22 21 20 19 18 17 16 15 14 13
10 9 8 7 6 5 4 3 2 1

The paper used in this publication meets the minimum requirements of ANSI/NISO Z39.48-1992 (R 2002) (*Permanence of Paper*). ⊗

CONTENTS

CONTRIBUTORS

ANDREW BACH is Associate Professor of Environmental Geography in Huxley College of the Environment, Western Washington University. His major research areas include geomorphology and soils, natural resources management, and climate change in the western United States. His research focuses on soil development, paleoecology, vegetation dynamics and fire history, the potential impacts of the removal of dams in the Olympic Peninsula of Washington, and glaciers as a water resource in the North Cascades.

EDWIN BERNBAUM is an author, mountaineer, and scholar of comparative religion and mythology who focuses on the relationship between culture and the environment. His book *Sacred Mountains of the World* won the Commonwealth Club of California's gold medal for best nonfiction work and an Italian award for literature of mountaineering, exploration, and the environment. As Director of the Sacred Mountains Program at The Mountain Institute, where he is a Senior Fellow, he initiated and implemented projects to develop interpretive materials with U.S. National Parks such as Mount Rainier, the Great Smoky Mountains, and Hawai'i volcanoes, based on the cultural and spiritual significance of different features of mountain environments in America and other cultures around the world.

KARL BIRKELAND is the Director and Avalanche Scientist for the U.S. Forest Service National Avalanche Center. In addition to conducting extensive research on avalanches, he works to transfer new and emerging technologies to field practitioners within the avalanche community. He is also Adjunct Professor of Earth Sciences at Montana State University, where he supervises a number of graduate students. His professional work with avalanches as a ski patroller, educator, backcountry forecaster, and researcher spans over 30 years. He founded the Gallatin National Forest Avalanche Center in Bozeman, Montana.

ALTON C. BYERS, Director of Science and Exploration at The Mountain Institute (TMI), is a mountain geographer and climber specializing in applied research, high-altitude conservation and restoration programs, climate change impacts in mountains, and highland–lowland interactive system approaches to conservation. He has worked with TMI since 1990 in Nepal, Peru, and the Appalachians. He is a National Geographic Society Explorer and grantee and recipient of the following: The Nature Conservancy's Annual Award for Outstanding Ecological Stewardship; the Association of American Geographers' Distinguished Career Award; the American Alpine Club's David Brower Conservation Award; and the Sir Edmund Hillary Mountain Legacy Medal for "remarkable service in the conservation of culture and nature in mountainous regions."

STEPHEN F. CUNHA is Chair and Professor of Geography at Humboldt State University. He writes on mountain issues in Central Asia, Alaska, and the Sierra Nevada. His teaching and research focus on environmental geography and mountain environments, particularly in Central Asia, Alaska, and California's Sierra Nevada. In 2007, he was named the California State University System's Outstanding Professor in the Social and Behavioral Sciences and Public Service.

LELAND R. DEXTER is an Emeritus Professor of Geography at Northern Arizona University. During his career there, he taught classes in mountain geography, snow and ice, physical geography, geomorphology, climatology, and geographic information systems. He conducted many winter mountain field camps in Colorado's San Juan Mountains with Melvin Marcus and Donald Friend. His research interests include snow and ice processes in high mountain environments, sandbar morphology and evolution in the Grand Canyon, microclimate energy balance studies, and GIS analysis of various environmental issues.

DONALD A. FRIEND is Professor and Chair of the Department of Geography at Minnesota State University. He is the U.S. Representative to the International Geographical Union Commission on Mountain Response to Global Change, and is past Chair and Founder of the Mountain Geography Specialty Group of the Association of American Geographers. From 2007 to 2010 he served as Associate Editor in Chief of the *Journal of Mountain Science,* published by the Chinese Academy of Sciences, and he sits on the advisory board of the Mountain Studies Institute. His research and teaching interests focus on physical geography, especially earth surface and atmospheric processes, their interaction, and human impacts in mountains.

JAMES S. GARDNER is Professor Emeritus, Natural Resources Institute, University of Manitoba, and Adjunct Professor, Department of Geography, University of Victoria, Canada. Formerly Provost and Professor at the University of Manitoba (1991–2001) and Professor and Dean of Graduate Studies at the University of Waterloo (1987–1991), he has pursued research and teaching in geomorphology, hydrology, glaciology, and resources and hazards management, with field studies in mountain environments in Canada, Europe, India, Pakistan, and China. He has published widely on alpine geomorphology and resources and hazards. Now retired, he continues to write, teach occasionally, and assist in supervision of graduate students at the Universities of Manitoba and Victoria.

VALERIUS GEIST is Professor Emeritus of Environmental Science in the Faculty of Environmental Design, the University of Calgary, Canada. He was a founding member and the first program director for Environmental Science in that graduate faculty. His interest in interdisciplinary scholarship resulted in 17 technical or popular books, some award-winning, the most important being *Life Strategies, Human Evolution, Environmental Design: Towards a Biological Theory of Health* (1978) and *Deer of the World* (1998). He has been very active in wildlife conservation.

GEORG GRABHERR is Professor Emeritus at the University of Vienna, Austria, having retired from his post as full Professor of Conservation Biology, Vegetation, and Landscape Ecology and head of the department in 2011. His research has related particularly to biodiversity and plant communities, effects of land use change on vegetation, and the use of models of vegetation to predict global change effects. He is founder and chair of the Global Observation Research Initiative in Alpine Environments (GLORIA). He was awarded the Austrian Conservation Award in 2011 and recognized as Austrian Scientist of the Year in 2012.

KEITH S. HADLEY is Associate Professor (retired) at the Department of Geography, Portland State University. A biogeographer with professional interests in alpine environments, vegetation dynamics, climate change, and dendroecology, his teaching and research interests focus on the spatial and temporal patterns of environmental change.

CAROL HARDEN is Professor of Geography at the University of Tennessee, Knoxville. She studies mountain watersheds to better understand the interrelationships between human activities and geomorphic processes. Her work in the Andes and Appalachians has focused on upland soil erosion and the delivery of sediment to streams; processes and rates of streambank erosion; effects of different

land uses on rainfall infiltration, runoff, and soil water-retention capacities; and trade-offs between management strategies for different environmental services. She is Editor-in-Chief of *Physical Geography,* and served as president of the Association of American Geographers, 2009–2010.

JASON R. JANKE is an Associate Professor of Environmental Science and lecturer with the Department of Earth and Atmospheric Sciences at the Metropolitan State University of Denver. His research interests include geographic information systems and geomorphology, in particular rock glaciers and permafrost.

THOMAS KOHLER is Associate Director in the Centre for Development and Environment and lecturer at the Department of Geography at the University of Bern, Switzerland. His field experience includes the mountains of Eastern Africa, Southeast Asia, the Caucasus, and Switzerland. He has a longstanding interest in mountain research and development, including policy advice, advocacy, and outreach, and has coedited and contributed to many publications for promoting mountains on the international development agenda. He is Managing Director of the International Mountain Society, the publisher of the international peer-reviewed journal *Mountain Research and Development.*

LARRY W. PRICE is Professor Emeritus in the Department of Geography at Portland State University, where he supervised large numbers of graduate students, particularly on mountain themes. In 1971, he instituted an internationally recognized summer mountain field camp. His best-known book is *Mountains and Man* (1981). He was president of the Association of Pacific Coast Geographers, 1984–1985.

MARTIN F. PRICE is Professor of Mountain Studies and Director of the Centre for Mountain Studies, Perth College, University of the Highlands and Islands, UK, where he holds the UNESCO Chair in Sustainable Mountain Development. He has been active in mountain research and policy development since the mid-1980s, including contributions to the development and implementation of the mountain chapter of Agenda 21 at the Rio Earth Summit, 1992, and the International Year of Mountains, 2002. He has coordinated three major assessments of Europe's mountains and has organized major conferences on global change and mountain regions. In 2012, his contribution to mountain science was recognized through the King Albert Mountain Award by the King Albert I Memorial Foundation.

ROBERT E. RHOADES was Professor of Anthropology at the University of Georgia, which recognized him as a Distinguished Research Professor in 2006. His research, particularly in the Ecuadorian Andes, challenged widespread assumptions that third-world farming systems are inefficient and demonstrated the need to incorporate traditional knowledge into policy. He passed away in March 2010.

JOHN F. SHRODER JR. is Emeritus Professor of Geography and Geology at the University of Nebraska at Omaha and still actively pursues research on landslides and glaciers in high mountain environments. He has written or edited some 30 books and more than 150 professional papers. He is a Fellow of the Geological Society of America and the American Association for the Advancement of Science and has received Distinguished Career awards from both the Mountain and Geomorphology Specialty Groups of the Association of American Geographers.

CHRISTOPH STADEL is Professor Emeritus at the University of Salzburg, Austria, and Adjunct Professor in the Department of Natural Resources at the University of Manitoba, Canada. His research and teaching have focused on comparative mountain geography, the Canadian prairies, and Andean and Alpine environments. He has served on various governmental and nongovernmental agencies relating to international development issues and has published extensively in English, German, Spanish, and French.

FOREWORD

JACK D. IVES

This major revision of Larry Price's book *Mountains and Man* (1981) is both timely and highly appropriate. The intervening three decades encompass a time of remarkable progress in our understanding of mountains from a purely academic point of view. Of even greater importance is that society at large is coming to realize that mountains and mountain people are no longer isolated from the mainstream of world affairs but are vital if we are to achieve an environmentally sustainable future.

The Foreword that I prepared for the 1981 book can yield comparisons between that book and this one and also allows us to emphasize some of the developments that have occurred within mountain geography and in the awareness of *mountains* during the last three decades. I wrote in the opening lines of the original: "This is a most important and timely book. It is destined to become a standard text." My undergraduate mountain geography classes alone sold well over a thousand copies, and I am sure this could be echoed by many faculty colleagues.

Nevertheless, my rereading of that old Foreword, after a lapse of more than a quarter century, prompted a sharp reawakening. The very title that Larry employed (*Mountains and Man*) would be unthinkable today. That may be a superficial assessment, although there are clear underlying implications. Women do account for somewhat more than half of humanity, and a comparable balance is being achieved in their contribution to mountain research, development, and overall adherence to the mountain cause. And we have learned of the overwhelming contribution of mountain women in the struggle for survival in these sometimes severe environments.

There is one particular lesson to be learned. In the late 1970s and early 1980s, world society was being made aware of a portending mountain catastrophe. For example, the early awakening to the relevance of mountains and mountain people by the world at large was prompted by a news media blitz that the mismanagement of their homeland by the mountain people themselves was pushing the Himalayas to the edge of environmental destruction, not only of the mountains but of the life-support systems of the hundreds of millions living on the plains below. The World Bank had predicted in 1979 that Nepal would be denuded of its forest cover by the year 2000. The United Nations Environment Programme speculated that Himalayan deforestation by 1980 was already compounding seasonal flooding in Bangladesh. The World Resources Institute claimed that "a few million subsistent hill farmers are undermining the life support of several hundred million people in the plains." And the mountain people were the culprits.

The preceding paragraph reflects the remarkable change that has occurred in our attitudes to, and understanding of, mountains and mountain people. The widespread change in understanding exemplified by this single example also provides part of the justification for production of this book.

During the late 1960s and early 1970s, under the leadership of the late Professor Carl Troll, the International Geographical Union's (IGU) Commission on High-Altitude Geoecology began to flourish. Larry Price was himself an active member of the original small group of geographers committed to high-mountain research. The commission changed its name to *Mountain Geoecology* so that it included study of the more inhabitable mountain environments and the people who lived there. It also extended its endeavors well beyond its initial academic arena, however, through the influence of UNESCO's Man and the Biosphere (MAB) Programme—Project 6: *Study of the Impact of Human Activities on Mountain Ecosystems* (1973). The move toward applied mountain research was subsequently accelerated by another leading German geographer, Professor Walther Manshard, as Vice-Rector of the then newly created United

Nations University (UNU), headquartered in Tokyo. There followed the founding of the International Mountain Society and its quarterly journal *Mountain Research and Development*. At that time (1981, coincidental with the publication of Larry's book), a small group of us realized that our efforts should be extended into the development field:

> *To strive for a better balance between mountain environment, development of resources, and the well-being of mountain peoples.*

The quarterly journal, together with UNU/IGU research in the Himalayas and elsewhere, became a prominent force that led to the Mohonk Mountain Conference of 1985. Maurice Strong was invited to serve as honorary chair of the conference, thereby establishing a valuable connection with the future Secretary General of the United Nations Conference on Environment and Development (UNCED 1992). In short, a group of mountain geographers and closely associated anthropologists and foresters set out to examine the validity of the catastrophe paradigm typified by the World Bank doomsday prediction. The Himalayan catastrophe theory was eventually overthrown, and mountain people were identified as part of the solution, decidedly not as the cause of environmental disaster.

As the process of academic and pragmatic mountain research gathered momentum, so institutional allegiances multiplied: IUCN, FAO, UNESCO, UNU, IGU, the East-West Center of the University of Hawaii; and nine different universities in eight countries. The growing debate, for instance, led to the UNU's mountain program being renamed as *Mountain Geoecology and Sustainable Development,* in response to Maurice Strong's request for assistance with the mountain cause in preparation for the United Nations Conference on Environment and Development (Rio 1992). Chapter 13, for mountains, was duly inserted into AGENDA 21. This, in conjunction with a rapidly enlarging group of institutions and individuals, culminated in the UN declaration of A.D. 2002 as *The International Year of Mountains* (IYM).

An important associated commitment throughout this period was the training of young scholars from developing mountain countries and their incorporation into the field research activities. The UNU/IGU collaboration, extensively assisted by the Swiss Development Cooperation (SDC), included research and training in western China, the Andes, Tajikistan, Thailand, Bangladesh, Ethiopia, East Africa, and Madagascar. The Mountain Institute was reborn of the Woodlands Mountain Institute in West Virginia and refocused. In Kathmandu, ICIMOD was revitalized. The Mountain Forum was established (1995), leading to worldwide electronic communication amongst all manner of mountain peoples. Following Rio (1992), FAO took on the vital role of nurturing mountain sustainable development activities. Further achievements since IYM 2002 are too many to be incorporated into this Foreword and would go far beyond my remit.

Coincident with these activities, the powerful idea that *mountains are the water towers of the world* became a virtual household saying. It was demonstrated that by *listening to the mountains,* major advances are possible (Rhoades 2007). FAO and the Centre for International Forestry Research (CIFOR) (2005) became the only major component of the UN system (excepting our collaborating institutions, UNU and UNESCO) to support the probability that the earlier paradigm of convenience—claiming that poor mountain peoples, through unthinking forest clearance, were causing downstream destruction—was nonsense. These are just a few indications of changes in attitudes and understanding in relation to mountains since *Mountains and Man* was published in 1981.

A recent theme which currently appears to eclipse all others, however, is global warming. It has been recognized during the last 10 years or so that high-altitude and high-latitude environments are experiencing the most pronounced effects of climate change. This reemphasizes the need for accelerated mountain research to ensure an adequate scientific basis for any attempted prognostication, particularly in the highest and more inaccessible regions where time-series data are almost entirely lacking. However, a new claim has arisen, with publicity equal to that which produced the original outcry of Himalayan catastrophe. Gross exaggeration, whether motivated by a felt need to obtain public acclaim or to increase research and consultancy funds, has pervaded much reporting by the news media (and, I am ashamed to say, academic and other research and commentary). This is not to say that future potential impacts of global warming are not serious; they are. And it has been clearly demonstrated that its impacts are especially noticeable in mountain regions. It is highly unlikely, however (to quote one current example), that glacial lake outburst floods will wash away major cities on the Ganges, or that the Himalayas will lose all their snow and glaciers within the next 20 or so years, so that the Ganges and similar great rivers will become seasonal streams with starvation and loss of hundreds of millions of lives. The dangers of this melodrama, aside from the moral implications, should not need emphasis; unfortunately, such is not the case.

By presenting such a timely and far-reaching reassessment of the current mountain *problematique,* the University of California Press has provided a sound base for ensuring that the next generation of students can better understand the processes occurring in the mountains of the world—processes both human and natural, if it is any longer possible to differentiate—and their impact on society as a whole. Of equal importance is the provision of so much new scientific information that our future students are better able to assess for themselves the exaggerations of a disaster-hungry world. I am sure that the coauthors of this book (including Larry himself, of course) have utilized the substantial original contribution to further strengthen the basis for mountain teaching and to enlarge the prospects for attainment of sustainable mountain development.

Jack D. Ives
Professor Emeritus, University of California, Davis
Honorary Research Professor, Department of Geography and Environmental Studies, Carleton University, Ottawa, Canada
412 Thessaly Circle, Ottawa, ON, K1H 5W5, Canada
e-mail:jack.ives@carleton.ca

References

FAO/CIFOR. 2005. *Forests and Floods: Drowning in Fiction or Thriving on Facts?* Forest Perspectives 2, Bangkok, Thailand, and Bogor Barat, Indonesia: FAO Regional Office for Asia and the Pacific, and Centre for International Forestry (CIFOR).

Rhoades, R. E. 2007. *Listening to the Mountains.* Dubuque, IA: Kendall/Hunt Publishing Co.

PREFACE

ALTON C. BYERS

As I sit in my study here in the quiet hills of West Virginia, my eyes wander to the bookshelf and a light-blue volume with the faded words "Mountains and Man" printed on its cracked and weathered spine. The glossy book jacket, showing two men silhouetted against a Hindu Kush–Himalayan skyline, has long been lost. Opening the book and flipping through its contents, I find that nearly every other sentence is underlined, every other page dog-eared, and the margins are blackened with notes and observations in my microscopic scrawl. Two muddy dog tracks grace page 1 from Chapter 1, "What Is a Mountain?"—a memento from fieldwork conducted in the Annapurna region of Nepal in the early 1980s. The book has traveled with me throughout the Hindu Kush–Himalayas; Yulongxue Shan Mountains of China; East African Highlands and Virunga Volcanoes; Drakensberg Mountains of South Africa; South American Andes; North American Rockies, Cascades, Adirondacks, and Appalachians; Russian and Mongolian Altai Mountains; and European, Australian, and New Zealand Alps. I have read and reread each chapter on buses, jeeps, airplanes, horseback, and in remote basecamps. It has helped me to plan my high mountain research, conservation, and climate change work throughout the mountain world; to design student and senior scientist field expeditions; and to develop a range of mountain education curricula, teacher training, and field methods courses.

The text I refer to is, of course, Dr. Larry Price's seminal book *Mountains and Man,* published in 1981, the first book devoted exclusively to undergraduate mountain education since Roderick Peattie's classic work from 1936, *Mountain Geography.* Several years ago I was discussing the status of mountain geography education and available textbooks with Don Friend during a teacher training workshop at The Mountain Institute's Spruce Knob Mountain Center in West Virginia. We realized that more than 20 years had passed since Dr. Price's innovative work, and that much had happened in the interim.

For example, dozens of new mountain organizations have been created, and new mountain-specific forums, publications, conferences, websites, nonprofits, partnerships, and development projects have been established. But while *Mountains and Man* still contained a wealth of solid information that had stood the test of time, no current undergraduate textbook on mountain geography was available. The solution? With his permission, we would revise and update Dr. Price's book by recruiting first-rate mountain geographers from around the world to revise each chapter, to be edited by a second team of mountain academics and specialists, including Martin Price, myself, Don Friend, and Thomas Kohler. Second, we would add new chapters on topics not addressed in the original edition, such as one devoted exclusively to mountain people and another on sustainable mountain development. Dr. Price was contacted and graciously agreed to the proposal. A contract with the prestigious University of California Press in Berkeley was signed, and the result is the product before you: *Mountain Geography: Physical and Human Dimensions.*

Alton C. Byers, Ph.D.
Director of Science and Exploration
The Mountain Institute
Elkins, West Virginia
March 2013

ACKNOWLEDGMENTS

We would like to extend our sincere thanks to Jack Dangermond, president of Environmental Systems Research Institute, Inc., and Ellen Stein of the Mountain Studies Institute for their generous contributions in support of this book during the initial project phases. Thomas Kohler, of the Centre for Development and Environment of the University of Bern, Switzerland, is also thanked for providing financial resources in support of the final revisions and manuscript preparation.

We gratefully acknowledge support of the Geography Departments at Humboldt State University (HSU) and Minnesota State University (MSU). All of the figures had to be redrafted electronically because they were hand-drawn for Larry Price's original text in 1980. This feat of re-creating some 200 figures was undertaken as part of two advanced cartography courses, one at HSU under the direction of MaryBeth Cunha and Dennis Fitzsimons, and the other at MSU under the direction of Kimberly Musser. Our appreciation goes to Mrs. Cunha and Ms. Musser for their devotion to the project and for their excellent supervision of student cartographers. We also wish to acknowledge the work of the student cartographers, most of whom have graduated and are now professionals themselves. MSU student cartographers include Derek Brown, Nicole Eppolito, Joe Holubar, Karly Klein, Mary Morgan, Chad Otto, and Tonya Rogers. HSU student cartographers include Andrew Allen, Christy Beard, Anita Bowen, Sean Canton, Miles Eggleston, Jo Erickson, Carla Esparza, Jeffrey Foster, David Gagner, Melinda Gentry, Breeanna Graydon, Kelley Hale, Michelle Hanna, Cassandra Hansen, Elizabeth Hausman, Jeffrey Johnson, Melissa Katz, Nathaniel Kelso, Samuel Levy, Brian Ludy, Josh Lyons-Tinsley, and Ross Nolan. In the final stages, Jana Cekalova provided further support to ensure that all figures met the necessary standards.

We thank the editors and authors of the individual chapters for agreeing to take on what at times seemed like a never-ending, but always important, contribution to the field of mountain geography. They include Andrew J. Bach, Edwin Bernbaum, Karl W. Birkeland, Stephen F. Cunha, Leland F. Dexter, Jim Gardner, Valerius Geist, Georg Grabherr, Keith S. Hadley, Carol P. Harden, Jason Janke, the late Robert Rhoades, John F. Shroder Jr., and Christoph Stadel. We would also like to thank Adean Lutton for her invaluable support in the very detailed process of preparing the text and figures of the final manuscript, as well as for securing permissions. Finally, we would like to thank the staff at University of California Press, especially Blake Edgar, Rich Nybakken, Franciso Reinking and Lynn Meinhardt, and Deepti Agarwal and her team at MPS Limited, for bringing this book to publication.

CHAPTER ONE

An Introduction to Mountains

ALTON C. BYERS, LARRY W. PRICE,
and MARTIN F. PRICE

Most people are familiar with the importance of oceans and rainforests (Byers et al. 1999), thanks in part to the dozens of books, documentaries, programs, and Internet sites developed by education and conservation groups over the past two decades. Yet there is at least as strong a case for arguing that mountains are also of critical importance to people in nearly every country of the world (Messerli and Ives 1997; Debarbieux and Price 2008).

For example, all of the world's major rivers have their headwaters in mountains, and more than half of humanity relies on the fresh water that accumulates in mountains for drinking, domestic use, irrigation, hydropower, industry, and transportation (Viviroli et al. 2007; Bandyopadhyay et al. 1997; this volume, Chapter 12). *Hydropower* from mountain watersheds provides 19 percent of the world's total electricity supply, roughly equivalent to all the electricity generated by alternative methods such as solar, wind, geothermal, and biomass (Schweizer and Preiser 1997; Mountain Agenda 2001). Mountain *forests* provide millions of people with both timber and non-timber forest products (e.g., mushrooms, medicinal plants) and play vital roles in downstream protection by capturing and storing rainfall and moisture, maintaining water quality, regulating river flow, and reducing erosion and downstream sedimentation (Price and Butt 2000; Price et al. 2011). Because the same geologic forces that have raised mountains have also helped concentrate assemblages of *minerals* useful to human society, the mines in today's mountains are the major source of the world's strategic nonferrous and precious metals (Fox 1997).

Many mountains are hotspots of *biodiversity* (Jeník 1997; Körner and Spehn 2002; Spehn et al. 2006; this volume, Chapters 7 and 8). With increasing altitude, changes in temperature, moisture, and soils can create a dense juxtaposition of differing ecological communities, sometimes ranging from dense tropical jungles to glacial ice within a few kilometers: This phenomenon is well illustrated by the six bioclimatic zones of the Makalu region of eastern Nepal that are found between 100 and 8,000 m over a mere 20 horizontal kilometers: Over 3,000 plant species are found within this range, including 25 species of rhododendrons, 50 of primroses, 45 of orchids, and 80 of fodder trees and shrubs (Shrestha 1989). Not only does such biodiversity have intrinsic value; it can also have great economic and health values. For instance, of the 962 species of medicinal plants that occur in the temperate to alpine zones of the Indian Himalaya, 175 are being used by herbal drug companies (Purohit 2002). Many mountains (e.g., Mount Kenya and Kilimanjaro in East Africa; Hedberg 1997) can be thought of as islands of biodiversity that rise above vast plains of human-transformed landscapes below. Mountains are often *sanctuaries* for plants and animals long since eliminated from these more transformed lowlands, such as the volcanoes of Rwanda and Uganda, where the last of the world's mountain gorillas—now numbering fewer than 300—survive (Weber and Vedder 2001). Many plant and animal species are *endemic* to mountain regions, having evolved over millennia of isolation to inhabit their specialized environments. Equally, many mountain ranges also function as *biological corridors,* connecting isolated habitats or protected areas and allowing species to migrate between them (Worboys et al. 2010).

Many of the most important *food staples* in the world— including potatoes, wheat, corn, and beans—were domesticated in mountains, and mountain peoples long ago developed elaborate agricultural production systems and strategies based on altitudinal and ecological zonation (Grötzbach and Stadel 1997; this volume, Chapter 11). Many other crops that have been cultivated for centuries in the

Andes have the potential to supply the increasing need for food as the world's population continues to grow (National Research Council 1989). Mountain people, particularly women, are exceptionally knowledgeable about, and make use of, the many *medicinal* and food plants found in mountain fields and forests (Daniggelis 1997). Of the hundreds of plants in the mountains of Nepal used for medicinal purposes, more than a hundred are undergoing commercial exploitation that can generate significant income for local people (Karki and Williams 1999; Guangwei 2002).

Biological and cultural *diversity* are often closely interrelated, and mountains contain an amazing diversity of human cultures and communities. For example, of the 1,054 languages spoken in New Guinea, 738 originate in mountainous regions, which cover only 33 percent of the island (Stepp et al. 2005). The late Anil Agarwal, founder and director of the Centre for Science and Environment in New Delhi, stated that "cultural diversity is not an historical accident. It is the direct outcome of the local people learning to live in harmony with the mountains' extraordinary biological diversity" (Centre for Science and Environment 1991, cited in Denniston 1995: 18). Mountains are also home to many *indigenous peoples,* the original inhabitants of a place before people of a different ethnic origin arrived—such as the Quechua people of Bolivia, Ecuador, and Peru; Naxi and Yi people of Yunnan Province, China; Batwa pygmies of the Ruhengeri Prefecture, Rwanda; and Rais and Sherpas of the eastern Himalaya and Mount Everest region.

The physical and cultural diversity found in many mountain countries is one of the major draws for world *tourism.* Tourism is the world's largest and fastest growing industry, and tourism to mountain areas represents a significant portion of this activity (Price et al. 1997; Godde et al. 2000; this volume, Chapter 12). Visitors go to the mountains for adventure, recreation, scenic beauty, solitude, and the opportunity to meet and interact with the people who live there. This large influx of visitors to mountain regions can have positive economic benefits for a community, helping to promote sustainable development and the capacity to balance human needs with the preservation of the environment. However, there is also the potential for negative environmental and cultural consequences, such as the impacts of large numbers of people and pack animals on fragile high-altitude environments (Byers 2005, 2007, 2008, 2009) and the loss of traditional cultural values (von Fürer-Haimendorf 1984; Mountain Forum 1998; Ortner 1999).

In many cultures, mountains have special *spiritual, cultural,* and *sacred significance.* Inspirational to most, mountains are held sacred by more than 1 billion people worldwide (Bernbaum 1997; Mathieu 2011; this volume, Chapter 9). As the highest and most impressive features of the landscape, mountains tend to reflect the highest and most central values and beliefs of cultures throughout the world. In the United States, mountain environments such as those found in the Rocky Mountains of the West or the Appalachians of the East enshrine cultural and spiritual values basic to American society, embodying what is interpreted as the original, unsullied spirit of the nation; others are sacred to native American peoples. The Japanese reverence for beauty in nature, an integral part of religious observance, bestows upon Mount Fuji a symbolic meaning for the entire nation. At 6,705 m (22,000 ft), Mount Kailash in Tibet is sacred to over a billion Hindus, Buddhists, Jains, and followers of the Bon religion.

Defining Mountains

Everyone can agree that every mountain has a summit. But how high should a feature be to be considered a "mountain," and how much of the Earth's surface do mountain areas cover? Such questions have long been discussed by geographers, explorers, mountain people, and mountaineers (Mathieu 2011).

During the 1930s, it became fashionable among members of various U.S. alpine clubs to climb the highest point in each of the continental 48 states. The highest of all was Mount Whitney in California at 4,418 m (14,496 ft); the lowest, Iron Mountain in Florida at 100 m (330 ft) (Sayward 1934). No one would doubt that Whitney is truly a mountain, but there is considerable question about Iron Mountain. *Merriam-Webster's* (Merriam-Webster 2013) defines a mountain as "a landmass which projects conspicuously above its surroundings and is higher than a hill." By this definition, Iron Mountain may be properly named, but most of us would judge this an exaggeration and regard it as a hill. At the opposite extreme, there is the story of a British climber in the Himalayas who asked his Sherpa guide the names of several of the surrounding 3,500 m (11,500 ft) peaks. The guide shrugged his shoulders, saying that they were just foothills with no name.

The difference between the two extremes is one of conspicuousness. The lesser peaks were lost in the majesty of the high Himalayas, whereas even a small promontory on a plain may be a "mountain" to the local people. Thus, Iron Mountain in Florida or landforms of only slightly larger stature, such as the Watchung Mountains in New Jersey, are important local landmarks to which the name "mountain" apparently seems appropriate even though they may not exceed 150 m (500 ft) in elevation. A similar pattern of place names can be found in South Africa (Browne et al. 2004). Nevertheless, calling a feature a mountain does not make it one.

Roderick Peattie, in his classic *Mountain Geography* (1936), suggests several subjective criteria for defining mountains: (1) mountains should be impressive, (2) they should enter into the imagination of the people who live within their shadow, and (3) they should have individuality. He cites Mount Fujiyama in Japan and Mount Etna in Italy as examples. Both are snowcapped volcanic cones that dominate the surrounding landscapes, and both have been immortalized in art and literature. They produce very different responses in the minds of the people who live near them, however. Fujiyama is benign and sacred, a symbol of peace and strength. Etna, on the other hand, is a devil, continually sending out boiling lava and fire to destroy farms and villages.

To a large extent, then, a mountain is a mountain because of the part it plays in the popular imagination. It may be hardly more than a hill, but if it has distinct individuality, or plays a symbolic role to the people, it is likely to be rated a mountain by those who live around its base (Peattie 1936). For similar symbolic reasons, mountains can come and go. For instance, the initial explorers who mapped the area around the Gulf of St. Lawrence in the seventeenth century identified the Wotchish Mountains, which presented a barrier to westward travel. As the region became more accessible, these low mountains, with summits just over 500 m, became recognized as just one part of the immense Labrador plateau (Debarbieux 2000).

It is difficult to include such intangibles in a workable definition. A more objective basis for defining mountains is elevation. For instance, a landform must attain at least a certain altitude (e.g., 300 m) to qualify. Although this is an important criterion, by itself it is still insufficient. The Great Plains of North America are over 1,500 m (5,000 ft) high, and the Tibetan Plateau reaches an elevation of 5,000 m (16,500 ft), but neither would generally be classified as mountainous. In Bolivia, the Potosí railway line reaches an elevation of 4,800 m (15,750 ft) near the station of El Condor, high enough to make your nose bleed, but it is situated in fairly level country with only occasional promontories exceeding 5,000 m (16,500 ft) (Troll 1972: 2). By contrast, western Spitsbergen in Norway, situated only a few hundred meters above sea level, has the appearance of a high mountain landscape, with its glaciers, frost debris, and tundra vegetation.

In addition to elevation, an objective definition of mountainous terrain should include local relief, steepness of slope, and the amount of land in slope. *Local relief* is the elevational distance between the highest and lowest points in an area. Its application depends upon the context in which it is applied. When compiling a global database of mountain protected areas (such as national parks), the United Nations Environmental Programme World Conservation Monitoring Centre, working with the World Conservation Union (IUCN), recognized only those that had at least 1,500 m (5,000 ft) of relief (Thorsell 1997). Several early European geographers believed that for an area to be truly mountainous there should be at least 900 m (3,000 ft) of local relief. If this standard is used, only the major ranges such as the Alps, Pyrenees, Caucasus, Himalayas, Andes, Rockies, Cascades, and Sierra Nevada qualify. Even the Appalachians would fail under this approach. On the other hand, American geographers working in the eastern and midwestern United States have thought that 300 m (1,000 ft) of local relief is sufficient to qualify as mountainous. Various landform

classifications have been proposed with specifications ranging between these figures (Hammond 1964).

Local relief by itself is, like elevation, an incomplete measure of mountains. A plateau may display spectacular relief when incised by deep valleys (e.g., the Grand Canyon). Such features are, essentially, inverted mountains, but we are accustomed to looking up at mountains, not down. (On the other hand, if one is at the bottom of the Grand Canyon looking up, the landscape can appear mountainous.) Still, this particular area of high local relief is of relatively limited extent and is surrounded on either side by primarily flat-lying surfaces. An opposite but comparable landscape is that of the Basin and Range Province in the western United States. Most of the area is in plains, but occasional ridges protrude 1,500 m (5,000 ft) above their surroundings. Such landscapes are problematic because they do not fit nicely into the category of either plain or mountain.

Mountains are usually envisaged as being both elevated and dissected landscapes. The land surface is predominantly inclined, and the slopes are steeper than those in lowlands. Although this is true as a generalization, the actual amount of steeply dissected land may be rather limited. Much depends upon geological structure and landscape history. In mountains such as the Alps or Himalayas, steep and serrated landforms are the dominant features; in other regions, these features may be more confined. The southern and middle Rocky Mountains display extensive broad and gentle summit uplands, and similar conditions exist in the Oregon Cascades. It is the young Pleistocene volcanoes sticking above the upland surface that give distinctiveness to the Cascades. The Sierra Nevada of California contains many strongly glaciated and spectacular features, but there are also large upland areas of only moderate relief. Yosemite Valley is carved into this undulating surface, and most of the impressive relief in this region derives from the occurrence of deep valleys rather than from the ruggedness of the upland topography. The world of mountains is basically one of verticality: Although slope angles of 10 to 30 degrees are characteristic, it is the intermittent cliffs, precipices, and ridges that give the impression of great steepness. Nevertheless, the horizontal distances between ridges and valleys, which establish the texture and framework for slope steepness, are just as fundamental to the delineation of mountains as the vertical distances that establish the relief.

Mountains may be delimited by geologic criteria—in particular, faulted or folded strata, metamorphosed rocks, and granitic batholiths (Hunt 1958; Ollier and Pain 2000). Most of the major mountain chains have these features, and they are also important in identifying former mountains. Good examples are found along the south shore of Lake Superior in Michigan and throughout much of southeastern Canada, where all of these characteristics are present, but erosion has long since removed the ancient peaks that once were mountains. Implicit in this definition is the idea that mountains are features of construction, built and produced by some internal force. This is certainly true of the major ranges, but mountainous terrain may also result from destructive processes, i.e., erosion. For example, a strongly dissected plateau may take on a mountainous character even though it contains none of the listed geologic characteristics. Certain areas of the southwestern United States do in fact display such dissection. Curiously, these landscapes are often perceived very differently from those of constructional origin. They are viewed as ruins, pathetic features, rather than as initial expressions of grand nature. They evoke "the sentiment of melancholy" (Tuan 1964).

Another basis for defining mountains is their climatic and vegetational characteristics. An essential difference between hills and mountains is that mountains have significantly different climates at successive levels (Barry 2008). This climatic variation is usually reflected in the vegetation, giving mountains a vertical change in plant communities, or *bioclimatic belts*, from bottom to top that hills lack (Jeník 1997; Körner 2003; Körner et al. 2011). It is argued that 600 m (2,000 ft) of local relief in most parts of the world suffices to bring about a distinct vegetation change. This is not always evident, because some plants, such as sagebrush (*Artemesia* spp.) in the western United States or heather (*Calluna vulgaris*) in Scotland, have great altitudinal range and may cover entire mountains. However, even if the vegetation is homogeneous, there are measurable climatic changes with altitude (Thompson 1964; Körner 2003; Barry 2008). The major advantage of this approach is that it recognizes ecology as well as topography. Clearly, one of the most distinctive characteristics of mountains, in addition to high relief and steepness of slope, is great environmental contrast within a relatively short distance (this volume, Chapters 7 and 11).

German-speaking peoples differentiate between the *Hochgebirge* (high mountains) and *Mittelgebirge* (medium mountains). The Harz Mountains and the Black Forest are *Mittelgebirge*, whereas the Alps are the classic example of *Hochgebirge* (Troll 1972: 2). French has the comparable terms *hautes montagnes* and *moyennes montagnes*; and in English we speak of the High Sierra or the High Cascades as opposed to the Sierra or the Cascades. The coastal ranges of the western United States are low mountains, and the Rockies are high mountains, but what distinguishes high mountains from low? Elevation alone is not sufficient: Compare the high plateaus of Tibet with the modest elevations of western Spitsbergen. High relief is not reliable, either: the California coastal ranges are on the whole probably more rugged than are most parts of the Rockies. Climate is the best determinant of where the alpine zone begins. For this reason, high mountain landscapes occur at different altitudes under different environmental conditions. In Java, the volcano Pangerango, which rises from sea level to 3,000 m (10,000 ft), is covered with tropical rainforest to its summit. "It is a high mountain without a high mountain landscape" (Troll 1972: 2).

The word "alpine" is European in origin, dating to pre-Roman times, with its roots in "alp" or "allo" meaning "mountain" (Körner 2003; Löve 1970). In Europe, New Zealand, and Japan, the term is commonly applied to whole mountain ranges that can include valleys, forests, and pastures. In biogeographical terms, however, the alpine life zone is confined to vegetation above the natural high-altitude forest or timberline. This is generally lowest in the polar regions, where "alpine" and "arctic" life zones with very similar geomorphological and ecological characteristics merge, and rises in elevation toward the equator. It is not a simple, straight-line relationship, however. The highest elevations at which trees grow occur at about 30° latitude in the arid zones of the Andes and Himalayas, rather than in the humid tropics (Körner 2003). Timberline also tends to rise from coastal areas toward the continental interiors. Thus, on Mount Washington in New Hampshire, the alpine zone begins at 1,500 m (5,000 ft); in the Rockies of Wyoming it occurs at over 3,000 m (10,000 ft); and in the Oregon Cascades, it drops again to 1,800 m (6,000 ft) (Daubenmire 1954: 121). Although the upper timberline is probably the major criterion for determining where the high mountain environment begins, it should not be the sole determinant. Because different tree species have different climatic requirements, contrasting abilities and potentials

TABLE I.I
Global Typology of Mountain Classes

Class (elevation in m)	Additional Criteria	% of Global Land Area
>4,500		1.2
3,500–4,499		1.8
2,500–3,499		4.7
1,500–2,499	>2° slope	3.6
1,000–1,499	>5° slope or LER >300 m	4.2
300–999	LER >300 m	8.8

SOURCE: Kapos et al. 2000.

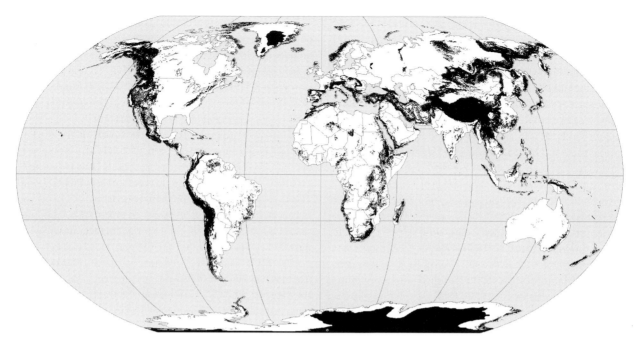

FIGURE 1.1 Mountains of the world (UNEP World Conservation Monitoring Centre 2002: 12–13). The boundaries and names shown and the designations used on maps do not imply official endorsement or acceptance by the United Nations Environment Programme or contributory organizations.

are involved in different regions. Geological or other natural factors may result in abnormally low timberlines. In addition, many timberlines have been greatly affected by human interference, especially through the agency of fire, cutting, and grazing, so they are not easily compared (Hedberg 1972, 1995; Braun et al. 2002; Broll and Keplin 2005).

Accordingly, a geoecological approach has been suggested for determining the lower limit of high mountain environments. There are three main criteria: High mountains should rise above the Pleistocene snow line, the zone of rugged and serrated topography associated with mountain glaciers and frost action; high mountains should extend above the regional (natural) timberline; and high mountains should display cryonival (i.e., cold climate) processes such as frost-heaving and solifluction (Troll 1972, 1973). Although each of these may exist at various altitudes, and one may be more important in some areas than in others, when considered together they provide a fairly good basis for delimiting high mountain environments. "According to this concept, high mountains are mountains which reach such altitudes that they offer landforms, plant cover, soil processes, and landscape character which in the classical region of mountain geography in the Alps is generally perceived as high-alpine" (Troll 1972: 4).

All of these approaches to defining mountains rely on a detailed analysis of one or more factors, usually based on fieldwork, ground-based topographic mapping, or both. More recently, modern technologies based on remote sensing have been applied to the definition of mountains at both regional and global scales. In 1996, the U.S. Geological Survey completed its GTOPO30 global digital elevation model. With a horizontal grid spacing of 30 arc seconds, the altitude of every square kilometer of the Earth's land surface was recorded in a database which could be used to derive a detailed typology of mountains based on not only altitude but also slope and terrain roughness (local elevation range, LER). Kapos et al. (2000) iteratively combined parameters from GTOPO30 to develop such a typology, starting from first principles and in consultation with scientists, policymakers, and mountaineers. First, 2,500 m, the threshold above which human physiology is affected by oxygen depletion, was defined as a limit above which all environments would be considered "mountain." Second, they considered that at middle elevations, some slope was necessary for terrain to be defined as "mountain," and that slopes should be steeper at lower elevations. Finally, to include low-elevation mountains, the LER was evaluated for a 7-km radius around each target cell: If the LER was at least 300 m, the cell was defined as "mountain." According to this typology, 35.8 million km^2 (24 percent of global land area) was classified as mountainous (Table 1.1; Fig. 1.1).

A further statistic of equal relevance is the global population in mountain areas, long estimated at about 10 percent (Ives and Messerli 1997). Using the mountain area defined by Kapos and others (2000), Huddleston and others (2003) estimated this to be approximately 720 million people (12 percent of the global population). Meybeck and others (2001), using an aggregated version of GTOPO30, also estimated that 26 percent of the global population live within or very close to mountain areas. Thus, with about a quarter of the Earth's land surface covered by mountains, and about a quarter of the global population living in or near them, mountain issues clearly have an important place on the global development and environment stage. Hence, mountains are on political agendas, and we should note that the definition of mountains is also a political process. For instance, decisions regarding the extent of mountains in Europe are closely linked to the availability of subsidies for mountain farmers (Price et al. 2004).

In summary, a universally accepted definition of what a mountain is will always remain elusive. For our purposes, however, a mountain can be defined as a conspicuous, elevated landform of high relative relief. Much of its surface has steep slopes, and it displays distinct variations in climate and vegetation zones from its base to its summit. A high mountain landscape is the area above the climatic timberline where glaciation, frost action, and mass wasting are dominant processes. Additionally, a landform is considered a mountain when local people rate it as such because it plays an important role in their cultural, spiritual, and working lives.

Mountain Challenges and Opportunities

Mountain people are typically *independent, innovative, resourceful, adaptive,* and *outstanding entrepreneurs.* At the same time, they include some of the *poorest, most remote,* and *disadvantaged* people in the world (Ives 1997; Huddleston et al. 2003). High elevations and cold climates exclude productive agriculture and limit animal husbandry. Poverty levels are often exceptionally high, and access to education, decision-making power, health services, financial resources, and land rights are inequitably distributed between upland and lowland communities (Pratt 2004; Körner and Ohsawa et al. 2005). Populations are scattered, and mountain peoples are typically distrustful of central governments and outsiders because of histories of exploitation with little compensation or long-term benefit (Libiszewski and Bächler 1997; Starr 2004; this volume, Chapter 12).

National governments typically do not connect with their citizens in peripheral mountain areas, despite the potential threats to the state that may emerge from neglect of those areas (Mountain Agenda 2002). At the end of the twentieth century, more than half of the world's 48 ongoing *wars and conflicts,* strongly linked to the poverty and historic marginalization of highland peoples, were taking place in mountains (Libiszewski and Bächler 1997). This trend continues, resulting not only in the tragic loss of human life but also in unprecedented levels of environmental degradation. In democratic states such as Germany, Italy, and Spain, however, the relative independence of mountain areas has often led to relative political autonomy.

The complex topography of mountain areas, and the often high frequency of natural hazards, such as landslides, avalanches, and floods, means that *communications* are typically poor, and roads and infrastructure marginal to nonexistent (Kohler et al. 2004; Chapter 12). However, although traditional means of communication may not be well developed, modern technologies such as the Internet and mobile phones are rapidly expanding in mountain areas (Ceccobelli and Machegiani 2006).

Because of their environmental diversity over small distances, mountains offer a huge potential for the local production of alternative energy through small-scale wind, solar, and hydroelectric power (Schweizer and Preiser 1997). Such small-scale facilities can allow for the development of local industries, bringing new sources of income—often from the rejuvenation of traditional crafts—and also provide light for children to study by (Banerji and Baruah 2006). Equally important, modern sources of energy can help to mitigate chronic *shortages of wood-based heating fuel,* the principal source of energy for the majority of mountain people (Schweizer and Preiser 1997).

High altitudes and cold climates can lead to *health problems* such as acute mountain sickness, caused by the body's inability to adapt to the decreased oxygen and pressure, with mild symptoms (headache, loss of appetite) occurring for some people as low as 3,000 m (West 2004). Care must be taken to *acclimate* (physically adjust) to higher altitudes by not ascending too quickly, drinking plenty of water, and "climbing high/sleeping low" during the rest or acclimatization period (Houston 1998). People who have lived at elevations above 4,000 m for generations, such as the Sherpa of Nepal or Quechua of Peru, seem to have a natural ability to live and work at elevations that are at least initially uncomfortable for the lower-altitude dweller. Goiter, a swelling of the thyroid gland in the neck, is caused by a lack of iodine in the diet that, until recently, was a relatively common occurrence in remote populations in the Hindu Kush–Himalaya and other mountain ranges (Fisher 1990). The use of fuelwood in improperly ventilated homes has been directly linked to high rates of bronchitis and other respiratory diseases in millions of mountain homes.

Work seasons are short in the mountains, and skilled labor forces often lacking, especially in areas characterized by the *out-migration* of young people (this volume, Chapters 10, 11, and 12). This is common in many mountain areas, from the Appalachians to the Hindu Kush, and many traditional mountain cultures are being rapidly assimilated into mainstream cultures as modern communication technologies and tourists reach even the most remote mountain villages. Yet emigrants often send remittances which can be vital sources of income and investment, and when they return, they may bring new ideas which can often be combined with traditional ideas and approaches, contributing to sustainable development. Examples include new means of producing and marketing the many high-quality and high-value products for which mountains are known, including textiles, food, and drink. In all of this, *women* are of critical importance to home life, the work force, farm maintenance, and as retainers of traditional biodiversity knowledge—yet they typically remain marginalized with regard to decision making, equity, and education (Byers and Sainju 1994; Ives 1997).

A key set of challenges derives from the fact that mountains are naturally *dynamic environments,* and low-frequency/high-magnitude events such as volcanic eruptions, landslides, debris flows, and glacial lake outburst floods (GLOFs) are capable of causing immense damage (Hewitt 1997; this volume, Chapter 5). One of the more extreme examples of catastrophic events was the 1970 Huascarán earthquake and debris flow in Peru, which buried the town of Yungay (population 20,000) within minutes and killed an estimated 80,000 people throughout the region (Browning 1973).

As many of the larger glaciers have melted in the Hindu Kush–Himalaya (Bajracharya and Shrestha 2011), hundreds of new glacier lakes, holding millions of cubic meters of water, have been created. Usually contained by dams of loose boulders and soil, these lakes present a risk of GLOFs. Triggering factors for GLOFs include "lake area expansion rate; up-glacier and down-valley expansion rate; dead-ice melting; seepage; lake water level change; and surge wave by rockfall and/or slide and ice calving" (Watanabe et al. 2009). GLOFs unleash stored lake water, often causing enormous devastation downstream that can include high death tolls as well as the destruction of valuable farmland and costly infrastructure (hydroelectric facilities, roads, and bridges). Examples include the 1941 outburst flood above Huaraz, Peru that killed nearly 6,000 people within minutes (Hambrey and Alean 2004; Carey 2005, 2010); the 1985 Langmoche outburst in the Sagarmatha (Mt. Everest) National Park, Nepal, which destroyed the $2 million Thami hydroelectric facility, hundreds of hectares of cropland, and dozens of bridges downstream (Vuichard and Zimmerman 1986; Byers et al. 2013; Byers 2013); and the 1998 outburst of the Sabai Tso in the Hinku valley, Makalu-Barun National Park, Nepal, that destroyed trails and seasonal settlements for nearly 100 km downstream (Cox, 1999; Osti and Egashira 2009).

A number of other physical attributes of mountains argue not only for their importance to policymakers in the immediate term but also for the special consideration necessary to ensure their sustainable use to meet the demands of both mountain people and those living downstream in the years to come (Ives et al. 1997; this volume, Chapter 12). For example, while mountains may seem indestructable— and are occasionally sources of great destruction—they include some of the *most fragile ecosystems in the world* (Jodha 1997; Hamilton and Bruijnzeel 1997; Körner 2003). Their steep slopes and young, thin soils make them particularly susceptible to accelerated soil erosion, gully formation, landslides, desertification, and downstream river siltation, particularly if the vegetation cover that protects their slopes is disturbed. Improper forest harvesting practices, overgrazing, mass tourism, the construction of ill-designed transport roads, and mining are the most frequent forms of land use leading to these advanced states of degradation and habitat destruction in the mountains. Particularly at higher elevations, off-road vehicle tracks, overgrazing, and the impacts of burning can take many

decades to heal (Byers 2005; Spehn et al. 2006). Knowledge—both scientific and traditional—of how to limit such damage and mitigate its consequences is often available but is not used effectively, if at all (Hamilton and Bruijnzeel 1997). The wide dissemination of case studies of good practice is therefore essential (Stocking et al. 2005).

The impacts of *climate change* on mountain ecosystems, especially regarding the retreat of most of the world's glaciers that has occurred over the past century and impacts on the world's freshwater supplies, has received considerable attention in the past decade (Barry and Thian 2011; National Research Council 2012). As temperatures increase, many alpine plants and animals may be at risk because of loss of habitat, or because they are not able to migrate upslope fast enough to cooler, more suitable habitats similar to those in which they evolved; thus, in particular, the design and location of protected areas may have to be reconsidered (Price 2008; Worboys et al. 2010). The melting of permafrost; increased risk of other high-magnitude events such as debris flows and landslides; accelerated erosion from increased glacial runoff; changes in agricultural patterns; the predicted depletion of glacier-fed freshwater for hydropower, agriculture, and drinking water; negative impacts of climate change on mountain tourism; and an increase in infectious diseases previously confined to the lowlands—all are real or predicted impacts of climate change in the mountains that are receiving increasing study and attention (Price and Barry 1997; Huber et al. 2005; Singh et al. 2011; this volume, Chapters 3, 4, 7, and 12). Finally, the impacts of acid rain, smog, and metal deposition from precipitation, all of which have lowland industrial sources, are often seen first in mountain regions. Some believe that mountains are "canaries in the coal mine" and among the most sensitive barometers of global climate change in the world (Hamilton 1997).

For these and other reasons, many geographers, development workers, and government officials feel that the future of the world's mountains, and the significant proportion of the global population who depend on them, will depend largely on achieving the same levels of international recognition and conservation efforts given to oceans, rainforests, wetlands, and deserts (Ives and Messerli 1997; Debarbieux and Price 2008; Rudaz 2011). Notable progress has been made during the past two decades to increase local and global awareness for the importance of mountain environments and their peoples (this volume, Chapter 12), but much remains to be done, particularly with regard to the concept of *ecosystem services* provided by mountain areas to wider populations. While this is increasingly being used as an argument for investing in mountain areas (e.g., Rasul et al. 2011), there is still debate about the utility of the concept (Grêt-Regamey et al. 2012). Education will be key to this process of awareness raising, and the objective of this book, an updated and expanded version of Larry Price's superb mountain geography textbook of 1981, is to facilitate the awareness-building process through the continued development and dissemination of mountain-related educational and resource materials.

Price, M. F., Lysenko, I., and Gloersen, E. 2004. Delineating Europe's mountains. *Revue de Géographie Alpine* 92(2) : 75–86.

Price, M. F., Moss, L. A. G., and Williams, P. W. 1997. Tourism and amenity migration. In B. Messerli and J.D. Ives, eds., *Mountains of the World: A Global Priority* (pp. 249–280). New York: Parthenon.

Purohit, A. N. 2002. Biodiversity in mountain medicinal plants and possible impacts of climatic change. In C. Körner and E. M. Spehn, eds., *Mountain Biodiversity: A Global Assessment* (pp. 267–273). New York and London: Parthenon.

Rasul, G., Chettri N., and Sharma, E. 2011. *Framework for Valuing Ecosystem Services in the Himalayas*. Kathmandu: International Centre for Integrated Mountain Development.

Rudaz, G. 2011. The causes of mountains: The politics of promoting a global agenda. *Global Environmental Politics* 11(4): 43–65.

Sayward, P. 1934. High points of the forty-eight states. *Appalachia* 20(78): 206–215.

Schweizer, P., and Preiser, K. 1997. Energy resources for remote highland areas. In B. Messerli and J.D. Ives, eds., *Mountains of the World: A Global Priority* (pp. 157–170). New York: Parthenon.

Shrestha, T. B. 1989. *Development Ecology of the Arun River Basin in Nepal*. Kathmandu: International Centre for Integrated Mountain Development.

Singh, S. P., Bassignan-Khadka, I., Karky, B. S., and Sharma, E. 2011. *Climate Change in the Hindu Kush-Himalayas: The State of Current Knowledge*. Kathmandu: International Centre for Integrated Mountain Development.

Spehn, E. M., Liberman, M., and Körner, C., eds. 2006. *Land Use Change and Mountain Biodiversity*. Boca Raton, FL: CRC Press.

Starr, S. F. 2004. Conflict and peace in mountain societies. In M. F. Price, L. Jansky, and A. A. Iatsenia, eds., *Key Issues for Mountain Areas* (pp. 169–180). Tokyo: UNU Press.

Stepp, J. R., Castaneda, H., and Cervone, S. 2005. Mountains and biocultural diversity. *Mountain Research and Development* 25: 223–227.

Stocking, M., Helleman, H., and White, R., eds. 2005. *Renewable Natural Resources Management for Mountain Communities*. Kathmandu: International Centre for Integrated Mountain Development.

Thompson, B. W. 1964. How and why to distinguish between mountains and hills. *Professional Geographer* 16(6): 6–8.

Thorsell, J. 1997. Protection of nature in mountain regions. In B. Messerli and J.D. Ives, eds., *Mountains of the World: A Global Priority* (pp. 237–248). New York: Parthenon.

Troll, C. 1972. Geoecology and the world-wide differentiation of high mountain ecosystems. In C. Troll, ed., *Geoecology of the High Mountain Regions of Eurasia* (pp. 1–13). Wiesbaden: Franz Steiner.

Troll, C. 1973. High mountain belts between the polar caps and the equator: Their definition and lower limit. *Arctic and Alpine Research* 5(3) (Pt. 2): 19–28.

Tuan Y. 1964. Mountains, ruins, and the sentiment of melancholy. *Landscape* 14(1): 27–30.

UNEP World Conservation Monitoring Centre. 2002. *Mountain Watch: Environmental Change and Sustainable Development in Mountains*. Cambridge, UK: UNEP World Conservation Monitoring Centre.

Viviroli, D., Dürr, H. H., Messerli, B., Meybeck, M., and Weingartner, R. 2007. Mountains of the world, water towers for humanity: Typology, mapping and global significance. *Water Resources Research* 43, W07447, doi:10.1029/2006WR005653.

von Fürer-Haimendorf, C. 1984. *The Sherpas Transformed*. New Delhi: Sterling Publishers Private Limited.

Vuichard, D., and Zimmerman, M. 1986. The Langmoche flash-flood, Khumbu Himal, Nepal. *Mountain Research and Development* 6(1): 90–94.

Watanabe, T., Lamsal, D., and Ives, J.D. 2009. Evaluating the growth characteristics of a glacial lake and its degree of danger of outburst flooding: Imja Glacier, Khumbu Himal, Nepal. *Norsk Geografisk Tidsskrift—Norwegian Journal of Geography* 63: 255–267.

Weber, B., and Vedder, A. 2001. *In the Kingdom of Gorillas: Fragile Species in a Dangerous Land*. New York: Simon and Schuster.

West, J. B. 2004. The physiologic basis of high-altitude diseases. *Annals of Internal Medicine* 141: 789–800.

Worboys, G. L., Francis, W. L., and Lockwood, M., eds. 2010. *Connectivity Conservation Management: A Global Guide*. London: Earthscan.

CHAPTER TWO

Origins of Mountains

JOHN F. SHRODER JR.
and LARRY W. PRICE

Views about mountain origins have changed considerably through time. During the Middle Ages and early modern Western Europe, mountains were regarded as "monstrous excrescences of nature"; a prevailing view was that they had been created as punishment after man's expulsion from the Garden of Eden. This idea apparently had its origin in the fact that the story of creation in Genesis makes no mention of mountains. Explanations of mountain creation differed: Some said that interior fluids ruptured the spherical surface and piled up in great heaps; others leaned toward the cataclysmic biblical flood. Although more advanced ideas had been developed by the ancient Greeks and medieval Arabs, such was the power of theology in Western Europe that science retrogressed. It was not until the bond between religion and science was severed in the 17th and 18th centuries that major scientific advances were made. Once begun, science progressed rapidly, and a number of plausible theories postulated the origin of mountains. Current attempts by enthusiasts of the oxymoronic "creation science" to set back the scientific clocks notwithstanding, advances have steadily increased understanding of the evolution of the Earth and its mountains (Turcotte and Schubert 2002; Owens and Slaymaker 2004; Owen 2004; Ollier 2004; Williams 2004). Earth's crust is known to be divided into moving rigid plates; where plates pull apart, new crustal material is produced at spreading center rifts, and where plates collide they descend back into the mantle in submarine trenches and subduction zones, or buckle upward into mountains.

Characteristics of Major Mountain Ranges

The most important clues to the origin of mountains are the distribution of mountain ranges over the globe and their structure and material composition. Both point to a fundamental association between oceans and mountains. Mountain ranges tend to run in long, linear belts along the margins of continents such as the Andes in South America, as well as in continental interiors such as the Urals or the Himalayas (Fig. 2.1). One long belt extends along the periphery of the Pacific Ocean, and another runs east–west along the "underbelly" of Eurasia. The distribution of mountains closely follows the distribution of earthquakes, fault zones, volcanic activity, ocean trenches, and certain curved chains of islands (*island arcs*).

Many coastal zones, as well as being less stable than most continental interiors, are made of much younger rock. Most continental interiors are composed of ancient crystalline cores of Precambrian granitic and metamorphic rock surrounded by patches of younger sedimentary rocks (Fig. 2.2). The primary process in continental growth or continental accretion is the addition of new land to continental cores by mountain building (Condie 2005).

The great mountain belts are chiefly marine sedimentary rocks (although the rocks are commonly metamorphosed and injected with igneous material during the course of mountain building). The sedimentary rocks of these mountains are undeniably marine, as proven by the presence of fossilized shallow-water seashells high on peaks. This has confounded everyone who has theorized about mountain genesis. How did fossils typical of shallow seas get to mountaintops? Marine fossils also occur deep in the Earth, far below the foot of the mountains. Indeed, sediments underlying mountains go far deeper than sedimentary accumulations under the surrounding lowlands. In the United States, for instance, it was observed more than a century ago that the marine sedimentary rocks in Iowa and Illinois (which were once covered by ancient shallow seas) were not as thick as those of the same age under the Appalachians in Pennsylvania and New York.

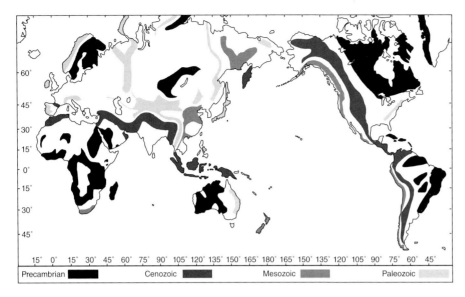

FIGURE 2.1 Precambrian shields (Achean and Proterozoic) age (>540 million years) consisting of crystalline rock are in black. Mountain belts are in shades of gray. Each shade represents a different age: Cenozoic age (the last 65 million years), Mesozoic age (65–225 million years ago), and Paleozoic age (225–540 million years ago) mountains. (Adapted from various sources.)

The presence of thicker accumulations of sedimentary rock under mountains than under surrounding lowlands reveals much. Geologists had long puzzled about whether the great weight of mountains was simply piled on the surface as excess load and was therefore in an unstable condition that would ultimately "pop up," or whether mountains have "roots" or foundations that are less dense than the surrounding strata so that a relatively balanced condition exists. In addition to the thicknesses of sedimentary rock, evidence for the presence of mountain roots came during the mid-1800s from northern India, where surveyors established precise land measurements. At survey stations near the Himalayas, they noticed that their plumb line (a string with a pointed weight attached to one end to make the string hang vertically) did not hang straight down: The line was deflected slightly toward the mountain, away from the Earth's center of gravity. This had been expected, considering the great mass of the Himalaya and its gravitational pull; in fact, the deflection was not as great as it should have been, according to their calculations. It was concluded, correctly, that the low degree of deflection was due to the mountain mass being composed of less dense, lighter rocks that extended deep into the Earth, displacing the more dense, heavier substrate (Holmes 1965: 28). In other words, mountains float on the Earth's surface somewhat like icebergs in water. The larger the mountain and the higher it rises, the deeper its roots extend into the Earth. This situation, called *isostasy* (from the Greek for "equal standing"), suggests that the Earth's crust always strives to maintain a state of gravitational equilibrium (Turcotte and Shubert 2002): A mountain mass rising high above the surrounding surface is compensated for by a deep root that displaces denser rocks at depth (Fig. 2.3).

The height mountains can achieve is limited by the nature of the physical processes that create mountains and by the ability of the base to support the accumulated mass. The type and intensity of tectonic activity or *orogenesis* (Greek for "mountain origin") determine the specific processes involved in mountain genesis, and the support capacity of the base is determined by the weight of the mountain mass and the pressure and melting point of the basal material—that is, at some point the basal material will begin to melt and flow (Weisskopf 1975: 608). The maximum possible height is not known, but from a quick survey of worldwide mountain elevations, it seems that ~6,000 m (~20,000 ft) may be close to the upper limit. Gravitational collapse and spreading can begin when the *orogenic welt* exceeds only half that height, and rock is forced downward and outward. The deeper rock is ductile or plastic and flows, but near the surface of the range the rock fractures in tensional normal faults. The rock is pushed outward and helps form fold and thrust-fault belts away from the core of the range. The Himalayas stand as high as they do because of the way in which they were formed. Portions of two continental masses were compressed together with the deposits of an ancient sea, as well as a volcanic island arc, caught between them. The resulting thick blanket of low-density sedimentary and some ancient volcanic rock also probably explains why there has been so little geologically recent volcanic activity in the Himalayas, although some has occurred in Tibet or where rapid erosion overhead has allowed decompression melting (Valdiya 1989; Harris et al. 1993; Schneider et al. 1999).

Just as the height and mass of a mountain range determine the depth of its roots, thereby maintaining a balance with the surroundings, so the Earth's crust continues to adjust to changing load conditions as

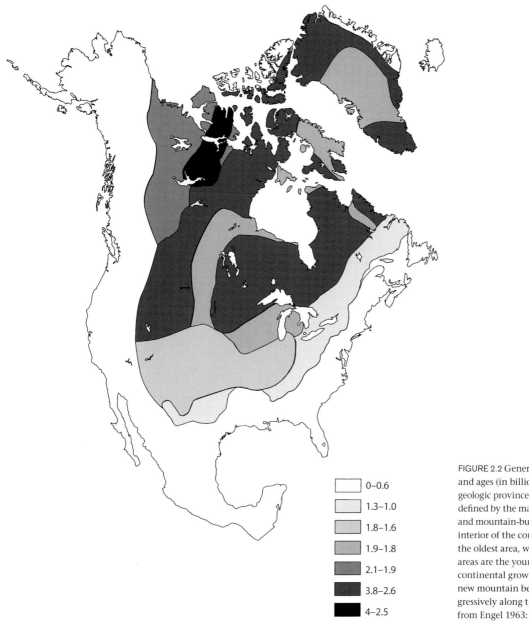

FIGURE 2.2 Generalized distribution and ages (in billions of years) of geologic provinces in North America, as defined by the major granite-forming and mountain-building events. The interior of the continent is generally the oldest area, whereas the coastal areas are the youngest, indicating continental growth by accretion as new mountain belts are formed progressively along the margins. (Adapted from Engel 1963: 145.)

Legend:
- 0–0.6
- 1.3–1.0
- 1.8–1.6
- 1.9–1.8
- 2.1–1.9
- 3.8–2.6
- 4–2.5

erosional unroofing or *gravity sliding* takes place. As the uppermost load is removed and the weight on the base is lightened, the mountain rises to compensate for the loss of weight. The isostatic reaction of the Earth's crust to vast applications of weight occurs in areas once covered by several kilometers by Pleistocene ice sheets, the weight of which was immense. For example, Hudson's Bay in Canada—the center for the accumulations of continental glaciation in North America—has been isostatically rebounding from being depressed under this great weight at a rate of more than 3.5–4 m (~12–13 ft) per century, although it is now slowing (Andrews 1970; Selby 1985; Goudie 1995).

The continents themselves are *isostatic:* They are composed of lighter rocks floating on a substrate of denser rocks. This was confirmed by the seismologist *Mohorovičić*, who observed that earthquake waves travel at low velocity until reaching a certain distance below the Earth's surface, where they speed up abruptly. He concluded this was due to denser rocks in Earth's mantle, and that the increased velocity marked the Earth's crustal base. The zone of density contrast between crust and mantle was subsequently named the *Mohorovičić discontinuity,* or simply the *Moho* (Fig. 2.3). The crust beneath the continents averages 20–30 km (12–19 mi) thick, but only 5–10 km (3–6 mi) under the ocean basins. The deep roots under the major

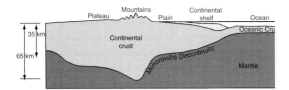

FIGURE 2.3 Idealized representation of crustal thickness beneath continents and oceans. The greatest thickness occurs under plateaus and mountains, the thinnest under the oceans. Continental shelves are largely underlain by geoclinal sediments and sedimentary rocks. (Adapted from Holmes 1965.)

mountain ranges, where the Earth's crust reaches its greatest thickness, descend to as much as 70 km (~44 mi) (Condie 2005).

A fundamental question at this point is: What accounts for the great accumulations of marine sediments that are present in the major mountain belts? Fossils collected near the summit of Mount Everest at 8,850 m (29,035 ft), as well as many other high mountain areas, indicate that these rocks were originally deposited in a shallow sea (Le Fort 1996). How is it possible for thick rock accumulations to have been continually deposited in shallow water? The answer seems to be that these sediments were eroded from the land and deposited in coastal areas, and as their weight increased, they gradually depressed the underlying rock, allowing sedimentation to continue in shallow water. Some deposits eventually reached thicknesses of up to 12,000 m (~40,000 ft) in huge linear, trough-like depressions. Such features, previously known as *geosynclines*, now *geoclines* or sedimentary basins, are known to be related to mountain building. Through the study of ancient folded mountains such as the Appalachians, a typical geocline was divided into two parallel components (King 1977): The inner continental part is composed of gently folded limestones and quartz-rich sandstones, termed the *miogeocline;* the outer seaward part, the *eugeocline,* has many turbidites from undersea landslides or turbidity currents. The eugeocline is more intensely faulted and folded, and perhaps has abundant volcanic or intrusive igneous material (Fig. 2.4). European researchers on alpine sedimentary basins in the Alps developed the term *flysch* for pre-orogenic shale and rock fragment–rich, sandstone turbidites originally deposited in submarine trenches, and *molasse* for postorogenic, feldspar-rich terrestrial sandstone deposited in nonmarine mountain basins (Einsele 2000).

Modern understandings of sedimentary basins that eventually become mountains consider two main divisions: basins that rest on either continental or oceanic lithosphere. Within the two classes are subclasses based on the tectonic processes that produced the basins. Further, using newer plate-tectonic terminology, shallow marine shelves without significant tectonism (miogeoclines) are recognized as passive continental margins; active continental

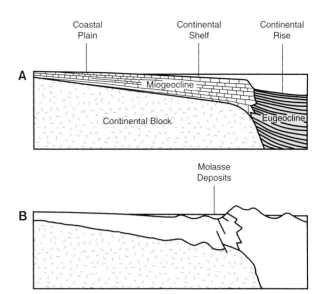

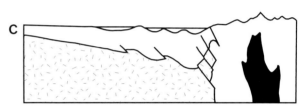

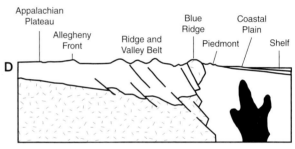

FIGURE 2.4 A simplified interpretation of the sequence of events that resulted in the active-margin crumpling of a former geoclinal or passive-margin couplet into the Appalachian Mountains. These rocks were deposited as sediments in Paleozoic time (540–225 million years ago). (A) The sandstones and limestones of the miogeocline, or landward continental shelf, were folded into a series of ridges between the Blue Ridge and the Allegheny Front. (B) The turbidites or flysch of the deeper-water eugeocline were altered by great heat and pressure associated with metamorphic and igneous activity as the passive margin became unstable and activated into a subduction zone. (C) Sediments eroded from the newly rising mountains were deposited as continental or terrestrial molasses deposits. The orogeny also produced extensive crystalline rocks that were elevated in a large mountain range whose crest lay to the east of the Blue Ridge. (d) Erosion continued to take its toll; all that remains are the present-day Appalachians, the deeply eroded roots of a range originally much like the Himalaya. (Adapted from Dietz 1972: 36.)

margins have offshore submarine trenches where oceanic plates descend into subduction zones, scraping off *accretionary wedges* and perhaps including island-arc volcanism, with forearc and backarc sedimentary basins. Five subclasses of continental basins are recognized, as well as four oceanic basins, with both modern and ancient examples known throughout the world (Strahler 1998).

Along the eastern seaboard of the United States, for example, a wedge of sediments some 250 km (155 mi) wide and 2,000 km (1,245 mi) long comprises the continental shelf and slope (Dietz and Holden 1966). This is a "living" miogeocline on a mature passive margin. The continental rise, which ascends to the continental slope from the abyssal plains of the oceanic deep, is thought to be a "living" eugeocline (Fig. 2.4a). The sediments on the continental shelf thicken seaward, attaining thicknesses of 3–5 km (1.8–3 mi), and consisting of heterogeneous deposits resulting from muddy suspensions of turbidity currents. These flow down the continental rise and deposit at its base great fans of turbidites having thicknesses of up to 10 km (6 mi). The recent sediments on the continental shelf and rise closely resemble ancient rocks occurring in the Appalachian fold belts (Fig. 2.4d). It is reasoned, therefore, that these deposits are the stuff of which mountains are made, and that similar examples of living geoclines exist off the coasts of the other major continents today; research continues on mechanisms causing their eventual deformation into mountains (Strahler 1998).

Theories of Mountain Origin

The ultimate causes of the origin of mountains has been one of the great enigmas of science. Even as late as the 1950s, geologists subscribed to at least half a dozen major theories. One of these was that mountains were created by the rise of *batholiths* (Greek for "deep rock"). Batholiths are masses of molten magma that intrude into rocks near the Earth's surface, where they cool, crystallize, and solidify. Because batholiths occur at the core of most mountain ranges, it was thought that they were also responsible for the uplift and deformation of the surface. Gravitational sliding of rock strata down mountain slopes was also recognized as important in folding, faulting, and distortion of overlying rocks. A major early theory was that Earth was contracting because it was thought to have been molten material ejected from the Sun and now gradually cooling. As the Earth cooled and contracted, the outer skin shriveled and wrinkled like a drying apple. An alternative theory was that Earth was actually expanding, which would cause rift valleys and tensional fault zones, as well as continental drift (Carey 1976). A still important early idea envisioned giant convection currents within the Earth; mountains develop where rising currents from two opposing convection systems converge, resulting in great compressional forces, which cause folding and deformation of the surface (Holmes 1931). Yet another theory held that mountain building took place as a result of continental drift, as the leading edge of a continent encountered resistance from the material through which it was moving and buckled under pressure.

Many of these theories are not mutually exclusive, and elements of each are still retained in one form or another, but no one believes they adequately explain primary initiating mechanisms for creating mountains. Revolutionary discoveries of the 20th century replaced them with the unifying concept of plate tectonics. This concept has proved to be one of the most significant developments in the history of science. Its immediate predecessor, continental drift, was not so well received.

Although the idea had been around in various forms for some time, the full theory of continental drift was proposed in 1910 by Alfred Wegener, a German meteorologist who noticed the complementarity of the continental outlines of the coasts of Europe and Africa with the Americas. It seemed to him, as it had to others in prior centuries, that if these continents were pushed together they would fit like the pieces of a jigsaw puzzle (Fig. 2.5). Wegener was the first to make this observation into a theory. He uncovered evidence of identical plant and animal fossils from coastal areas in Brazil and Africa. This strengthened his conviction that there had been land connections, and from that point on he devoted his life to searching for evidence to support ideas of continental drift. Similarly in the southern hemisphere, South African Alexander du Toit compiled evidence of a former supercontinent that later broke into the separate units that now exist as individual continents (du Toit 1937).

Wegener's and du Toit's evidence for former land connections included matching fossils and unusual rock types, alignment of fault zones, matching former areas of glaciation, and the presence of similar mountain types in Europe and North America. Most geologists, especially in the northern hemisphere, rejected the theory of continental drift as too fantastic. The similarities across oceans were all explained by other means. The major objection, however, was the mechanism Wegener offered for the movement of entire continents—the tidal attractions of the sun and moon. In succeeding years, interest in continental drift waxed and waned as new theories were proposed and symposia were held, but by and large the scientific community remained unconvinced. Discoveries in the last three decades of the 20th century, however, piled one type of evidence upon another so heavily that the conclusion became inescapable—the continents had indeed moved apart (Strahler 1998).

One of the first modern discoveries was made in the early 1950s by geophysicists studying rock paleomagnetism. When molten volcanic rocks with traces of iron solidify, the rocks are slightly magnetized in the direction of Earth's magnetic field. It follows that molten rocks formed at about the same time on different continents should be like compasses with their needles frozen in the direction of the Earth's magnetic pole, but it was discovered the orientation of similar ancient rocks on different continents did not line up with the Earth's present magnetic field.

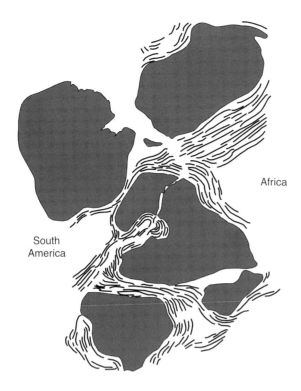

 Archean Crust    Proterozoic Mountain Belts

FIGURE 2.5 The "fit of the South American and African continents around the south Atlantic Ocean," showing close conformity of coastlines, continental shelves, Achean crustal blocks, and trends to Proterozoic mountain belts. (Adapted from various sources.)

This suggested that either the poles had "wandered," or the continents had moved with respect to the magnetic pole. Because it is generally believed that the Earth's magnetic and rotational poles have always been close to each other (within 15°), the continents were assumed to have moved, but the mechanism for this movement was still unknown.

Plate Tectonics

The basis of plate tectonics is the idea that the Earth's surface is broken into six large and many smaller rigid plates, like a huge cracked sphere (Fig. 2.6). The plates consist of portions of both continents and ocean basins, each about 100 km (~60 mi) thick, that are moving in various directions at rates varying from ~1 to 10 cm (~¼–4 in.) per year, or about as fast as fingernails grow. Where plates pull apart, new volcanic material fills the void, but where they come together, one oceanic plate dives beneath the other and is absorbed back into the Earth. If the second plate is a continent, its rocks are commonly squeezed and buckled up into mountains.

The concept of plate tectonics was born in the 1960s by combining two preexisting ideas—continental drift and *sea-floor spreading* (Fig. 2.7). Initial evidence for plate tectonics came from the sea floor. Systematic surveys in the late 1950s and 1960s revealed that the globe is virtually encircled by spectacular undersea mountain ranges or oceanic ridges. The development of sophisticated seismic, sonar, and computerized equipment allowed detailed sampling and analysis of the ocean floor. It became evident that, similar to those on land, the undersea mountain ranges are the loci of frequent volcanic and earthquake activity, and that they are areas of abnormally high heat flow. The largest and best-known undersea range is the Mid-Atlantic Ridge, which extends north–south for several thousand kilometers, roughly parallel to the coastlines of Europe, Africa, and the Americas. The active volcanic island of Iceland is a place where the ridge is large enough to protrude above sea level.

A corollary discovery based on dating of rocks from the sea floor revealed that the youngest volcanic rocks are near the centers of the undersea ridges; rock ages increase with distance outward from the ridge in both directions.

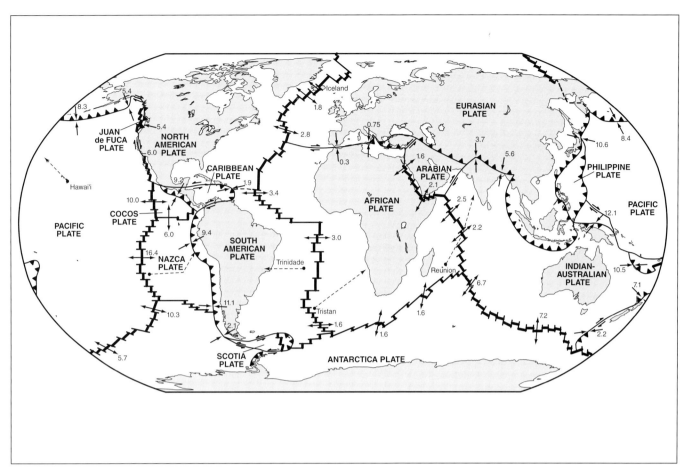

FIGURE 2.6 Distribution of major plates, spreading centers (diverging oceanic ridges), and subduction zones of convergence. Heavy, dark, offset lines represent ridge axes; connecting single lines with parallel but opposing arrows indicate displacement along transform faults. Lighter toothed lines with triangular symbols represent subduction zones. Dashed lines show uncertain plate boundaries and areas of extension within continents. A dot with a dashed arrow indicates a hot spot and its trace. Divergence direction and spreading rate (cm/yr) are shown with diverging arrows and convergence direction and rate with head-to-head converging arrows. High rates of convergence on land represent places where the mountains are growing most actively. (Adapted from various sources.)

This led to the discovery of *sea-floor spreading,* the idea that new volcanic crust is created at mid-ocean ridges and then spreads outward by dividing between the two plates moving in opposite directions. In addition, other studies revealed distinct magnetic strips of volcanic rock paralleling the axis of the mid-ocean ridges. These strips occur in symmetrically distributed pairs, with the pattern on one side of the ridge forming a mirror image of the other. The magnetic orientation of iron traces contained in the formerly molten rock in these strips can reveal opposite headings. It is well known that the Earth's magnetic field has reversed itself repeatedly throughout geologic time—north has become south and south has become north (although the magnetic poles have stayed in about the same locations). When lava was erupted along the mid-ocean crest, it became magnetized in the direction of the prevailing magnetic field. As the sea floor spread, the rock was carried away from the center in both directions like

two gigantic conveyor belts. Later eruptions deposited new material, which was in turn magnetized in the direction of the new magnetic field and then transported away from the ridge axis. By dating the rocks in these strips, it is possible to calculate the rates of sea-floor spreading. Rates of plate movement are also measured over short time periods by using astronomical and satellite laser ranging techniques, as well as repeat measurements of continental plate displacement by the global positioning navigation satellites (Gripp and Gordon 1990; Herring 1996; Turcotte and Schubert 2002).

As our understanding of sea-floor spreading developed, it became apparent that a characteristic sequence of igneous and sedimentary rocks formed at the spreading centers on the bottom of the sea. These *ophiolites* (Greek for "snakeskin," after their characteristic formations of shiny green metamorphic serpentine rock) are characterized by layers of the dense, heavy *ultramafic peridotite* that seems to

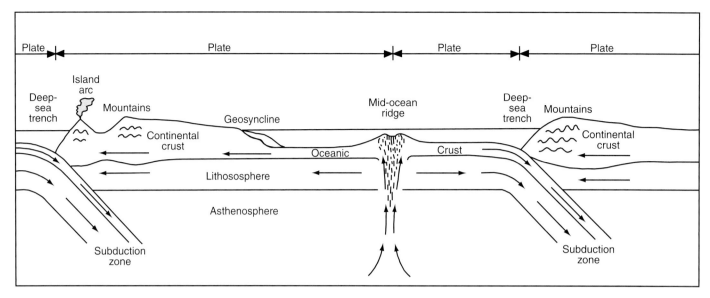

FIGURE 2.7 Schematic view of plate movement through sea-floor spreading and subduction. New crustal material is being created at the mid-ocean ridge (diverging and accreting plate margin), whereas oceanic crust and lithosphere are being consumed back into the Earth at the deep-sea trenches (converging plate boundary). Accordingly, the central ocean in this sketch is increasing in size while the one on the left is closing. (Adapted from various sources.)

represent parts of the upper mantle. Ophiolites include gabbro batholith intrusions, multiple sheeted dike complexes that probably served as feeder dikes to the upper pillow lavas of basalt that form in the erupting submarine spreading centers. They are overlain by fine-grained deep-sea muddy oozes that harden into cherts and limestones. Such ophiolite suites are commonly incorporated into landmasses and represent bits of ancient sea floor now caught up in mountain ranges.

Although the discoveries about sea-floor spreading fitted into the context of continental drift as it was conceived in the 1960s, they also raised new questions. If new material was rising and moving away from the mid-ocean ridges that formed the divergent margin of a plate, what was happening at the other end—the convergent margin of a plate? It was soon recognized that where plates came together a deep-sea trench developed and one plate sank, or subducted, beneath the other, to be consumed deep in the mantle. Subduction involes dense oceanic lithosphere sinking back into the mantle, largely under its own weight (Fig. 2.7). The evidence for this process comes mainly from seismology (Tatsumi 2005).

Most volcanic and earthquake activity occurs along plate edges as direct by-products of rifting, plate movement, and subduction. Four basic types of areas have earthquakes (Turcotte and Schuster 2002). Along the mid-ocean ridges, high heat flow and volcanic activity is caused by the stretching of the Earth's surface (Fig. 2.8). Mid-ocean earthquakes there are generally of shallow focus, originating at depths of less than 70 km (~44 mi).

Second, shallow earthquakes occur along *transform faults,* where one section of the Earth is sliding past another; examples include the San Andreas Fault in California and the Dead Sea–Jordan Fault in the Middle East, along which so many earthquakes occurred in antiquity. Third, a belt of shallow- and intermediate-focus earthquakes extends from the Alps to the Himalayas; it is apparently associated with compressive forces responsible for the creation of these mountains. In general, shallow-focus earthquakes pose the greatest danger because they are the most numerous and closest to where people live. Other earthquake areas are the deep-sea trenches and volcanic island arcs (Fig. 2.9). The foci of earthquakes occurring in these regions may be shallow, intermediate, or as deep as 700 km (~440 mi), depending upon their exact location in the subduction zone (Strahler 1998; Lliboutry 1999). Independent tracing of the different depths of earthquake foci in these areas by geophysicists in the mid-20th century finally allowed the naming of the downward-slanting seismic features, known as Wadati–Benioff zones, that define where a descending subduction slab is grinding beneath the upper plate (Wadati 1935; Benioff 1949, 1954; Oliver and Isaacs 1967). Therefore, by using sensitive seismographs to locate the foci of these earthquakes, it is possible to establish the position and steepness of the subduction zone. Earthquakes do not occur below 700 km (~440 mi) because descending lithosphere has heated up sufficiently to be ductile, thus allowing deformation of the crust without earthquakes. Some subduction zones do descend any further than this;

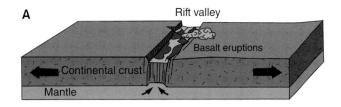

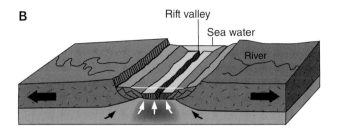

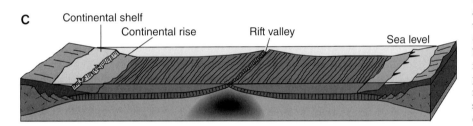

FIGURE 2.8 Generalized block diagrams that show three stages of: (A) the uplift, rifting, and volcanic activity along a new spreading center beneath a continent; (B) formation of a new narrow ocean basin, floored with new basaltic oceanic crust; and (C) the widening of the new ocean basin as it rides down off the spreading center and subsides as stable or passive continental margins that receive miogeoclinal and eugeoclinal sediments from the continents. (After Plummer et al. 2003.)

subduction slabs are known to travel all the way through the 2,900 km (~1,800 mi) thick mantle to the core.

This conveyor-belt system involves creation of new crust at the divergent mid-ocean ridges and reincorporation of the plate's leading edge back into the mantle at the convergent subduction zones. The material in between is slowly transported from one point to another. Continents drift, but they do not, as formerly envisioned, plow across a viscous sea bottom; rather, they are passive passengers on lithospheric plates moving outward from oceanic spreading centers. Mechanisms that drive plate tectonics are thought to be largely due to convective flow in the mantle—in which warm, less dense rock rises and cooler, more dense material sinks. In this context, *ridge-push* results from the elevated position of the oceanic ridge system, so that oceanic lithosphere slides gravitationally down the flanks of the ridge, and *slab-pull* occurs where old oceanic crust, which is relatively cool and dense, sinks into the mantle and "pulls" the trailing lithosphere along. *Slab-suction* occurs at a plate margin where the descending plate pulls at the upper plate, and as a result the subduction zone can migrate in the direction opposite that of the lower plate's movement. The suction force pulling on the overriding lithosphere can move it toward the trench, or even rift it apart.

The lithospheric plate system consists of twelve major plates. Six of the twelve are "great plates." The remaining six are intermediate to comparatively small. The great Pacific Plate occupies most of the Pacific Ocean basin and consists almost entirely of oceanic lithosphere. Its relative motion is northwesterly, so that it has convergent subduction zones along much of its northern and western edges, and huge island arcs of volcanoes have developed on the peripheral parts of the plates above the subduction zones. A spreading boundary occurs along much of its eastern and southern edges. The North American Plate consists of the continent and the western half of the Atlantic Ocean; the entire unit is moving to the west, where it is colliding with the Pacific Plate. The eastern side of the Atlantic Ocean basin is part of the vast Eurasian Plate, which is generally moving east and colliding with the western edge of the Pacific Plate. Thus the Atlantic Ocean basin is opening and the Pacific Ocean basin is closing. Because continental crust is composed of low-density rock and is more buoyant than oceanic crust, it cannot be subducted. Consequently, the Pacific Plate, which is composed of oceanic crust, is descending beneath the North American and Eurasian Plates, which carry continental crust on them, and it is undergoing subduction in the deep-sea trenches that occur on its margin—the Aleutian, Kuril, Japan, and Marianas trenches (Fig. 2.10).

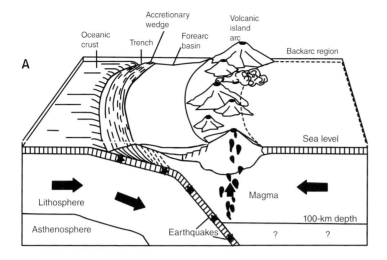

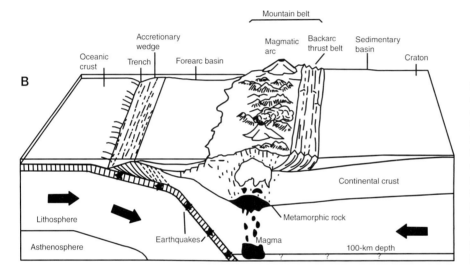

FIGURE 2.9 Generalized view of subduction in: (A) a deep-sea trench in ocean-to-ocean convergence, with an island arc of composite volcanoes moving along above it; and (B) a deep-sea trench in ocean-to-continent convergence, where an accretionary wedge forms on the edge of the overlying plate and magmatic arc forms in the mountain belt. The buoyant magma plumes that rise from the subduction zones in both cases are responsible for building the island arcs.

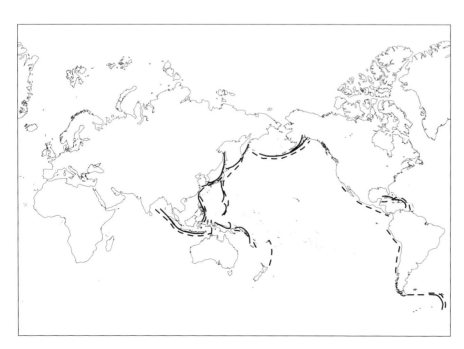

FIGURE 2.10 Distribution of island arcs (solid lines) and deep-sea trenches (dashed lines). Both phenomena occur mostly around the Pacific, a closing ocean. (Adapted from various sources.)

Plate tectonics provides a broad and unifying framework into which virtually all aspects of Earth science fit. Although details are still being added, especially in the more complex and older continental areas, the theory explains the movement of continents, as well as such formerly perplexing problems as the youthfulness of the bottoms of the ocean basins. The origin of the oceans has long been a great enigma: They are thought to have formed soon after initial *accretion,* when the crust had cooled enough to allow liquid water. The Earth is over 4 billion years old, yet the sea-floor rocks are less than 180 million years old. This disparity can be explained easily in plate tectonic theory. New sea-floor crust is continually produced at the mid-ocean ridges and destroyed at the subduction zones; as a result, it can never be very old. The continents, on the other hand, can be broken apart and reassembled in various ways as continental plates are rifted apart or sutured together in collisions, and because of their low density they are not subducted easily; the basic amount of continental material stays the same or increases through time. Erosion may wear down the land and transport the low-density sediments to the sea, but much of this material is reconstituted by being carried down subduction zones, melted, and then rising up as igneous intrusions and added to existing continental masses through mountain building. Continents thus display accretion of younger material and active tectonism at their edges, whereas their interiors are commonly older and more stable.

Mountain Building and Plate Tectonics

The essential feature in mountain building through plate tectonics is the plate boundary, of which there are three main types, each having different characteristics. Divergent rifting occurs where spreading centers develop beneath a lithospheric plate, such as the rift valleys of East Africa (Wright et al. 2006). As magma rises beneath a continent, the overlying crust is weakened and domes upward above the hot, rising plumes. Such uplift stretches and thins the crust, which ruptures along linear rift valleys, allowing formation of multiple normal fault scarps, down-dropped *grabens,* upraised *horst* mountains, and extrusive volcanoes from escaping magmas. Where such orogenic processes continue for a long time, the rift valley will eventually open to an invasion by the sea. Further progressive rifting and slow movement of the continental plate down from upraised spreading centers will cause subsidence of the fault mountains and volcanoes beneath the sea, where they become part of a new passive-margin accumulation wedge of thick, shallow-water sandstones, limestones, and shales, similar to the Atlantic Ocean and its coasts. Over time, some passive margins can become unstable and begin to subside into a newly developed subduction zone, where the oceanic part of the plate now begins to dive beneath the continental portion.

In some cases, however, a number of passive margins seem to have maintained their fault or warped mountain relief for a long time, perhaps as a result of enigmatic swelling forces (isostasy, thermal expansion, intrusion, renewed uplift) on passive continental margins. The Western Ghats of southern India, the Great Escarpment of eastern Australia, and the Drakensberg of South Africa constitute erosional mountains of passive-margin plateaus (Ollier and Pain 2000).

In a transform plate boundary, plates slide past each other; here, small volcanoes can occur and other small mountains are produced by buckling, especially where faults change direction. Convergent plate margins, where plates come together and subduction takes place, are the most important for mountain building. Subduction convergence of oceanic lithosphere produces two basic different kinds of orogenic belts, depending upon the nature of the overriding plates: the Aleutian-type island arc or the Andean-type mountain belt (Kearey and Vine 1996).

Aleutian-Type Island Arc

Island arc chains of volcanic islands curve in a convex arc toward the open ocean, with a deep-sea trench on the descending plate. In this ocean–ocean convergence, subduction beneath an oceanic plate generates a magmatic arc of volcanoes and associated features on the opposing plate. Related processes include partial melting of the mantle wedge located above the subducting plate; sporadic volcanic activity; emplacement of magmas as batholiths at depth; high-temperature metamorphism of the deeper surrounding rock; and the accumulation of an *accretionary wedge* of sediment scraped off the subducting plate into the chaos of folded and faulted marine sediment, along with pieces of oceanic crust that are subjected to low-temperature, high-pressure metamorphism.

Regional extension or spreading can also occur behind or within island arcs. Such backarc spreading can tear an arc in two, move an arc away from a continent, or split off a small part of a continent, which seems to be how Japan formed. This spreading also forms oceanic crust similar to that at the crest of mid-ocean spreading ridges, which seems to be the type of crust occurring in most ophiolites. The causes of backarc spreading are diverse and much debated. In some cases the overlying plate seems to move away from the subducting plate, leaving the island arc in place with a new spreading area behind. Alternatively, the down-going subducting plate may either generate a rising and spreading mantle *diapir* of molten rock above it, or drag on the overlying mantle, causing it to form secondary convection cells that fracture the overlying oceanic crust.

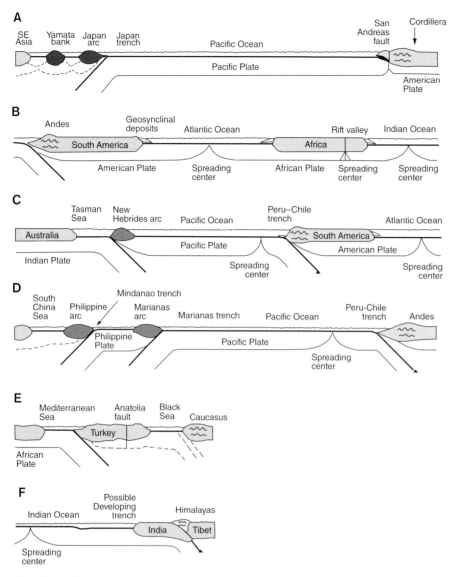

FIGURE 2.11 Schematic sections at various locations showing relationships among lithospheric plates, continents, oceans, spreading centers, island arcs, deep-sea trenches (subduction zones), and mountain building. (Adapted from various sources.)

Island-arc volcanoes grow larger over time as material is brought into the subduction zone, consumed by melting, and reinjected upward as magmas. Island arcs contain many of the world's most destructive volcanoes. These are primarily the large *composite cones* built from silica-rich andesite and rhyolite lavas, which are cooler and more viscous than the low-silica basalt types, with the result that the silica-rich types choke their vents and cause huge explosions of *pyroclastics* (*blocks, bombs, cinders,* and *ash*). Island arcs, and the volcanoes they contain, are especially common in the Pacific Ocean (Fig. 2.11), where they occur around the periphery in a zone known as the "ring of fire." A few island arcs also occur along the American coasts, including the Aleutians of the Alaskan coast, the Lesser Antilles of the Caribbean,

and the South Sandwich Islands in the Scotia Arc at the tip of South America.

At different times in the geologic past, island arcs have occurred off the east and west coasts of North America and the west coast of South America, but these have been incorporated into the main continental blocks by plate-tectonic collisions. Ophiolite sequences commonly mark the suture of their collisions. On the west coast of North America, the Cascades and the Sierra Nevada mountains are made in part of ancient island arcs, and in the east, the low Taconic and Berkshire Mountains along the borders of Vermont, Massachusetts, and New York State are the remains of a very old island arc. Numerous other island arc fragments are also known to be incorporated into the Appalachian Mountains (Fig. 2.12).

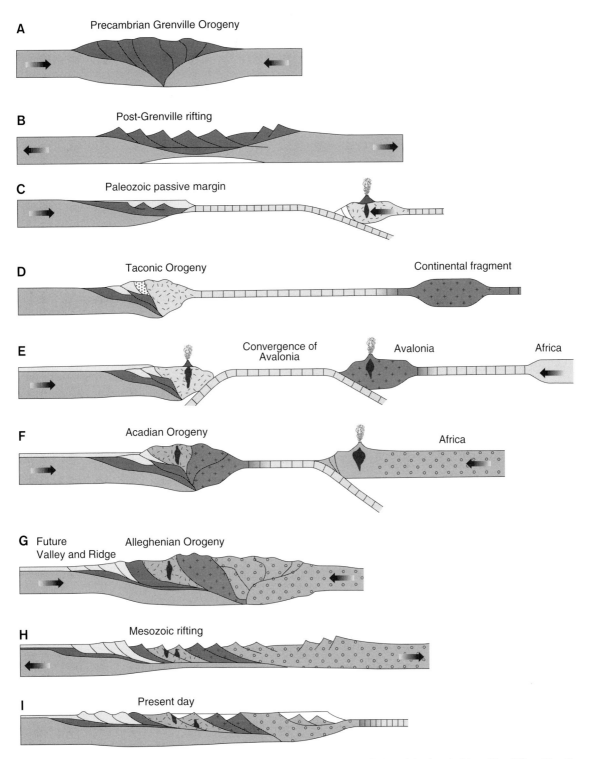

A Precambrian Grenville Orogeny

B Post-Grenville rifting

C Paleozoic passive margin

D Taconic Orogeny   Continental fragment

E Convergence of   Avalonia   Africa
        Avalonia

F Acadian Orogeny   Africa

G Future   Alleghenian Orogeny
       Valley and Ridge

H Mesozoic rifting

I Present day

FIGURE 2.12 Idealized and schematic interpretation of events showing the tectonic evolution of the Appalachians. The different locations and dip directions of all the subduction zones are not known precisely. Although mountains do form in rifting events, geologists generally only assign names to convergent or collisional events when significant igneous and metamorphic rocks are produced. (A) A continent-to-continent collision in the Late Proterozoic produced the crystalline rocks of the Grenville Orogeny. (B and C) A rift subsequently developed between the two continents and the ancient Atlantic Ocean of the Paleozoic was formed, complete with continental shelves (geoclines) and offshore island arcs. (D) The Taconic Orogeny resulted when island arcs collided and were sutured onto the continent, together with geoclinal sediments. (E) The continental fragment of Avalonia and parts of northern Europe next collided and were sutured onto the continent in the (F) Acadian Orogeny. This was followed in the Alleghenian Orogeny (G) by collision with Africa. (H and I) In Mesozoic time rifting took place once again, leaving part of Africa behind under what is now Florida, and a new complex of geoclinal sediments was deposited as a continental shelf in the newest Atlantic Ocean. That this process is continuing is demonstrated by the fact that the Atlantic is still growing wider at the rate of ~3 cm/yr (1.2 in./yr). (Adapted from various sources.)

Andean-Type Mountain Belt

The first stage in the development of ocean–continent convergence and formation of an idealized Andean-type mountain cordillera occurs prior to the formation of a later subduction zone. During this earliest stage, the continental edge exists as a passive margin which accumulates sandstones, limestones, and shales in the shelf miogeocline and the deeper water eugeocline. At some point the passive continental margin can become active. A subduction zone then develops beneath the continental plate and gives rise to a new magmatic arc together with a suite of associated features similar to the Aleutian-type of island arc. This includes andesitic volcanoes of intermediate silica composition on the overriding plate margin. Partial melting occurs in the mantle wedge located above the subducting plate, magma is emplaced at depth as batholiths, high-temperature metamorphism affects the deeper surrounding rock, and an accretionary wedge accumulates. Prolonged subduction can lead to an accretionary wedge that is large enough to rise above sea level (Strahler 1998; Brandon et al. 1998).

Another reason for the growth of the cordilleran mountain belt is the stacking up of fold and thrust-fault sheets on the continental or backarc side of the magmatic arc. Slices of the mountain belt are moved landward over the continental interior or craton, in some cases as the cold, rigid craton is pushed beneath the hot, mobile core of the mountain belt. These active continental margins, with the landward chain of volcanoes and the seaward accretionary wedge, are subject to extensive, long-term deformation and metamorphism. Both Aleutian-type island arcs and Andean-type mountain belts develop progressively larger size through long-continued subduction and associated sedimentary, igneous, and metamorphic activities.

Collisional Mountain Ranges

Collisional mountain ranges develop from Andean-type orogenesis, and are complex because of the variety of low-density lithosphere that can be brought into the subduction trench. Two kinds of island arc–continental collisions are possible: one where the subduction dips beneath the continent and one where it dips away. Where the subduction zone is initially beneath the continent, as the oceanic plate moves toward the continent, it can carry an older inactive or mature island arc on it to eventually collide with the continental margin. Alternatively, where the subduction zone dips away from the continent and beneath the island arc, the closing of the small oceanic basin between continent and island arc will eventually suture the island arc to the continental edge. In both cases, because of its buoyancy, the island arc will not subduct beneath the continental plate but will instead plow into the continent, deforming both blocks. When no further convergence is possible, a new trench and subduction zone can develop on

the seaward side of the accreted island arc (Dewey and Bird 1970; Strahler 1998).

In other cases of accretion of continental fragments, both small and large, their low density and buoyancy prevent their descent into the subduction zone. The ensuing collision of buoyant crustal fragments with the continent forces the stacking of crustal slabs into thrust slices embedded in the Andean-type margin. Small microcontinents can be diminutive islands or as large as Madagascar; there can even be submerged continental fragments known as oceanic plateaus. There may also be submerged volcanic plateaus created by massive outpourings of lava from hot-spot activity, or extinct volcanic islands such as the Hawai'ian Island–Emperor Seamount chain. Some smaller crustal fragments can ultimately become terranes accreted to the main continental mass; these commonly have a geologic history quite distinct from that of the adjoining regions. Alternatively, large continental masses, such as the well-known example of the Indian plate—which broke from the Gondwanaland supercontinent in Mesozoic time and moved north through the Indian Ocean to dock with Asia in Cenozoic time—represent massive continent-to-continent collisions (Fig. 2.13). The majestic Himalaya and the massive Tibetan Plateau resulted from this continent-to-continent collisional convergence. In collisional convergence, the remains of ancient spreading centers, sea-floor basalts, and overlying deep-sea sediments can be caught up in the collision and sutured into the collisional mountain range. These ophiolite suites are a clue to the presence of the former ocean bottom and help to decipher the sequence of collisional events, although whether most ophiolites are typical sea floor, or due only to backarc spreading, is much debated.

Erosional Mountains

Mountain building as a result of plate-tectonic divergent and convergent stress is the traditional understanding of orogenesis. Nevertheless, at crustal and lithospheric scales, gravitationally induced stress is as important as tectonic stress, with the result that syntectonic deformation can be modified significantly by erosion (Beaumont et al. 1992). Paradoxically, the shaping of mountains can depend as much on the destructive forces of erosion as on the constructive power of tectonics (Pinter and Brandon 1997). In fact, after more than a century of viewing erosion as the weaker brother of tectonics, many scientists now elevate erosion to the head of the family. As Hoffman and Grotzinger (1993) noted, "savor the irony should those orogens most alluring to hard-rock geologists owe their metamorphic muscles to the drumbeat of tiny raindrops."

Erosion actually builds mountains in a variety of ways. Isostasy represents a kind of crustal buoyancy in which the crust of the Earth floats on top of the denser, deformable rocks of the mantle. Mountain belts stand high above the surrounding lowlands because they have buoyant, less dense crustal roots that extend deep into the upper mantle supporting them from below. Analogously, any floating

A Ocean–continent convergence

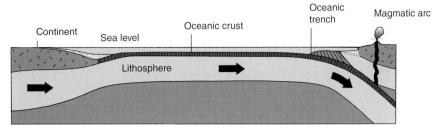

B Ocean–continent convergence

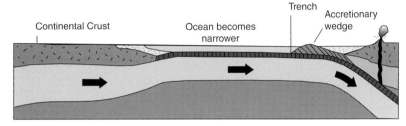

C Continent–continent collision

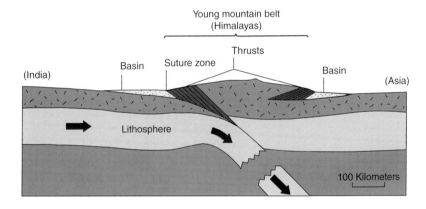

FIGURE 2.13 One interpretation of the sequence of events in continent–continent collision. A subduction zone develops between two plates, each carrying continental material, allowing them to converge and combine, with ophiolite rock remnants of the former ocean bottom caught in a suture between them. Because lighter, less dense continental rock cannot be subducted, the entire mass is compressed, metamorphosed and intruded with igneous rocks, and buckled into high mountains (e.g., the Himalayas). The former subduction zone eventually becomes inactive; another can form somewhere else. (Adapted from various sources.)

body creates the addition of mass and loading. For example, increased glacial ice or a new lake from a dammed river will cause subsidence of the crust, whereas removal of mass through deglaciation or massive erosion of rock will cause isostatic uplift of a mountain mass. If a high plateau undergoes overall erosion without significant dissection, isostasy will uplift the whole plateau, but not quite to its original level. If, however, erosion of the plateau carves many deep canyons and leaves only a few mountain ridges and peaks behind, the entire regional block will have less mass and will float isostatically upward. Its average surface level, falling somewhere between ridge and peak top and canyon valley bottom, will rise toward the original plateau surface level, but this will carry the residual mountain peaks considerably higher than they were before, when they were part of the plateau (Montgomery 1994). In this fashion, dissection of a high plateau will generate mountain peaks well above the original plateau surface. This seems to have happened in the eastern Himalaya, where deep glacial erosion

during the Pleistocene seems to have raised the peaks above their pre-Pleistocene tectonic levels.

The literature abounds with many other minor examples of erosional mountains, in which a variety of special features are produced by erosional isolation of particular rock structures. Fairbridge (1968) divided them into (a) structural, tectonic, or constructional forms, and (b) denudational, subsequent, destructional, or sequential mountain forms. Recent understanding of how erosion actually builds mountains recognizes the intricate linkages and feedbacks between tectonics, isostasy, climate, and erosion (Hoffman and Grotzinger 1993; Pinter and Brandon 1997; Koons et al. 2002; Stewart et al. 2008). Orogeny leads to orography, the perturbation of the regional climate by topography. An asymmetric pattern of precipitation results, wherein a dominant wind direction or storm track has moisture drawn out from the rising, cooling air mass on the windward side of the range, thus producing a rain shadow to the leeward, where the air is descending and warming. Erosion is enhanced on the windward side and reduced on the lee-ward (Fig. 2.14). In ranges near oceans, where prevailing winds are in the same direction as subduction, erosion denudes the coastal side of the range, which effectively "pulls" or unloads the rocks so that the compressive forces of collision, coupled with isostasy, move the buried rocks toward the surface. Where prevailing winds blow offshore and opposite to the direction of subduction, erosion is concentrated on the inland side of the range, exposing the deepest and most deformed and metamorphosed rocks in that area. The control of erosional agents—gravity, water, wind, and ice—acting upon any particular mountain landscape depends largely upon the local climate, the steepness of slopes, and local rock types (Pinter and Brandon 1997). In addition, high-altitude/high-latitude mountain areas tend to be protected by cold-based glacier ice frozen to its bed, whereas at lower altitudes and latitudes, warm-based ice with basal meltwater is highly erosive. Similarly, other factors that decrease erosion in mountain areas are rock strength, greater aridity, and gentle slopes.

In recognition of the unusual juxtaposition of numerous massive Himalayan peaks over 7–8 km high that are directly adjacent to major high-order rivers at low altitudes at their base, Zeitler et al. (2001a, 2001b; Finnegan et al. 2008) discovered that some actively uplifting metamorphic massifs are a direct consequence of erosion. Intense localized river erosion initiated as a result of capture of a major river by a steeper but smaller one is capable of focusing deformation into small areas being rapidly eroded. Intense metamorphism at depth results in a "tectonic aneurysm" that is accompanied by a bowing up of the brittle–ductile transition zone, decompression melting, and intrusion (Fig. 2.15). New crustal mass is advected in at depth to replace the continued loss of surficial mass from the local high elevation and high relief that maintains the rapid surficial exhumation in a closed feedback loop. Passage of the rising mountain top though high altitude causes formation of cold-based, protective ice that helps maintain steep slopes against mass failure or further glacier erosion. Upward movement of rock at depth into the tectonic aneurysm is thought to produce the distinctive petrology and structure of mantled gneiss domes. These features are known from elsewhere in the Himalaya where deep erosion is known or suspected (Finnegan et al. 2008; Shroder et al. 2011), as well as in other ancient collision mountain zones, where these initially deep internal dome structures are ultimately exposed at the surface by deep erosional unroofing (An Yin 2004; Fletcher and Hallet 2004; Mahéo et al. 2004).

Erosional Relict or Residual Mountains

In many continental interior cratons, where tectonic stability has existed for millions of years, long-continued, deep weathering to saprolites and their erosion has left behind residual bedrock mountains. Examples occur on all continents, but those of Australia, Africa, and South America

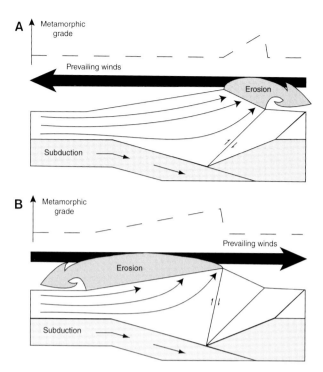

FIGURE 2.14 The direction of prevailing winds over a mountain range can have profound effects because of the effects of orography. Mountains lift the air flowing over them, causing strongest precipitation on the windward side. This causes maximum erosion on that side, with the consequent exposure of the most deformed and metamorphosed rocks there. The growth and exhumation of mountains are thus affected differently, depending upon whether the direction of subduction and prevailing winds are in the same or the opposite direction. (After Willet et al. 1993; Pinter and Brandon 1997; Keller and Pinter 2002.)

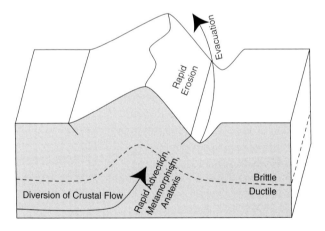

FIGURE 2.15 Diagram illustrating dynamics of a tectonic aneurysm, shown at a mature state. The feature begins with a capture or diversion of a major river, producing a deep gorge that weakens the crust and draws advective flow of crustal rock at depth toward the topographic gap. This builds relief, leading to further rapid erosion rates, which further weaken the crust, allowing updoming of the brittle–ductile transition, as well as a steepening of the thermal gradient. Hot crustal rock can experience decompression melting and intrusion (*anatexis*), as well as low-pressure/high-temperature metamorphism as it moves rapidly toward the surface. (After Zeitler et al. 2001b.)

are the best known. The famous Ayers Rock (Uluru to the Aboriginal Australians) is such a remnant, consisting of Precambrian sandstone. In Africa and South America *inselberg* (German for island mountain) landscapes abound. Where near-vertical regional joints in the bedrock are widely spaced, curved sheeting joints that crudely parallel the Earth's surface can cause isolated, solid-rock inselbergs to develop smooth, domal *bornhardt* or sugarloaf shapes. Half Dome in Yosemite National Park and the great Sugarloaf Mountain with the colossal Christ figure in the harbor of Rio de Janeiro, Brazil, are examples. Where regional and sheeting joints are closely spaced and strongly weathered, corestones develop; tors or castle koppies (kopje) also result from stripping away of weathered saprolites.

Principal Mountain Types

Individual mountains or small mountain ranges are also formed in a variety of ways. Turning to Fairbridge's (1968) pre–plate tectonic, generic classification of mountains, and leaving aside the most minor types and those discussed above, we have various categories of structural, tectonic, or constructional forms: volcanic mountains, fold and nappe types, fault-block mountains, and dome mountains, all with various degrees of erosion.

Volcanic Mountains

Volcanoes result from both molten and solid material that is ejected from the Earth to accumulate on the surface in various sizes and forms (Decker and Decker 2006). Volcanic mountains are classified into four types by differing eruptive habits and constituent materials (Fig. 2.16). Volcanic activity commonly begins as gas-rich magma moves forcefully to the surface through a circular conduit pipe, and more rarely through a linear fissure in the Earth. Successive eruptions of different lavas or of semisolid or solid particles known as pyroclastics (Greek for "fire-broken"), or a combination of these, commonly separated by long periods of inactivity, build the structure of a volcano.

Small, steep-sided volcanic mountains include *ash* and *cinder cones* and silica-rich rhyolite *lava domes*. Ash and cinder cones result from explosive eruptions of pyroclastics. Volcanic domes are built of slow-flowing, cooler lava, which, because of high viscosity, also tend to plug volcanic conduits and lead to later explosions. Among the largest of volcanoes are *shield volcanoes*, gentle-sided, massive accumulations of fluid basaltic lava, such as those of the Hawai'ian Islands, resembling a gigantic warrior's shield emerging from the sea. *Composite cones* or *stratovolcanoes* are equally huge, steep-sided accumulations of interspersed lava flows and pyroclastics that pile up into a major mountain form.

Volcanic eruptions provide spectacular evidence of their ongoing construction. In some cases they can develop with incredible speed. The Paracutín ash and cinder cone in

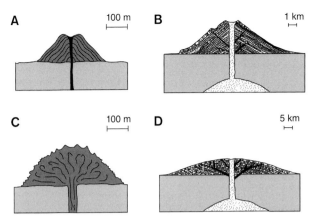

FIGURE 2.16 Volcanic mountain features shown in schematic cross sections arranged according to relative size and typical explosiveness of eruption. (A) A cinder cone or small steep-sided volcano formed by explosive eruption of pyroclastics. (B) A small, quiet effusion of high-viscosity, rhyolitic lava forming a dome with steep sides. (C) A composite or stratovolcano formed from multiple, alternating eruptions of lava and pyroclastic deposits, whose interlayering and imbrication provide greater strength and result in a large, steep-sided cone. (D) A shield volcano, built from many outpourings of low-viscosity basaltic lava which produce a large, gently arched surface. (Adapted from various sources.)

Mexico emerged from a small depression in a farmer's field in 1943 and grew to 40 m (130 ft) the first day and 140 m (460 ft) in a week. By the end of the first year it had grown to a height of 325 m (1,065 ft). After nine years of sporadic pyroclastic eruptions and nearly continuous discharge of lava from vents at its base, activity ceased and the mountain went into dormancy or extinction (Anonymous 1982; Luhr and Simkin 1993; Decker and Decker 2006).

Volcanoes occur on all continents and make up a substantial percentage of the world's mountains. The number becomes even more impressive if we include those hidden by oceans—most islands are nothing more than the summits of undersea volcanoes. If one includes the submerged as well as the visible part, the world's tallest and largest peak is a volcano—Mauna Loa in Hawai'i. The above-water portion of this gently sloping shield volcano, at 4,169 m (13,680 ft), is not particularly impressive, but it rises a total of 9,100 m (30,030 ft) from its sea-floor base, which is over 225 km (~140 mi) in diameter. The highest volcano on land, at 6,267 m (20,556 ft), is Mount Chimborazo in the Andes of Ecuador. The highest known mountain in the solar system is the giant shield volcano of Olympus Mons on Mars. Its basal diameter is 600 km (375 mi), more than the width of Texas, and the mountain is about 27 km (~17 mi) high, more than three times higher than Mount Everest. Mars has no plate tectonics, so Olympus Mons was fixed in place and grew larger by long-continued eruptions.

Volcanoes are linked to three basic regions: rift-valley spreading centers, such as Iceland on the Mid-Atlantic rift, or Mount Kilimanjaro and many other volcanoes on the

East African rift system; on the plate above subduction zones; and above intraplate "hot spots" or mantle-plume thermal centers, where plumes of solid yet mobile mantle rock rise toward the surface like the blobs in a lava lamp (Strahler 1998; Macdonald 1972; Ollier 1988; Scarth 1994; Decker and Decker 2006). In the latter case, the mantle plume has a bulbous head that draws out into a narrow stalk as it rises. As the head nears the top of the mantle, decompression melting generates basaltic magmas, some of which can pour out on the surface in vast floods; or, if the plate carries the prior solidified outpourings along from the source hotspot, a chain of volcanoes can result. The Hawai'ian Islands are a northwest–southeast string, with the oldest to the northwest and the youngest to the southeast (Macdonald and Abbott 1970). This strung-out pattern is the result of the Pacific Plate moving to the northwest and the islands continually being moved away from the hot spot (Condie 2005; Decker and Decker 2006). The youngest island, Hawai'i, with the Mauna Loa volcano on it, now overlies the hot spot. Hawai'i continues to drift to the northwest, with the result that, although the Kiluaea volcano there is still active, already a new eruptive center is developing underwater offshore to the southeast. Eventually the island of Hawai'i will become entirely extinct as it drifts away from the hot spot and the new underwater volcano rises above the waves. The volcanoes of the Snake River Plain and Yellowstone National Park represent a similar situation and result from lava outpourings from the Yellowstone Plume and hot spot (Condie 2005; Greeley and King 1977; Decker and Decker 2006).

VOLCANIC MOUNTAIN HAZARDS

People have lived around volcanic mountains for thousands of years, in part because of the fertile soils there and limited other available land, but also because, most of the time, volcanoes are not hazardous. But as anyone knows who has watched the common presentations of volcanic disasters on television in recent years, a wide variety of deadly and destructive events are possible—even probable in some areas. Volcanoes kill and destroy in several different ways: lava flows, ash falls, pyroclastic flows, *lahars*, and *phreatic* explosions. Pyroclastic flows (also known as fiery clouds, glowing avalanches, or *nuées ardentes*) are mixtures of hot gases infused with incandescent ash and larger fragments. Because the pyroclastics are buoyed up by the hot gases, the whole is heavier than air, and they race down the steep slopes of volcanoes commonly at speeds greater than 200 km/hr (125 mi/hr). In 1902 a small volcano, Mt. Pelée, on the island of Martinique in the Lesser Antilles, produced a fiery cloud that destroyed the city of St. Pierre, along with 28,000 of its inhabitants (Anonymous 1982; Decker and Decker 2006). Lahars, or volcanic mud and debris flows, are common on volcanoes that have lakes or glaciers, which can melt to provide plentiful water to mobilize loose

pyroclastics into a devastating slurry of rapidly flowing, wet debris. A lahar on the volcano Nevado del Ruiz in Colombia killed 23,000 people in 1985. The volcanic heat melted glacier ice, which mobilized mud and pyroclastics, which then raced down the steep slopes of the volcano and overwhelmed the town of Armero (Wright and Pierson 1992; Decker and Decker 2006). Phreatic explosions occur when underground water comes in contact with rising magma, flashes into steam, and bursts out of the ground. The crater mountain of Diamond Head in Honolulu, Hawai'i, is the result of a phreatic eruption. In the year 79 C.E., the people of ancient Pompeii, at the foot of the mountain now called Vesuvius near Naples, Italy, discovered to their horror that catastrophic ash falls and pyroclastic flows cannot be escaped if one lives in the direct vicinity. Failure to heed the warnings of an impending eruption—common ground tremors and precursory minor eruptions—too often has led to major loss of life in eruptions.

The explosive eruption of the stratovolcano Mount St. Helens in the Cascade Range of Washington State, USA, in 1980 (Lipman and Mullineaux 1981; Decker and Decker 2006) occurred when new magma moved slowly up into the volcano after several centuries of dormancy. A bulge formed on the north side of the mountain, which grew at a rate of ~1.5 m/day (~5 ft) until after an intermediate-sized earthquake it collapsed abruptly in a massive landslide. This released the pressure on the gas- and steam-filled magma, which then burst forth in a superheated (500°C/932°F), supersonic lateral blast, devastating a zone 20 km (13 mi) outward from the volcano and 30 km (19 mi) wide. Within seconds a large vertical eruption sent a cloud of volcanic ash to an altitude of about 18 km (11 mi). More than 60 people were killed. Study of the volcanic landslide that triggered the eruption enabled recognition of many prior events elsewhere involving volcanoes that had collapsed and sent out large rock and debris avalanches. This newly recognized hazard is now a part of the disaster planning for the monitoring and warning systems that exist around many volcanic mountains in the developed parts of the world.

Faulted and Folded Mountains

The major mountain ranges of the world are eroded out of rocks that were broken and crumpled by the forces of plate motion. Most geologists differentiate between intense deformational forces that cause faulting and folding, and the more general uplift that elevates landscapes. Many deformation structures are formed deep within the Earth and then uplifted to produce mountains. In most cases, the geologic structures and the landforms eroded from them are separate and distinct. In some regions, however, faults and folds are forced up out of the ground in active deformation, and the structure and the landform are commonly the same. Such *neotectonic* activity generally produces a distinctive suite of landforms that enable interpretation of the sequence of structural and geomorphic events that

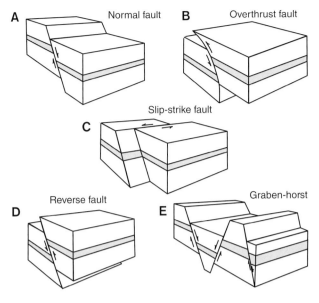

FIGURE 2.17 Schematic and simplified representation of several major types of faulting associated with mountain building. The history of erosion of many of these scarp types is important in understanding their characteristics and history. (After Price 1981, adapted from California Geology 1971.)

produced them. It is important to distinguish between previously deeply buried, old rock structures that have later been exposed and fashioned into landforms by erosion, and those structures that are landforms by virtue of their recent deformation at the surface.

FAULTED MOUNTAINS

Faulted mountains result from displacement of the Earth's surface along fracture zones (Fig. 2.17). This movement is generally very slow, taking place in short jerks over thousands to millions of years. Displacements may be in any direction whatsoever, but commonly one block is raised past another. Features associated with faulting include juxtaposition of different types of material, raised escarpments, local alteration or crushing of rocks, and various sorts of surface disturbances, such as the damming or rearrangement of stream flow. The forces that stretch (tension) the crust and forces that squeeze (compress) it each produce their own characteristic forms of faulting. Tension tends to create abrupt and spectacular fault scarps, dropping or raising crustal blocks with respect to one another; this is *normal faulting* (Fig. 2.18). If a central block drops in relation to the land on both sides of it, a fault trough or graben (German for "grave") results. Normal faults and grabens are characteristic of rift valleys associated with plate-tectonic spreading centers. Fault-scarp mountains along such rift valleys that have been invaded by the sea can rise in a spectacular array of steep slopes to over 3,000 m (~10,000 ft asl) high, as along the coast of the Red Sea. A block that is raised with normal faults on both sides, such as between two grabens, is called a horst (German for "eagle's nest"). The Rhine Valley in Germany is a graben; the Ruwenzori Mountain Range or "Mountains of the Moon" in East Africa is an example of a horst rising from the grabens of the African Rift system. Down-faulted grabens in rift valleys are among the lowest places on Earth outside of ocean

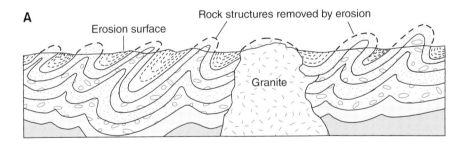

A
Erosion surface
Rock structures removed by erosion
Granite

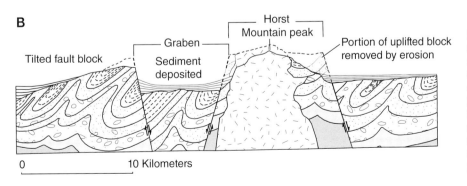

B
Tilted fault block
Graben
Sediment deposited
Horst
Mountain peak
Portion of uplifted block removed by erosion
0   10 Kilometers

FIGURE 2.18 Progressive development of fault-block mountain ranges. (A) Before faulting: The folds and intrusion of a granitic pluton during a mountain-building orogeny have been followed by erosion. (B) After block faulting: The tilted fault-block and horst mountains have been further eroded to produce sedimentation into the graben valley and nearby lowlands. (After Plummer et al. 2003.)

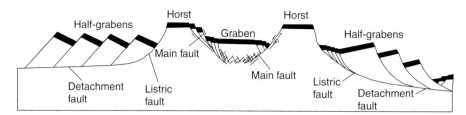

FIGURE 2.19 Systems of tensional normal faults commonly are characterized by a main fault with subsidiary faults and by low-angle detachment faults with imbricate fault blocks on the upper, hanging-wall block. Curved listric detachment faults that are gravity driven can move rapidly and may even resemble giant landslides. All such complex fault systems can produce a variety of surficial structural landforms. (After Twiss and Moores 1992.)

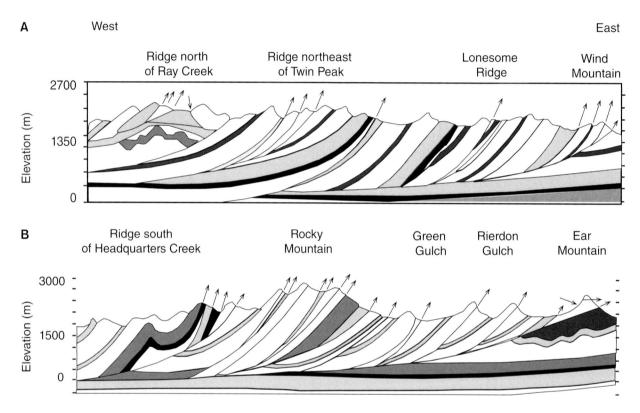

FIGURE 2.20 Cross sections of rock structures resulting from low- to high-angle thrust faulting in the Sawtooth Range of Montana. This commonly results in the juxtaposition of rocks of varying resistance and produces an extremely rugged landscape. Length of cross section is 16 km (10 mi). (Adapted from Deiss 1943, in Thornbury 1965: 391.)

basins, with the Dead Sea between Israel and Jordan being the world's deepest place on land below sea level. Death Valley in California is a graben that lies 86 m (282 ft) below sea level between two horsts, the Panamint and the Black Mountains of the Basin and Range geomorphic province.

Most of the Basin and Range in the western United States resulted from tensional forces. This region, centered in Nevada, has been broken into a series of fault zones oriented roughly northwest–southeast, and the resulting blocks have been tilted, lifted, and dropped relative to one another to create more or less parallel ridges and valleys. Many of the blocks occur as asymmetric massifs with one

side, a fault scarp, rising as much as 1,500 m (~5,000 ft) above the adjacent surface, with the other side dipping gently into the surrounding lowlands. From the Wasatch Range in Utah to the Sierra Nevada in California, the complexly eroded fault scarps of these ranges, and most of the others in the Basin and Range, are the result, depending primarily upon rock type, either of erosional downwearing of the original fault scarps to reduced angles, or of parallel erosional retreat at approximately the same angle. All the fault scarps are the result of abrupt incremental fault offsets of a few meters at a time, generally at intervals of centuries between earthquake events. Over millions

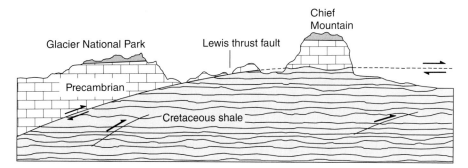

FIGURE 2.21 Generalized east–west cross section of the Lewis Overthrust in northern Montana. Older (Precambrian) rocks have been thrust over younger (Cretaceous) rocks along the low-angle fault for at least 24 km (15 mi). Chief Mountain is an erosional remnant (*outlier* or *klippe*) of the overthrust sheet, which at one time was much more extensive. (After Plummer et al. 2003.)

of years the recurrent offsets generate the mountain ranges of the present day. Bull (2007) assessed the typical landforms of these faulted mountain fronts in the southwestern United States to establish more than five classes of relative tectonic activity, from active maximal to minimal, as well as various types of inactive mountain fronts. He used various criteria to establish his classification, including mountain-front sinuosity, widths of valleys in the ranges, dissection characters of fault scarp facets, and other features.

The Sierra Nevada consists of several individual segments that have tilted more or less in unison to present a block ~650 km (~400 mi) long by ~80 km (~50 mi) wide. This huge block slopes gently to the west, whereas east-facing scarps rise abruptly along eroded fault scarps, presenting the highest mountain front in the contiguous United States. In a horizontal distance of less than 10 km (~6 mi), a drop of ~3,350 m (~11,000 ft) occurs between the higher peaks of the Sierra Nevada and the graben floor of Owens Valley. Similarly, the spectacular fault scarps of the Wasatch Range in Utah or the Grand Tetons of Wyoming display abrupt mountain fronts rising as much as 1,800 m (~6,000 ft) above the surrounding plains.

The cause of the tensional breakup of the Basin and Range Province after the prior lengthy regime of compressional uplift and thrust faulting that had produced the North American Cordillera has long been enigmatic. At first the application of a plate tectonics model seemed to offer an explanation. The plate in the western Pacific, which had long been subducting beneath western North America, was ultimately consumed and the transform faults of its spreading center overridden. Disappearance of this slab allowed hot mantle to rise and spread out beneath the province to cause horizontal extension of the crust. Later evidence of the exact timing of all this, however, did not corroborate the theory, and other explanations were sought. It is now thought that gravitational collapse of the Basin and Range area occurred after the strong compressional phase produced an overthickened collisional orogen that was gravitationally unstable. The best model of collapse of the orogen involves production of simple horsts and grabens, tilted blocks, and listric or domino-style faulting on a basal detachment surface with ductile deformation at depth that would fill in voids otherwise required if the blocks behaved in an entirely rigid fashion (Fig. 2.19; Strahler 1998; Stockli et al. 2003; Colgan et al. 2006).

Although normal faulting can be complex, the structures and landforms produced through displacement and deformation of strata are relatively simple compared with the bewildering disarray of high-angle *reverse* and gentle-angle *thrust faults*. These are caused by compressive forces that tend to squeeze and push one segment of the Earth's surface over another. Low-angle faults can pass upward to the surface at high angles. A series of high-angle faults or an imbricated thrust system will produce a more or less vertical stacking of beds that can erode differentially into a series of parallel ridges and valleys. The result is sharp and rugged topography, as in the Sawtooth Range of Montana, where thrust and reverse faulting produced juxtaposed harder and softer rocks (Fig. 2.20).

Low-angle thrust faulting tends to produce nearly horizontal displacement as low-lying older rocks are pushed over younger strata. The rock units involved may vary from a few tens to hundreds of meters in thickness, and they can override the adjoining surface for hundreds of kilometers. Erosion can then remove the weaker strata, leaving the more resistant strata as isolated remnants. Extremely confusing stratigraphic relationships are produced: A mountain peak can be composed of rocks older than underlying strata and not found elsewhere in the vicinity; alternatively, a mountain peak can be composed of rocks identical to those exposed far below in an adjacent valley, while sandwiched between are several hundred meters of younger rock.

Many overthrust faults occur in major mountain ranges. In the Lewis Overthrust near Glacier National Park, Montana, ancient Precambrian rocks were uplifted and thrust a distance of 24 km (15 mi) over younger shales and sandstones of Mesozoic age (King 1977). Chief Mountain is an *erosional outlier* remnant, or *klippe,* of that overthrust sheet (Fig. 2.21). Results similar to overthrust faulting can also be accomplished through other processes, such as gravity sliding and overturned folding (discussed below). All three of these processes can be interrelated, and it is commonly difficult to identify the relative influence of each in a given situation.

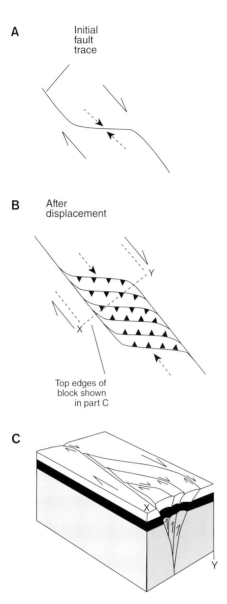

A Initial
fault
trace

B After
displacement

Y

X

Top edges of
block shown
in part C

C

X

Y

FIGURE 2.22 (A) At a contractional or restraining jog in a strike-slip fault, the resulting unequal compression of the rocks can force multiple sheaves of rock (B) up out of the ground. (C) The result is a transpressive, *flower* or *palm-tree* structure. (After Twiss and Moores 1992.)

In another major type of faulting, one piece of land surface is displaced horizontally past another to create *strike-slip* or transform faults. This faulting is less important to mountain building than those producing primarily vertical displacement, but is nevertheless an important process. The San Andreas Fault of California, extending in a northwesterly direction from Mexico to the northern Pacific for 3,000 km (~1,900 mi), is a highly active strike-slip fault that is the source of the region's most violent earthquakes (Keller and Pinter 2002). The western section (the Pacific Plate, on which Los Angeles is located) is moving northward at the rate of ~5.8 cm (~2.3 in.) per year. Past movement is estimated to have been ~1,100 km (~700 mi), and if the present trend continues, Los Angeles will one day—several million years from now—be a suburb of San Francisco. Although movement along the fault is primarily horizontal, some buckling and vertical displacements have taken place, resulting in mountainous terrain (e.g., the San Gabriel and San Bernardino Mountains).

The manner in which strike-slip or transform faults produce mountains is called *transpression,* and the result may be a *flower structure* (Fig. 2.22). The uplift mechanism occurs when two plates move past each other along a fault that has a curve or a jog in it. Where the jog causes the moving rock to be restrained, rock is then squeezed up and out along associated high-angle reverse faults on the strike-slip fault. The Alpine Fault of New Zealand is a primary example of transpression, wherein the Southern Alps are the uplifted result of the oblique convergence of the Indian (Australian) and Pacific Plates (Kneupfer 1992; Tippett and Hovius 2000).

A similar sort of faulting that can produce major mountains is also exemplified by Nanga Parbat (8,125 m, 26,660 ft), a pop-up structure in the western Himalaya (Schneider et al. 1999). In this case, major erosional unroofing caused a tectonic aneurysm in which the mountain rose like a cork along two inward-facing, high-angle, reverse faults to produce the ninth highest mountain in the world (Zeitler et al. 2001a, 2001b).

FAULT AND FAULT-LINE SCARPS

Fault and *fault-line scarps* are characteristic of most mountain ranges, yet they are notoriously difficult to interpret. Most of the features have a long history of formation, with concomitant erosion that may so alter the original structural landform as to obscure its origins (Fig. 2.23). Once a fault scarp is produced on the landscape it undergoes erosion by either downwearing or parallel retreat, generally as a function of lithology. For example, crystalline rocks tend to decline in slope angle through time, whereas some sedimentary rocks will tend to promote erosional maintenance of similar slope angles. In either case, if the fault undergoes renewed movement, a *composite fault scarp* (upper eroded, lower uneroded) results. If erosion is continued it can remove the original fault scarp and produce a new, entirely erosional escarpment, near the location of the original fault. This is a *fault-line scarp,* and it can occur with the scarp facing in the same direction as the original (*resequent fault-line scarp*), or in the opposite direction (*obsequent fault-line scarp*). A fault scarp buried in sediment and then exposed at a later time is an *exhumed fault scarp.* Deciphering the sequence of fault and erosional events in any particular complex faulted area is commonly a difficult task.

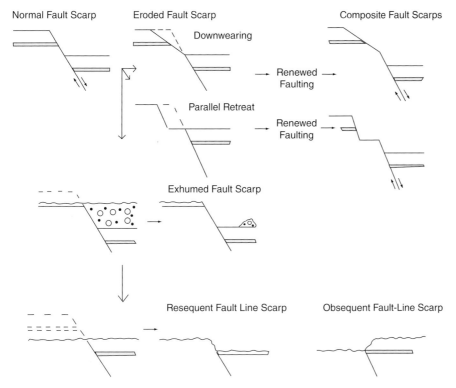

Normal Fault Scarp   Eroded Fault Scarp   Composite Fault Scarps

Downwearing

→ Renewed →
Faulting

Parallel Retreat

Renewed
Faulting

Exhumed Fault Scarp

Resequent Fault Line Scarp   Obsequent Fault-Line Scarp

FIGURE 2.23 Fault scarps and fault-line scarps where the erosional histories of the structure can be quite different. Differentiating the various types is important when reconstructing histories of past and assessing potential hazardous seismic events.

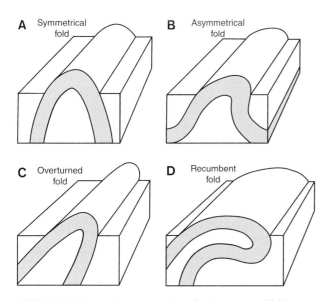

A Symmetrical fold
B Asymmetrical fold
C Overturned fold
D Recumbent fold

FIGURE 2.24 Schematic representation of various types of folded rock structures. Such features can occur in a neotectonic, uneroded fashion, as in this idealized view, where internal rock structure and external landscape form or geomorphology are coincident or congruent. More commonly, such rock structures are eroded into and, because the rocks in the axes of the anticlines are under more tension than the commonly adjacent synclines, a breached anticline results. (Adapted from Putnam 1971: 392.)

FOLDED MOUNTAINS

Folding compresses rock strata into wave-like troughs and ridges without fracturing the rock (Fig. 2.24). This plastic deformation takes place slowly over millions of years. Where the deforming forces are mildly compressive, the amplitude of the spacing of the downwarped *synclines* and upwarped *anticlines* resembles a gently rippled sea, but under extreme pressures the rock strata can take on the appearance of a tempest frozen at the point of greatest fury, as in most complex mountains where the strata have been deformed, overturned, and wrapped into the configuration of a ribbon candy. Because folded mountains commonly stretch rocks under tension in the axes of anticlines and compress axes of intervening synclines, the uparched anticlines are subject to greater erosion and synclines are protected. The result can be *topographic inversion*, such that rock structure and topography are incongruent and instead, anticlinal valleys and synclinal ridges result (Fig. 2.25).

The major orogenic belts are folded mountains, consisting of thick accumulations of marine sediments (geoclines) that have been altered and deformed (Fig. 2.26). This is thought to result from large-scale compressive forces shortening sections of the Earth's crust so that the strata become

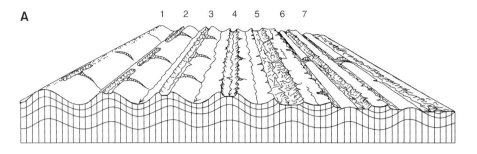

A

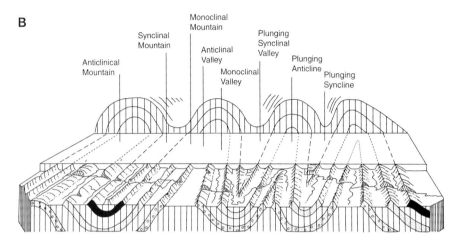

B

Anticlinical Mountain
Synclinal Mountain
Monoclinal Mountain
Anticlinal Valley
Monoclinal Valley
Plunging Synclinal Valley
Plunging Anticline
Plunging Syncline

FIGURE 2.25 (A) Progressive erosion of nonplunging anticlines and synclines (1–7) that show topographic reversal from: (1) anticlinal mountain and synclinal valley; (2 and 3) breached anticlines; (4 and 5) anticlinal valleys; and synclinal ridges between 5 and 6 and between 6 and 7. (B) On left are nonplunging fold landforms; on right are plunging fold landforms that produce zigzag ridges where differential erosion has left resistant rocks in relief. (After Lobeck 1939.)

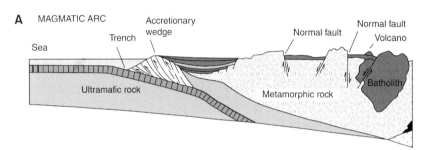

A   MAGMATIC ARC

Accretionary wedge
Trench
Sea
Normal fault
Normal fault
Volcano
Batholith
Ultramafic rock
Metamorphic rock

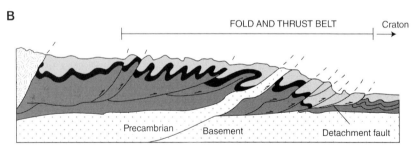

B

FOLD AND THRUST BELT   Craton

Precambrian   Basement   Detachment fault

FIGURE 2.26 Cross section of convergent mountain belt at continent edge in continuous section from beginning of (A) to end of (B). (A) Geoclinal sediments are metamorphosed, intruded with granite batholiths, and volcanoes and horsts and grabens are produced on the surface. An accretionary wedge of material scraped up at the submarine trench and above the subduction zone, a forearc basin, and sedimentation in the grabens add to the complexity. (B) Further inland, past the crystalline core of the mountain range, a fold and thrust belt has developed and forced geoclinal rocks over the stable interior craton. (After Plummer et al. 2003.)

thickened and pile up as mountainous accumulations of deformed rock. Similar effects, however, can result from local processes; if an area is domed upward, for example, the beds on either side can gravity slide downslope under their own weight (Fig. 2.27). Plentiful evidence supports both processes; strongly metamorphosed rocks indicate that folding took place at considerable depth within the Earth under great heat and pressure; rocks with little or no metamorphism suggest that folding occurred at the surface

under normal atmospheric temperatures, and may be due to gravity sliding.

Gravitational collapse of a range is much more important in the deformation of mountains than previously thought, displaying a wide range of phenomena. In fact, degradation of mountains caused by gravity is scale-dependent, with a range of such phenomena encompassing gravitational extension of a wide range, from the largest, slowest end member; through *sackung* (German for "sagging") collapse

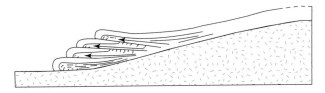

FIGURE 2.27 Graphic representation of gravity sliding. Rock beds slide slowly downslope and deform under their own weight to produce local folding and faulting.

of a single mountain in the middle range of the phenomena; down to smaller, rapid landslides (Shroder and Bishop 1998). Evidence of gravitational extension includes the fact that, whereas the most intense rock metamorphism occurs near the axis of the range (commonly occupied by a granitic-core batholith), the most complex folding is on the margins (Fig. 2.28). As rock masses are transported downslope through gravity sliding (Fig. 2.29), the central area typically undergoes isostatic uplift to compensate for loss of weight. When this happens, the central core attains the highest point of relief, both because it undergoes greater uplift and because the crystalline rocks of which it is composed are more resistant to erosion. Mount Denali (McKinley) in Alaska, for example, the highest mountain in North America (6,188 m/20,300 ft), is carved from a granite batholith that underlies the Alaska Range (Hunt 1974). Gravity sliding can also account for displacement of older strata over younger rock in overthrust faults. The Heart Mountain detachment fault (Fig. 2.29), located in northwestern Wyoming near Yellowstone National Park, for example, is a series of huge rock blocks that slid down a slight incline, probably as a result of ancient volcanism in early Cenozoic time, to lie unconformably on younger rocks (Thornbury 1965). Chief Mountain, mentioned above as an example of an overthrust remnant, may actually have been formed by gravity sliding. Another of the most famous examples of displacement from gravity sliding is the Rock of Gibraltar in the Mediterranean; it consists of

older limestones and shales overlying younger sedimentary rocks. This huge monolith was carved from a larger mass that moved laterally from the east to its present location (Garner 1974: 194).

At its simplest, in folded rocks, a number of common mountain shapes can form in high mountains of sedimentary rocks (Cruden 2003). Castellate and Matterhorn-type mountains occur in subhorizontal to gently inclined beds, cuestas form in gently dipping beds up to about 10° dip, monoclinal ridges form between ~10° and 45° of dip, and hogback ranges form in steeply dipping beds >45°. Cataclinal slopes on mountains in sedimentary rocks have the dip in the same direction as the slope, whereas anaclinal slopes have a dip in the opposite direction.

The most spectacular feature associated with folded mountains is the *nappe* fold (from the French for "tablecloth"; Fig. 2.30). Nappes are enormous slab-like masses of rock, bulbous protrusions that have been extruded into and over other rock strata for considerable distances, in some cases up to 100 km (~60 mi). Closely associated with *overthrust faults*, or *breakthrusts* where the nappe's lower overturned strata fracture and thrust forward, nappes are generally attributed to *overturned* and *recumbent folding*, wherein the axis of maximum curvature of the anticline is strongly asymmetrical and is commonly forced laterally over the adjacent strata for some distance. Nappes are formed as a result of several possible mechanisms, including horizontal compression, or gravity glide and gravitational collapse. In horizontal compression, nappes are emplaced by a force applied to the rear of the nappe. This so-called "push from behind" involves ductile shearing of a nappe over its base by a combination of simple shear and shortening parallel to the basal fault. In gravity models, the nappe deforms under gravity by gliding down a gently inclined base.

Nappes, which occur in most mountain ranges, are an integral part of orogenesis. The entire superstructure of the European Alps is dominated by a series of overlapping nappes that were displaced from their point of origin near present-day Italy as Africa collided with Europe, causing

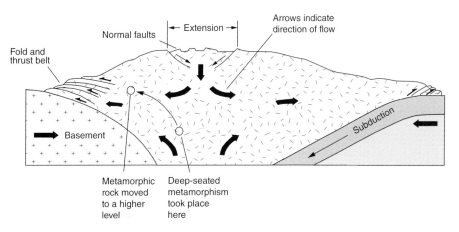

FIGURE 2.28 Cross section of a large collisional mountain range in which gravitational collapse by internal fracture and flow allow lateral movement of rock outward from the range. Faulting occurs in surficial brittle rock, whereas rock metamorphosed at depth flows up and out at a higher level in the mountain belt to replace the mass lost through folding, thrust faulting, and surface erosion. (After Plummer et al. 2003.)

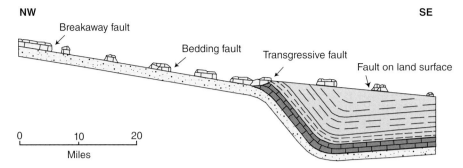

FIGURE 2.29 Heart Mountain detachment fault, located in northwestern Wyoming near Yellowstone National Park. Blocks up to 8 km (5 mi) across have broken away from strata at upper left and slid down the gentle incline so that older strata now rest upon younger rock. The origin of this structure has been controversial for decades but may have been associated with the massive Absaroka volcanoes and their sector collapses in early Tertiary time. (Adapted from Pierce 1957 and Garner 1974: 194.)

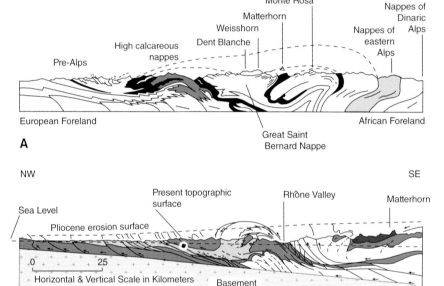

FIGURE 2.30 Generalized cross sections of eastern Alps, based upon the idea that they were formed from recumbent folds and nappes driven northward from the ancient Tethyan (Mediterranean Sea) geocline by the closing of Africa and Europe. (A) An older interpretation of nappe folds without significant thrust faulting (after Holmes 1965). (B) Recent interpretation of thrust-faulted detachment blocks and nappe folds (after Boyer and Elliot 1982; Plummer et al. 2003).

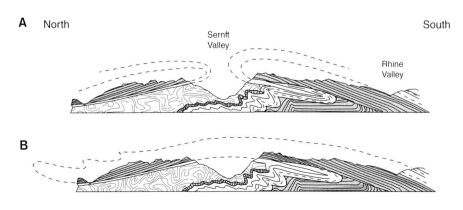

FIGURE 2.31 Two interpretations of the mechanism responsible for emplacing older strata over younger in the Alps near Glarus, Switzerland. The lower illustration, of one over-thrust sheet or nappe, is the accepted view. Length of section is 40 km (25 mi). (After Holmes 1965.)

the creation of these mountains from the sediments of the ancient Tethys Seaway (Strahler 1998; De Graciansky et al. 2010). Millions of years of erosion have removed many of the displaced rock sheets entirely, leaving only remnants. The famous Swiss Matterhorn in the Alps is a remnant of one of these nappes. It consists of older, more resistant rocks overlying younger rocks.

The Alps today are only the skeletal remains of a formerly more robust mountain body. Theoretical reconstruction of the original is exceedingly complex, and several schools of thought have arisen from different interpretations of nappes. Early workers had difficult rationalizing the great horizontal displacements involved. A study in 1841 near Glarus, Switzerland, revealed that for a distance of over 50 km (~30 mi) the mountain peaks are composed of rocks older than the valleys. The original explanation for this anomalous situation was that the older rock had been transported to its present location by two recumbent folds coming from opposite directions (Fig. 2.31a). Even this relatively cautious speculation was so revolutionary at the time that its author would not allow publication of the work: "No one would believe me if I published my sections; they would put me in an asylum" (Vanderlinth 1841, cited in Holmes 1965: 1162). In 1884, one gigantic overthrust was postulated for the entire region (Fig. 2.31b). This eventually became the mostly widely accepted view, although some controversy has occurred.

Dome Mountains

Dome mountains can be simple igneous intrusives such as laccoliths, which are mushroom-shaped *plutonic* bodies that dome up overlying sedimentary rocks. The Henry and La Sal Mountains of Utah (Hunt 1974) are a series of simple to complex laccoliths and irregularly shaped igneous stocks that have uparched surrounding strata. Complex domal uplifts, such as the Black Hills of South Dakota or the Ozarks of Missouri, involve a wide variety of tectonically uplifted igneous and metamorphic rock with overlying arches of sedimentary rock (Thornbury 1965; Hunt 1974). Purely sedimentary rock domes of small to large size can result from various compressive or uplifting rock stresses. Salt-dome mountains form because rock salt is highly mobile when loaded with other sedimentary rock; it bulges upward, in some cases piercing the rock overhead and flowing slowly out of the ground. But because salt is so soluble in water, only desert areas, such as the Zagros Mountains in southern Iran, retain surficial salt mountains.

Topographic Inversion

Rock structure deep underground and landforms on the surface are generally mutually independent. Close to the surface, rock structure may or may not be reflected in the landform. Anticlinal mountain and normal fault scarps are examples of the convergence of rock structure with topography. *Topographic inversion* is an example of differential erosion, whereby the attitude of the rock structure and the topography have opposite character, that is, an anticlinal valley or an obsequent fault-line scarp. Proper interpretation of the correct sequence of geologic events that led to the existing structure and topography of the mountain system in question is essential.

Conclusion

Geologic interpretation of mountain structure requires not only careful study and mapping of rock strata, but also an ability to visualize the larger picture with enough imagination. The importance of these qualities can be seen in the progression of ideas concerning the origin of mountains; the theory of plate tectonics is a fantastic and spectacular culmination of human ingenuity. Although much remains to be learned, especially in the most complex continental areas, it has provided a logical framework into which most observations fit. The new recognition that erosion can actually produce mountain uplift in focused isostasy or tectonic aneurysms is equally exciting because it is so counterintuitive. The perspectives gained from these examinations of the origin of mountains should be retained as we move to the central concern of this book—the processes that characterize present-day mountain landscapes.

References

Andrews, J. T. 1970. A geomorphological study of post-glacial uplift with particular reference to Arctic Canada. *Institute of British Geographers Special Publication* 2.

Anonymous (Editors of Time-Life Books). 1982. *Volcano.* Alexandria, VA: Time-Life Books.

An Yin. 2004. Gneiss domes and gneiss dome systems. In D. L. Whitney, C. Teyssier, and C. S. Siddoway, eds., *Gneiss Domes in Orogeny. Geological Society America Special Paper* 380: 1–14.

Beaumont, C., Fullsack, P., and Hamilton, J. 1992. Erosional control of active compressional orogens. In K. R. McCay, ed., *Thrust Tectonics* (pp. 1–18). London: Chapman and Hall.

Benioff, H. 1949. Seismic evidence for the fault origin of the oceanic deeps. *Geological Society of America Bulletin* 60: 1837–1856.

Benioff, H. 1954. Orogenesis and deep crustal structure: Additional evidence from seismology. *Geological Society of America Bulletin* 65: 385–400.

Boyer, S. E., and Elliot, D. 1982. Thrust systems. *American Association of Petroleum Geologists Bulletin* 66(9): 1196–1230.

Brandon, M. T., Roden-Tice, M. K., and Graver, J. I. 1998. Late Cenozoic exhumation of the Cascadia accretionary wedge in the Olympic Mountains, northwest Washington. *Geological Society America Bulletin* 110(8): 985–1009.

Bull, W. B. 2007. *Tectonic Geomorphology of Mountains: A New Approach to Paleoseismology.* Malden, MA: Blackwell Publishing.

Carey, S. W. 1976. *The Expanding Earth.* Amsterdam: Elsevier.

Colgan, J. P., Dumitru, T. A., Miller, E. L., and Reiners, P. W. 2006. Cenozoic tectonic evolution of the Basin and Range

Province in northwestern Nevada. *American Journal of Science* 306: 616–654.

Condie, K. C. 2005. *Earth as an Evolving Planetary System*. Burlington, MA: Elsevier Academic Press.

Cruden, D. M. 2003. The shapes of cold, high mountains in sedimentary rocks. *Geomorphology* 55: 249–261.

Decker, R., and Decker, B., 2006. Volcanoes. 4th ed. New York: W. H. Freeman.

De Graciansky, P.-C., Roberts, D. G., Tricart, P., and Shroder, J. F., Jr. (series ed.). 2010. *The Western Alps, from Rift to Passive Margin, to Orogenic Belt: An Integrated Geoscience Overview*. Developments in Earth Surface Processes 14. Philadelphia: Elsevier.

Dewey, J. F., and Bird, J. M. 1970. Mountain belts and the new global tectonics. *Journal of Geophysical Research* 75(14): 2625–2647.

Dietz, R. S., and Holden, J. C. 1966. Miogeoclines (miogeosynclines) in space and time. *Journal of Geology* 74(5): 566–583.

Dietz, R. S. 1972. Geosynclines, mountains, and continent building. *Scientific American* 226(3): 30–38.

du Toit, A. L. 1937. *Our Wandering Continents*. Edinburgh: Oliver and Boyd.

Einsele, G. 2000. *Sedimentary Basins, Evolution, Facies, and Sediment Budget*. 2nd ed. Berlin: Springer.

Engel, A. E. J. 1963. Geological evolution of North America. *Science* 140: 143–152.

Fairbridge, R. W. 1968. Mountain types. In *Encyclopedia of Geomorphology* (pp. 751–760). New York: Reinhold.

Finnegan, N. J., Hallet, B., Montgomery, D. R., Zeitler, P. K., Stone, J. O., Anders, A. M., and Liu Yuping. 2008. Coupling of rock uplift and river incision in the Namche Barwa–Gyala Peri massif, Tibet. *GSA Bulletin* 120(1/2): 142–155; doi: 10.1130/B26224.1.

Fletcher, R. C., and Hallet, B., 2004. Initiation of gneiss domes by necking, density instability, and erosion. In D. L. Whitney, C. Teyssier, and C. S. Siddoway, eds., *Gneiss Domes in Orogeny. Geological Society America Special Paper* 380.

Garner, H. F. 1974. *The Origin of Landscapes*. New York: Oxford University Press.

Goudie, A., 1995. *The Changing Earth: Rates of Geomorphological Processes*. Oxford, UK: Blackwell.

Greeley, R., and King, J. S. 1977. *Volcanism of the Eastern Snake River Plain, Idaho*. Washington, DC: NASA.

Gripp, A. E., and Gordon, R. G. 1990. Current plate velocities relative to hotspots incorporating NUVEL-1 global plate motion model. *Geophysical Research Letters* 17(8):1109–1112.

Harris, N., Inger, S., and Massey, J., 1993. The role of fluids in the formation of the High Himalayan leucogranites. In P. J. Treloar and M. P. Searle, eds., *Himalayan Tectonics. Geological Society (London) Special Publication* 74: 391–400.

Herring, T. 1996. The global positioning system. *Scientific American* 274(2): 44–50.

Hoffman, P. F., and Grotzinger, J. P. 1993. Orographic precipitation, erosional unloading, and tectonic style. *Geology* 21: 195–198.

Holmes, A. 1931. Radioactivity and earth movements. *Geological Society Glasgow Transactions* 18: 559–606.

Holmes, A. 1965. *Principles of Physical Geology*. New York: Ronald Press.

Hunt, C. B. 1974. *Natural Regions of the United States and Canada*. San Francisco: W. H. Freeman.

Kearey, P., and Vine, F. J. 1996. *Global Tectonics*. London: Blackwell Science.

Keller, E. A., and Pinter, N. 2002. *Active Tectonics: Earthquakes, Uplift and Landscape*. Upper Saddle River, NJ: Prentice-Hall.

King, P. B. 1977. *The Evolution of North America*. rev. ed. Princeton, NJ: Princeton University Press.

Knuepfer, P. L. K. 1992. Temporal variations in latest Quaternary slip across the Australian–Pacific Plate boundary, northeast South Island, New Zealand. *Tectonics* 11: 449–464.

Koons, P. O., Zeitler, P. K., Chamberlain, C. P., Craw, D., and Meltzer, A. S. 2002. Mechanical links between erosion and metamorphism in Nanga Parbat, Pakistan Himalaya. *American Journal of Science* 302: 749–773.

Le Fort, P. 1996. Evolution of the Himalaya. In An Yin and T. M. Harrison, eds., *The Tectonic Evolution of Asia* (pp. 95–109). Cambridge, UK: Cambridge University Press.

Lliboutry, L., 1999. *Quantitative Geophysics and Geology*. Chichester, UK: Springer Praxis.

Lipman, P. W., and Mullineaux, D. R., eds. 1981. The 1980 eruptions of Mount St. Helens, Washington. *U.S. Geological Survey Professional Paper* 1250.

Lobeck, A. K. 1939. *Geomorphology: An Introduction to the Study of Landscapes*. New York: McGraw-Hill.

Luhr, J. F., and Simkin, T. 1993. *The Volcano Born in a Mexican Cornfield*. Phoenix, AZ: Geoscience Press.

Macdonald, G. A. 1972. *Volcanoes*. Englewood Cliffs, NJ: Prentice-Hall.

Macdonald, G. A., and Abbott, A. T. 1970. *Volcanoes in the Sea: The Geology of Hawaii*. Honolulu: University of Hawaii Press.

Mahéo, G., Pêcher, A., Guillot, S., Rolland, Y., and Delacourt, C. 2004. Exhumation of Neogene gneiss domes between oblique crustal boundaries in south Karakoram (northwest Himalaya, Pakistan). In D. L. Whitney, C. Teyssier, and C. S. Siddoway, eds., *Gneiss Domes in Orogeny. Geological Society America Special Paper* 380: 141–154.

Montgomery, D. R. 1994. Valley incision and the uplift of mountain peaks. *Journal of Geophysical Research* 99(B7): 13913–13921.

Oliver, J., and Isaacs, B. 1967. Deep earthquake zones, anomalous structures in the upper mantle, and the lithosphere. *Journal of Geophysical Research* 72: 4259–4275.

Ollier, C. D. 1988. *Volcanoes*. Oxford, UK: Basil Blackwell Ltd.

Ollier, C. 2004. The evolution of mountains on passive margins. In P. N. Owens and O. Slaymaker, eds., *Mountain Geomorphology* (pp. 59–88). London: Arnold.

Ollier, C., and Pain, C., 2000. *The Origin of Mountains*. London: Routledge.

Owen, L. A. 2004. Cenozoic evolution of global mountain systems. In P. N. Owens and O. Slaymaker, eds., *Mountain Geomorphology* (pp. 33–58). London: Arnold.

Owens, P. N., and Slaymaker, O. 2004. An introduction to mountain geomorphology. In P. N. Owens and O. Slaymaker, eds., *Mountain Geomorphology* (pp. 3–29). London: Arnold.

Pierce, W. G. 1957. Heart Mountain and South Fork detachment thrusts of Wyoming. *Bulletin American Association of Petroleum Geologists* 41: 591–626.

Pinter, N., and Brandon, M. T. 1997. How erosion builds mountains. *Scientific American* (April): 74–79.

Plummer, C. C., McGreary, D., and Carlson, D. H., 2003. *Physical Geology*. New York: McGraw-Hill.

Price, L. W. 1981. *Mountains and Man: A Study of Process and Environment*. Berkeley: University of California Press.

Putnam, W. C. 1971. *Geology*. 2nd ed. New York: Oxford University Press.

Scarth, A., 1994. *Volcanoes: An Introduction*. College Station: Texas A&M University Press.

Schneider, D. A., Edwards, M. A., Kidd, W. S., Khan, M. A., Seeber, L., and Zeitler, P. K. 1999. Tectonics of Nanga Parbat, western Himalaya: Synkinematic plutonism within the doubly vergent shear zones of a crustal-scale pop-up structure. *Geology* 27: 999–1002.

Selby, M. J. 1985. *Earth's Changing Surface: An Introduction to Geomorphology*. Oxford, UK: Clarendon Press.

Shroder, J. F., Jr., and Bishop, M. P. 1998. Mass movement in the Himalaya: New insights and research directions. *Geomorphology* 26: 13–35.

Shroder, J. F., Jr., Owen, L. A., Seong, Y. B., Bishop, M. P., Bush, A., Caffee, M. W., Copland, L., Finkel, R. C., and Kamp, U. 2011. The role of mass movements on landscape evolution in the Central Karakoram: Discussion and speculation. *Quaternary International* 236: 34–47, doi:10.1016/j.quaint.2010.05.024.

Stewart, R. J., Hallet, B., Zeitler, P. K., Malloy, M. A., Allen, C. M., and Trippett, D. 2008. Brahmaputra sediment flux dominated by highly localized rapid erosion from the easternmost Himalaya. *Geology* 36: 711–741, doi: 10.1130/G24890A.1.

Stockli, D. F., Dumitru, T. A., McWilliams, M. O., and Farley, K. A. 2003. Cenozoic tectonic evolution of the White Mountains, California and Nevada. *Geological Society of America Bulletin* 115: 788–816.

Strahler, A. N. 1998. *Plate Tectonics*. Cambridge, MA: Geobooks.

Tatsumi, Y. (2005). The subduction factory: How it operates on Earth. *GSA Today* 15(7): 4–10.

Thornbury, W. D. 1965. *Regional Geomorphology of the United States*. New York: John Wiley and Sons.

Tippett, J. M., and Hovius, N. 2000. Geodynamic processes in the Southern Alps, New Zealand. In M. A. Summerfield, ed., *Geomorphology and Global Tectonics* (pp. 109–134). Chichester, UK: John Wiley and Sons.

Turcotte, D. L., and Schubert, G. 2002. *Geodynamics*. 2nd ed. Cambridge, UK: Cambridge University Press.

Twiss, R. J., and Moores, E. M. 1992. *Structural Geology*. New York: W. H. Freeman.

Valdiya, K. S. 1989. Trans-Himadri intracrustal fault and basement upwarps south of the Indus–Tsanpo Suture Zone. In L. M. Malinconico Jr. and R. J. Lillie, eds., *Tectonics of the Western. Geological Society of America Special Paper* 232: 153–168.

Wadati, K. 1935. On the activity of deep-focus earthquakes in the Japan Islands and neighborhoods. *Geophysical Magazine* (Tokyo) 8: 305–325.

Weisskopf, V. F. 1975. Of atoms, mountains, and stars: A study in qualitative physics. *Science* 187(4177): 605–612.

Willett, S., Beaumont, C., and Fullsack, P. 1993. Mechanical model for the tectonics of doubly vergent compressional orogens. *Geology* 21: 371–374.

Williams, P. W. 2004. The evolution of the mountains of New Zealand. In P. N. Owens and O. Slaymaker, eds., *Mountain Geomorphology* (pp. 89–106). London: Arnold.

Wright,T .J., Ebinger, C., Biggs,, J., Ayele, A., Gezahegan, Y., Yirgu, D., Keir, D., and Stork, A. 2006. Magma-maintained rift segmentation at continental rupture in the 2005 Afar dyking episode. *Nature* 442: 291–294.

Wright, T. L., and Pierson. T. C. 1992. Living with volcanoes. *U.S. Geological Survey Circular* 1073.

Zeitler, P. K., Koons, P., Bishop, M. P., Chamberlain, C. P., Craw, D., Edwards, M., Hamidullah, S., Jan, M. Q., Khan, M. A., Khattak, U. K., Kidd, W. S. F., Mackie, R. L., Meltzer, A. S., Park, S. K., Pecher, A., Poage, M. A., Sarker, G., Schneider, D. A., Seeber, L., and Shroder, J. F., Jr. 2001a. Crustal reworking at Nanga Parbat, Pakistan: Metamorphic consequences of thermal-mechanical coupling facilitated by erosion. *Tectonics* 20: 712–728.

Zeitler, P. K., Meltzer, A. S., Koons, P. O., Craw, D., Hallet, B., Chamberlain, C. P., Kidd, W. S. F., Park, S. K., Seeber, L., Bishop, M., and Shroder, J. F., Jr. 2001b. Erosion, Himalayan geodynamics, and the geomorphology of metamorphism. *GSA Today* 11: 4–9.

CHAPTER THREE

Mountain Climate

ANDREW J. BACH and LARRY W. PRICE

Climate is the fundamental factor in establishing a natural environment, setting the stage on which all physical, chemical, and biological processes operate. This becomes especially evident at the climatic margins of the Earth, namely desert and tundra. Under temperate conditions, the effects of climate are often muted and intermingled, so that the relationships between stimuli and reactions are difficult to isolate, but under extreme conditions such relationships become more evident. As extremes constitute the norm in many areas within high mountains, a basic knowledge of climatic processes and characteristics is key to understanding the mountain milieu.

In mountain areas, great environmental contrasts occur within short distances as a result of the diverse topography and highly variable nature of the energy and moisture fluxes. While in the mountains, have you ever sought refuge from the wind in the lee of a rock? If so, you have experienced the kind of difference that can occur over a small distance. Near the margin of a species' distribution, such differences may decide between life and death; thus, plants and animals reach their highest elevations by taking advantage of microhabitats. Great variations also occur within short time spans. When the sun is shining, it may be quite warm, even in winter, but if a passing cloud blocks the sun, the temperature can drop rapidly. Therefore, areas exposed to the sun undergo much greater and more frequent temperature contrasts than those in shade. This is true for all environments, but the difference is much greater in mountains because the thin alpine air does not hold heat well and allows more solar radiation to reach the surface.

More generally, the climate of a slope may be very different from that of a ridge or valley. When these basic differences are compounded by the infinite variety of combinations created by the orientation, spacing, and steepness of slopes—along with the presence of snow patches, shade, vegetation, and soil—the complexity of climatic patterns in mountains becomes truly overwhelming. Nevertheless, predictable patterns and characteristics are found within this heterogeneous system; for example, temperatures normally decrease with elevation while cloudiness and precipitation increase; moreover, it is usually windier in mountains, the air is thinner and clearer, and the sun's rays are more intense.

The dynamic characteristics of mountains also have a major impact on regional and local airflow patterns that affect the climates of adjacent regions (Xu et al. 2010). Their influence may be felt for hundreds or thousands of kilometers, making surrounding areas warmer or colder, or wetter or drier, than they would otherwise be. The effect of the mountains depends upon their location, size, and orientation with respect to the moisture source and the direction of prevailing winds. The 2,400 km (1,500 mi)-long barrier of the Himalaya permits tropical climates to extend farther north in India and Southeast Asia than anywhere else in the world (Xu et al. 2010). One of the wettest places on Earth is Cherrapunji, near the base of the Himalaya in Assam, with an annual rainfall of 10,871 mm (428 in.; Cerveny et al. 2007); the record for a single day is 1,041 mm (41 in.), as much as Chicago or London receives in an entire year! On the north side of the Himalaya, however, there are extensive deserts, and temperatures are abnormally low for the latitude. This contrast in environment between north and south derives almost entirely from the presence of the mountains, whose east–west orientation and great height prevent the invasion of warm air into central Asia just as surely as they prevent major invasions of cold air into India. It is no wonder that the Hindus pay homage to Shiva, the great god of the Himalaya.

External Climatic Controls

Mountain climates occur within the framework of the surrounding regional climate and are controlled by the same factors, including latitude, altitude, continentality, and regional circumstances such as ocean currents, prevailing wind direction, and the location of semipermanent high- and low-pressure cells. Mountains themselves, by acting as a barrier, affect regional climate and modify passing storms. Our primary concern is in the significance of all these more or less independent controls for mountain weather and climate.

Latitude

Distance from the equator governs the angle at which the sun's rays strike the Earth, day length, and thus the amount of solar radiation arriving at the surface. In the tropics, the sun is always high overhead at midday, and days and nights are of nearly equal length throughout the year. There is no winter or summer; one day differs from another only in the amount of cloud cover. With increasing latitude, however, the height of the sun changes throughout the year, and days and nights become longer or shorter depending on the season (Fig. 3.1).

Although the highest latitudes receive the lowest annual amounts of heat energy, middle latitudes frequently experience higher temperatures in summer than the tropics. This is because of the moderate sun heights and longer days of the middle latitudes. Mountains in these latitudes may experience even greater solar intensity than lowlands, both because the atmosphere is thinner and because the sun's rays strike slopes oriented toward the sun at a higher angle than level surfaces. A surface inclined 20° toward the sun in middle latitudes receives about twice as much radiation during the winter as a level surface. Thus, slope angle and orientation with respect to the sun are very important and may partially compensate for latitude.

The basic pattern of global atmospheric pressure systems—the equatorial low (0°–20° latitude), subtropical high (20°–40° latitude), polar front and subpolar lows (40°–70° latitude), and polar high (70°–90° latitude)—reflects the role of latitude in determining climatic patterns. The equatorial and subpolar lows are zones of relatively heavy precipitation, while the subtropical and polar highs are areas of low precipitation. These pressure zones create the global circulation system (Fig. 3.2), which in turn dictates the prevailing wind direction and types of storms that occur latitudinally. The easterly *trade winds* have warm, very moist, convective (tropical) storms. The subtropical highs have slack winds and clear skies year round. The subpolar lows and polar front are embedded in the *westerlies*, bringing cool, wet cyclonic storms and large seasonal temperature fluctuations. The cold and dry *Polar easterlies* develop seasonally, dissipating in summer.

Jet streams are high-velocity, relatively narrow air currents found at about 11 km (36,000 ft) above the surface (Archer and Caldeira 2008). They form at the boundaries of air masses with significant differences in temperature, such as at the polar front. Thus they develop seasonally, and their strength and position are variable. While jets form in several locations, the best developed are the polar jets, westerly winds in both hemispheres at latitudes 30–70°. The path of a jet typically has a meandering shape (Rossby waves), and these meanders propagate eastward, at lower speeds than that of the actual wind within the flow. Jet streams are important in steering storms, particularly midlatitude cyclones. The altitude of jets is above most mountains; however, at higher elevations, wind velocities generally increase.

Where mountains are located within the global circulation system greatly influences their climate. Mountains near the equator—such as Kilimanjaro in East Africa, Kinabalu in Borneo, or Cotopaxi in Ecuador—are under the influence of the equatorial low and receive precipitation almost daily on their east-facing windward slopes. By contrast, mountains located around 30° latitude may experience considerable aridity, as do the northern Himalayas, Tibetan highlands, the Puna de Atacama in the Andes, the Atlas Mountains of North Africa, and the mountains of the southwestern United States and northern Mexico (Troll 1968). Farther poleward, the Alps, Rockies, Cascades, southern Andes, and the Southern Alps of New Zealand receive heavy precipitation on westward slopes facing the prevailing westerlies. Leeward-facing slopes and lands downwind are notably arid. Polar mountains are cold and dry year round.

Altitude

Fundamental to mountain climatology are the changes that occur in the atmosphere with increasing altitude, especially decreases in temperature, air density, water vapor, carbon dioxide, and impurities. The sun is the ultimate source of energy, but little heating of the atmosphere takes place directly. Rather, solar radiation passes through

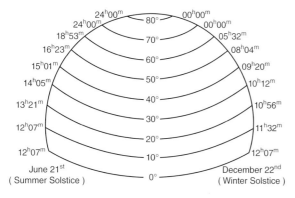

FIGURE 3.1 Length of daylight received at each latitude during summer (left) and winter (right) solstice in the northern hemisphere. (After Rumney 1968: 90.)

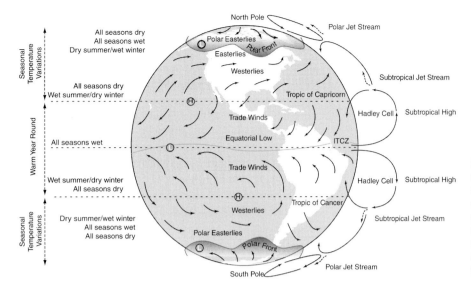

The diagram labels:

North Pole
Polar Jet Stream
Polar Easterlies
Polar Front
Easterlies
Westerlies
Subtropical Jet Stream
Tropic of Capricorn
Hadley Cell
Subtropical High
Trade Winds
Equatorial Low
ITCZ
Trade Winds
Hadley Cell
Subtropical High
Westerlies
Tropic of Cancer
Subtropical Jet Stream
Polar Easterlies
Polar Front
Polar Jet Stream
South Pole

Left-side labels:
Seasonal Temperature Variations
All seasons dry
All seasons wet
Dry summer/wet winter
All seasons dry
Wet summer/dry winter
Warm Year Round
All seasons wet
Wet summer/dry winter
All seasons dry
Seasonal Temperature Variations
Dry summer/wet winter
All seasons wet
All seasons dry

FIGURE 3.2 The general distribution of global atmospheric pressure systems and general circulation of the atmosphere. These winds dictate global climatic patterns associated with latitude. The general latitudinal climatic zones are shown along the right side of the diagram. (Adapted from several sources.)

the atmosphere and is absorbed by the Earth's surface. The Earth itself becomes the radiating body, emitting long-wave energy that is readily absorbed by CO_2, H_2O, and other greenhouse gases in the atmosphere. Thus, as the atmosphere is heated directly by the Earth rather than by the sun, temperatures are usually highest near the Earth's surface and decrease with altitude. Mountains are part of the Earth, too, but present a smaller land area at higher altitudes within the atmosphere, and so are less able to modify the temperature of the surrounding air. A mountain peak is analogous to an oceanic island. The smaller the island and the farther it is from large land masses, the more its climate will be like that of the surrounding sea. By contrast, the larger the island or mountain area, the more it modifies its own climate. This *mountain mass effect* is a major factor in the local climate.

The density and composition of the air control its ability to absorb and hold heat. The weight or density of the air at sea level (standard atmospheric pressure) is generally expressed as 1,013 mb (millibars), or 760 mm (29.92 in.) of mercury. Near sea level, pressure decreases at approximately 1 mb per 10 m (30 mm/300 m or 1 in./1,000 ft) of increased altitude. Above 5,000 m (20,000 ft) atmospheric pressure begins to fall off exponentially. Thus, half the weight of the atmosphere occurs below 5,500 m (18,000 ft) and pressure is halved again in the next 6,000 m (Fig. 3.3).

The ability of air to hold heat is a function of its molecular structure. At higher altitudes, as molecules are farther apart, there are fewer in a given parcel of air to receive and hold heat. Similarly, water vapor, carbon dioxide, and suspended particulate matter decrease rapidly with altitude (Tables 3.1, 3.2). These constituents, important in determining the ability of the air to absorb heat, are all concentrated in the lower atmosphere. Water vapor is the chief heat-absorbing constituent. Half of the water vapor in the air occurs below 1,800 m (6,000 ft), diminishing rapidly above this elevation and barely detectable above 12,000 m (40,000 ft).

The importance of water vapor as a reservoir of heat can be seen by comparing the daily temperature ranges of deserts and humid areas. Both may heat up equally during the day but, because of the relative absence of water vapor to absorb and hold the heat energy, deserts cool down much more at night. The mountain environment responds in a fashion similar to that of a desert, but is even more accentuated. As the thin, pure air of high altitudes does not effectively intercept radiation, that radiation is lost to space. Mountain temperatures respond almost entirely to radiation fluxes, not to the temperature of the surrounding air, although some mountains receive considerable heat from precipitation. The sun's rays heat the high thin air very little. Consequently, although the temperature at 1,800 m (6,000 ft) in the free atmosphere changes very little between day and night, a mountain peak intercepts and absorbs the sun's rays. The soil surface may be quite warm, but the envelope of heated air is usually only a few meters thick and displays a steep temperature gradient.

In theory, every point along a given latitude receives the same amount of sunshine; in reality, clouds interfere. Cloudiness is controlled by distance from the ocean, direction of prevailing winds, dominance of pressure systems, and altitude. Precipitation normally increases with elevation, but only up to a certain point, and is generally heaviest on middle slopes, where clouds first form and cloud moisture is greatest. Precipitation then decreases at higher elevations. Thus, lower slopes can be wrapped in clouds while higher slopes are sunny. In the Alps, for example, the outer ranges receive more precipitation and less sunshine than the higher interior ranges. The herders in the Tien Shan and Pamir Mountains of Central Asia traditionally take their flocks higher in winter than in summer to take advantage of the lower snowfall and sunnier conditions at higher elevations. High mountains have another advantage with respect to possible sunshine: In effect, they lower the

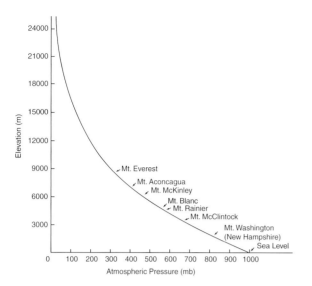

FIGURE 3.3 Generalized profile showing the decrease of atmospheric pressure with altitude. (Adapted from several sources.)

horizon. The sun shines earlier in the morning and later in the evening on mountain peaks than in lowlands. The same peaks, however, effectively raise the horizon for adjacent lands, thus delaying sunrise or creating early sunsets.

Continentality

The relationship between land and water strongly influences the climate of a region. Generally, the more water-dominated an area is, the more moderate its climate. An extreme example is a small oceanic island, on which the climate is essentially that of the surrounding sea. At the other extreme is a central location on a large land mass such as Eurasia, far from the sea. Water heats and cools more slowly than land, so temperature ranges between day and night and between winter and summer are smaller in marine areas than in continental areas.

The same principle applies to alpine landscapes, but is intensified by the barrier effect of mountains. We have already noted this effect of the Himalaya between India and China. Another good example is the Cascades in the Pacific Northwest of the United States, extending north–south at right angles to the prevailing westerly wind off the Pacific Ocean. Consequently, western Oregon and Washington have a marine-dominated climate with moderate temperatures, cloudiness, and persistent winter precipitation (Mass 2008). The eastern side of the Cascades, however, experiences a continental climate with hot summers and cold winters with low precipitation. In less than 85 km (50 mi) across the Cascades, the vegetation changes from lush green forests to dryland shrubs and grasses (Price 1978). This spectacular transect provides eloquent testimony to the great differences in climate that may occur within a short horizontal distance. The presence of the mountains increases the precipitation

in western Oregon and Washington at the expense of that received on the east side. Additionally, the Cascades inhibit the invasion of cold continental air to the Pacific side and, by obstructing the mild Pacific air, allow the continental climate to extend much closer to the ocean than it otherwise would (Whiteman et al. 2001). It must be stressed that the significance of mountains in accentuating continentality depends upon their orientation with respect to the ocean and prevailing winds. Western Europe has a climate similar to the Pacific Northwest, but the east–west orientation of the mountains allows the marine climate to extend far inland, resulting in a milder climate throughout Europe.

Barrier Effects

While the Himalaya and Cascades are both outstanding climatic divides that create unlike conditions on their windward and leeward sides, all mountains serve as barriers to some extent, depending on their size, shape, orientation, and relative location. Specifically, the barrier effects of mountains can be grouped under the following subheadings: (1) damming, (2) deflection and funneling, (3) blocking and disturbance of the upper air, (4) forced ascent, and (5) forced descent.

DAMMING

Damming of stable air occurs when mountains are high enough to prevent the passage of an air mass across them. A steep pressure gradient may develop between the windward and leeward sides of the range. The effectiveness of the damming depends on the depth of the air mass and the elevation of the lowest valleys or passes (Smith 1979). A shallow, ground-hugging air mass may be effectively dammed, but a deep one is likely to flow through higher gaps and transverse valleys to the other side. In the Los Angeles Basin of southern California, for example, the San Gabriel, San Bernardino, and San Jacinto Mountains act as dams for marine air from the Pacific Ocean. As the automobile-based culture of southern California pollutes the air, the pollution can only be vented as far east as the towns of San Bernardino and Riverside at the base of the mountains. In the absence of a strong wind system, the pollution can build up to dangerous levels as the air stagnates at the mountain barrier (Jerrett et al. 2005).

DEFLECTION AND FUNNELING

When a mountain range dams an air mass, the winds can be deflected around the mountains if topographic gaps exist. Deflected winds can have higher velocities as their streamlines are compressed, the so-called "Bernoulli effect" (Lin et al. 2005). In winter, polar continental air coming from Canada across the central United States is channeled to the south and east by the Rocky Mountains. Similarly, as

TABLE 3.1
Average Density of Suspended Particulate Matter in the Atmosphere
with Changing Elevation

Elevation	Number of Particles (cu. cm)	Number of Particles (cu. in)
0–500	25,000	435,000
500–1,000	12,000	209,000
1,000–2,000	2,000	34,800
2,000–3,000	800	14,000
3,000–4,000	350	6,100
4,000–5,000	170	3,000
5,000–6,000	80	1,400

SOURCE: Landsberg 1962: 114.

the cold air progresses southward, the Sierra Madre Oriental prevents it from crossing into the interior of Mexico. The east coast of Mexico also provides an excellent example of deflection in summer: The northeast trade winds blowing across the Gulf of Mexico cannot cross the mountains and are deflected southward through the Isthmus of Tehuantepec, where they become northerly winds of unusual violence (Brennan et al. 2010). Maritime air from the northeastern Pacific is deflected north and south around the Olympic Mountains (Fig. 3.4). To their north, where wind is also deflected south from the Vancouver Island Ranges, these winds converge into the topographic funnel of the Strait of Juan de Fuca, resulting in much higher wind speeds (Mass 2008). A similar phenomenon occurs around the Southern Alps of New Zealand, with winds funneled through the Cook Strait between the islands (Sturman and Tapper 1996). These perturbations of the local airflow influence transiting storms, making local forecasts difficult (Lin et al. 2005). The same funneling effect occurs over mountain passes as winds are deflected around peaks or ridges on either side of the pass.

BLOCKING AND DISTURBANCE OF THE UPPER AIR

High-pressure areas prevent the passage of storms. Large mountain ranges such as the Rockies, Southern Alps, Andes, and Himalayas are very efficient at blocking storms, since they are often the foci of anticyclonic systems (because the mountains are a center of cold air), and thus storms detour around the mountains (Insel et al. 2010). Mountains also cause other perturbations to upper-air circulation, with consequent effects on clouds and precipitation (Epifanio and Rotunno 2005; Lin et al. 2005; Neiman et al. 2010; Xu et al. 2010). This occurs on a variety of scales: locally, with the wind immediately adjacent to the mountains; on an intermediate scale, creating large waves in the air; and on a global basis, with the

larger mountain ranges actually influencing the motion of planetary waves and the transport momentum of the total circulation (Park et al. 2010). Disturbance of the air by mountains generally creates a wave pattern much like that found in the wake of a ship, and may result in the kind of clear-air turbulence feared by airline pilots, or it may simply produce lee waves with their beautiful lenticular (standing-wave) clouds, associated with mountains the world over (Scorer 1961).

Mountains also influence the location and intensity of jet streams, which greatly influence mountain weather (Insel et al. 2010). The jet streams may split to flow around the mountains, rejoining to the lee of the range, where they often intensify and produce storms (Neiman et al. 2010). In North America, such storms, known as "Colorado Lows" or "Alberta Lows," reach their greatest frequency and intensity in spring, sometimes causing heavy blizzards on the Great Plains and Prairie provinces (Sato and Kimura 2003). The tornadoes and violent squall lines that form in the American Midwest also result from the great contrasts in air masses which develop in the confluence zone in the lee of the Rockies (Jeglum et al. 2010).

The splitting of the jet streams by the Himalaya has the effect of intensifying the range's barrier effect and produces a stronger climatic divide. The presence of the Himalaya also reverses the direction of the jet streams in early summer. The Tibetan Highlands act as a "heat engine" in the warm season, with a giant chimney in their southeastern corner through which heat is carried upward into the atmosphere. During the spring, this causes a gradual warming of the upper air above the Himalaya, which weakens and finally eliminates the subtropical westerly jet. The easterly tropical jet then replaces the subtropical jet during the summer. Thus, the Himalayas are intimately connected with the complex interaction of the upper air and the development of the Indian monsoon (Xu et al. 2010).

TABLE 3.2
Average Water-Vapor Content of Air with Elevation
in the Middle Latitudes

Elevation	Volume
0	1.30
500	1.16
1,000	1.01
1,500	0.81
2,000	0.69
2,500	0.61
3,000	0.49
3,500	0.41
4,000	0.37
5,000	0.27
6,000	0.15
7,000	0.09
8,000	0.05

SOURCE: Landsberg 1962: 110.

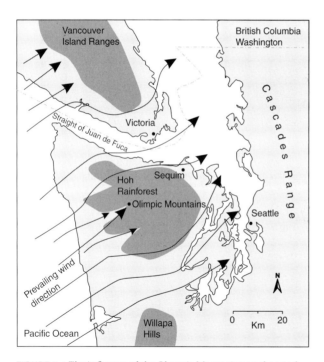

FIGURE 3.4 The influence of the Olympic Mountains on the wind field and precipitation. The arrows are flow lines indicating wind direction. Distance between the flow lines indicates relative speed: The closer they are to one another, the faster the wind in that region. Notice that the flow lines are evenly spaced over the Pacific Ocean. As they are deflected through the Strait of Juan de Fuca, the wind speed increases. Also notice that winds are funneled up the western valleys of the Olympics, concentrating moist air and increasing precipitation at the Hoh Rain Forest (3,800 mm), while Sequim only receives 430 mm in the rainshadow. (A. J. Bach.)

FORCED ASCENT

Air that blows perpendicular to a mountain range is forced to rise, and is cooled adiabatically. Eventually the dew point is reached, condensation occurs, clouds form, and precipitation results. This increased cloudiness and precipitation on the windward slope is known as the *orographic effect* (Lin et al. 2001). The rainiest places in the world are mountain slopes in the path of winds blowing off relatively warm oceans. There are many examples from every continent, but the mountainous Hawai'ian Islands serve as an illustration. The precipitation over the water around Hawai'i averages about 650 mm (25 in.) per year, while the islands average 1,800 mm (70 in.) per year. This is largely because of the presence of mountains, many of which receive over 6,000 mm (240 in.) per year (Nullet and McGranaghan 1988). At Mount Waialeale on Kaua'i, the average annual rainfall reaches the extraordinary total of 12,344 mm (486 in.), that is, 12.3 m (40.5 ft.), one of the highest recorded annual averages in the world (Blumenstock and Price 1967). In the continental United States, the heaviest precipitation occurs at the Hoh Rain Forest on the western side of the Olympic Mountains in Washington, where an average of 3,800 mm (150 in.) is received annually as storms are funneled up valleys oriented toward winter storm tracks (Fig. 3.4; Mass 2008).

FORCED DESCENT

Atmospheric-pressure conditions determine whether the air, after passing over a mountain barrier, maintains its altitude or is forced to descend; it is then heated by compression (adiabatic heating), resulting in clear, dry conditions. This characteristic phenomenon in the lee of mountains is responsible for the famous foehn or Chinook winds. The important point is that the descent of the air is induced by the barrier effect, resulting in clear, dry conditions that allow sunshine to reach the ground with much greater intensity and frequency than it otherwise would. This can produce "climatic oases" in the lee of mountain ranges, as in the Po Valley of Italy (Hoggarth et al. 2006).

Although heavy precipitation may occur on the windward side of mountains where the air is forced to rise, the leeward side may receive considerably less precipitation because the air is no longer being lifted (it is descending) and much of the moisture has already been removed. This *rainshadow effect* creates an arid area on the leeward or downwind side of mountains. In the lee of Mount Waialeale, Kaua'i, precipitation decreases at the rate of 3,000 mm (118 in.) per 1.6 km (1 mi) along a 4-km (2.5 mi) transect to Hanalei Tunnel (Blumenstock and Price 1967). In the Olympic Mountains, precipitation decreases from totals approaching 6,350 mm (250 in.) on the windward side to less than 430 mm (17 in.) at the town of Sequim on the leeward, a distance of only 48 km (30 mi) (Fig. 3.4; Mass 2008). Since both of these leeward areas are maritime, they are still quite cloudy; under more continental conditions, there would be a corresponding increase in sunshine as precipitation decreases, especially where the air is forced to descend on the leeward side.

Major Climatic Elements

The discussion so far has covered the more or less independent climatic controls of latitude, altitude, continentality, and the barrier effect of mountains. These factors, along with ocean currents, pressure conditions, and prevailing winds, control the distribution of sunshine, temperature, humidity, precipitation, and local winds. The climatic elements of sunshine, temperature, and precipitation are essentially dependent variables reflecting the major climatic controls (Barry 2008). They interact in complex ways to produce the day-to-day weather conditions experienced in different regions. In mountains, these processes frequently occur on scales small enough to be invisible to standard measurement networks used in weather forecasting, though their impact can be serious.

Solar Radiation

The effect of the sun becomes more exaggerated and distinct with elevation. The time lag, in terms of energy flow, between stimulus and reaction is greatly compressed in mountains. Looking at the effect of the sun in high mountains is like viewing its effects at lower elevations through a powerful magnifying glass. The alpine environment has perhaps the most extreme and variable radiation climate on Earth (Emck and Richter 2008). The thin, clean air allows very high solar intensities, and the topographically complex landscape provides surfaces with a range of different exposures, shadowing, and reflection from nearby peaks. Although the air next to the ground may heat up very rapidly during daylight, it may cool just as rapidly if the sun's rays are blocked. Thus, in the sun's daily and seasonal march through the sky, mountains experience a continually changing pattern of sunshine and shadow, influencing the energy flux in the ecosystem (Germino and Smith 2000). The factors to consider are the amount of sunlight received, the quality or kinds of radiation, and the effect of slopes upon this energy.

AMOUNT OF SOLAR RADIATION

The most striking aspect of the vertical distribution of solar radiation in the atmosphere is the rapid depletion of short-wavelength energy at lower elevations. This *attenuation* results from the increased density of the atmosphere at lower altitudes (Tables 3.1, 3.2). The atmosphere acts as a filter, reducing the intensity of some wavelengths and screening out others altogether. Consequently, the amount of energy reaching the surface at sea level is only about half that at the top of the atmosphere (Fig. 3.5). High mountains protrude through the lower atmospheric blanket and thus have the potential to receive much higher levels of

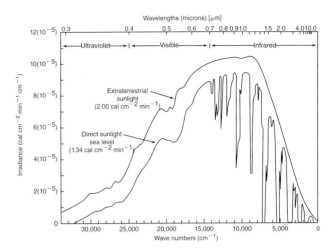

FIGURE 3.5 Spectral distribution of direct solar radiation at the top of the atmosphere and at sea level. Calculations are for clear skies with the sun directly overhead. Also shown is the spectral distribution of cloud light and sky light. The graph is plotted on a wave number scale in cm^{-1}—that is, the reciprocal of the wavelength, directly proportional to the frequency of light—to allow display of the full spectrum (a wavelength plot has difficulty including the visible and infrared together). The total area under the upper curve is the solar constant, 2.0 cal. cm^{-2} min^{-1} (1,365 W/m^2). (After Gates and Janke 1966: 42.)

solar radiation, as well as cosmic-ray and ultraviolet radiation (Emck and Richter 2008).

The first, and very vital, screening of solar energy takes place in the stratosphere, where ozone absorbs most of the ultraviolet radiation from the sun. Greenhouse gases absorb infrared solar radiation, but visible light passes through to the surface except when there is cloud cover. The visible light is scattered as it strikes molecules of air, water, and dust. Scattering is a selective process, principally affecting the wavelengths of blue light. The sky looks much bluer or darker in high mountains than at lower elevations, where more water and pollutants scatter light of other wavelengths, diluting the blue color. Clouds, of course, are the single most important factor controlling the receipt of solar energy at any given latitude and in mountains (Germino and Smith 2000). Since mountains stand above the lower reaches of the atmosphere, the solar radiation is much more intense since it has passed through less atmosphere (Fig. 3.6).

The solar constant is defined as the average amount of total radiation energy received from the sun at the top of the atmosphere on a surface perpendicular to the sun's rays (Fig. 3.5). This is approximately 1,365 Wm^{-2} (2 calories per square centimeter per min). At midday under clear skies, the total energy flux from the sun in high mountains may approach the solar constant, but several field investigations have recorded readings even slightly above the solar constant (Gates and Janke 1966; Terjung et al. 1969). Astonishingly high values of irradiance (up to 1,832 Wm^{-2}, 134 percent of the solar constant) incident upon

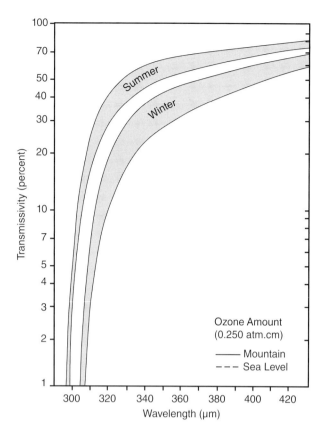

FIGURE 3.6 Spectral transmissivity of the atmosphere at 4,200 m (14,000 ft) and at sea level for latitude 40°N at summer and winter solstice. The attenuation shown here is for clear skies and is due entirely to ozone absorption. When the effects of dust, water vapor, and other impurities are included, the difference in transmissions between high and low elevations becomes considerably greater. (After Gates and Janke 1966: 45.)

a horizontal surface have been observed in the southern Ecuadorian Andes (4°S, 1,500 and 3,400 m). Cloud radiative effects were identified as the exclusive source of the "superirradiance" (Emck and Richter 2008).

QUALITY OF SOLAR RADIATION

The alpine environment receives considerably more ultraviolet radiation (UV) than low elevations. If only wavelengths shorter than 320 μm are considered, alpine areas receive 50 percent more UV during the summer solstice than areas at sea level. During the winter season, when the sun is lower in the sky (and therefore passes through denser atmosphere), alpine areas receive 120 percent more UV than sea level (Gates and Janke 1966). Many local influences, especially solar elevation, cloud coverage and type, albedo, and aerosol properties strongly affect the amount of UV received (Pfeifer et al. 2006). Clouds are a major source of variation in solar radiation and UV received, and can even increase UV levels compared to a cloud-free day (Parisi and Downs 2004). The relatively greater quantity of UV at high elevations has special significance for human

comfort and biological processes. A proverb in the Andes says, *"Solo los gringos y los burros caminan en el sol"* ("Only foreigners and donkeys walk in the sunshine"), indicating the respect that local people give to the strength of the sun at high altitudes (Prohaska 1970). UV has been cited for a number of harmful effects, ranging from the retardation of growth in tundra plants (Körner 2003) to cancer and eye damage in humans (Parisi and Downs 2004; Lichte et al. 2010). UV is mainly responsible for the deep tans of mountain dwellers and the painful sunburns of neophytes who expose too much of their skin (Lichte et al. 2010). The wavelengths responsible for sunburn occur primarily between 280 and 320 μm, while those responsible for darkening the skin occur between 300 and 400 μm. Wavelengths less than 320 μm are known to cause skin cancer and weaken the immune system.

EFFECT OF SLOPES ON SOLAR RADIATION

The play of the sun on the mountain landscape is like a symphony. As the hours, days, and seasons follow one another, the sun bursts upon some slopes with all the strength of fortissimo while the shadows lengthen and fade into pianissimo on others. The melody is continuous and ever changing, with as many scores as there are mountain regions, but the theme remains the same. It is a study of slope angle and orientation.

The closer to perpendicular the sun's rays strike a surface, the greater their intensity. The longer the sun shines on a surface, the greater the heating that takes place (Anderson 1998). In mountains, every slope has a different potential for receiving solar radiation, which can be measured using the following variables: latitude, time of year (height of sun), time of day, elevation, slope angle, and slope orientation (Huo and Bailey 1992). The basic characteristics of solar radiation on slopes are illustrated in Figure 3.7. This very useful diagram shows the situation for one latitude at four times of the year, at four slope orientations. It does not include the effects of clouds, diffuse sky radiation, the receptiveness of different slopes to the sun's rays, or shadow effects.

Most mountain slopes receive fewer hours of sunshine than a level surface, although slopes facing the sun may receive more energy than a level surface (especially at higher latitudes). In the tropics, level surfaces usually receive a higher solar intensity than slopes because the sun is always high in the sky. Whatever the duration and intensity of sunlight, the effects are generally clearly evident in the local ecology (Fig. 3.8). In the northern hemisphere, south-facing slopes are warmer and drier than north-facing slopes and, under humid conditions, are more favorable for life. Timberlines go higher on south-facing slopes, and the number and diversity of plants and animals are greater (Germino and Smith 2000). Humans take advantage of the sunny slopes. In the east–west valleys of the Alps, most settlements are located on south-facing slopes.

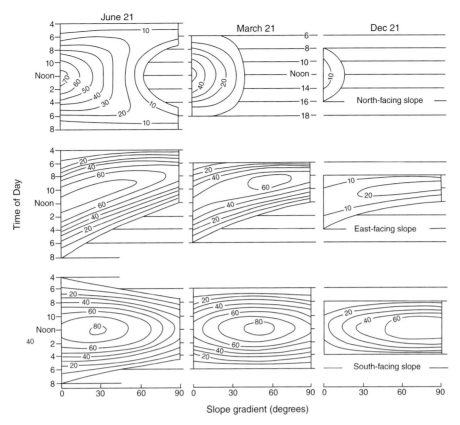

FIGURE 3.7 Direct solar radiation (cal. cm⁻² hr⁻¹) received on different slopes during clear weather at 50°N latitude. Three slopes are shown: north-, south-, and east-facing (west would be a mirror image of east), for summer and winter solstice and equinox (vernal is a mirror image of autumnal). The left-hand side of each diagram shows the distribution of solar energy on a horizontal surface (0° gradient) and is therefore identical for each set of three in the same column. The right-hand side of each diagram represents a vertical wall (90° gradient). The top of each diagram shows sunrise and the bottom shows sunset. As can be seen, the north- and south-facing slopes experience a symmetrical distribution of energy, while the east- and west-facing slopes reveal an asymmetrical distribution. Thus, on the east-facing slope during summer solstice the sun begins shining on a vertical cliff at about 4:00 A.M. and highest intensity occurs At 8:00 A.M. By noon the cliff passes into shadow. The opposite would hold true for a west-facing wall: It would begin receiving the direct rays of the sun immediately past noon.

The bottom row of diagrams illustrates a south-facing slope. During equinox days and nights are equal, so the distribution of energy is equal. During winter solstice the sun strikes south-facing slopes of all gradients at the same time (sunrise), but during summer the sun rises farther to the northeast, so some time elapses before it can shine on a south-facing slope. This difference in time increases with steeper slopes; for example, a 30° south-facing slope would receive the sun at about 5:00 A.M. and would pass into shadow at about 6:30 P.M., while a 60° south-facing slope would receive the sun at 6:30 A.M. (1½ hrs later) and the sun would set at 5:30 P.M. (1 hr earlier). On a north-facing slope (top row of diagrams) during summer, slopes up to 60° receive the sun at the same time, but if the slope is greater than 60 percent the sun cannot shine on it at noon; hence the "neck" cut out of the right-hand margin. Steep north-facing slopes at this latitude would only receive the sun early in the morning and late in the evening. During the winter solstice only north-facing slopes with gradients of less than 15° would receive any sun at all. (After Geiger 1965: 374.)

In spring, north-facing slopes may still be deep in snow while south-facing slopes are clear. As a result, north-facing slopes have traditionally been left in forest while south-facing slopes are used for high pastures (Fig. 3.9).

East- and west-facing slopes are also affected differently by solar radiation. Soil and vegetation surfaces are often moist in the morning, owing to the formation of dew or frost. On east-facing slopes, the sun's energy has to evaporate this moisture before the slope can heat appreciably. By the time the sun reaches the west-facing slope, however, the moisture has already evaporated, so the sun's energy more effectively heats the slope. The driest and warmest slopes, therefore, face southwest rather than south.

Cloud cover, which varies latitudinally, seasonally, and according to time of day, can greatly influence the amount of solar energy received on slopes. During storms, the entire mountain may be wrapped in clouds; even during relatively clear weather, mountains may still experience local clouds. In winter, stratus clouds and fog are characteristic on intermediate slopes and valleys, but these frequently burn off by midday. In summer, mornings are typically clear but convection clouds (cumulus) build by mid-afternoon from thermal heating. Consequently, convection clouds result in east-facing slopes receiving greater sunlight while stratus clouds, as described above, allow more sun on west-facing slopes. As clouds move over mountains, build,

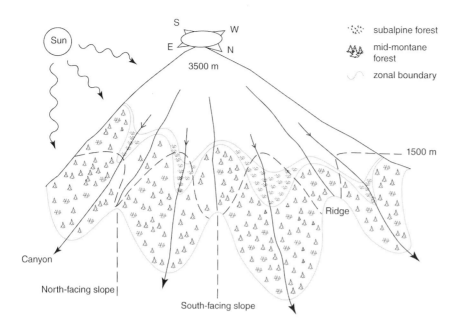

FIGURE 3.8 Topo- and microclimatic influences of slope and aspect on vegetation types. The northern hemisphere example is given where more solar receipt on south-facing slopes warms temperatures to where forest is replaced by grass. North-facing slopes are shaded and cooler with more soil moisture retention and thicker forests. On a larger scale, forests move down valleys following moisture and cooler temperatures created by cold air drainage. (After Kruckeberg 1991.)

and dissipate, they have a marked effect upon the amount and character of radiation received.

The extensive variety of surface characteristics of mountains—snow, ice, water, pastures, extensive forests, shrubs, soils, bare bedrock—affects the absorption of incoming solar radiation (Pomeroy et al. 2006). The effects of two factors—groundcover and topographic setting—illustrate this. Dark-colored features, including vegetation, absorb rather than reflect radiation, receiving increased amounts of energy. Snowfields, glaciers, and light-colored rocks have a high reflectivity (*albedo*), so that much of the incoming short-wave energy is reflected back into the atmosphere. If the snow is in a valley or on a concave slope, reflected energy may bounce from slope to slope, increasing the energy budget of the upper slopes. The opposite occurs on a mountain ridge or convex slope, where the energy is reflected back into space. Consequently, valleys and depressions are areas of heat build-up and generally experience greater temperature extremes than do ridges and convex slopes. Reflected energy can be an important source of heat for trees in the high mountains (Pomeroy et al. 2006). Snow typically melts faster around trees because the increased heat is transferred, as long-wave thermal energy, to the adjacent surface. On a larger scale, the presence of forests adds significantly to the heat budget of snow-covered areas. The shortwave energy from the sun can pass through a coniferous forest canopy, but very little of it escapes again. The absorbed energy heats the tree foliage and produces higher temperatures than in open areas. This results in rapid melting rates in the snowpack (Pomeroy et al. 2006).

Variation in the components of the surface energy budget provides the main driving force for regional differences in climate. In particular, the relative magnitude of sensible and latent heat fluxes reflects the influence of prevailing weather systems, and plays an important role in determining atmospheric temperature and moisture content (McCutchan and Fox 1986; Kelliher et al. 1996). These factors in turn influence the development of local wind systems. Surface energy budgets can vary significantly in mountains due to the effects of both complex topography and surface characteristics. When snow or ice is present, energy must first be partitioned to ablation before temperatures rise and, once the snow melts, there are large changes in albedo (Cline 1997). These variations affect the distribution of incoming and outgoing radiation, influencing net radiation, soil heat flux, and sensible and latent heat, producing a range of topo- and microclimates (Germino and Smith 2000).

Temperature

The decrease of temperature with elevation is one of the most striking and fundamental features of mountain climate. Those of us who are fortunate enough to live near mountains are constantly reminded of this, either by spending time in the mountains or by viewing the snow-capped peaks from a distance. Nevertheless, many characteristics of the nature of temperature in mountains are subtle and poorly understood. Alexander von Humboldt was so struck by the effect of temperature on the elevational zonation of climate and vegetation in the tropics that he proposed the terms *tierra caliente, tierra templada,* and *tierra fria* for the hot, temperate, and cold zones, respectively. These terms, commonplace in the tropics today, are still valid there. Their extension to higher latitudes by others, however, under the mistaken assumption that the same

FIGURE 3.9 View of an east–west valley near Davos, Switzerland, showing settlement and clearing on the sunny side (south-facing), while the shady side (north-facing) is left in forest. (L. W. Price.)

basic kinds of temperature conditions occur in belts from the equator to the poles, has been unfortunate, though some textbooks still use this simplistic approach.

VERTICAL TEMPERATURE GRADIENT

Change of temperature with elevation is called the *environmental* or *normal lapse rate*. De Saussure, who climbed Mont Blanc in 1787, was one of the first to measure temperature at different elevations. Many temperature measurements have since been made in mountains around the world. The lapse rate varies spatially according to many factors. Nevertheless, by averaging the temperatures at different levels, as well as those measured in the free air by balloon, radiosonde, and aircraft, average lapse rates have been established, ranging from 1°C to 2°C (1.8°F to 3.6°F) per 300 m (1,000 ft) (Minder et al. 2010). Aside from purposes of gross generalization, however, average lapse rates have little value in mountains. There is no constant relationship between altitude and temperature across space or time. Instead, the lapse rate changes continually with changing conditions, particularly the diurnal heating and cooling of the Earth's surface. For example, the vertical temperature gradient is normally greater during the day than at night, and during the summer than in winter. The gradient is steeper under clear than under cloudy conditions, on sun-exposed slopes than on shaded ones, and on continental mountains than on maritime mountains (Peattie 1936; Yoshino 1975). Also, the characteristics of free-air temperature differ from those on a mountain slope (Pepin and Losleben 2002), though the higher and more isolated a mountain peak is, the more closely its temperature will approach that of the free atmosphere.

Table 3.3 presents the average decrease of temperature with changing elevation in the Alps, and Figure 3.10 illustrates temperature changes with elevation in the Cascade Mountains of the United States. The temperatures shown are averages, with some interpolation between stations; the actual decrease with elevation is much more variable, and is influenced by whether a station is on a sunny or a shaded slope, the disposition of winds and clouds, and the nature of the slope surface. Snowcover and snowmelt each act to lower atmospheric temperatures (Takechi et al. 2002). A convex slope has qualities of heat retention different from those of a concave slope. A high valley will heat up more during the day (and cool down more at night) than an exposed ridge at the same elevation. Nevertheless, broad averages smooth out the extremes and individual differences, generally showing a steady and progressive decrease in temperature with increased elevation (Minder et al. 2010).

MOUNTAIN MASS (*MASSENERHEBUNG*) EFFECT

Large mountain systems create their own surrounding climate. The greater the surface area or landmass at any given elevation, the greater the effect on its own environment. Mountains serve as elevated heat islands where solar radiation is absorbed and transformed into long-wave heat energy, resulting in much higher temperatures than those found at similar altitudes in the free air (Chen et al. 1985). Accordingly, the larger the mountain mass, the more its climate will vary from the free atmosphere at any given altitude. This is particularly evident on high plateaus where treeline and snowline often occur at higher elevations than on isolated peaks at the same elevation. On the broad general level of the Himalaya, at 4,000 m (13,100 ft) it seldom freezes during summer,

TABLE 3.3
Temperature Conditions with Elevation in the Eastern Alps

Mean Air Elevation (m)	Temperature				Annual Number of		
	January	July	Year	Annual Range	Frost-free Days	Frost Alternation Days	Continuous Frost Days
200	−1.4	19.5	9.0	20.9	272	67	26
400	−2.5	18.3	8.0	20.8	267	97	1
600	−3.5	17.1	7.1	20.6	250	78	37
800	−3.9	16.0	6.4	19.9	234	91	40
1,000	−3.9	14.8	5.7	18.7	226	86	53
1,200	−3.9	13.6	4.9	17.5	218	84	63
1,400	−4.1	12.4	4.0	16.5	211	81	73
1,600	−4.9	11.2	2.8	16.1	203	78	84
1,800	−6.1	9.9	1.6	16.0	190	76	99
2,000	−7.1	8.7	0.4	15.8	178	73	114
2,200	−8.2	7.2	−0.8	15.4	163	71	131
2,400	−9.2	5.9	−2.0	15.1	146	68	151
2,600	−10.3	4.6	−3.3	14.9	125	66	174
2,800	−11.3	3.2	−4.5	14.5	101	64	200
3,000	−12.4	1.8	−5.7	14.2	71	62	232

SOURCE: After Geiger 1965: 444.

while on the isolated peaks at 5,000 m (16,400 ft) it seldom thaws (Peattie 1936).

In establishing the relationships between mountain mass and the heat balance, continentality, latitude, amount of cloud cover, winds, precipitation, and surface conditions must all be considered. Persistent cloud cover during summer can prevent a large mountain mass from showing substantial warming. A heavy snowcover can also retard the warming of a mountain area in spring because of surface reflectivity and the amount of initial heat required to melt the snow. The high Sierra Nevada of California is relatively warm compared with other mountain areas, in spite of heavy snowfalls, partially because the extreme clarity of the skies over this region in late summer allows maximum reception of solar energy. In general, the effect of greater mountain mass on climate is somewhat like that of increasing continentality. The ranges of temperature are greater than on small mountains (i.e., the winters are colder and the summers warmer), but the average of these temperatures will generally be higher than the free air at the same altitude. The effective growing climate, especially, is more favorable at the soil surface than in the free air, owing to higher soil temperatures, particularly when there is a high percentage of sunshine (Yoshino 1975).

Generally, the larger the mountain mass, the higher the elevation at which vegetation grows. The most striking example is in the Himalaya, where plants reach their absolute highest altitude (Chen et al. 1985). In the Alps (where the influence of mountain mass, *Massenerhebung*, was first observed), the timberline is higher in the more massive central part than on the marginal ranges (Imhof 1900, in Peattie 1936: 18). More locally, the effects of mountain mass on vegetation development can be observed in the Oregon Cascades. Except for Mount McLoughlin in southern Oregon, timberline is highest and alpine vegetation reaches its best development in the Three Sisters Wilderness area, where three peaks join to form a relatively large landmass above 1,800 m (6,000 ft) (Price 1978). On the higher but less massive peaks of Mount Hood and Mount Washington, a few kilometers to the north, the timberline is 150–300 m (500–1,000 ft) lower and the alpine vegetation is considerably more impoverished. The development of vegetation involves more than climate, of course, since plant adaptations and species diversity are related to the size of the gene pool and other factors (Körner 2003). Nevertheless, as discussed further in Chapter 7, vegetation is a useful indicator of environmental conditions, and a positive correlation between vegetation development and mountain mass can be observed in most mountain areas.

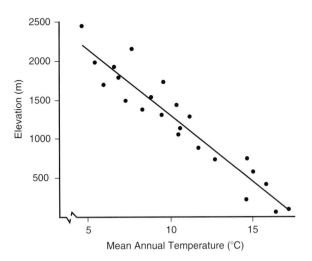

FIGURE 3.10 Mean annual temperature with altitude in the Cascade Mountains. Dots represent U.S. Weather Bureau First Order Stations in Washington State. Temperatures were calculated for period 1971–2000.

TEMPERATURE INVERSION

Temperature inversions are ubiquitous in landscapes with marked relief, and anyone who has spent time in or around mountains will have experienced their effects. Inversions are the exception to the general rule of decrease in temperature with elevation. During a temperature inversion, temperatures are lowest in the valley and increase upward along the mountain slope. Eventually, however, temperatures decrease again, so that an intermediate zone, the *thermal belt,* will experience higher night temperatures than either the valley bottom or the upper slopes. Inversions typically break up by thermal heating about three hours after sunrise (Whiteman et al. 2004).

Cold air is denser and therefore heavier than warm air. As slopes cool at night, colder air begins to slide downslope, flowing underneath and displacing warm air in the valley. Temperature inversions are best developed under calm, clear skies, where there is no wind to mix and equalize the temperatures, and the transparent sky allows the surface heat to be rapidly radiated and lost to space (Rucker et al. 2008; Daly et al. 2010). Consequently, the surface becomes colder than the air above it, and the air next to the ground flows downslope. The cold air will continue to collect in the valley until equilibrium between the temperatures of the slopes and the valleys has been established. The speed of the winds is controlled by valley width, increasing through narrows (i.e., Bernoulli's Theorem), as well as by differential heating along the valley (Rucker et al. 2008). If the valley is enclosed, a pool of relatively stagnant colder air may collect, but, if the valley is open, there may be continuous movement of air to lower levels, leading to air pollution (Gohm et al. 2009). The depth of the inversion depends on the characteristics of the local topography, airshed size, mixing with air aloft,

and general weather conditions, but is generally not more than 300–600 m (1,000–2,000 ft) (Whiteman et al. 2001; Smith et al. 2010).

Figure 3.11 demonstrates a temperature inversion in Gstettneralm, a small enclosed basin at an elevation of 1,270 m (4,165 ft) in the Austrian Alps, about 100 km (62 mi) southwest of Vienna. Because of the local topographic situation and the "pooling" of cold air, this valley experiences some of the lowest temperatures in Europe, even lower than the high peaks (Schmidt 1934). The lowest temperature recorded there is −51°C (−59.8°F), while the lowest temperature recorded at Sonnblick at 3,100 m (10,170 ft) is −32.6°C (−26.7°F).

As might be expected, distinct vegetation patterns are associated with these extreme temperatures. Normally, valley bottoms are forested and trees become stunted on higher slopes, being replaced by shrubs and grasses still higher up, but the exact opposite occurs here. The valley floor is covered with grass, shrubs, and stunted trees, while the larger trees occur higher up. An inversion of vegetation matches that of temperature (Schmidt 1934). A similar vegetative pattern has been found in the arid mountains of Nevada, where valley bottoms support sagebrush, while higher up is a zone of pinyon and juniper woodland. Higher still, the trees again disappear (Bradley and Fleishman 2008). The pinyon/juniper zone, the thermal belt, is sandwiched between the lower night temperatures of the valley bottom and those occurring higher up.

Human populations have taken advantage of thermal belts for centuries, particularly to cultivate frost-susceptible crops such as vineyards and orchards. In the southern Appalachians of North Carolina, the effect of temperature inversions is clearly displayed by the distribution of the fruit orchards (Dunbar 1966). During winter, the valleys are often brown with dormant vegetation, while the mountain tops at 1,350 m (4,430 ft) may be white with snow. In between is a strip of green that marks the thermal belt. Frost is common in the valley but, in the thermal belt, the sensitive Isabella grape has apparently grown for years without danger from frost (Peattie 1936). A similar situation exists in the Hood River Valley of Oregon, on the north side of Mount Hood, where cherries are grown in a sharply delimited thermal belt between the river and the upper slopes.

TEMPERATURE RANGE

The temperature difference between day and night and between winter and summer generally decreases with elevation (Fig. 3.12). This is because of the relatively greater distance from the heat source: the broad level of the Earth's surface. Temperature in mountains is largely a response to solar radiation. The free air, however, essentially does not respond to the heating effects of the sun, particularly at higher altitudes. A mountain becomes heated at the surface but there is a rapid temperature gradient in the

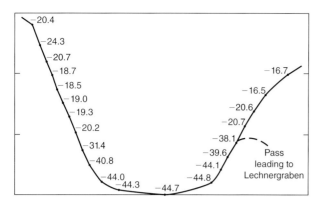

FIGURE 3.11 Cross section of an enclosed basin, Gstettneralm, in the Austrian Alps, showing a temperature inversion in early spring. Elevation of valley bottom is 1,270 m (4,165 ft). Note increase in temperature (°C) with elevation above valley floor, especially the rapid rise directly above the pass. This results from the colder air flowing into a lower valley at this point. (After Schmidt 1934: 347.)

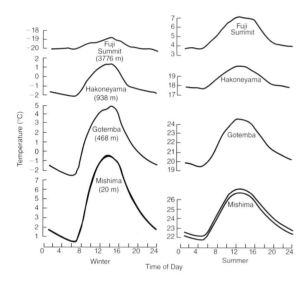

FIGURE 3.12 Diurnal temperature range at different elevations on Mount Fuji, Japan. The difference between high and low altitudes is much more exaggerated in winter (left) than in summer (right). (After Yoshino 1975: 193.)

surrounding air. Consequently, only a thin boundary layer or thermal shell surrounds the mountain, its exact thickness depending on a variety of factors (e.g., solar intensity, mountain mass, humidity, wind velocity, surface conditions, and topographic setting).

Ambient temperatures are normally measured at a standard instrument-shelter height of 1.5 m (5 ft). Such measurements generally show a progressive decline in temperature and a lower temperature range with elevation (Table 3.3; Figs. 3.10, 3.12). Yet there is a vast difference between the temperature at a height of 1.5 m (5 ft) and that immediately next to the soil surface. Paradoxically, the soil surface in alpine areas may experience higher temperatures

(and therefore a greater temperature range) than at low elevations, due to the greater intensity of the sun at high elevations (Anderson 1998). At an elevation of 2,070 m (6,800 ft) in the Alps, temperatures up to 80°C (176°F) were measured on a dark humus surface near timberline on a southwest-facing slope with a gradient of 35° (Terjung et al. 1969). This is comparable to the maximum temperatures recorded in hot deserts! At the same time, the air temperature at a height of 2 m (6.5 ft) was only 30°C (86°F), a difference of 50°C (90°F). Such high surface temperatures may occur infrequently and only under ideal conditions, but somewhat less extreme temperatures are characteristic, demonstrating the vast differences that may exist between the surface and the overlying air (Fig. 3.13). The soil surface in alpine tundra will almost always be warmer during the day than the air above it (Pomeroy et al. 2006). It may also become colder at night, although the differences are far less than during the day. The low growth of most alpine vegetation may be viewed as an adaptation to take advantage of these warmer surface conditions. In fact, several studies have shown that tundra plants may suffer more from high temperatures than from low temperatures (Körner 2003).

Temperature ranges vary not only with elevation, but latitudinally. The contrast in daily and annual temperature ranges is one of the most important distinguishing characteristics between tropical and midlatitude or polar climates. The average annual temperatures of high tropical mountains and polar climates are similar. The average annual temperature of El Misti in Peru at 5,850 m (19,193 ft) is −8°C (18°F), comparable to many polar stations. The use of this value alone is grossly misleading, given the vast differences in the temperature regimes. Tropical mountains experience a temperature range between day and night that is relatively greater than other mountain areas, due to the strongly positive heating effect of the sun in the tropics. On the other hand, changes in average temperature from month to month, or between winter and summer, are minimal. This is in great contrast to midlatitude and polar mountains, which experience lower daily temperature ranges but are increasingly dominated by strong seasonal gradients. Knowledge of the differences between these temperature regimes is essential to understanding the nature and significance of the physical and biological processes at work at each latitude.

Figure 3.14a depicts the temperature characteristics of Irkutsk, Siberia, a subpolar station with strong continentality. The most striking feature of this temperature regime is its marked seasonality. The daily range is only 5°C (9°F), while the annual range is over 60°C (108°F). Thus in winter, from October to May, temperatures are always below freezing, while in summer they are consistently above freezing. The period of stress for organisms, then, is concentrated into winter. An alpine station at this latitude would have essentially the same temperature regime, except for a relatively longer period with negative temperatures and a

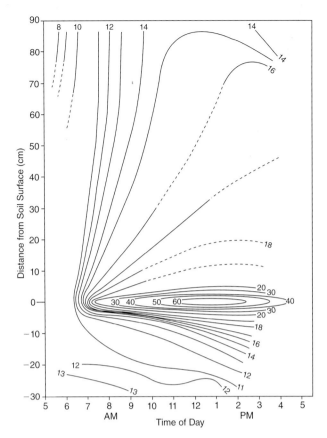

FIGURE 3.13 Vertical profile of soil and air temperatures (°C) under clear skies on a well-drained alpine tundra surface at 3,580 m (11,740 ft) in the White Mountains of California. Note the tremendous gradient occurring immediately above and below the soil surface. The slightly higher temperatures at a depth of 25–30 cm (10–12 in.) are a result of the previous day's heating and are out of phase with present surface conditions. (After Terjung et al. 1969: 256.)

shorter period with positive temperatures. More poleward stations would show an even smaller daily temperature range (Troll 1968).

Such a temperature regime stands in great contrast to that of tropical mountains. Figure 3.14b shows the temperature characteristics of Quito, Ecuador, on the equator at an elevation of 2,850 m (9,350 ft). The isotherms on the graph are oriented vertically, indicating very little change between winter and summer, but a marked contrast between day and night. The average annual range is less than 1°C (1.8°F), while the average daily range is approximately 11°C (19.8°F). This beautifully demonstrates the saying, "Night is the winter of the tropics." This is particularly true if the station is high enough for freezing to occur.

The lower limit of frost is determined principally by latitude, mountain mass, continentality, and the local topographic situation. In the equatorial Andes, it is at about 3,000 m (10,000 ft). The elevation of this limit decreases with latitude; the point where frost begins to occur in the lowlands is normally taken as being the outer limit of the tropics. In North America, the frost line runs through the middle of Baja California and eastward to the mouth of the Rio Grande, although it is highly variable from year to year. The frost line in tropical mountains is much more sharply delineated. In Quito, at 2,850 m (9,350 ft), frost is practically unknown. The vegetation consists of tropical evergreen plants which blossom continuously; farmers plant and harvest crops throughout the year. By an elevation of 3,500 m (11,500 ft), however, frost becomes a limiting factor (Troll 1968). At an elevation of 4,700 m (15,400 ft) on El Misti in southern Peru, it freezes and thaws almost every day.

The fundamental relationships between these disparate freeze–thaw regimes are demonstrated in Figure 3.15. Each site has a similar average annual temperature of −8°C to −2°C (18°F to 28°F), but the daily and annual ranges are markedly different. Yakutsk, Siberia, experiences strong seasonality, with a frost-free summer period of 126 days, but in winter the temperatures remain below freezing for 197 days. Alternating freezing and thawing take place for 42 days in the spring and fall. At Sonnblick in the Austrian Alps, the winter season is much longer (276 days), with a very short summer during which freezing and thawing can occur at any time. El Misti, however, is dominated by an almost daily freeze–thaw regime. This type of weather has been characterized as "perpetual spring": The sun melts the night frost every morning, and the days are quite pleasant. The 12-hour day adds to the impression of spring. As discussed in Chapter 7 and 8, these different systems provide greatly contrasting frameworks for the survival of plants and animals, as well as for the development of landscapes.

Humidity and Evaporation

Water vapor constitutes less than 5 percent of the atmosphere but is by far the single most important component with regard to weather and climate. It is highly variable in space and time. Water vapor provides energy for storms, and its abundance is an index of the potential of the air for yielding precipitation. It absorbs infrared energy from the sun and reduces the amount of short-wave energy reaching the Earth; serves as a buffer from temperature extremes; and is important biologically, since it controls the rate of chemical reactions and the drying power of the air. The moisture content of the atmosphere decreases rapidly with increasing altitude. At 2,000 m (6,600 ft) it is only about 50 percent of that at sea level, at 5,000 m (16,400 ft) it is less than 25 percent, and at 8,000 m (26,200 ft) less than 1 percent of that at sea level (Table 3.2). Within this framework, however, the presence of moisture is highly variable. This is true on a temporal basis, between winter and summer, day and night, or within a matter of minutes when the saturated air of a passing cloud shrouds a mountain peak (McCutchan and Fox 1986; Huntington et al. 1998; Xie et al. 2010). It is also true on a spatial basis, between high and low latitudes, marine and continental locations, the windward and leeward sides of a mountain range, or north- and south-facing slopes. The general upward decrease in water-vapor content, and the variations that

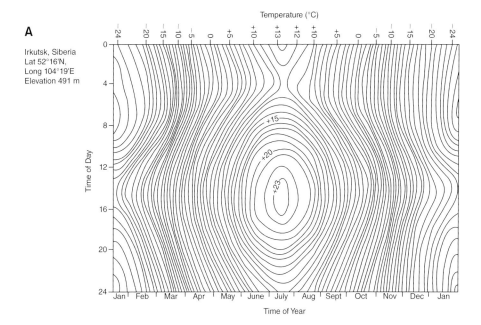

A

Irkutsk, Siberia
Lat 52°16'N,
Long 104°19'E
Elevation 491 m

Temperature (°C)

Time of Day

Time of Year

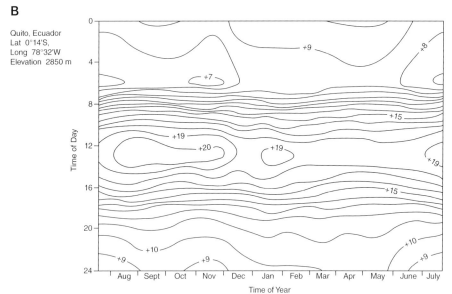

B

Quito, Ecuador
Lat 0°14'S,
Long 78°32'W
Elevation 2850 m

Time of Day

Time of Year

FIGURE 3.14 Daily and seasonal temperature distribution in a subarctic continental (A) and alpine tropical (B) climate. The opposite orientation of the isotherms reflects the fundamental differences in daily and seasonal temperature ranges in the two contrasting environments. The subarctic continental station (A) experiences a small daily temperature range (read vertically) but a large annual range (read horizontally). Conversely, the high-altitude tropical station (B) experiences a much greater daily temperature range than the annual range. (Adapted from Troll 1958: 11.)

occur, are illustrated by the east and west sides of the tropical Andes (Fig. 3.16). The contrast in absolute humidity between these two environments is immediately apparent, although the difference decreases with elevation and probably disappears altogether above the mountains. Imata, Salcedo, and Arequipa on the west have only about half the water-vapor content of stations on the east (Cerro do Pasco, Pachachaca, Huancayo, Bambamarca). Values similar to those at Arequipa occur at elevations 2,000 m (6,600 ft) higher on the east side (e.g., at Pachachaca). During the wet season, however, the absolute humidity at Arequipa may be two to three times higher than during the dry season, corresponding to an elevational difference of up to 3,000 m (10,000 ft).

The decrease in water vapor with altitude may seem somewhat difficult to explain, since precipitation increases with elevation. The two phenomena are not directly related, however. Precipitation results from the lifting of moist air from lower elevations upward into an area of lower temperature. Increasing precipitation does create a more humid environment in mountains, at least for part of the year and up to certain elevations, but eventually signs of aridity increase. Aridity at high elevations is related, in part, to lower barometric pressure, stronger winds, porous well-drained soils, and the intense sunlight.

The greater aridity of high-elevation environments is evident from the plants and animals, many of which

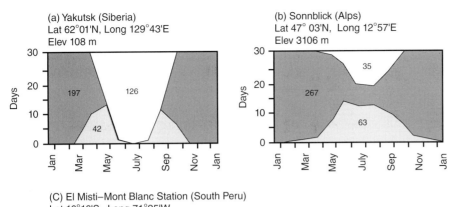

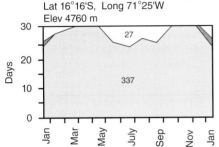

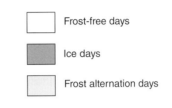

(a) Yakutsk (Siberia)
Lat 62°01'N, Long 129°43'E
Elev 108 m

197   126
42

(b) Sonnblick (Alps)
Lat 47° 03'N, Long 12°57'E
Elev 3106 m

267   35
63

(C) El Misti–Mont Blanc Station (South Peru)
Lat 16°16'S, Long 71°25'W
Elev 4760 m

27
337

☐ Frost-free days

▨ Ice days

▨ Frost alternation days

FIGURE 3.15 Freeze–thaw regimes at different latitudes and altitudes. Frost-free days indicate the number of days when freezing did not occur, ice days are those when the temperature was continually below freezing, and frost alternation days are the days when both freezing and thawing occurred. Note that the greatest number of the latter occur in tropical mountains. (Adapted from Troll 1958: 12–13.)

have adapted to a dry environment. Thick, corky bark and waxy leaves are common in alpine plants (Körner 2003). Mountain sheep and goats and their cousins, the llama, guanaco, alpaca, chamois, and ibex, can all live for prolonged periods on little moisture. Geomorphologically, eolian processes become increasingly important in higher landscapes, and the low availability of moisture is reflected in soil development (Bach 1995). One physiological stress reported by climbers on Mount Everest is a dryness of the throat and a general desiccation. Air-dried meat is a provincial dish in the high Engadine of Switzerland, and pemmican and jerky were both important in the mountains of western North America. In the Andes, an ancient method exists for producing dried potatoes (chuño) in the high, dry air above 3,000 m (10,000 ft). Permanent settlement at higher elevations apparently depended upon the development of this technique of food preservation (Troll 1968). Most mummification of the dead in Peru, Bolivia, and Chile occurred naturally, resulting from the aridity of the coastal foothill regions. Although rare, the Incas performed human sacrifices and buried bodies on mountain summits where the freezing conditions allowed mummies to be exceptionally well preserved (Reinhard 2005).

The lower absolute humidity and the tendency toward aridity at higher altitudes suggest greater evaporation rates with elevation. However, this may not be true, since the few studies of alpine evaporation have conflicting results (see Barry 2008). Several studies do indicate increased evaporation with elevation (Matthes 1934; Henning and Henning 1981; Sturman and Tapper 1996). Two years of water balance data from a high-elevation (2,800–3,400 m)

lake in California's Sierra Nevada show that evaporation accounts for 19–32 percent of the ablation (Kattelman and Elder 1991). Snowfall contributed 95 percent of the precipitation, and 80 percent of the evaporative (sublimation) losses came from snowcover. However, other studies have shown that evaporation does not exceed 10 percent of the total ablation (Hock 2005). Whichever of these tendencies is accepted as being the more general, it should be noted that these particular alpine areas are exceptionally, if not uniquely, dry environments, with high solar intensities, strong winds, and persistent subfreezing temperatures (Terjung et al. 1969). Most investigations on snowfields and glaciers in other regions have tended to show that evaporation is relatively unimportant in total ablation and may actually inhibit ablation, owing to the heat it extracts (Hock 2005).

Evaporation and the factors that control it in a natural environment are exceedingly complex (Penman 1963; Calder 1990). The rate depends upon temperature, solar intensity, atmospheric pressure, the available quantity of water (soil moisture), the degree of saturation of the air, and wind. One problem in measuring the rate of evaporation is the availability of moisture. In a lake or evaporating pan, the available moisture is, for all practical purposes, unlimited, but this is not true for most surfaces in high mountains. Rainfall is generally lost from the surface by drainage through porous soil or by runoff on steep slopes. As a result, there is frequently little surface moisture available for evaporation, no matter how great the measured rates are from an evaporation pan. For this reason, the determination of *evapotranspiration*, the loss of water to the air

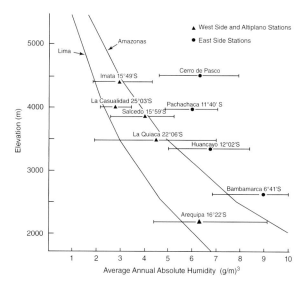

FIGURE 3.16 Average annual absolute humidity (mass of water vapor per unit volume, g/m³) with elevation on the humid eastern and arid western side of the tropical Andes. Horizontal lines provide a measure of the annual range of the monthly means of absolute humidity. The extremes are largely a reflection of the wet and dry seasons. Profiles are calculated as a function of height, according to starting values at Lima and Amazonas, based on empirical formulas obtained from observations in the Alps. The tropical-station data indicate that the decrease in vapor density with height is less pronounced than in middle latitudes. (Adapted from Prohaska 1970: 3.)

from both plant and soil surfaces, has become increasingly attractive (Penman 1963; Henning and Henning 1981).

The single most important factor controlling the decrease of evaporation with elevation is temperature, both of the evaporation surface and of the air directly above it (Huntington et al. 1998). While soil surfaces exposed to the sun at high elevations may reach exceptionally high temperatures, this is a highly variable condition (Anderson 1998; Germino and Smith 2000). During periods of high sun intensity and high soil temperatures, the potential for evaporation may be considerable, especially when the wind is blowing.

Precipitation

The increase of precipitation with elevation is well known from every country, even if the landforms involved are only small hills. In many regions, an isohyetal map with its lines of equal precipitation will look similar to a topographic map composed of lines of equal elevation (Fig. 3.17). Of course, the data on which most precipitation maps are based are scanty, so that considerable interpolation may be necessary, particularly in areas of higher relief (Kyriakidis et al. 2001; Scherrer et al. 2010). Precipitation does not always correspond to landforms. In some cases, maximum precipitation may occur at the foot or in advance of the mountain slopes (Barry 2008). In some regions and under certain conditions such as topographic funneling, valleys may receive more rainfall than the nearby mountains

(Sinclair et al. 1997; Mass 2008). In many higher alpine areas, precipitation decreases above a certain elevation, with the peaks receiving less than the lower slopes. Wind direction, temperature, moisture content, storm and cloud type, depth of the air mass and its relative stability, orientation and aspect, and configuration of the landforms are all contributing factors in determining the location and amount of precipitation (Ferretti et al. 2000; McGinnis 2000; Drogue et al. 2002; Lin et al. 2001; Houze and Medina 2005; Roe and Baker 2006). The complex topographic arrangement and often high relief of mountains create complex meso- and microscale three-dimensional circulation and cloud formations, leading to complex spatial patterns of precipitation (Garreaud 1999; Germann and Joss 2001, 2002; Roe and Baker 2006). Great variations in precipitation occur within short distances; one slope may be excessively wet while another is relatively dry. The terms "wet hole" and "dry hole" may be used in this regard. For instance, while the Grand Tetons of Wyoming receive 1,400 mm (55 in.), Jackson Hole, in a protected site at their base, only 16 km (10 mi) away, receives 380 mm (15 in.).

The most fundamental reason for increased precipitation with elevation is that landforms obstruct the movement of air and force it to rise, due to the *orographic effect*. Forced ascent of air is most effective when mountains are oriented perpendicular to the prevailing winds; the steeper and more exposed the slope, the more rapidly air will be forced to rise. As air is lifted over the mountains, it is cooled by expansion and mixing with cooler air at higher elevations (i.e., adiabatic processes). The ability of air to hold moisture depends primarily upon its temperature: Warm air can hold much more moisture than cold air. The temperature, the pressure, and the presence of hygroscopic nuclei in the atmosphere tend to concentrate the water vapor in its lower reaches. This is why most clouds occur below 9,000 m (30,000 ft), and why those that do develop higher than this are usually thin and composed of ice particles, yielding little or no precipitation.

When the air holds as much moisture as it can (i.e., relative humidity is 100 percent), it is considered saturated. The temperature at which condensation takes place is called the *dew point*. Ground forms of condensation (i.e., fog, frost, and dew) are caused by cooling of the air in contact with the ground surface, but condensation in the free atmosphere (i.e., clouds) only results from rising air. This may occur in one of several ways. Convection (thermal heating) takes place where the sun warms the Earth's surface and warm air rises until clouds begin to form. Such clouds may grow to great size since they are fed from below by relatively warm, moist rising air. Convectional rainfall is best displayed in the humid tropics, where water vapor is abundant, but it occurs in all climates. Air may also be forced to rise by the passage of cyclonic storms, where warm and cold fronts lift moist, warm air over cooler, denser air. This takes place primarily in the middle latitudes in association with the polar front (Fig. 3.2). Although both of these processes can operate without the presence of mountains, their

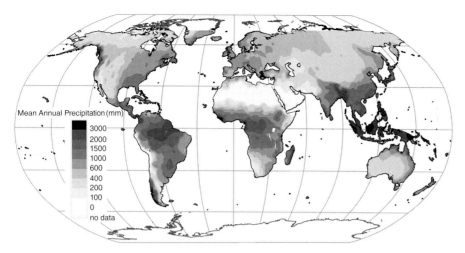

FIGURE 3.17 Annual average precipitation. (From several sources.)

effectiveness is greatly increased on the windward sides of mountains and decreases on the leeward sides.

One has only to compare the distribution of world precipitation with the location of mountains to see the profound influence of those mountains (Fig. 3.17). Almost every area of heavy rainfall is associated with mountains. In general, any area outside the tropics receiving more than 2,500 mm (100 in.), and any area within the tropics receiving more than 5,000 mm (200 in.), is experiencing a climate affected by mountains. In addition to the examples of Cherrapunji, Assam, Mount Waialeale, Hawai'i, and the Olympic Mountains, given earlier, many others could be added, including Mount Cameroon, West Africa, the Ghats along the west coast of India, the Scottish Highlands, the Blue Mountains of Jamaica, and the Southern Alps of New Zealand. The reverse is also true, for in the lee of each of these ranges is a rainshadow in which precipitation decreases drastically (Sklenář and Lægaard 2003; Smith et al. 2009). The western Ghats receive over 5,000 mm (200 in.), but immediately to their lee, on the Deccan Plateau, the average precipitation is only 380 mm (15 in.). The windward slopes of the Scottish Highlands receive over 4,300 mm (170 in.), but the lowlands around the Moray Firth receive 600 mm (24 in.). The Blue Mountains of northeast Jamaica face the trade winds and receive over 5,600 mm (220 in.), while Kingston, 56 km (35 mi) to the leeward, receives only 780 mm (31 in.). Mountains, therefore, not only cause increased precipitation, but also have the reciprocal effect of decreasing precipitation.

Despite these useful generalities, many local and regional variations occur within mountains. The complex local topography creates funneling effects that can increase atmospheric moisture content and precipitation, even downwind from the funnel (Sinclair et al. 1997). High peaks or ridges can create "mini-rainshadow" zones, even in the center of a range (Garreaud 1999). Significant quantities of precipitation can fall on the lee side of mountains because of spillover effects (Sinclair et al. 1997). Many precipitation maps of mountainous regions do not indicate

this internal variability because of the lack of weather stations and interpolation between existing station data using generalized elevation–precipitation relationships (Kyriakidis et al. 2001; Minder et al. 2010; Scherrer et al. 2010). Additionally, seasonal and interannual variability of storm tracks and storm intensities can create nonelevational precipitation patterns.

The movement of air up a mountain slope, creating clouds and precipitation, may be due simply to the wind, but is usually associated with convection and frontal activity. Rising air cools at a rate of 3.05°C (5.5°F) per 300 m (1,000 ft) (dry adiabatic rate) until the dew point is reached and condensation occurs (Fig. 3.28). Thereafter, the air cools at a slightly lower rate (wet adiabatic rate) because of the release of the latent heat of condensation. If, upon being lifted, the air has a high relative humidity, it may take only slight cooling to reach saturation but, if it has a low relative humidity, it may be lifted considerable distances without reaching the dew point. Conversely, if the air is warm, it often takes considerable cooling to reach dew point but then may yield copious amounts of rainfall; cool air usually needs only slight cooling to reach dew point but also yields far less precipitation. After the air has passed over the mountains, precipitation decreases or may cease as the air descends. Descending air gains heat at the same rate at which it was cooled initially: 3.05°C per 300 m (5.5°F per 1,000 ft), since it is being compressed and is moving into warmer air (Fig. 3.28). Such conditions are not conducive to precipitation.

In relation to precipitation, the orographic effect involves several distinct processes: (1) forced ascent, (2) blocking (or retardation) of storms, (3) the triggering effect, (4) local convection, (5) condensation and precipitation processes, and (6) runoff.

FORCED ASCENT

Forced ascent is the most important precipitation process in mountains. Steeper slopes produce more precipitation than gentle slopes (Lin et al. 2001). The process may be most clearly seen in coastal mountains, like the Olympics, that lie athwart moisture-laden winds. Other processes contribute to the total precipitation, and differentiation among them is difficult. To explain the amount and distribution of rainfall caused strictly by forced ascent, it is necessary to consider the atmospheric conditions from three different perspectives.

The first perspective involves the large-scale synoptic pattern that determines the characteristics of the air mass crossing the mountains: its depth, stability, moisture content, wind speed, and direction (McGinnis 2000). Second is the microphysics of the clouds, the presence of hydroscopic nuclei, the size of the water droplets, and their temperature, which determine whether the precipitation will fall as rain or as snow or evaporate before reaching the ground (Uddstrom et al. 2001). Third, and most important, is the motion of the air with respect to the mountain (Tucker and Crook 1999). Will it blow over the mountain, or around it? This determines to what depth and extent the air mass at each level is lifted. It is not realistic, for example, to assume that the air will be lifted the same amount at all levels. The solution to these problems involves atmospheric physics and the construction of dynamic models (Drogue et al. 2002).

The simplest system is that of coastal mountains with moisture-laden winds approaching from the ocean. As the air is lifted from sea level, the resulting precipitation is clearly due to the landforms (Mass 2008). Exceptions may occur where the mountains are oriented parallel to the prevailing winds and/or where the frontal systems resist lifting. In southern California, for example, precipitation is often heavier in the Los Angeles coastal lowlands than in the Santa Inez and San Gabriel Mountains, due to the blocking of storms. The orographic component of precipitation increases only when the approaching air mass is unstable. Under stable conditions, the wind will flow around the mountains (which are oriented east–west), so there is no significant orographic lifting and the precipitation is related entirely to frontal lifting. The mountains apparently receive less rainfall than the lowlands under these conditions, because the shallow cloud development does not allow as much depth for falling precipitation particles to grow by collision and coalescence with cloud droplets before reaching the elevated land.

The situation becomes more complex in interior high-elevation areas where there is more than one source region and storms enter the area at various levels in the atmosphere. For instance, in the Wasatch Mountains of Utah, precipitation is highly variable; the valleys may receive greater amounts than the mountains during any given storm or season (Sassen and Zhao 1993). The average over a period of years, however, shows an increase with elevation. The greater precipitation in valleys is apparently associated with certain synoptic situations, particularly when a "cold low" is observed on the upper-air charts. These occur as closed lows on the 500-millibar pressure chart (i.e., at a height of about 5,500 m [18,000 ft]), and are associated with large-scale upward (vertical) movement of air, which is not displayed in normal cold- or warm-front precipitation (Schultz et al. 2002).

BLOCKING OF STORMS

By hindering the free movement of storm systems, mountains can cause increased precipitation. Storms often linger for several days or weeks as they slowly move up and over mountains, producing a steady downpour (Mass 2008). This is best displayed in the middle latitudes with high barrier mountains. Winter storms linger with amazing persistence in the Cascades and the Gulf of Alaska before they pass across the mountains, or are replaced by another storm. Storms of similar character in the Great Plains travel much more rapidly, since there are no restrictions to their movement. In northern Italy, between the Alps and the Apennines, heavy and persistent rains are associated with the "lee depressions" caused by the interception of polar air by the Alps (Hoggarth et al. 2006).

THE TRIGGERING EFFECT

Another important variable influencing the amount of precipitation is the stability of the air, that is, its resistance to vertical displacement. This is controlled primarily by temperature. When there is a low environmental lapse rate, that is, less than 1.4°C per 300 m (2.5°F per 1,000 ft), as there often is at night in mountains, the air is stable. During the day, when the sun warms the slopes and the surface air is heated, the environmental lapse rate increases and the air will tend to rise, frequently producing afternoon clouds. When the lapse rate exceeds the dry adiabatic rate of 3.05°C per 300 m (5.5°F per 1,000 ft), a condition of absolute instability prevails. Under these conditions, even a slight lifting of the air by a landform is enough to "trigger" it into continued lifting of its own accord. If it then begins to feed upon itself through the release of latent heat of condensation, it can yield considerable precipitation (Smith et al. 2009). As a result, thunderstorms can develop, even on small hills in the path of moist unstable air (Schaaf et al. 1988).

LOCAL CONVECTION

Clouds commonly form over mountains during the day, especially in summer, when nights and early mornings are clear, but by midmorning clouds begin to build, often culminating in thunderstorms with hail and heavy rain (Kirshbaum and Durran 2004). This has been well documented for the base of the Colorado Rockies, where the higher peaks of the Front Range create a "heated chimney effect" in the initiation of thunder and hailstorms (Banta and Schaaf 1987). Mountains serve as elevated heat islands during the day, since their surfaces can be warmed to a temperature similar to surrounding lowlands. As a consequence, the air at a given altitude is much warmer over the mountains than over the valleys. The lapse rate above the peaks, therefore, is considerably greater than in the surrounding free air, resulting in actively rising air. Glider pilots have long taken advantage of this (Ludlam, 1980). Airline pilots, on the other hand, make every effort to avoid the turbulence associated with unstable air over mountains. Rarely, given weak synoptic conditions, local mountain convection can become organized into a meso-scale convective complex (Tucker and Crook 1999). These strong storms can reinforce themselves, spawning severe thunderstorms and even tornadoes ("mountainadoes").

Clouds and thunderstorms initiated in Colorado's Front Range frequently drift eastward, continuing to develop as they move onto the plains, and producing locally heavy precipitation (Sato and Kimura 2003; Jeglum et al. 2010). A study in the San Francisco Mountains north of Flagstaff, Arizona, suggests that clouds may increase in volume by as much as 10 times after drifting away from a mountain source (Banta and Schaaf 1987). Most clouds observed in this area were small cumuli that eventually dissipated once removed from their supply of moist, rising air, but a large cumulonimbus could maintain itself independently of the mountains and result in storms at some distance away. Fujita (1967) found a ring of low precipitation about 24 km (15 mi) in diameter encircling these mountains, with an outer ring of heavier precipitation. During the day, the rainfall is over the mountains, but at night it falls over the lowlands because the mountains are relatively cold. A "wake effect" due to wave action created by airflow over the mountains may be partly responsible for the inner ring of light precipitation (Fujita 1967; Brady and Waldstreicher 2001; Epifanio and Rotunno 2005).

The height of the cloud base is very important to the development of convection in mountains since, once the sun is blocked, the positive effect of solar heating is eliminated. The height of the cloud base is also critical to the distribution of precipitation. If the cloud base is below the level of the peaks, as it usually is in winter, when forced ascent occurs, cloud growth and precipitation will take place mainly on the windward side. In summer, however, the base of convection clouds is generally much higher.

Mountains, as sites of natural atmospheric instability, are ideal areas for artificial stimulation of precipitation. The considerable efforts that have been made in this regard have met with varied success, depending upon technique and local atmospheric conditions (Xing et al. 2005; Noppel and Beheng 2009). Most projects have been aimed at increasing the snowpack for runoff during the summer. This appears to be a desirable objective, but the ecological implications of such undertakings are far reaching (MacCracken 2006). For example, the Portland General Electric Company of Portland, Oregon, hired a commercial firm during the winter of 1974/75 to engage in cloud seeding on the eastern side of the Cascades. The objective was

to increase the snowpack in the Deschutes River watershed, where the company has two dams and power-generating plants. Considerable success was apparently achieved, but problems arose when residents of small towns at the base of the mountains were suddenly faced with a marked increase in snow. There were new problems of transportation, snow removal, and other hardships for the local people. Greater snowfall meant greater profits for the power company, but also greater expenses for local people. Objections were raised in the courts, and the project was eventually halted. The positive effects of such programs must always be balanced against the negative. In our efforts to manipulate nature, we are made increasingly aware of how little we understand the effects of our actions on natural systems (MacCracken 2006).

CONDENSATION PROCESSES

The presence of fog or clouds near the ground may increase moisture delivery to the ground. Water droplets in fog and clouds are usually so small that they remain suspended, and even a slight wind will carry them through the air until they strike a solid object and condense upon it. You have experienced this if water droplets have ever formed on your hair and eyebrows as you passed through a cloud or fog. Fog drip and rime deposits, which form at subfreezing temperatures, are responsible for an appreciable amount of the moisture in mountains, since elevated slopes are often in contact with clouds.

CLOUDS Cloud cover is generally more frequent and thicker over mountains than over surrounding lowlands (Uddstrom et al. 2001). Forced lifting of a moist air mass over the topographic barrier is the primary cause, although this may be augmented by convective processes. A slowing of storm movement by the blocking effect also leads to an increase in cloud water content (Mass 2008). Cloud type in mountain areas is primarily determined by synoptic characteristics. In middle and high latitudes, stratiform clouds are common, especially during winter in the absence of convection. These clouds often envelop the ground as hill fog. Cumulus clouds are typically associated with convection in continental, subtropical, and tropcal areas, and during the summer in midlatitudes. A problem relating to cloud data from mountain stations is that the clouds often engulf the observer, obstructing the view of the cloud forms. Likewise, cloud tops can be below the station.

A number of cloud forms are unique to mountain environments (Ludlam 1980). All are stationary clouds, which continually dissipate on the lee edge of the cloud and re-form on the upwind edge, thus appearing to remain in the same location for long periods. A cap or crest cloud forms over the top of an isolated peak or ridge, resembling a cumulus cloud, although often streamlined, or with streamers of cirrus forms. They sit near or just below the summit level, appearing like a hat atop the peak. Banner

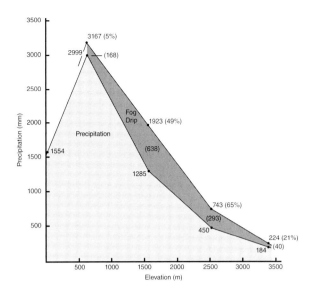

FIGURE 3.18 Contribution of fog drip to precipitation during 28-week study period (October 1972–April 1973) on the forested northeast slopes of Mauna Loa, Hawai'i. Numbers show precipitation totals in millimeters. Those in parentheses indicate fog drip. Percentages are the relative amounts contributed to the total by fog drip at each station. (After Juvik and Perreira 1974: 24.)

clouds are cap clouds which extend downwind from the peak like a flag waving in the wind, and are sometimes difficult to distinguish from streamers of snow blowing from summits. Lenticular clouds are lens-shaped clouds formed in regular spaced bands parallel to the mountain barrier on the lee side. These streamlined cloud features form through the interaction of high-velocity winds with the mountain barriers. Stratification of humidity in the atmosphere can result in multistoried lenticular clouds, forming a "pile of plates" or "pile of pancakes." These sometimes eerie-looking clouds might be responsible for the "flying saucer" scare of the 1950s, which originated from a sighting of "a disc-shaped craft skimming along the crest of the Cascades Range in Washington" (Arnold and Palmer 1952).

FOG (CLOUD) DRIP   Fog drip is most significant in areas adjacent to oceans with relatively warm, moist air moving across the windward slopes. The moisture yield from fog drip may exceed that of mean rainfall by as much as 462 percent (Juvik et al. 2011). The potential of clouds for yielding fog drip depends primarily upon their liquid content, the size of the cloud-droplet spectrum, and the wind velocity (Vermeulen et al. 1997). The amount depends upon the nature of the obstacles encountered and their exposure to the clouds and wind. For example, a tree will yield more moisture than a rock, and a needle-leaf tree is more efficient at "combing" the moisture from the clouds than a broadleaf tree (Vermeulen et al. 1997). A tall tree will yield more moisture than a short one, and a tree with frontline exposure will yield more than one surrounded by other trees (Juvik et al. 2011). The tiny fog droplets are intercepted by the leaves and branches and grow by

coalescence until they become heavy enough to fall to the ground, thereby increasing soil moisture and feeding the groundwater table. If the trees are removed, this source of moisture is also eliminated.

Many tropical and subtropical mountains sustain "cloud forests," which are largely controlled by the abundance of fog drip (Jarvis and Mulligan 2011). Along the east coast of Mexico in the Sierra Madre Oriental, luxuriant cloud forests occur between 1,300 and 2,400 m (4,300–7,900 ft). The coastal lowlands are arid by comparison, as is the high interior plateau beyond the mountains. Globally, 477 cloud forests have been identified, at elevations between 220 and 5005 m (720–16,420 ft), with an average elevation of 1,700 m (5,600 ft) (Jarvis and Mulligan 2011). Such forests were once much more extensive, but they have been severely disturbed by people and are now in danger of being eliminated. The climate of cloud forests is highly variable from site to site, with an average rainfall of ~2,000 mm per year and an average temperature of 17.7°C (Jarvis and Mulligan 2011). On the northeast slopes of Mauna Loa, Hawai'i, at 1,500–2,500 m (5,000–8,200 ft.), above the zone of maximum precipitation, fog drip is a major ecological factor in the floristic richness of the forests. Over 28 weeks, fog drip was found to provide 638 mm (25.3 in.) of moisture at an elevation of 1,500 m (5,000 ft) and, at 2,500 m (8,200 ft), 293 mm (11.5 in.), 65 percent of the direct rainfall (Fig. 3.18; Juvik and Perreira 1974). The study was replicated in 2006–2008 with automated instrumentation, measuring annual fog drip totals of 901–2883 mm (35.5–113.5 in.): 181–462 percent of measured rainfall (Juvik et al. 2011).

The contribution of fog drip on middle and upper mountain slopes at lower latitudes is clearly a major factor in the moisture regime. The relationship between the cloud forest and fog drip is essentially reciprocal. The trees cause additional moisture in the area. At the same time, they need the fog drip in order to survive, especially in areas with a pronounced dry season, when fog drip provides the sole source of moisture for the plants. In middle latitudes, fog drip is less critical to the growth of trees, but can still be important (Jarvis and Mulligan 2011). This can be seen in the mountains of Japan, where there is heavy fog at intermediate altitudes (Fig. 3.19).

RIME DEPOSITS   Rime is formed at subfreezing temperatures when supercooled cloud droplets are blown against solid obstacles, freezing on contact (Krzysztof et al. 2002). Rime deposits tend to accumulate on the windward side of objects. The growth rate is directly related to wind velocity. Rime deposits can reach spectacular dimensions and, by their weight, cause considerable damage to tree branches, especially if followed by snow or freezing rain. Trees at the forest edge and at timberline frequently have their limbs bent and broken by this process; power lines and ski lifts are also greatly affected. One study in Germany measured a maximum hourly growth of 230 g per m (8.1 oz per 3.3 ft) on a power-line cable (Geiger 1965). The stress caused by this added weight may cause a power failure

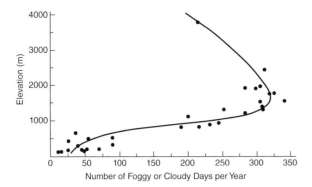

FIGURE 3.19 Relationship between the number of foggy or cloudy days and elevation in the mountains of Japan. (Dots represent data from weather stations at various altitudes.) The elevation of greatest cloudiness is 1,500–2,000 m (5,000–6,000 ft), where clouds develop almost daily, especially in August. This is caused by the inflow of cool marine air at these levels. The actual height of maximum cloudiness varies from one season to another and from one mountain range to another. (After Yoshino 1975: 205.)

if the supporting structures are not properly engineered. Rime accumulation is a severe obstacle to the maintenance of mountain weather stations because instruments become coated, making accurate measurements extremely difficult. Some instruments can be heated or enclosed in protected housing, but the logistical problems of accurately monitoring the alpine environment are very great. The U.S. Weather Bureau Station on Mount Washington, New Hampshire, where rime-forming fogs are frequent and the wind is indefatigable, exemplifies the problems encountered. Rime occurs 149 days/year on average on Mt. Szrenica, Poland, where daily accumulation rates of 0.56–2 kg per 200 cm² have been measured (Krzysztof et al. 2002).

Few investigations have been made concerning the moisture contribution of rime. It is known to be generally somewhat less than fog drip, but may nevertheless be significant. Rime is found primarily in midlatitude and polar mountains, although it also occurs at the highest elevations in the tropics. Like fog drip, it is most effective on forest-covered slopes that provide a large surface area for its accumulation. At very high altitudes and at latitudes where total precipitation is low, rime deposits on glaciers and snowfields may constitute the primary source of the water taken from the air.

ZONE OF MAXIMUM PRECIPITATION

Precipitation is generally thought to increase only up to a certain elevation, beyond which it decreases (Lauer 1975). The argument is that the greatest amount of precipitation will usually occur immediately above the cloud base level because most of the moisture is concentrated there. As the air lifts and cools further, the amount of precipitation will eventually decrease, because a substantial percentage of the moisture has already been released below (Miniscloux et al. 2001). In addition, the decreased temperature and pressure

at higher elevations reduce the capacity of the air to hold moisture. The water-vapor content at 3,000 m (10,000 ft) is only about one-third that at sea level. Forced ascent also plays a part, since the air, seeking the path of least resistance, will generally move around the higher peaks rather than over them.

The concept of a zone of maximum precipitation was developed over a century ago from studies in tropical mountains and the Alps (Hann 1903). Other studies seemed to confirm the concept and its application to other areas (Peattie 1936; Lauer 1975). The elevation of maximum precipitation varies geographically, depending upon the synoptic setting (McGinnis 2000; Barry 2008). Tropical mountains tend to have precipitation maxima at lower elevations, with the maximum zone rising with decreasing annual totals. In middle latitudes, the general trend is for precipitation to increase with elevation, often to the highest observation station (Mass 2008). The existence of such a zone has been challenged, as calculations of the amount of precipitation necessary to maintain active glaciers in high mountains and observations of relatively heavy runoff from small alpine watersheds seem to call for more precipitation in certain mountain areas than climatic station data would indicate (Smith et al. 2009; Daly et al. 2010).

Currently, the question is moot, the problem being one of measurement. There are very few weather stations in high mountains, and even where measurements are available, their reliability is questionable (Scherrer et al. 2010). As one author says, "Precipitation in mountain areas is as nearly unmeasureable as any physical phenomenon" (Anderson 1972: 347). This is particularly true at high altitudes with strong winds. Not surprisingly, many studies have shown that wind greatly affects the amount of water collected in a rain gauge (Fig. 3.20; Zhihua and Li 2007). Considerable effort has been made to alleviate this problem by the use of shields on gauges, location in protected sites, use of horizontal or inclined gauges, the use of radar techniques, and statistical corrections (Peck 1972; Zhihua and Li 2007).

Measuring snow is even more difficult, since the wind not only drives falling snow but redistributes it on the ground (Hiemstra et al. 2002). Correction factors have been developed for certain types of gauges (Kyriakidis et al. 2001). There are also problems in storage and melting of snow for water equivalency, as well as losses due to evaporation. The major problem, however, is accurate monitoring of snowfall. Small clearings are used in conifer forests and, above timberline, snow fences are increasingly used to enclose and shield the gauges. This still does not guarantee accurate measurements, but shielded gauges (whether for rain or snow) record greater amounts of precipitation than unshielded gauges in the same location. For example, the University of Colorado has operated a series of weather stations in the Front Range of the Rocky Mountains since 1952. The measured precipitation amounts from the two highest sites above treeline increased abruptly in 1964

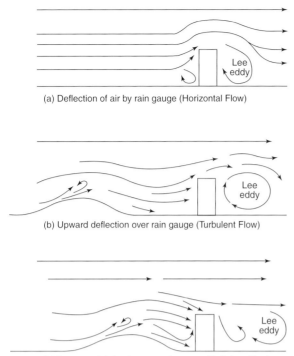

(a) Deflection of air by rain gauge (Horizontal Flow)

(b) Upward deflection over rain gauge (Turbulent Flow)

(c) Downward deflection over rain gauge (Turbulent Flow)

FIGURE 3.20 The effects of a precipitation gauge on surface wind-flow. In the first case (A) the wind may tend to speed up next to the gauge since it must travel farther to get around the obstacle. The lower illustrations (B and C) show that turbulence caused by surface roughness may result in upflow or downflow at the gauge orifice, depending on its location with respect to surrounding topography and wind direction. The lee eddy created in each situation is a location of snow and dust deposition due to slow (reversed) wind speeds. (Adapted from Peck 1972: 8.)

TABLE 3.4

Average Annual Precipitation at Four Ridge Sites in a Transect Up the Front Range of the Colorado Rockies during 1965–1970

Station	Elevation (m)	Amount (mm)
Ponderosa	2,195	579
Sugarloaf	2,591	578
Como	3,048	771
Niwot Ridge	3,750	1,021

SOURCE: Barry 1973: 96.

when snow fences were erected around the recording gauges. Before the gauges were shielded, the average annual amount was 655 mm (25.8 in.); it jumped to 1,021 mm (40.2 in.) after the snow fence was installed. The data now show an absolute increase in precipitation with increasing elevation (Table 3.4). More reliable instrumentation in the Alps has led to similar results, at least up to an elevation of 3,000 m (10,000 ft) (Schmidli et al. 2002).

Snow accumulation in alpine watersheds can be investigated more thoroughly by collecting depth and density data from snow pits, which can be converted to water equivalent (Perkins et al. 2009). In North America, an extensive network of over 1,200 snow courses is surveyed monthly by the Natural Resources Conservation Service (NRCS, formerly the Soil Conservation Service) of the U.S. Department of Agriculture (Perkins et al. 2009). In snow courses, depth and density are measured manually to estimate annual water availability, spring runoff, and summer streamflows. In more remote locations in the western United States, the SNOTEL (Snow Telemetry) network of over 730 automated snow reporting stations uses an air-filled pillow attached to a pressure gauge to measure snowpack weight, which is transmitted via VHF signals to a data-collection station (Perkins et al. 2009). Combining field measurements of snow–water-equivalency with topographic data (slope and aspect) and net radiation, estimates of watershed snowpack water content can be modeled (Clow 2010).

Another problem with precipitation analysis in mountains is that many weather stations are located in valleys. Uncritical use of these data may lead to erroneous results (Kyriakidis et al. 2001). Valleys oriented parallel to prevailing winds may receive as much or more precipitation than the mountains on either side, while valleys oriented perpendicular to the prevailing winds may be "dry holes" (Neiman et al. 2002; Mass 2008). In addition, local circulation systems between valleys and upper slopes may result in valleys being considerably drier than the ridges. For example, in parts of the Hindu Kush, Karakoram, Himalaya, and Hengduan Mountains, many valleys are distinctly arid (Troll 1968).

In the tropics, decreasing precipitation above a certain elevation is much better established (Fig. 3.21; Lauer 1975). The precipitation falls principally as rain, with snow or rime on the highest peaks; tropical mountains experience considerably less wind than in those middle latitudes. As a result, simple rainfall measurements are more dependable. The zone of maximum precipitation varies according to location. In the tropical Andes and in Central America, it lies at 900–1,600 m (3,000–5,300 ft) (Weischet 1969). Mount Cameroon in West Africa, near the Gulf of Guinea, receives an annual rainfall of 8,950 mm (355 in.) on the lower slopes but less than 2,000 mm (80 in.) at the summit; its zone of maximum precipitation occurs at 1,800 m (6,000 ft). In East Africa, measurements on Mount Kenya and Kilimanjaro show an increase up to the montane forest belt at 1,500 m (5,000 ft), and then a sharp decrease (Fig. 3.21). The maximum zone receives about 2,500 mm (100 in.), but less than 500 mm (20 in.) falls on the summits. The effects of low rainfall, high sun intensity, and porous soils give the alpine belt a desert-like appearance, although both summit areas support small glaciers (Kaser et al. 2004; Hastenrath 2010). Desert-like conditions exist at the summits of many tropical mountains. On the islands of Indonesia and Sri Lanka,

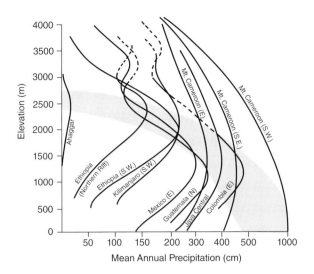

FIGURE 3.21 Generalized profiles of mean annual precipitation (cm) vs. elevation (m) in the tropics. The shaded area shows the zone of maximum precipitation. (Adapted from Lauer 1975.)

the zone of maximum precipitation is at 900–1,400 m (3,000–4,600 ft) (Weischet 1969). It is at 600–900 m (2,000–3,000 ft) in Hawai'i (Blumenstock and Price 1967; Juvik and Perreira 1974; Nullet and McGranaghan 1988), where the decrease above is counteracted somewhat by the presence of fog drip, since this is a zone of frequent cloudiness (Fig. 3.18).

The vertical distribution of precipitation illustrates yet another environmental distinction between tropical and extratropical mountains. The presence of a zone of maximum precipitation is well established for the tropics, but less defined in middle latitudes. Although there are insufficient measurements to settle the question categorically, evidence from mass-balance studies on glaciers, runoff from mountain watersheds, and improved methods of instrumentation seem to indicate that precipitation continues to increase with altitude in middle latitudes, at least up to 3,000–3,500 m (10,000–11,000 ft). The decrease beyond moderate elevations in the tropics is explained by the dominance there of convection rainfall, which means that the greatest precipitation occurs near the base of the clouds. Where forced ascent is important, the level may be somewhat higher, but it does not vary more than a few hundred meters. In many tropical areas, an upper air inversion composed of dry, stable air tends to restrict the deep development of clouds. This is the case on Mount Kenya and Kilimanjaro, as well as on Mauna Loa and Mauna Kea in Hawai'i (Juvik and Perreira 1974; Kaser et al. 2004; Hastenrath 2010).

RUNOFF

Mountain surface runoff is related to the topographic, biotic, pedologic, and, predominantly, climatic characteristics of a watershed (Stewart 2009). In particular, seasonality of precipitation inputs, temperatures, snowpack characteristics, and nonprecipitation water sources (i.e., groundwater and glaciers) are important variables in determining the amount of water flowing down a mountain stream (Peterson et al. 2000). Globally, mountain runoff displays significant temporal and spatial variation. The temporal heterogeneity arises from the intraannual, interannual, and secular changes in temperature, precipitation, and other climatic factors (Xie et al. 2010; Clow 2010). Spatial heterogeneity is due to climatic, topographic, biotic, land-use, and pedologic variability within and between mountains, making generalities about mountain hydrology difficult. Two unique aspects of mountain hydrology are the generation of flood events and the influences of snowpack meltwater.

Snow meltwater has four principal impacts on watershed hydrology: (1) lowering stream temperature; (2) sudden contributions to discharge resulting from rapid melting (rain on snow events); (3) an increase in melt-season discharge and decrease in snow-accumulation season discharge; and (4) a decrease in annual and especially seasonal variations in runoff (Wohl 2004). The changes in average seasonal discharge because of snowmelt are illustrated in Figure 3.22. Differences in discharge in these side-by-side mountainous watersheds of the same size are related to the elevational effects on snowfall (Bach 2002). During October, the beginning of the wet season, both basins have similar discharges. Between November and about April, two differences become apparent in the flow characteristics: The higher-elevation basin discharge becomes smaller in volume and less variable than the lower basin (Fig. 3.22). The decrease in volume is due to more precipitation falling as snow and accumulating in the upper basin, while rain falls throughout the winter in the lower basin and runs off. The greater temporal variability in the lower basin is related to the chaotic timing of storms. In March, temperatures begin to warm and the snowmelt season begins—a few weeks earlier in the lower basin (Fig. 3.22). The variability of discharge decreases in both basins (lines become smoother), indicating a change from storm event–dominated runoff to temperature-driven snowmelt runoff (Peterson et al. 2000). The peak in the snowmelt flood of the lower basin occurs about one month earlier and is only 70 percent the size of the upper basin, reflecting the difference in snowpack volume. These runoff characteristics are further confounded by the presence of glaciers in a watershed (Clow 2010).

Besides the daily to weekly variations caused by storm events, and the seasonal variations throughout a year, streamflow regimes in mountains are prone to interannual and secular variations related to large-scale patterns of climate variations, such as the El Niño/Southern Oscillation (Wohl 2004). The type of climatic variations vary around the globe, but generally result in extreme weather and climate conditions such as flooding, droughts, different storm frequency and precipitation, or changes in snowmelt

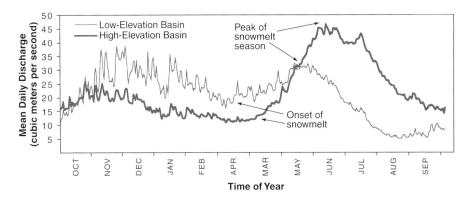

FIGURE 3.22 Influence of snowpack and glacier cover on runoff is illustrated by data from side-by-side basins of the same size but different elevations. Water year (Oct.–Sept.) hydrographs showing mean daily discharge (1938–1999) for high-elevation and low-elevation subbasins of the Nooksack River, Washington. The high-elevation North Fork has a mean elevation of 1,311 m, 6 percent glacier cover, and a mean annual discharge of 22.0 m³/s. The lower-elevation South Fork has a mean elevation of 914 m, no glacier cover, and a mean annual discharge of 20.8 m³/s. (Daily discharge data from USGS, from Bach 2002.)

season (Viles and Goudie 2003). Global warming is likely to increase the frequency and magnitude of these climatic variations and their impacts on the hydrological system (Xie et al. 2010).

High-elevation snowsheds are important to regional water supplies, as they provide water for domestic, industrial, and agricultural users, as well as for recreation, hydroelectric power, and habitat. They also produce flood hazards. Rapid population growth, increasing environmental concerns, and resulting changes in the character of demands for water have led to increased competition for water even under normal flow conditions. Water management practices, storage infrastructure, and patterns of use are tuned to the expected range of variation in surface runoff and groundwater availability. The abundant surface water supply from mountainous regions has promoted a historic reliance on this resource in adjacent lowlands. Effective water development planning and policy making must recognize how changes in upper watershed conditions will impact lowland water resources (Wohl 2004). New reservoirs and water transfer systems require considerable lead time to plan and construct. As discussed in Chapter 12, such structures will be necessary to deal with changing water supplies in the future.

Winds

Mountains are among the windiest places on Earth. They protrude into the high atmosphere, where there is less friction to retard air movement. While there is not a constant increase in wind speed with altitude, measurements from weather balloons and aircraft show a persistent increase at least up to the tropopause, where, in middle latitudes, the wind culminates in the jet streams. Similar increases occur in mountains, although the conditions at any particular site are highly variable. Wind speeds are greater in middle latitudes than in tropical or polar areas, in marine than in continental locations, in winter than in summer, and during the day than at night—and, of course, the velocity of the wind is dependent on the local topographic setting and the overall synoptic conditions (Smith 1979; Gallus and Klemp 2000). The wind is usually greatest in mountains oriented perpendicular to the prevailing wind, on the windward rather than the leeward side, and on isolated, unobstructed peaks rather than on those surrounded by other peaks. The reverse situation may exist in valleys, since those oriented perpendicular to the prevailing winds are protected while those oriented parallel to the wind may experience even greater velocities than the peaks, owing to funneling and intensification (Mass 2008). Table 3.5 lists the mean monthly wind speeds during the winter for several representative mountain stations in the northern hemisphere.

Mountains greatly modify the normal wind patterns of the atmosphere (Smith 1979; Insel et al. 2010). Their effect may be felt for many times their height in both horizontal and vertical distance. The question of whether the wind speed is greater close to mountains or in the free air has long been problematic. The two basic factors that affect wind speeds over mountains operate in opposition to on another. The vertical compression of airflow over a mountain causes acceleration of the air, while frictional effects cause a slowing. Frictional drag in the lowest layers of the atmosphere is caused by the interaction of air with individual small-scale roughness elements (i.e., vegetation, rocks, buildings, or landforms of <10 m dimensions) and by the influence of larger topographic features and vegetation canopies (Walmsley et al. 1989).

The sharpest gradient in wind speed usually occurs immediately above the surface. Wind speed doubles or triples within the first few meters, but the vegetation and surface roughness make a great difference in the absolute velocity (Fig. 3.23). The low-lying foliage of alpine vegetation

TABLE 3.5
Mean Monthly Wind Speeds during Winter at Selected Mountain Weather Stations
In Order of Decreasing Velocity
Readings were taken above treeline or in treeless areas, but anemometers were located at various heights above the ground.

Location	Elevation (m)	Monthly Windspeed (mph)					
		Nov.	Dec.	Jan.	Feb.	Mar.	Apr.
Mount Fujiyama, Japan	3,776	42	42	47	37	43	34
Mount Washington, N.H.	1,909	25	36	39	49	41	36
Jungfraujoch, Switzerland	3,575	27	29	25	24	26	25
Niwot Ridge, Colorado	3,749	21	25	26	24	22	21
Pic du Midi, France	2,860	15	19	20	17	20	17
Sonnblick, Austria	3,106	22	16	18	15	18	15
Berthoud Pass, Colorado	3,621	15	15	17	17	16	17
Mauna Loa, Hawaii	3,399	15	12	19	15	13	10

SOURCE: After Judson 1965: 13.

does not produce much frictional drag on the wind, so the wind can reach quite high velocities close to the ground. There is nevertheless a sharp gradient within the first few centimeters of the surface, and most alpine plants escape much of the wind (Liptzin and Seastedt 2009). A reciprocal and reinforcing effect operates here: Taller vegetation tends to reduce the wind speed and provide a less windy environment for plants, while low-lying alpine vegetation provides little braking effect, so the wind blows freely and becomes a major factor of stress in the environment. Under these conditions, the presence of microhabitats becomes increasingly important.

Surface roughness caused by clumps of vegetation and rocks creates turbulence and hence great variability in wind speed near the surface (Fig. 3.24). In the illustration, wind speed at a height of 1 m (3.3 ft) above the grass tussock is 390 cm/s, while closer to the ground it is 50 cm/sec on the exposed side of the tussock and 10 cm/s on the lee side (Fig. 3.24a). Similar conditions exist with the eroded soil bank, except that wind speeds are higher on the exposed side and there is more eddy action and reverse flow on the lee. The restriction of the vegetation to the lee of the soil bank is largely due to the reduced wind speed there (Fig. 3.24b). The wind follows a similar pattern across the rock, with small eddies developing in depressions and to the lee (Liptzin and Seastedt 2009). A mat of vegetation occupies the center depression where wind speeds are lower (Fig. 3.24c).

Wind is clearly an extreme environmental stress; in many cases, it serves as the limiting factor to life. What may be the two most extreme environments in mountains are caused by the wind: late-lying snowbanks, where the growing season is extremely short, and windswept, dry ridges. Both of these

environments become more common and more extreme with elevation, until eventually the only plants are mosses and lichens—or nothing at all. Trees on a windswept ridge may be "flagged" with the majority of branch growth on the protected lee side (Yoshino 1975). In the extreme conditions within the krummholz (crooked wood) zone, trees take on a prostate cushion form (see Chapter 7).

The redistribution of snow by the wind is a major feature of the alpine environment (Hiemstra et al. 2002). The wind speed necessary to pick snow up from the surface and transport it depends upon the state of the snowcover, including temperature, size, shape and density of the snow particles, and the degree of intergranular bonding (Liston et al. 2007). For loose, unbound snow, the typical velocity is about 5 m/s (16 ft/s), while a dense, bonded snowcover requires velocities in excess of 25 m/s (82 ft/s). Blowing snow can abrade surfaces, causing erosion to snowcover and flagging trees. Once the wind velocity lowers, the snow is deposited into dune-like features called *drifts*. Drifts are found in the lee eddy of obstacles of all sizes (e.g., trees, ridges, fences; Fig. 3.20). To control blowing snow, snow fences and other barriers are specially engineered to maximize deposition, and carefully placed to reduce the hazard of blowing snow or drifts (Liston et al. 2007). In mountains, snow redistribution by wind is strongly affected by meso- and microscale topography and vegetation (Pomeroy et al. 2006; Löffler 2007; Liptzin and Seastedt 2009). Topographic traps fill in with snow, where it may survive late into spring or summer because of its depth and because of temperature inversions. Many glaciers receive a significant component of their accumulation from snow blown over crests (Pelto 1996).

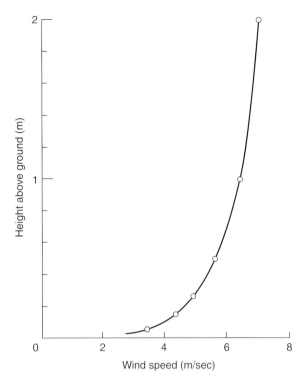

FIGURE 3.23 Wind velocity with height above a tundra surface. Note how wind speed increases with distance above the ground, one reason why alpine plants grow so close to the ground. (From Warren-Wilson 1959: 416.)

There are two overall groups or types of winds associated with mountains. One type originates within the mountains themselves. These are local, thermally induced winds given distinct expression by the topography. The other type is caused by obstruction and modification of winds originating from outside the mountain area. The first type is a relatively predictable, daily phenomenon, while the second is more variable, depending on the vagaries of changing regional wind and pressure patterns.

LOCAL WIND SYSTEMS IN MOUNTAINS

Winds that blow upslope and upvalley during the day and downslope and downvalley at night are common. Albrecht von Haller, author of *Die Alpen,* observed and described these during his stay in the Rhône Valley of Switzerland from 1758 to 1764. Since then, many studies have been made on thermally induced winds (see Barry 2008). The driving force for these winds is differential heating and cooling, which produces air density differences between slopes and valleys, and between mountains and adjacent lowlands (McGowan and Sturman 1996a). During the day, slopes are warmed more than the air at the same elevation in the center of the valley; the warm air, being less dense, moves upward along the slopes. Similarly, mountain valleys are warmed more than the air at the same elevation over adjacent lowlands, so the air begins to move up the

valley. These are the same processes that give rise to convection clouds over mountains during the day and provide good soaring for glider pilots and birds. At night, when the air cools and becomes dense, it moves downslope and downvalley under the influence of gravity. This is the flow responsible for the development of temperature inversions. Although they are interconnected and part of the same system, a distinction is generally made between slope winds, and larger mountain and valley wind systems (Fig. 3.25).

SLOPE WINDS

Slope winds consist of thin layers of air, usually less than 100 m (330 ft) thick. In general, the upslope movement of warm air during the day is termed *anabatic* flow, and the downslope movement of cold air during the night is referred to as *katabatic* flow, or a gravity or drainage wind. The upslope flow of air during the day is associated with surface heating and the resulting buoyancy of the warm air (Rucker et al. 2008). The wind typically begins to blow uphill about half an hour after sunrise and reaches its greatest intensity shortly after noon (Fig. 3.25a). By late afternoon, the wind abates, and within half an hour after sunset reverses to blow downslope (Fig. 3.25c). Katabatic winds in the strict sense are local downslope gravity flows caused by nocturnal radiative cooling near the surface under calm, clear-sky conditions, or by the cooling of air over a cold surface such as a lake or glacier (Whiteman and Zhong 2008). The extra weight of the stable layer, relative to the ambient air at the same altitude, provides the mechanism for the flow. Since slope winds are entirely thermally induced, they are better developed in clear weather than in clouds, on sun-exposed rather than on shaded slopes, and in the absence of overwhelming synoptic winds. Local topography is important in directing these winds; greater wind speeds will generally be experienced in ravines and gullies than on broad slopes (Defant 1951; Whiteman and Zhong 2008).

Downslope winds form better at night and during the winter, when radiative cooling dominates the surface energy system (Whiteman and Zhong 2008). The downslope flow of cold air is analogous to that of water, since it follows the path of least resistance and always tends toward equilibrium, although water has a density 800 times greater than air (Porch et al. 1989). Even with a temperature difference of 10°C (18°F), the density of cold air is only 4 percent greater than warm air; and unlike the rapid flow of water because of gravity, the displacement of warm air by cold air is a relatively slow process (Geiger 1965). Katabatic winds begin periodically when the layer of air just above the surface cools, then slides downslope (Papadopoulos and Helmis 1999). The cycle is repeated when the radiative cooling rebuilds the downslope pressure gradient. This pulsating downslope flow depends on the temperature difference between the katabatic layer and valley temperature (Whiteman and Zhong 2008). Surges of cold air,

A

B

C

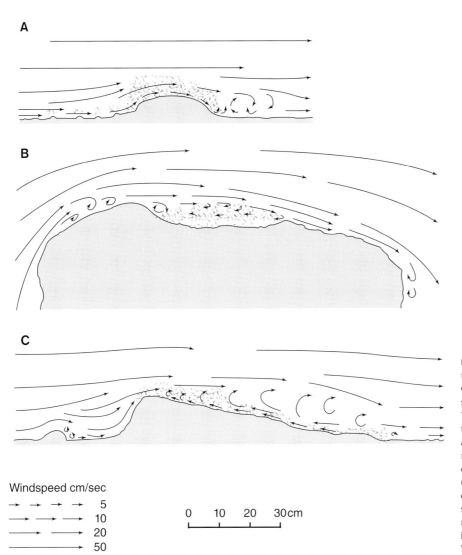

Windspeed cm/sec

→ → → →   5

——→ ——→ ——→   10

————→ ————→   20

—————————→   50

0   10   20   30cm

FIGURE 3.24 Wind behavior in relation to microtopography in the Cairngorm Mountains, Scotland. The stippled area represents vegetation. Vertical scale is roughly equivalent to the horizontal. (A) Air movement across a grassy tussock. (B) The movement of air over a rock with a depression occupied by vegetation. (C) A wind-eroded bank. Note the eddies that develop to the lee of small obstacles: Wind speed is greatly reduced in these areas and vegetation is better developed. (Adapted from Warren-Wilson 1959: 417–418.)

termed "air avalanches," are commonly observed on slopes greater than 10° (Geiger 1965). A final steady velocity will be achieved once a certain temperature has been reached (Papadopoulos and Helmis 1999). Further down a drainage basin, a steady velocity will be reached, maintained, and increased throughout the night as individual slope winds accumulate down basin in a fashion similar to tributaries in a stream (Porch et al. 1989).

Closed basins, even those created by dense forest cover, can trap the cold air, creating cold pockets or "frost hollows" (Clements et al. 2003; Daly et al. 2010). These temperature inversions can reach 30°C (77°F) below the ambient atmosphere and can persist for weeks or months, effectively trapping atmospheric contaminants until strong enough winds can clear the air (Geiger 1965; Iijima and Shinoda 2000). These have obvious significance for a number of human activities, including agriculture, forestry, tourism, and air pollution. In the wine-producing regions of Germany, hedges are frequently planted above the vineyards to deflect cold air coming from upslope (Geiger 1965).

Upslope winds form best during the day and during the summer when surfaces are radiatively warmed (Rampanelli et al. 2004). The upslope wind does not rise far above the ridge tops since it is absorbed and overruled by the regional prevailing wind (Yu et al. 2009). The upward movement of two slope winds establishes a small convection system in which a return flow from aloft descends in the center of the valley (Figs. 3.25, 3.26; cf. Fig. 3.25a). This descending flow brings from aloft drier air that has been heated slightly by compression, and thus is strongly opposed to cloud formation. For this reason, the dissipation of low-lying fog and clouds generally takes place first in the center of the valley (Fig. 3.26). If the valley is deep enough, the dry descending air can produce markedly arid zones. In the dry gorges and deep valleys of the Andes of Bolivia and in the Himalaya, the vegetation ranges from semidesert shrubs in valley bottoms to lush forests on the upper slopes where clouds form (Troll 1968).

(a) Upslope wind (daytime)

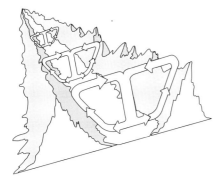

(b) Valley wind (daytime)

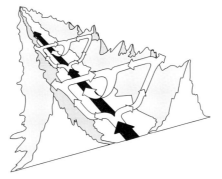

(c) Downslope wind (night)

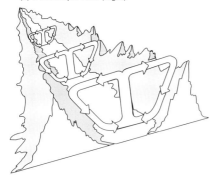

(d) Mountain wind (night)

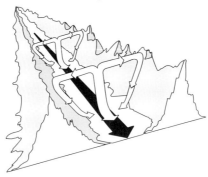

FIGURE 3.25 Schematic representation of slope winds (open arrows) and mountain and valley winds (black arrows). (A and B) Day conditions. (C and D) Night conditions. (After Defant 1951: 665 and Hindman 1973: 199.)

MOUNTAIN AND VALLEY WINDS

The integrated effects of slope-generated flows produce mountain and valley winds, blowing longitudinally up and down the main valleys, essentially at right angles to the slope winds (Yu et al. 2009). They are all part of the same system, however, and are controlled by similar thermal responses. The valley wind (blowing from the valley toward the mountain) is interlocked with the upslope winds, and both begin after sunrise (Buettner and Thyer 1965; Rucker et al. 2008; Fig. 3.25b). Valley winds involve greater thermal contrast and a larger air mass than slope winds, however, so they attain greater wind speeds. In the wide and deep valleys of the Alps, the smooth surfaces left by glaciation allow maximum development of the wind. The Rhône Valley has many areas where the trees are shaped by the wind and flagged in the upvalley direction. Mountain winds (blowing from the mountains downvalley) are associated with the nocturnal downslope winds and can be very strong and quite cold in the winter (Porch et al. 1989; Fig. 3.25d).

As with slope winds, a circulation system is established in mountain and valley winds. The return flow from aloft (called an anti-wind) can frequently be found immediately above the valley wind (Defant 1951; McGowan 2004), as shown by a study near Mount Rainier, Washington, using weather balloons (Fig. 3.26; Buettner and Thyer 1965). This wind system beautifully demonstrates the three-dimensional aspects of mountain climatology: Next to the surface are the slope and mountain-valley winds; above them is the return flow or anti-wind; and above this is the prevailing regional gradient wind (McGowan and Sturman 1996a; Rampanelli et al. 2004). During clear weather, all of these may operate at the same time, each moving in a different direction.

OTHER LOCAL MOUNTAIN WINDS

An important variant of the thermal slope wind is the *glacier wind*, which arises as the air adjacent to the icy surface is cooled and moves downslope due to gravity. The glacier wind has no diurnal period but blows continuously, since the refrigeration source is always present. It reaches its greatest depth and intensity at midafternoon, however, when the thermal contrast is greatest. At these times, the cold air may rush downslope like a torrent. During the day, the glacier wind frequently collides with the valley wind and slides under it (Fig. 3.27). At night, it merges with the mountain wind that blows in the same direction (Defant 1951). Glacier winds have a strong ecological effect, since the frigid temperatures are transported downslope with authority, and the combined effect of wind and low temperatures can make the area they dominate quite inhospitable.

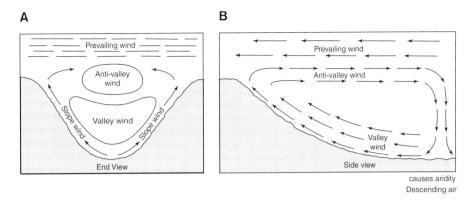

A

Prevailing wind

Anti-valley wind

Slope wind   Valley wind   Slope wind

End View

B

Prevailing wind

Anti-valley wind

Valley wind

Side view

causes aridity
Descending air

FIGURE 3.26 Graphic representation of slope and valley winds. The view on the left is looking upvalley at midday. Slope winds are rising along the slopes, while the valley wind and anti-wind are moving opposite each other, up and down the valley. The illustration on right provides a vertical cross section of the same situation, viewed from the side. The valley wind and anti-wind essentially establish a small convection system. The regional gradient wind is shown blowing above the mountains. If the regional wind is very strong, of course, it may override and prevent development of the slope and valley winds. (Adapted from Buettner and Thyer 1965: 144.)

In a valley with a receding glacier, these winds can entrain the unconsolidated till, sand-blasting vegetation and rocks (into ventifacts) downvalley (Bach 1995).

Another famous local wind in mountains is the *Maloja wind*, named after the Maloja Pass in Switzerland between the Engadine and Bergell Valleys (Hann 1903; Defant 1951; Whipperman 1984). This wind, which blows downvalley both day and night, results from the mountain wind of one valley reaching over a low pass into another valley, where it overcomes and reverses the normal upvalley windflow. This anomaly occurs in the valley with the greater temperature gradient and the ability to extend its circulation into the neighboring valley across the pass. Thus, the wind ascends from the steep Bergell Valley and extends across the Maloja Pass downward into the Engadine Valley to St. Moritz and beyond. Other examples of local winds could be given, since every mountainous country has its own peculiarities, but those mentioned suffice to illustrate their general nature.

MOUNTAIN WINDS CAUSED BY BARRIER EFFECTS

As mentioned earlier in this chapter, mountains can act as barriers to the prevailing general circulation of the atmosphere. The barrier effect introduces turbulence to the winds, increasing and decreasing speeds, changes directions, and modifies storms (reviewed earlier). Once the wind passes over the mountain crest, however, it will do one of two things: flow down the lee side, or stay lifted in the atmosphere (Durran 1990; Mass 2008). Most commonly, the wind will fall down the lee side under the influence of gravity. These surface winds are sometimes collectively termed *fall winds*, but they are known by a variety of local terms because they have long been observed in many regions downwind from mountains and have distinct weather phenomena associated with them. When

the winds leave the surface in a hydraulic jump, they often travel through the atmosphere in a wave motion, producing unique cloud forms.

FOEHN WIND

Of all the transitory climatic phenomena of mountains, the *foehn wind* (pronounced "fern" and sometimes spelled föhn) is the most intriguing. Legends, folklore, and misconceptions have often arisen about this warm, dry wind that descends with great suddenness from mountains. The foehn, known in the Alps for centuries, is common to all major mountain regions, formed as synoptic winds blow over a mountain crest and down the lee side (Barry 2008). In North America, it is called the "Chinook"; in the Yakima Valley of eastern Washington, the "winawaay"; in the Argentine Andes, the "zonda"; in southern California, the "Santa Ana"; in New Zealand, the "Canterbury northwester"; in New Guinea, the "warm braw"; in Japan, the "yamo oroshi"; in the Barison Mountains of Sumatra, the "bohorok"; in Poland, the "halny wigtr"; and in Romania, the "autru." Other mountain regions have their own local names for it as well (Brinkman 1971; Forrester 1982).

The foehn produces distinctive weather: gusts of wind, high temperatures, low humidity, and very transparent and limpid air (Brinkman 1971; McGowan 2004). When viewed through the foehn, mountains frequently take on a deep blue or violet tinge and seem unnaturally close and high, because light rays are refracted upward through layers of cold and warm air. The bank of clouds that typically forms along the crest line is associated with the precipitation failing on the windward side. This bank of clouds remains stationary in spite of strong winds and is known as the *foehn wall* (when viewed from the lee side).

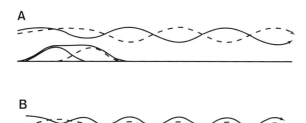

FIGURE 3.29 Air current over mountains, showing superposition of lee-wave trains. The mountain ridge (indicated by dashed line of mountain form) produces a certain wave pattern (dashed streamline) and the other mountain (solid line) produces a different wave pattern (continuous streamline). Together, the mountains have the effect of creating an obstacle (indicated by the continuous line). In the upper diagram the wavelength is such that the wave trains cancel out; in the lower diagram the amplitude is doubled. Since the wavelength is determined by the flow of air across the ridge, the same airstream could produce either large-amplitude lee waves or none at all, depending on its direction. (After Scorer 1967: 76.)

LEE WAVES

The behavior of airflow over an obstacle depends largely on the vertical wind profile, the stability structure, the shape of the obstacle, and the surface roughness (Barry 2008; Doyle et al. 2011). When wind passes over an obstacle, its normal flow is disrupted, and a train of waves may be created that extends downwind for considerable distances. The major mountain ranges produce large-amplitude waves that extend around the globe (Vosper and Parker 2002). On a smaller scale, these waves take on a regional significance reflected in their relationship to the foehn, in distinctive cloud forms, in upper-air turbulence, and in downwind climate (Scorer 1961, 1967; Durran 1990; Reinking et al. 2000; Brady and Waldstreicher 2001; Doyle et al. 2011). The amplitude and spacing of lee waves depend on the wind speed and the shape and height of the mountains, among other factors (Epifanio and Rotunno 2005). An average wavelength is between 2 and 40 km (1–25 mi), the vertical amplitude is usually between 1 and 5 km (0.6–3 mi), and they occur at altitudes of 300–7,600 m (1,000–25,000 ft) (Durran 1990). Wind speeds within lee waves are quite strong, frequently exceeding 160 km (100 mi) per hour (Scorer 1961; Doyle et al. 2011).

The most distinctive visible features of lee waves are the *lenticular* (lens-shaped) or *lee-wave* clouds that form at the crests of waves. These are created when the air reaches dew point and condensation occurs as the air moves upward in the wave (Ludlam 1980). The clouds do not form in the troughs of the waves, since the air is descending and warming slightly (Fig. 3.29). The relatively flat cloud bottoms represent the level of condensation, and the smoothly curved top follows the outline of the wave crest. The clouds are restricted in vertical extent by overlying stable air (Vosper

and Parker 2002). Lenticular clouds are relatively stationary (hence the name "standing-wave clouds"), although the wind may be passing through them at high speeds. Lee-wave clouds frequently develop above one another, as well as in horizontal rows. They typically consist of one to five clouds and extend only a few kilometers or miles downwind, but satellite photography has revealed series of 30 to 40 clouds extending for several hundred kilometers (Bader et al. 1995).

Much of the early knowledge about lee waves was acquired by glider pilots who found, to their surprise, that there was often greater lift to the lee of a hill than on the windward side. The pilots had long made use of upslope and valley winds, but by this method could never achieve a height of more than a couple of hundred meters above the ridges. In southern England, the members of the London Gliding Club had soared for years in the lift of a modest 70 m (230 ft) hill, never achieving more than 240 m (800 ft). After discovering the up-currents in the lee wave, however, one member soared to a height of 900 m (3,000 ft), 13 times higher than the hill producing the wave (Scorer 1961). German pilots were the first to explore and exploit lee waves fully. In 1940, one pilot soared to 11,300 m (37,400 ft) in the lee of the Alps. The world's altitude record of 15,460 m (49,213 ft) was set in 2006 in the lee of the Andes Mountains of Argentina.

Another aspect of lee waves is the development of rotors, awesome roll-like circulations that develop to the immediate lee of mountains, usually forming beneath the wave crests (Fig. 3.29). The rotor flow moves toward the mountain at the base and away from it at the top (Doyle and Durran 2002). It is marked by a row of cumulus clouds but, unlike ordinary cumulus, they may contain updrafts of 95 km (60 mi) per hour. The potential of such a wind for damage to an airplane can well be imagined. The height of the rotor clouds is about the same as that of the crest cloud or foehn wall. The rotating motion is thought to be created when the lee waves reach a certain amplitude and frictional drag causes a roll-like motion in the underlying air (Fig. 3.29; Scorer 1961, 1967).

Several other kinds of turbulence may be associated with lee waves, particularly when the wave train produced by one mountain is augmented by that of another situated in the right phase relationship. In some cases, they cancel each other; in others, they reinforce each other. Wind strength and direction are also important, since a small change in either can alter the wavelength of two superposed wave trains so that they become additive and create violent turbulence (Scorer 1967; Neiman et al. 2001).

Microclimates

In addition to the climatic characteristics reviewed above, it should be emphasized that there are substantial variations in climates over very short distances within mountains. Mountain environments are exceedingly spatially complex in terms of vegetation types and structures, geology, hydrology, soils, and topography. All vary in composition (i.e., plant species, canopy characteristics, or rock types), and variations occur across a range of slopes and aspects. The climate over each of these surfaces, or *microclimate,* can differ significantly because of variations in net radiation, soil and air temperature, humidity, precipitation accumulation (amount and form), soil moisture, and winds (Pomeroy et al. 2006; Löffler 2007; Daly et al. 2010).

Large differences in temperature, moisture, and wind can be found within a few meters, or even centimeters (McCutchan and Fox 1986). The thin atmosphere at high elevation means that surfaces facing the sun on a clear day can warm dramatically, while shaded surfaces remain cold (Germino and Smith 2000). Other effects may arise according to valley orientation with respect to the mountain range, valley cross profile, and the effect of winds and cold air drainage. The effect of aspect in generating slope winds can exceed the influence of elevation on wind velocity and temperature (McCutchan and Fox 1986).

The mosaic of microclimates determines local variability in ecosystem processes, forcing plants and animals to a high degree of specialization and adaptation (Körner 2003). The distribution of vegetation zones and individual species often follow the distribution of microclimates (Fig. 3.8; Löffler 2007; Daly et al. 2010; see Chapter 7).

Vegetation creates its own microenvironment by creating shade and windbreaks (Fig. 3.24). In association with the wind regime is a recurring pattern of snow accumulation in the lee of obstacles. These snowdrifts add to soil moisture during the melt season, and protect trees from freezing in winter (Pomeroy et al. 2006). The distributions of snow and vegetation are closely linked. Plant species in these areas exhibit distributions related to the duration of snowcover as well as relief (Körner 2003; Erschbamer 2007; Löffler 2007).

The spatial resolution of most weather station networks in mountains is far too coarse to capture the spatial variability of their microclimates. Maps of climatic variables are often interpolated from existing meager data sets, using assumed or empirical relationships with elevation (Kyriakidis et al. 2001; Horsch 2003; Minder et al. 2010). These models are unable to demonstrate local deviations in trends, and when combined with map scale, microclimates are typically eliminated from most maps of mountains. Likewise, vegetation maps of mountainous areas rarely show the small patches of vegetation that occur in microclimatic habitats. While these features may be unmappable, they are certainly observable in mountains, adding to the splendor of the multifaceted mountain environment.

Climate Change and Variability

The variability of climatic phenomena is an important natural component of the Earth's climate system. Climatic variability (i.e., occurrence of certain climatic events) is different than climate change, which is a permanent change in climatic conditions. However, changes in variability are one likely result of climatic change. The middle and high latitudes inherently have especially variable climates since they are influenced by large seasonal changes in energy. The equatorial region experiences little variability, as it has nearly the same energy fluxes year round. Reflecting the complexity of the climate system, most regions of the world show different patterns and magnitudes of variability and trends through time (Viles and Goudie 2003).

All temporal climate records demonstrate some degree of interannual variability (e.g., Beniston et al. 1997; Liu and Chen 2000; Kane 2000; Viles and Goudie 2003; Xie et al. 2010). Every mountain location has its record high and low temperature, snowfall, rain event, drought, and wind speed (Cerveny et al. 2007). While these extreme events are rare, they often occur with a greater frequency, and with more extreme magnitudes in mountainous regions, than in lowlands (Frei and Schär 2001). Extreme storm events are exacerbated by the topographic setting of mountains, producing even higher precipitation totals, lower temperatures, and higher wind velocities. As discussed in Chapter 10, extreme precipitation events in mountains are of significance because they lead to hazards such as downstream flooding, soil erosion, and mass movements on slopes. Temperature, precipitation, and resulting runoff are often related to distant forcing mechanisms, such as the El Niño/Southern Oscillation (Cayan et al. 1999; Diaz et al. 2001, 2003; Clare et al. 2002; Rowe et al. 2002). Several other periodic, yet chaotic, perturbations to the climate system have been linked to increased climatic variability (Fowler and Kilsby 2002; Viles and Goudie 2003).

Among the regional differences in variability, the following consistent temporal trends emerge in data sets over the last century: (1) The number of extremely warm summer temperatures has increased; (2) the number of extremely cold winter temperatures has decreased (with fewer frost days); and (3) mean summer season precipitation has increased—in particular, an increase in heavy precipitation events (Viles and Goudie 2003). All of these general trends have temporally reversed during the period of record, so that while variability is expected, changes in the frequency of occurrence of extreme events are recognized as a signal of ongoing climate change (Diaz et al. 2003). The variations in the climate system, especially decadal- to millennial-scale oscillations, can be mistaken for climate trends. An often-cited example is the question of whether the current global temperature trend is the result of recovery from the Little Ice Age or the result of anthropogenic greenhouse gas increases (Meier et al. 2003).

Mountain and glacier environments are especially sensitive to climate changes and variability (Meier et al. 2003; Kohler et al. 2010; Kaltenborn et al. 2010). Many climate changes have been detected in mountain records (e.g., Cayan et al. 2001; Pepin and Losleben, 2002; Huber et al. 2005; Kaser et al. 2004; Hastenrath 2010). Climate changes are well documented to have occurred in the geologic past, as illustrated by the glacial and interglacial climates of the Pleistocene and over the period of instrumental records (Beniston et al. 1997; Thompson et al. 2003). Current scientific consensus holds that the climate is, in fact, in the process of changing, primarily warming because of anthropogenic inputs of greenhouse gases to the atmosphere (IPCC 2007). Different magnitudes of warming, and even cooling, are predicted for different mountainous regions of the world (Thompson et al. 2003; IPCC 2007). Landmasses are expected to warm more than the oceans, and northern, middle, and high latitudes more than the tropics (Lean and Rind 2009). Precipitation, in particular, is predicted to both increase and decrease depending upon region because of changes in general circulation (IPCC 2007). Climate models in mountainous regions, however, tend to be rather poor because of coarse spatial resolution, topographic smoothing, and local effects not captured by the models (Beniston 2003; Suklitsch et al. 2010). In mountains, higher temperatures would cause both a higher percentage of annual precipitation to fall as rain (i.e., higher snowlines) and an acceleration of summer ablation (Stewart 2009). Longer snow-free periods will increase evaporative demands and lower soil moisture, increasing the dominance of drought-tolerant species (Erschbamer 2007). Characterizing the exact climatic impacts on any mountain site is difficult. However, past and likely future climatic changes and variations are likely to have major impacts in mountain environments.

Mountains provide freshwater to half of the global population, and climate change will affect its availability (Kohler et al. 2010; Kaltenborn et al. 2010). Changes in winter precipitation and summer temperatures will alter the rate and extent to which snowlines migrate up or downslope, and contribute to glacier mass balance and runoff (Clare et al. 2002; Diaz et al. 2003). Seasonal snowpacks in the northern hemisphere have significantly declined over recent years (Diaz et al. 2003; Pielke et al. 2004; Mote et al. 2005; Adam et al. 2009; Stewart 2009). Glaciers are likely to experience loss of mass, which will contribute more water to melt-season runoff and cause the glacier to thin and retreat. Glacier recession will have an impact on local climatic conditions, such as energy and moisture exchanges and the generation of local winds. Measurements of alpine glacier mass balances globally have documented retreats in recent decades (Pelto 1996; Meier et al. 2003, Oerlemans 2005; Dyurgerov and McCabe 2006; Hoelzle et al. 2007; Barry 2008; Nesje et al. 2008; Zemp et al. 2009; Hastenrath 2010). A comprehensive survey of global glacier coverage is not yet possible due to limited spatial and temporal observations (Barry 2008; Zemp et al. 2009). However, the vast majority of studies demonstrate a major trend of glacier recession, with global estimates of glacier area (and volume) lost ranging from 0.9 to 1.5 percent during the period 1961–1990 and accelerated loss since then (Dyurgerov and Meier 1997; Meier et al. 2003; Barry 2008; Zemp et al. 2009). The recession

rates observed in most mountains are considerably greater, with some experiencing 60–80 percent loss in small glacier cover over the last century (Kaab et al. 2002; Kaser et al. 2004; Barry 2008). Small glaciers in many mountains have already disappeared and more are feared to become extinct if global warming trends continue (Dyurgerov 2005; Zemp et al. 2006).

As a result of temperature changes, glacial recession, and changing snowpacks, downstream runoff characteristics (i.e., seasonality and magnitude) may be altered appreciably over the next several decades (Adam et al. 2009; Stewart 2009; Clow 2010; Kohler et al. 2010; Xie et al. 2010; Kaltenborn et al. 2010). If glaciers entirely disappear from mountains, then melt-season, especially late-melt-season, discharge will decrease substantially (Fig. 3.22). Glaciers are estimated to provide 6–30 percent of annual runoff in some rivers (Bach 2002). Changes to mountain hydrology will have significant consequences not only in the mountains themselves, but also in the populated lowland regions that depend on the runoff for domestic, agricultural, energy, and industrial uses (Beniston 2006; Kohler et al. 2010). Even climate changes in nonglaciated, low mountains can have a significant impact on municipal water supplies (Frei et al. 2002; Mote et al. 2003). Runoff from glacier melt is responsible for a substantial component of the current rise in eustatic sea level (Dyurgerov and Meier 1997; Raper and Braithwaite 2006).

Climate change is considered a major threat to mountain ecosystems (McCarty 2001; Grace et al. 2002; Parmesan and Yohe 2003; Root et al. 2003; Beniston 2006; Nogués-Bravo et al. 2008; Kohler et al. 2010). Since many organisms living in mountains survive near their tolerance range for climatic conditions, even minor climatic changes could have a significant impact on alpine ecosystems (Grabherr et al. 1994; Thuiller 2004; Erschbamer 2007). Vegetation zones will migrate altitudinally and latitudinally in response to warming temperatures, possibly eliminating some biomes, although the adaptations will likely be more complex (Thuiller 2004; Colwell et al. 2008). Because mountain tops are smaller than their bases, biomes shifting upslope will occupy smaller and more fragmented areas, reducing populations and increasing genetic and environmental pressures (Thuiller 2004; Beniston 2006). However, because different species respond uniquely to environmental pressures, ecological communities may disassemble as individual species shift their ranges in different directions and at different rates (Colwell et al. 2008). Migrants and hibernators may experience problems as a consequence of these changes in phenology, and may no longer be stimulated by the same environmental cues (Inouye et al. 2000). Complex topography will result in habitat fragmentation and the creation of barriers to migration, making it difficult for some species to adapt and allowing others, often invasive species, to expand their range. There is a chance that the quaking aspen (*Populus tremuloides*), Engleman spruce (*Picea englemannii*), and lodgepole pine (*Pinus contorta*) of the western

North American mountains might not survive under projected climate changes (Hansen et al. 2001; Coops and Waring 2011).

With warmer temperatures and elevated CO_2 concentrations, forests may begin growing earlier in the spring and throughout the growing season; however, changes in moisture could slow growth (McKenzie et al. 2001). Treelines in many mountainous regions have been responding to recent temperature changes by migrating upslope (Kullman and Kjallgren 2000; Pallatt et al. 2000; Klasner and Fagre 2002; Grace et al. 2002; Graumlich et al. 2005). However, in some topographic settings cold air drainage will equal or exceed the warming rate, potentially providing refugia for some speices (Daly et al. 2010). These responses are largely but not entirely related to rising freezing levels and reduced snowcover (Grace et al. 2002; Diaz et al. 2003). Many alpine plants and tree seedlings not only respond to temperature changes, but are sensitive to day length at the beginning of the growing period and increased CO_2 concentrations (Keller et al. 2005). A warming of seasonal mean temperatures of 3–5°C can reduce snowcover and duration by more than a month on average (Keller et al. 2005). Some mountain forests will experience changes to disturbance regimes, such as fire frequency and intensity, blowdown (wind), drought, insects, and disease, caused by climate change and variability. The changing character of these systems with changing climates is poorly understood, and few generalizations can be made for all mountain systems.

In response to the habitat changes, wildlife also migrates to find appropriate climatic niches (Hansen et al. 2001; Nogués-Bravo et al. 2008; Mantyka-Pringle et al. 2012). Because of microclimatic complexity, populations or single species could readily be insolated on individual slopes or peaks, as the mountain environment increases in fragmentation (Fig. 3.8; Beniston 2006). Since this climate shift is occurring rapidly, some species may not be able to adapt or migrate quickly enough (Grabherr et al. 1994; McCarty 2001). Not only must prey species adapt to new environmental conditions, but behaviors will need to change as new predators migrate in or out of range (McCarty 2001). It is probable that some alpine and cold-water fish species will not survive climatic changes, and new water temperatures will allow for the invasion of nonnative fish species (Mohseni et al. 2003). Pacific salmon, which migrate to and spawn in some mountains, have experienced population fluctuations related to climate (Downton and Miller 1998). In the Columbia River system, the projected impacts of global warming are warmer water temperatures and earlier snowmelt peak flows, which are likely to further impact the beleaguered salmon populations and related ecosystems (Miller 2000; Mote et al. 2003).

Land-use changes in mountains, especially urbanization, logging, and hydrolake (i.e., lakes created for power generation) development can have significant impacts on regional and microclimates in mountains (McGowan and Sturman 1996a; Godde et al. 2000; Mote et al. 2003). These environmental disturbances can have long-term influences on climates since they change the surface characteristics as well as energy and moisture fluxes. Hydrolakes have been found to moderate temperatures, increase atmospheric water-vapor content and precipitation, and increase windiness by decreasing surface roughness and developing their own wind systems (McGowan and Sturman 1996a). Snowmaking is an adaptation strategy for ski resorts facing declining snowpacks, but it can alter local hydrology and climate (Steiger and Meyer 2008).

Because of their slope, aspect, verticality, mass, and altitude, mountains are particularly sensitive to changes in climate; they have been called some of the best natural "barometers" and predictors of global climate change consequences in the world (Beniston 2006; Rhoades 2006; Barry 2008). Changes in mountain environments are important not only within the mountains, but also in terms of the resources and services they provide to adjacent lowlands. Frequently cited in the press and growing climate change literature is the retreat of the world's glaciers that has occurred over the past century, most noticeably in the tropics and subtropics. This rapid melting of snow and ice has resulted in an increase in the formation of high-altitude glacial lakes, sometimes too fast to monitor accurately, increasing the potential for catastrophic down-valley floods that can destroy everything in their paths (e.g., Mark et al. 2010). Other potential impacts on people include the increased risk of other high-magnitude events such as debris flows and landslides (Jomelli et al. 2004); accelerated erosion from increased glacial runoff (Mark et al. 2010); changes in agricultural patterns (Chhetri et al. 2010); depletion of glacier-fed rivers for power, agriculture, and drinking water (Mote et al. 2003; Kaltenborn et al. 2010); increases in conflicts over diminishing water supplies for irrigation (Rhoades 2006); negative impacts on mountain tourism (Godde et al. 2000); and even an increase in infectious diseases previously confined to the lowlands (Beniston 2003), also fall within the realm of current speculation, documentation, and discussion.

References

Adam, J. C., Hamlet, A. F., and Lettenmaier, D. P. 2009. Implications of global climate change for snowmelt hydrology in the 21st century. *Hydrological Processes* 23: 962–972.

Anderson, H. W. 1972. Water yield as an index of lee and windward topographic effects on precipitation. In *Distributions of Precipitation in Mountainous Areas*, Vol. II (pp. 341–358). Geilo, Norway: World Meteorological Organization.

Anderson, R. S. 1998. Near-surface thermal profiles in alpine bedrock: Implications for frost weathering of rock. *Arctic and Alpine Research* 30: 362–372.

Archer, C. L., and Caldeira, K. 2008. Historical trends in the jet streams. *Geophysical Research Letters* 35: L08803.

Arnold, K., and Palmer, R. 1952. *The Coming of the Saucers.* Amherst, WI: Palmer Publications.

Bach, A. J. 1995. Aeolian Modifications of glacial moraines at Bishop Creek, eastern California. In V. P. Tchakerian, ed., *Desert Aeolian Geomorphology* (pp. 179–197). London: Chapman and Hall.

Bach, A. J. 2002. Snowshed contributions to the Nooksack River watershed, North Cascades, Washington. *The Geographical Review* 92: 192–212.

Bader, M. J., Forbes, J. R., Grant, J. R., Lilley, R. B., and Waters, A. J. 1995. *Images in Weather Forecasting: Practical Guide for Interpreting Satellite and Radar Data.* Cambridge, UK: Cambridge University Press.

Banta, R. M., and Schaaf, C. L. B. 1987. Thunderstorm genesis zones in the Colorado Rocky Mountains as determined by traceback of geosynchronous satellite images. *Monthly Weather Review* 115: 463–476.

Barry, R. G. 1973. A climatological transect along the east slope of the Front Range, Colorado. *Arctic and Alpine Research* 5: 89–110.

Barry, R. G. 2008. *Mountain Weather and Climate.* 3rd ed. Cambridge, UK: Cambridge University Press

Beniston, M. 2003. Climatic changes in mountain regions: A review of possible impacts. *Climatic Change* 59: 5–31.

Beniston, M. 2006. Mountain weather and climate: A general overview and a focus on climatic change in the Alps. *Hydrobiologia* 562: 3–16.

Beniston, M., Diaz, H. F., and Bradley, R. S. 1997. Climatic changes at high elevation sites: An overview. *Climatic Change* 36: 233–251.

Blumenstock, D. I., and Price, S. 1967. The climate of Hawaii. In *Climates of the States,* Vol. 2: *Western States* (pp. 481–975). Port Washington, NY: Water Information Center.

Bradley, B. A., and Fleishman, E. 2008. Relationships between expanding pinyon-juniper cover and topography in the central Great Basin, Nevada. *Journal of Biogeography* 35: 951–964.

Brady, R. H., and Waldstreicher, J. S. 2001. Observations of mountain wave-Induced precipitation shadows over northeast Pennsylvania. *Weather and Forecasting* 16: 281–300.

Brennan, M. J., Cobb, H. D., and Knabb, R. D. 2010. Observations of Gulf of Tehuantepec gap wind events from QuikSCAT: An updated event climatology and operational model evaluation. *Weather and Forecasting* 25: 646–658.

Brinkman, W. A. R. 1971. What is a foehn? *Weather* 26: 230–239.

Buettner, K. J. K. and Thyer, N. 1965. Valley winds of the Mt. Rainer area. *Archives for Meteorology, Geophysics, and Bioclimatology, Series B* 14: 9–148.

Calder, I. R. 1990. *Evaporation in the Uplands.* Chichester, UK: John Wiley and Sons.

Cayan, D. R., Kammerdiener, S., Dettinger, M. D., Caprio, J. M., and Peterson, D. H. 2001. Changes in the onset of spring in the western United States. *Bulletin of the American Metrological Society* 82: 399–415.

Cayan, D. R., Redmond, K. T., and Riddle, L. G. 1999. ENSO and hydrologic extremes in the western United States. *Journal of Climate* 12: 2881–2893.

Cerveny, R. S., Lawrimore, J., Edwards, R., and Landsea, C. 2007. Extreme weather records: Compilation, adjudication and publication. *Bulletin of the American Meteorological Society* 88: 853–860.

Chen, L., Reiter, E. R., and Feng, Z. 1985. The atmospheric heat source over the Tibetan Plateau; May–August 1979. *Monthly Weather Review* 113: 1771–1790.

Chhetri, N. B., Easterling, W. E., Terando, A., and Mearns, L. 2010. Modeling path dependence in agricultural adaptation to climate variability and change. *Annals of the Association of American Geographers* 100: 894–907.

Clare, G. R., Fitzharris, B. B., Chinn, T. J. H., and Salinger, M. J. 2002. Interannual variation in end-of-summer snowlines of the Southern Alps of New Zealand, and relationships with Southern Hemisphere atmospheric circulation and sea surface temperature patterns. *International Journal of Climatology* 22: 107–120 .

Clements, C. B., Whiteman, C. D., and Horel, J. D. 2003. Cold-air-pool structure and evolution in a mountain basin: Peter Sinks, Utah. *Journal of Applied Meteorology* 42: 752–768.

Cline, D. W. 1997. Snow surface energy exchanges and snowmelt at a continental, mid-latitude Alpine site. *Water Resources Research* 33: 689–702.

Clow, D. W. 2010. Changes in the timing of snowmelt and streamflow in Colorado: A response to recent warming. *Journal of Climate* 23: 2293–2306.

Colwell, B. K., Brehm, G., Cardelús, C. L., Gilman, A. C., and Longino, J. T. 2008. Global warming, elevational range shifts, and lowland biotic attrition in the wet tropics. *Science* 322: 258–261.

Cooke, L. J., Rose, M. S., and Becker, W. J. 2000. Chinook winds and migraine headache. *Neurology* 54: 302–307.

Coops, N. C., and Waring, R. H. 2011. A process-based approach to estimate lodgepole pine (*Pinus contorta* Dougl.) distribution in the Pacific Northwest under climate change. *Climatic Change* 105: 313 328.

Daly, C., Conklin, D. R., and Unsworth, M. H. 2010. Local atmospheric decoupling in complex topography alters climate change impacts. *International Journal of Climatology* 30:1857–1864.

Defant, F. 1951. Local winds. In T. F. Malone, ed., *Compeddium of Meteorology* (pp. 655–672). Boston: American Meteorology Society.

Deisenhammer, E.A. 2003. Weather and suicide: The present state of knowledge on the association of meteorological factors with suicidal behaviour. *Acta Psychiatrica Scandinavica* 108: 402–409.

De La Rue, E. A. 1955. *Man and the Winds.* New York: Philosophical Library.

Diaz, H. F., Eischeid, J. K., Duncan, C., and Bradley, R. S. 2003. Variability of freezing levels, melting season indicators, and snow cover for selected high-elevation and continental regions in the last 50 years. *Climatic Change* 59: 33–52.

Diaz, H. F., Hoerling, M. P., and Eischeid, J. K. 2001. ENSO variability, teleconnections and climate change. *International Journal of Climatology* 21: 1845 1862.

Downton, M. W., and Miller, K. A. 1998. Relationships between salmon catch and north Pacific climate on interannual and decadal time scales. *Canadian Journal of Fisheries and Aquatic Sciences* 55: 2255–2265.

Doyle, J. D., and Durran, D. R. 2002. The dynamics of mountain-wave-induced rotors. *Journal of the Atmospheric Sciences* 59: 186–201.

Doyle, J. D., Jiang, Q., Smith, R. B., and Grubiši, V. 2011. Three-dimensional characteristics of stratospheric mountain waves during T-REX. *Monthly Weather Review* 139: 3–23.

Drogue, G., Humbert, J., Deraisme, J., Mahr, N., and Freslon, N. 2002. A statistical-topographic model using an omnidirectional parameterization of the relief for mapping orographic rainfall. *International Journal of Climatology* 22: 599–613.

Dunbar, G. S. 1966. Thermal belts in North Carolina. *Geographical Review* 56: 516–526.

Durran, D. R. 1990. Mountain waves and downslope winds. In W. Blumen, ed., *Atmospheric Processes over Complex Terrain* (pp 59–81). Meteorological Monograph 23(45). Boston: American Meteorological Society.

Dyurgerov, M. B. 2005. Mountain glaciers are at risk of extinction, In U. M. Huber, H. K. M. Bugmann, and M. A. Reasoner, eds., *Global Change and Mountain Regions: An Overview of Current Knowledge* (pp. 177–184). Dordrecht: Springer.

Dyurgerov, M., and McCabe, G. J. 2006. Associations between accelerated glacier mass wastage and increased summer temperature in coastal regions. *Arctic, Antarctic and Alpine Research* 38: 190–197.

Dyurgerov, M. B., and Meier, M. F. 1997. Year-to-year fluctuations of global mass balance of small glaciers and their contributions to sea level changes. *Arctic and Alpine Research* 29: 392–402.

Emck, P., and Richter, M. 2008. An upper threshold of enhanced global shortwave irradiance in the troposphere derived from field measurements in tropical mountains. *Journal of Applied Meteorology and Climatology* 47: 2828–2845.

Epifanio, C. C., and Rotunno, R. 2005. The dynamics of orographic wake formation in flows with upstream blocking. *Journal of the Atmospheric Sciences* 62: 3127–3150.

Erschbamer, B. 2007. Winners and losers of climate change in a central alpine glacier foreland. *Arctic, Antarctic and Alpine Research* 39: 237–244.

Ferretti, R., Paolucci, T., Zheng, W., Visconti, G., and Bonelli, P. 2000. Analyses of the precipitation pattern on the alpine region using different cumulus convection parameterizations. *Journal of Applied Meteorology* 39: 182–200.

Forrester, F. H. 1982. Winds of the world. *Weatherwise* 35: 204–210.

Fowler, H. J., and Kilsby, C. G. 2002. Precipitation and the North Atlantic Oscillation: A study of climatic variability in northern England. *International Journal of Climatology* 22: 843–866.

Frei, A., Armstrong, R. L., Clark, M. P., and Serreze, M. C. 2002. Catskill Mountain water resources: Vulnerability, hydroclimatology, and climate-change sensitivity. *Annals of the Association of American Geographers* 92: 203–224.

Frei, C., and Schär, C. 2001. Detection probability of trends in rare events: Theory and application to heavy precipitation in the alpine region. *Journal of Climate* 14: 1568–1584.

Fujita, T. 1967. Mesoscale aspects of orographic influences on flow and precipitation patterns. In E. R. Reiter and J. L. Rasmussen, eds., *Proceedings of the Symposium on Mountain Meteorology* (pp. 131–146). Atmospheric Science Paper 122. Fort Collins: Colorado State University.

Gade, D. W. 1978. Windbreaks in the lower Rhône Valley. *Geographical Review* 68: 127–144.

Gaffin, D. M. 2007. Foehn winds that produced large temperature differences near the southern Appalachian Mountains. *Weather and Forecasting* 22: 145–159.

Gallus, W. A., and Klemp, J. B. 2000. Behavior of flow over steep orography. *Monthly Weather Review* 128: 1153–1164.

Garreaud, R. D. 1999. Multiscale analysis of the summer time precipitation over the central Andes. *Monthly Weather Review* 127: 901–921.

Gates, D. M., and Janke, R. 1966. The energy environment of the alpine tundra. *Oecologia Plantarum* 1: 39–62.

Geiger, R. 1965. *The Climate Near the Ground*. Cambridge, MA: Harvard University Press.

Germann, U., and Joss, J. 2001. Variograms of radar reflectivity to describe the spatial continuity of alpine precipitation. *Journal of Applied Meteorology* 40: 1042–1059.

Germann, U., and Joss, J. 2002. Mesobeta profiles to extrapolate radar precipitation measurements above the Alps to the ground level. *Journal of Applied Meteorology* 41: 542–557.

Germino, M. J., and Smith, W. K. 2000. Differences in microsite, plant form, and low-temperature photoinhibition in alpine plants. *Arctic, Antarctic, and Alpine Research* 32: 388–396.

Godde, P., Price, M. F., and Zimmermann, F. M., eds. 2000. *Tourism and Development in Mountain Regions*. Wallingford, UK: CABI Publishing.

Gohm, A., Harnisch, F., Vergeiner, J., Obleitner, F., Schnitzhofer, R., Hansel, A., Fix, A., Neininger, B., Emeis, S., and Schafer, K. 2009. Air pollution transport in an alpine valley: Results from airborne and ground-based observations. *Boundary-Layer Meteorology* 131: 441–463.

Grabherr, G., Gottfried, M., and Pauli, H. 1994. Climate effects on mountain plants. *Nature* 369: 448.

Grace, J., Berninger, F., and Nagy, L. 2002. Impacts of climate change on the tree line. *Annals of Botany* 90: 537–544.

Graumlich, L. J., Waggoner, L. A., and Bunn, A. G. 2005. Detecting global change at Alpine treeline: Coupling paleoecology with contemporary studies. In U. M. Huber, H. K. M. Bugmann, and M. A. Reasoner, eds., *Global Change and Mountain Regions: An Overview of Current Knowledge* (pp. 501–508). Dordrecht: Springer.

Hann, J. 1903. *Handbook of Climatology*. London: Macmillan.

Hansen, A. J., Neilson, R. P., Dale, V., Flather, C., Iverson, L., Currie, D. J., Shafer, S., Cook, R., and Bartlein, P. J. 2001. Global change in forests: Responses of species, communities, and biomes. *BioScience* 51: 765–779.

Hastenrath, S. 2010. Climatic forcing of glacier thinning on the mountains of equatorial East Africa. *International Journal of Climatology* 30: 146–152.

Henning, D., and Henning, D. 1981. Potential evapotranspiration in mountain ecosystems on different altitudes and latitudes. *Mountain Research and Development* 1: 267–274.

Hiemstra, C. A., Liston, G. E., and Reiners, W. A. 2002. Snow redistribution by wind and interactions with vegetation at upper treeline in the Medicine Bow Mountains, Wyoming, U.S.A. *Arctic, Antarctic and Alpine Research* 34: 262–273.

Hindman, E. E. 1973. Air currents in a mountain valley deduced from the break-up of a stratus deck. *Monthly Weather Review* 101: 195–200.

Hock, R. 2005. Glacier melt: A review of processes and their modeling. *Progress in Physical Geography* 29: 362–391.

Hoelzle M., Chinn T., Stumm D., Paul F., Zemp M., and Haeberli, W. 2007. The application of glacier inventory data for estimating past climate change effects on mountain glaciers: A comparison between the European Alps and the Southern Alps of New Zealand. *Global and Planetary Change* 56: 69–82.

Hoggarth, A. M., Reeves, H. D., and Lin, Y.-L. 2006. Formation and maintenance mechanisms of the stable layer over the Po Valley during MAP IOP-8. *Monthly Weather Review* 134: 3336–3354.

Horsch, B. 2003. Modelling the spatial distribution of montane and subalpine forests in the central Alps using digital elevation models. *Ecological Modelling* 168: 267–282.

Houze, R. A., and Medina, S. 2005. Turbulence as a mechanism for orographic precipitation enhancement. *Journal of the Atmospheric Sciences* 62: 3599–3623.

Huber, U. M., Bugmann, H. K. M., and Reasoner, M. A., eds. 2005. *Global Change and Mountain Regions: An Overview of Current Knowledge*. Dordrecht: Springer.

Huntington, C., Blyth, E. M., Wood, N., Hewer, H. E., and Gant, A. 1998. The effect of orography on evaporation. *Boundary-Layer Meteorology* 86: 487–504.

Huo, Z., and Bailey, W. G. 1992. Evaluation of models for estimating net radiation for alpine sloping surfaces. *Acta Meteorologica Sinica* 6: 189–197.

Iijima, Y., and Shinoda, M. 2000. Seasonal changes in the cold-air pool formation in a subalpine hollow, central Japan. *International Journal of Climatology* 20: 1471–1483.

Inouye, D. W., Barr, B., Armitage, K. B., and Inouye, B. D. 2000. Climate change is affecting altitudinal migrants and hibernating species. *Proceedings of the National Academy of Sciences* 97: 630–1633.

Insel, N., Poulsen, C. J., and Ehlers, T. A. 2010. Influence of the Andes Mountains on South American moisture transport, convection, and precipitation. *Climate Dynamics* 35: 1477–1492.

Intergovernmental Panel on Climate Change (IPCC). 2007. *Contribution of Working Group I to the Fourth Assessment Report of the Intergovernmental Panel on Climate Change.* S. Solomon, D. Qin, M. Manning, Z. Chen, M. Marquis, K.B. Averyt, M. Tignor, and H.L. Miller, eds. Cambridge, UK: Cambridge University Press.

Jansa, A. 1987. Distribution of the mistral: a satellite observation. *Meteorology and Atmospheric Physics* 36: 201–214.

Jarvis, A., and Mulligan, M. 2011. The climate of cloud forests. *Hydrological Processes* 25: 327–343.

Jeglum, M. E., Steenburgh, W. J., Lee, T. P., and Bosart, L. F. 2010. Multi-reanalysis climatology of intermountain cyclones. *Monthly Weather Review* 138: 4035–4053.

Jerrett, M., Burnett, R. T., Ma, R., Pope, C. A., Krewski, D., Newbold, K. B., Thurston, G., Shi, Y., Finkelstein, N., Calle, E. E., and Thun, M. J. 2005. Spatial analysis of air pollution and mortality in Los Angeles. *Epidemiology* 16: 727–736.

Jomelli, V., Pech, V. P., Chochillon, C., and Brunstein, D. 2004. Geomorphic variations of debris flows and recent climatic change in the French Alps. *Climatic Change* 64: 77–102.

Judson, A. 1965. The weather and climate of a high mountain pass in the Colorado Rockies. U.S. Forest Service Research Paper RM-16.

Juvik, J. O., DeLay, J. K., Kinney, K. M., and Hansen, E. W. 2011. A 50th anniversary reassessment of the seminal "Lana'i fog drip study" in Hawai'i. *Hydrological Processes* 25: 402–410.

Juvik, J. O., and Perreira, D. J. 1974. Fog interception on Mauna Loa, Hawaii. *Proceedings of the Association of American Geographers* 8: 22–25.

Kaab, A., Maisch, M., Kellenberger, T., and Haeberli, W. 2002. The new remote sensing-derived Swiss glacier inventory, II: First results. *Annals of Glaciology* 34: 362–366.

Kaltenborn, B. P., Nellemann, C., Vistnes, I. I., eds. 2010. High mountain glaciers and climate change: Challenges to human livelihoods and adaptation. Arendal, Norway: United Nations Environment Programme, GRID-Arendal.

Kane, R. P. 2000. El Niño/La Niña relationship with rainfall at Huancayo, in the Peruvian Andes. *International Journal of Climatology* 20: 63–72.

Kaser, G., Hardy, D. R., Mölg, T., Bradley, R. S., and Hylera, T. M. 2004. Modern glacier retreat on Kilimanjaro as evidence of climate change: Observations and facts. *International Journal of Climatology* 24: 329–339.

Kattelman, R., and Elder, K. 1991. Hydrologic characteristics and balance of an alpine basin in the Sierra Nevada. *Water Resources Research* 27: 1553–1562.

Keller, F., Goyette, S., and Beniston, M. 2005. Sensitivity analysis of snow cover to climate change scenarios and their impact on plant habitats in alpine terrain. *Climatic Change* 72: 299–319.

Kelliher, F. M., Owens, I. F., Sturman, A. P., Byers, J. N., Hunt, J. E., and McSeveny, T. M. 1996. Radiation and ablation on the névé of Franz Josef Glacier. *New Zealand Journal of Hydrology* 35: 131–145.

Kirshbaum, D. J., and Durran, D. R. 2004. Factors governing cellular convection in orographic precipitation. *Journal of the Atmospheric Sciences* 61: 682–698.

Klasner, F. L., and Fagre, D. B. 2002. A half century of change in alpine treeline patterns at Glacier National Park, Montana, U.S.A. *Arctic, Antarctic, and Alpine Research* 34: 49–56.

Kohler, T., Giger, M., Hurni, H., Ott, C., Wiesmann, U., Wymann von Dach, S., and Maselli, D. 2010. Mountains and climate change: A global concern. *Mountain Research and Development* 30: 53–55.

Körner, C. 2003. *Alpine Plant Life: Functional Plant Ecology of High Mountain Ecosystems*. Berlin: Springer.

Kruckeberg, A. R. 1991. *The Natural History of Puget Sound Country*. Seattle: University of Washington Press.

Krzysztof, M., Liebersbach, J., and Sobik, M. 2002. Rime in the Giant Mts. (the Sudetes, Poland). *Atmospheric Research* 64: 63–73.

Kullman, L., and Kjällgren, L. 2000. A coherent postglacial tree-limit chronology (*Pinus sylvestris* L.) for the Swedish Scandes: Aspects of paleoclimate and "recent warming," based on megafossil evidence. *Arctic, Antarctic and Alpine Research* 32: 419–428.

Kyriakidis, P. C., Kim, J., and Miller, N. L. 2001. Geostatistical mapping of precipitation from rain gauge data using atmospheric and terrain characteristics. *Journal of Applied Meteorology* 40: 1855–1877.

Landsberg, H. 1962. *Physical Climatology.* DuBois, PA: Grey Printing.

Lauer, W. 1975. Klimatische grundzüge der höhenstufung tropischer gebirge. In F. Steiner, ed., *Tagungsbericht und Wissenschaftliche Abhandlungen, 40 Deutscher Geographentag, Innsbruck* (pp. 79–90). Innsbruck: F. Steiner.

Lean, J. L., and Rind, D. H. 2009. How will Earth's surface temperature change in future decades? *Geophysical Research Letters* 36, L15708.

Lichte, V., Dennenmoser, B., Dietz, K., Häfner, H.-M., Schlagenhauff, B., Garbe, C., Fischer, J., and Moehrle, M. 2010. Professional risk for skin cancer development in male mountain guides: A cross-sectional study. *Journal of the European Academy of Dermatology and Venereology* 24: 797–804.

Lin, Y. L., Chen, S.Y., Hill, C. M., and Huang, C. Y. 2005. Control parameters for the influence of a mesoscale mountain range on cyclone track continuity and deflection. *Journal of the Atmospheric Sciences* 62: 1849–1866.

Lin, Y. L., Chiao, S., Wang, T., Kaplan, M. L., and Weglarz, R. P. 2001. Some common ingredients for heavy orographic rainfall. *Weather and Forecasting* 16: 633–660.

Liptzin, D., and Seastedt, T. R. 2009. Patterns of snow, deposition, and soil nutrients at multiple spatial scales at a Rocky Mountain tree line ecotone. *Journal of Geophysical Research* 114: 1–13.

Liston, G. E., Haehnel, R. B., Sturm, M., Hiemstra, C. A., Berezovskaya, S., and Tabler, R. D. 2007. Simulating complex snow distributions in windy environments using SnowTran-3D. *Journal of Glaciology* 53: 241–256.

Liu, X., and Chen, B. 2000. Climatic warming in the Tibetan Plateau during recent decades. *International Journal of Climatology* 20: 1729–1742.

Löffler, J. 2007. The influence of micro-climate, snow cover, and soil moisture on ecosystem functioning in high mountains. *Journal of Geographical Sciences* 17: 3–19.

Ludlam, F. H. 1980. *Clouds and Storms.* University Park: Pennsylvania State University Press.

MacCracken, M. C. 2006. Geoengineering: Worthy of cautious evaluation? *Climatic Change* 77: 235–243.

Mantyka-Pringle, C. S., Martin, T. G., and Rhodes, J. R. (2012). Interactions between climate and habitat loss effects on biodiversity: A systematic review and meta-analysis. *Global Change Biology* 18: 1239–1252.

Mark, B. G., Bury, J., McKenzie, J. M., French, A., and Baraer, M. 2010. Climate change and tropical Andean glacier recession: Evaluating hydrologic changes and livelihood vulnerability in the Cordillera Blanca, Peru. *Annals of the Association of American Geographers* 100: 794–805.

Mass, C. 2008. *Weather of the Pacific Northwest.* Seattle: University of Washington Press.

Matthes, F. E. 1934. Ablation of snow-fields at high altitudes by radiant solar heat. *Transactions of the American Geophysical Union,*15: 380–385.

McCarty, J. P. 2001. Ecological consequences of recent climate change. *Conservation Biology* 15: 320–331.

McCutchan, M. H., and Fox, D. G. 1986. Effect of elevation and aspect on wind, temperature and humidity. *Journal of Climate and Applied Meteorology* 25: 1996–2013.

McGinnis, D. L. 2000. Synoptic controls on upper Columbia River basin snowfall. *International Journal of Climatology* 20: 131–149.

McGowan, H. A. 2004. Observations of anti-winds in a deep alpine valley, Lake Tekapo, New Zealand. *Arctic, Antarctic and Alpine Research* 36: 495–501.

McGowan, H. A., and Sturman, A. P. 1996a. Interacting multi-scale wind systems within an alpine basin, Lake Tekapo, New Zealand. *Meteorology and Atmospheric Physics* 58: 165–177.

McGowan, H. A., and Sturman, A. P. 1996b. Regional and local scale characteristics of foehn wind events over the South Island of New Zealand. *Meteorology and Atmospheric Physics* 58: 151–164.

McKenzie, D., Hessl, A. E., and Peterson, D. L. 2001. Recent growth in conifer species of western North America: Assessing regional and continental spatial patterns of radial growth trends. *Canadian Journal of Forest Research* 31: 526–536.

Meier, M. F., Dyurgerov, M. B., and McCabe, G. J. 2003. The health of glaciers: Recent changes in glacier regime. *Climatic Change* 59: 123–135.

Miller, K. A. 2000. Pacific salmon fisheries: Climate, information and adaptation in a conflict-ridden context. *Climatic Change* 45: 36–61.

Minder, J. R., Mote, P. W., and Lundquist, J. D. 2010 Surface temperature lapse rates over complex terrain: Lessons from the Cascade Mountains. *Journal of Geophysical Research* 115: D14122, doi:10.1029/2009JD013493.

Miniscloux, F., Creutin, J. D., and Anquetin, S. 2001. Geostatistical analysis of orographic rainbands. *Journal of Applied Meteorology* 40: 1835–1854.

Mohseni, O., Stefan, H. G., and Eaton, J. G. 2003. Global warming and potential changes in fish habitat in U.S. streams. *Climatic Change* 59: 389–409.

Mote, P. W., Hamlet, A. F., Clark, M. P., and Lettenmaier, D. P. 2005. Declining mountain snowpack in western North America. *Bulletin of the American Meteorological Society* 86: 39–49.

Mote, P. W., Parson, E. A., Hamlet, A. F., Keeton, W. S., Lettenmaier, D. P., Mantaua, N., Miles, E. L., Peterson, D. W., Peterson, D. L., Slaughter, R., and Snover, A. K. 2003. Preparing for climatic change: The water, salmon, and forests of the Pacific Northwest. *Climatic Change* 61: 45–88.

Neiman, P. J., Ralph, F. M., Weber, R. L., Uttal, T., Nance, L. B., and Levinson, D. H. 2001. Observations of nonclassical frontal propagation and frontally forced gravity waves adjacent to steep topography. *Monthly Weather Review* 129: 2633–2659.

Neiman, P. J., Ralph, F. M., White, A. B., Kingsmill, D. E., and Persson, P. O. G. 2002. The statistical relationship between upslope flow and rainfall in California's Coastal Mountains: Observations during CALJET. *Monthly Weather Review* 130: 468–1492.

Neiman, P. J., Sukovich, E. M., Ralph, F. M., and Hughes, M. 2010. A seven-year wind Profiler–based climatology of the windward barrier jet along California's northern Sierra Nevada. *Monthly Weather Review* 138: 1206–1233.

Nesje, A., Bakke, J., Dahl, S. O., Lie, O., and Matthews, J. A. 2008. Norwegian mountain glaciers in the past, present and future. *Global and Planetary Change* 60: 10–27.

Nogués-Bravo, D., Araújo, M. B., Romdal, T., and Rahbek, C. 2008. Scale effects and human impact on the elevational species richness gradients. *Nature* 453: 216–219.

Noppel, H., and Beheng, K. D. 2009. Effects of intentional and inadvertent hygroscopic cloud seeding. In *High Performance Computing in Science and Engineering '08, Part 6* (pp. 443–457). Proceedings of the High Performance Computing Center, Stuttgart. New York: Springer-Verlag.

Nullet, D., and McGranaghan, M. 1988. Rainfall enhancement over the Hawaiian Islands. *Journal of Climatology* 1: 837–839.

Oerlemans, J. 2005. Extracting a climate signal from 169 glacier records. *Science* 308: 675–677.

Papadopoulos, K. H., and Helmis, C. G. 1999. Evening and morning transition of katabatic flows. *Boundary-Layer Meteorology* 92: 195–227.

Parisi, A. V., and Downs, N. 2004. Cloud cover and horizontal plane eye damaging solar UV exposures. *International Journal of Biometeorology* 49: 130–136.

Park, H. S., Chiang, J. C. H., and Son S. W. 2010. The role of the Central Asian mountains on the midwinter suppression of North Pacific storminess. *Journal of the Atmospheric Sciences* 67: 3706–3720.

Parmesan, C., and Yohe, G. 2003. A globally coherent fingerprint of climate change impacts across natural systems. *Nature* 421: 37–42.

Peattie, R. 1936. *Mountain Geography.* Cambridge, MA: Harvard University Press.

Peck, E. L. 1972. Discussion of problems in measuring precipitation in mountainous areas. *Distribution of Precipitation in Mountainous Area.* Proceedings of the Geilo Symposium. World Meterological Organization Publication 326, Vol. 1 (pp. 5–15).

Pellatt, M. G., Smith, M. J., Mathewes, R. W., Walker, I. R. and Palmer, S. L. 2000. Holocene treeline climate change in the subalpine zone near Stoyoma Mountain, Cascade Mountains, southwestern British Columbia, Canada. *Arctic, Antarctic, and Alpine Research* 32: 73–83.

Pelto, M. 1996. Annual net balance of North Cascade glaciers 1984–1994. *Journal of Glaciology* 42: 3–9.

Penman, H. L. 1963. *Vegetation and Hydrology.* Commonwealth Bureau of Soils Technical Communication 53. Farnham Royal: Commonwealth Agricultural Bureaux.

Pepin, N., and Losleben, M. 2002. Climate change in the Colorado Rocky Mountains: Free air versus surface temperature trends. *International Journal of Climatology* 22: 311–329.

Perkins, T. R., Pagano, T. C., and Garen, D. C. 2009. Innovative operational seasonal water supply forecasting technologies. *Journal of Soil and Water Conservation* 64: 15–17.

Peterson, D. H., Smith, R. E., Dettinger, M. D., Cayan, D. R., and Riddle, L. 2000. An organized signal in snowmelt runoff over the Western United States. *Water Resources Research* 36: 421–432.

Pfeifer, M. T., Koepke, P., and Reuder, J. 2006. Effects of altitude and aerosol on UV radiation. *Journal of Geophysical Research*, 111, D01203, doi:10.1029/2005JD006444.

Pielke, R. A., Liston, G. E., Chapman, W. L., and Robinson, D. A. 2004. Actual and insolation-weighted Northern Hemisphere snow cover and sea-ice between 1973–2002. *Climate Dynamics* 22: 591–595.

Pomeroy, J. W., Bewley, D. S., Essery, R. L. H., Hedstrom, N. R., Link, T., Granger, R. J., Sicart, J. E., Ellis, C. R., and Janowicz, J. R. 2006. Shrub tundra snowmelt. *Hydrological Processes* 20: 923–941.

Porch, W. H., Fritz, R. B., Coulter, R. L., and Gudiksen, P. H. 1989. Tributary, valley, and sidewall air flow interactions in deep valley. *Journal of Applied Meteorology* 28: 578–589.

Price, L. W. 1978. Mountains of the Pacific Northwest: A study in contrast. *Arctic and Alpine Research* 10: 465–478.

Prohaska, F. 1970. Distinctive bioclimatic parameters of the subtropical-tropical Andes. *International Journal of Biometeorology* 14: 1–12.

Rampanelli, G., Zardi, D., and Rotunno, R. 2004. Mechanisms of up-valley winds. *Journal of Atmospheric Science* 61: 3097–3111.

Raper, S. C. B., and Braithwaite, R. J. 2006. Low sea level rise projections from mountain glaciers and icecaps under global warming. *Nature* 439: 311–313.

Reid S. J., and Turner, R. 1997. Wind storms. *Tephra* 16: 24–32.

Reinhard, J. 2005. *The Ice Maiden: Inca Mummies, Mountain Gods, and Sacred Sites in the Andes.* Washington, DC: National Geographic Society.

Reinking, R. F., Snider, J. B., and Coen, J. L. 2000. Influences of storm-embedded orographic gravity waves on cloud liquid water and precipitation. *Journal of Applied Meteorology* 39: 733–759.

Rhoades, R., ed. 2006. *Development with Identity: Community, Culture and Sustainability in the Andes.* Oxford, UK: CAB International Publishing.

Roe, G. H., and Baker, M. B. 2006. Microphysical and geometrical controls on the pattern of orographic precipitation. *Journal of Atmospheric Sciences* 63: 861–880.

Root, T. L., Price, J. T., Hall, K. R., Schneider, S. H., Rosenzweig, C., and Pounds, J. A. 2003. Fingerprints of global warming on wild animals and plants. *Nature* 421: 57–60.

Rowe, H. D., Dunbar, R. B., Mucciarone, D. A., Seltzer, G. O., Baker, P. A., and Fritz, S. 2002. Insolation, moisture balance and climate change on the South American altiplano since the last glacial maximum. *Climatic Change* 52: 175–199.

Rucker, M., Banta, R. M., and Steyn, D. G. 2008. Along-valley structure of daytime thermally driven flows in the Wipp Valley. *Journal of Applied Meteorology and Climatology* 47: 733–751.

Rumney, G. R. 1968. *Climatology of the World's Climates.* New York: Macmillan.

Sassen, K., and Zhao, H. 1993. Supercooled liquid water clouds in Utah winter mountain storms: Cloud-seeding implications of a remote-sensing dataset. *Journal of Applied Meteorology* 32: 1548–1558.

Sato, T., and Kimura, F. 2003. A two-dimensional numerical study on diurnal cycle of mountain lee precipitation. *Journal of Atmospheric Sciences* 60: 1992–2003.

Schaaf, C. L. B., Wurman, J., and Banta, R. M. 1988. Thunderstorm-producing terrain features. *Bulletin of the American Meteorological Society* 69: 272–277.

Scherrer, D., Schmid, S., and Körner, C. 2010. Elevational species shifts in a warmer climate are overestimated when based on weather station data. *International Journal of Biometeorology,* doi:10.1007/s00484-010-0364-7.

Schmidli, J., Schmutz, C., Frei, C., Wanner, H., and Schär, C. 2002. Mesoscale precipitation variability in the region of the European Alps during the 20th century. *International Journal of Climatology* 22: 1049–1074.

Schmidt, W. 1934. Observations on local climatology in Austrian mountains. *Quarterly Journal of the Royal Meteorological Society* 60: 345–351.

Schultz, D. M., Steenburgh, W. J., Trapp, R. J., Horel, J., Kingsmill, D. E., Dunn, L. B., Rust, W. D., Cheng, L., Bansemer, A., Cox, J., Daugherty, J., Jorgensen, D. P., Meitín, J., Showell, L., Smull, B. F., Tarp, K., and Trainor, M. 2002. Understanding Utah winter storms: The Intermountain Precipitation Experiment. *Bulletin of the American Meteorological Society* 83: 189–210.

Scorer, R. S. 1961. Lee waves in the atmosphere. *Scientific American* 204: 124–134.

Scorer, R. S. 1967. Causes and consequences of standing waves, In E. R. Reiter and J. L. Rassmussen, eds. *Proceedings of the Symposium on Mountain Meteorology* (pp. 75–101). Fort Collins: Department of Atmospheric Sciences, Colorado State University 1.

Sinclair, M. R., Wratt, D. S., Henderson, R. D., and Gray, W. R. 1997. Factors affecting the distribution and spillover of precipitation in the Southern Alps of New Zealand: A case study. *Journal of Applied Meteorology* 36: 428–442.

Sklenář, P., and Lægaard, S. 2003. Rain-shadow in the high Andes of Ecuador evidenced by Páramo vegetation. *Arctic, Antarctic and Alpine Research* 35: 8–17.

Smith, R. B. 1979. The influence of mountains on the atmosphere. *Advances in Geophysics* 21: 87–230.

Smith, R. B., Schafer, P., Kirshbaum, D. J. and Regina, E. 2009. Orographic precipitation in the tropics: Experiments in Dominica. *Journal of the Atmospheric Sciences* 66: 1698–1716.

Smith, S. A., Brown, A. R., Vosper, S. B., Murkin, P. A., and Veal, A. T. 2010. Observations and simulations of cold air pooling in valleys. *Boundary-Layer Meteorology* 134: 85–108.

Steiger, S., and Meyer, M. 2008. Snowmaking and climate change: Future options for snow production in Tyrolean ski resorts. *Mountain Research and Development* 28: 292–298.

Stewart, I. T. 2009. Changes in snowpack and snowmelt runoff for key mountain regions. *Hydrological Processes* 23: 78–94.

Sturman., A. P., and Tapper, N. J. 1996. *The Weather and Climate of Australia and New Zealand.* Melbourne: Oxford University Press.

Suklitsch, M., Gobiet, A., Truhetz, H., Awan, N. K., Göttel, H., and Jacob, D. 2010. Error characteristics of high resolution regional climate models over the Alpine area. *Climate Dynamics,* DOI 10.1007/s00382-010-0848-5.

Takechi, Y., Kodama, Y., and Ishikawa, N. 2002. The thermal effect of melting snow/ice surface on lower atmospheric temperature. *Arctic, Antarctic and Alpine Research* 34: 20–25.

Terjung, W. H., Kickert, R. N., Potter, G. L., and Swarts, S. W. 1969. Energy and moisture balances of an alpine tundra in mid-July. *Arctic and Alpine Research* 1: 247–266.

Thompson, L. G., Mosley-Thompson, E., Davis, M. E., Lin, P.-N., Henderson, K., and Mashiotta, T. A. 2003. Tropical glacier and ice core evidence of climate change on annual to millennial time scales. *Climatic Change* 59: 137–155.

Thuiller, W. 2004. Patterns and uncertainties of species' range shifts under climate change. *Global Change Biology* 10: 2020–2027.

Troll, C. 1958. *Soil Structures, Solifluction, and Frost Climates of the Earth.* U.S. Army Snow, Ice and Permafrost Research Transactions 43.

Troll, C. 1968. The cordilleras of the tropical Americas: Aspects of climatic, phytogeographical and agrarian ecology. In *Geoecology of the Mountainous Regions of the Tropical Americas* (pp. 15–56). Bonn: Dummlers Verlag.

Tucker, D. F., and Crook, N. A. 1999. The generation of a mesoscale convective system from mountain convection. *Monthly Weather Review* 127: 1259–1273.

Uddstrom, M. J., McGregor, J. A., Gray, W. R., and Kidson, J. W. 2001. A high-resolution analysis of cloud amount and type over complex topography. *Journal of Applied Meteorology* 40: 16–33.

Vermeulen A. T., Wyers, G. D., Romer, F. G., van Leewen, N. F. M., Draaijers, N. F. M., and Erimim, J. W. 1997. Fog deposition on a coniferous forest in the Netherlands. *Atmospheric Environment* 31: 388–396.

Viles, H. A., and Goudie, A. S. 2003. Interannual, decadal and multidecadal scale climatic variability and geomorphology. *Earth-Science Reviews* 61: 105–131.

Vosper, S. B., and Parker, D. J. 2002. Some perspectives on wave clouds. *Weather* 57: 3–7.

Walmsley, J. L., Taylor, P. A., and Salmon, J. R. 1989. Simple guidelines for estimating wind speed variations due to small-scale topographic features: An update. *Climatological Bulletin* 23: 3–14.

Warren-Wilson, J. 1959. Notes on wind and its effects in arctic alpine vegetation. *Journal of Ecology* 47: 415–427.

Weischet, W. 1969. Klimatologische Regeln zur Vertikalverteilung der Niederschläge in Tropengebirgen. *Der Erde* 100: 287–306.

Whipperman, F. 1984. Airflow over and in broad valleys: channeling and counter-current. *Contributions to Atmospheric Physics* 57: 184–195.

Whiteman, C. D., Pospichal, B., Eisenbach, S., Weihs, P., Clements, C. B., Steinacker, R., Mursch-Radlgruber, E., and Dorninger, M. 2004. Inversion breakup in small Rocky Mountain and Alpine basins. *Journal of Applied Meteorology* 43: 1069–1082.

Whiteman, C. D., and Zhong, S. 2008. Downslope flows on a low-angle slope and their interactions with valley inversions, Part I: Observations. *Journal of Applied Meteorology and Climatology* 47: 2023–2038.

Whiteman, C. D., Zhong, S., Shaw, W. J., Hubbe, J. M., Bian, X., and Mittelstadt, J. 2001. Cold pools in the Columbia Basin. *Weather and Forecasting* 16: 432–447.

Wohl, E. 2004. *Disconnected Rivers: Linking Rivers to Landscapes.* New Haven, CT: Yale University Press.

Xie, H., Ye, J., Liu, X. and Chongyi, E. 2010. Warming and drying trends on the Tibetan Plateau (1971–2005). *Theoretical and Applied Climatology* 101: 241–253.

Xing, Y., Jin, D., Hengchi, L., and Peng, F. 2005. Comparison between computer simulation of transport and diffusion of cloud seeding material within stratiform cloud and the

NOAA-14 satellite cloud track. *Advances in Atmospheric Sciences* 22: 133–141.

Xu, Z., Qian, Y., and Fu, C. 2010. The role of land-sea distribution and orography in the Asian monsoon, Part II: Orography. *Advances in Atmospheric Sciences* 27: 528–542.

Yoshino, M. M. 1975. *Climate in a Small Area*. Tokyo: University of Tokyo Press.

Yu, R., Li, J., and Chen, H. 2009. Diurnal variation of surface wind over central eastern China. *Climate Dynamics* 33: 1089–1097.

Zemp, M., Haeberli, W., Hoelzle, M., and Paul, F. 2006. Alpine glaciers to disappear within decades? *Geophysical Research Letters* 33: L13504.

Zemp, M., Hoelzle, M., and Haeberli, W. 2009. Six decades of glacier mass balance observation: A review of the worldwide monitoring network. *Annals of Glaciology* 50: 101–111.

Zhihua, R., and M. Li 2007. Errors and correction of precipitation measurements in China. *Advances in Atmospheric Sciences* 24: 449–458.

CHAPTER FOUR

Snow, Ice, Avalanches, and Glaciers

LELAND R. DEXTER, KARL W. BIRKELAND,
and LARRY W. PRICE

The presence of frozen water in several forms is fundamental at high altitudes and provides the essential ingredient for the development of avalanches and glaciers. These interrelated phenomena, which contribute much to the distinctiveness of high mountain landscapes, offer a considerable challenge to the inhabitants, both plant and animal, of these regions.

Snow and Ice

Snowfall and New Snow

Snow is precipitation in the solid form that originates from the freezing of water in the atmosphere. This leads to one of the great mysteries of nature: Why should snow fall in the form of delicate and varying lacy crystals rather than as frozen raindrops? The commonly held assumption that water must freeze at 0°C (32°F) is incorrect. The freezing temperature can range as low as −40°C (−40°F), which, coincidentally, is the crossover point of the two temperature scales. Water that remains liquid when cooled below 0°C is referred to as *supercooled* water. The actual freezing point of water in the atmosphere depends not only on ambient temperature but also on water droplet size, droplet purity, and mechanical agitation. Smaller droplets are more resistant to freezing. Very small droplets may resist freezing to the −40°C value mentioned above. Dissolved salts will retard freezing, but certain particulates will enhance freezing (i.e., promote freezing at temperatures closer to 0°C) (Knight 1967; Hobbs 1974; Pruppacher and Klett 1987; Tabazedeh et al. 2002).

Clouds form most readily when certain contaminants are present in the atmosphere. These contaminants can be divided into two classes, depending on their ability to promote either condensation or freezing. *Condensation nuclei* are hygroscopic materials that attract water, such as salt

and smoke. *Freezing* (more properly called *deposition*) *nuclei* generally are particles that mimic the hexagonal crystal structure of ice, although dry ice (frozen CO_2) is also an effective freezing nucleator based on its low temperature. Effective freezing nuclei include clays, certain bacteria, and silver iodide. In nature, most clouds contain a mixture of water droplets formed around condensation nuclei and small ice crystals formed around freezing nuclei. At typical cloud temperatures of −10°C (14°F), the freezing nuclei are effective in overcoming the *activation energy* and hence allow the surrounding water to freeze. Droplets formed around the condensation nuclei are too small or too salty to freeze directly at this temperature. Most storm clouds, therefore, are a three-phase mixture of water vapor, supercooled liquid droplets, and small ice crystals. The affinity of ice surfaces for attracting water vapor is slightly greater than that of the supercooled liquid surface (stated another way, saturation vapor pressure is lower over ice than over liquid water at the same temperature). Therefore, water vapor molecules have a tendency to deposit more rapidly on small ice seed crystals (hence drying the air near them), whereas water vapor tends to evaporate from supercooled droplets (thus moistening the air near them). The net result is a vapor flow from the supercooled droplets to the ice crystals, causing shrinkage of the former and growth of the latter (Fig. 4.1). Thus, it can be seen that snow crystals grow molecule by molecule (analogous to bricks placed one by one in a complex building project), and this helps explain why snowflakes can be so delicate and varied. This mechanism is referred to as the *Wegener–Bergeron–Findeisen* process, named after persons involved in the development of the theory (Knight 1967; Hobbs 1974; Pruppacher and Klett 1987; Nelson and Baker 1996; Wood et al. 2001; Barry and Gan 2011).

Snow and ice crystals grow in some variation of the hexagonal (six-sided) crystal system (Fig. 4.2). Once formed,

FIGURE 4.1 The Wegener–Bergeron–Findeisen mechanism for snow crystal growth in the atmosphere. Greater saturation vapor pressure over liquid water than over ice causes supercooled droplets to evaporate and ice crystals to grow. (Image courtesy of L. R. Dexter and K. Birkeland.)

ice crystals and snowflakes are subject to continual change. They may grow through deposition and accretion or diminish through sublimation and melting, and they may be fragmented and recombined in numerous ways. The variations on the basic hexagonal pattern display almost infinite variety. We are taught from childhood on that every snowflake is different! In absolute terms this is true, but most often snow crystals falling from homogeneous cloud conditions resemble one another closely in basic shape. Snow crystals are generally small and simple when first formed in the cold, dry air of high altitudes. As they fall, snow crystals can become larger and more complex when they encounter warmer or more moisture-laden atmospheric layers, often becoming large enough to earn the name *snowflakes*. Thus, snow received at the summits

FIGURE 4.2 A classic example of the six-sided hexagonal crystal structure of ice I. (Image courtesy of K. Libbricht.)

of mountains is often quite different from that received on the mid-elevation slopes; in fact, it may melt to rain by the time it reaches the valley bottoms. Most rainfall outside the tropics begins as snowfall at high altitudes (Knight 1967; Hobbs 1974, Barry and Gan 2011).

For 40 years around the turn of the twentieth century, a dedicated photographer named Wilson Bentley took thousands of photographs of newly fallen snowflakes while braving the outdoor conditions of New England winters (Figs. 4.3, 4.4). Bentley cataloged his snowflake photographs into different types based on similar form characteristics (Bentley and Johnson 1931). During the 1930s to 1950s, a patient scientist from Japan spent a great deal of time studying the seemingly infinite varieties, trying to make some physical sense of snow crystal form. Ukichiro Nakaya (1954) grew snow crystals indoors in a cold chamber where temperature and humidity could be carefully controlled. He grew snow crystals from small "ice seeds" frozen onto a strand of rabbit hair and noted the form results for varying temperatures and amounts of supersaturation. Nakaya's original results are shown in Figure 4.5 and are summarized in Table 4.1.

The crystal form changes in a consistent manner depending on cloud temperature and degree of supersaturation. It is most typical for one type of crystal to fall from a given cloud, rather than having a mix of types all falling at once. The bottom line is that if you can identify the basic form of the snow crystal at the ground, you can tell what the conditions are in the clouds above. Nakaya poetically referred to this connection between crystal form and cloud conditions as "letters from the sky."

The principal forms of snow crystals falling from the atmosphere are generally grouped into eight to ten main types. The newer *International Commission on Snow and Ice (ICSI) classification scheme* has nine different types (Fierz et al., 2009) and includes rime, which is formed on terrestrial surfaces when supercooled water is deposited directly on object. The major atmospheric forms are shown in Figure 4.6. This and older classification schemes are applicable only to falling

FIGURE 4.3 A sample of the thousands of snow crystal photographs taken by Wilson Bentley. (Image courtesy of Wilson Bentley Digital Archives of the Jericho Historical Society, http://snowflakebentley.com.)

TABLE 4.1
Crystal Form Results for Varying Temperatures

Temperature (°C)	Ice Crystal Habit	Temperature (°F)
0 to −3	Thin hexagonal plates	32 to 27
−3 to −5	Needles	27 to 23
−5 to −8	Hollow prismatic columns	23 to 18
−8 to −12	Hexagonal plates	18 to 10
−12 to −16	Dendritic, fern-like crystals	10 to 3
−16 to −25	Hexagonal plates	3 to −13
−25 to −50	Hollow prisms	−13 to −58

SOURCE: Nakaya 1954.

FIGURE 4.4 Wilson Bentley at work photographing snow crystals outdoors. (Photo courtesy of photos: Wilson Bentley Digital Archives of the Jericho Historical Society, http://snowflakebentley.com.)

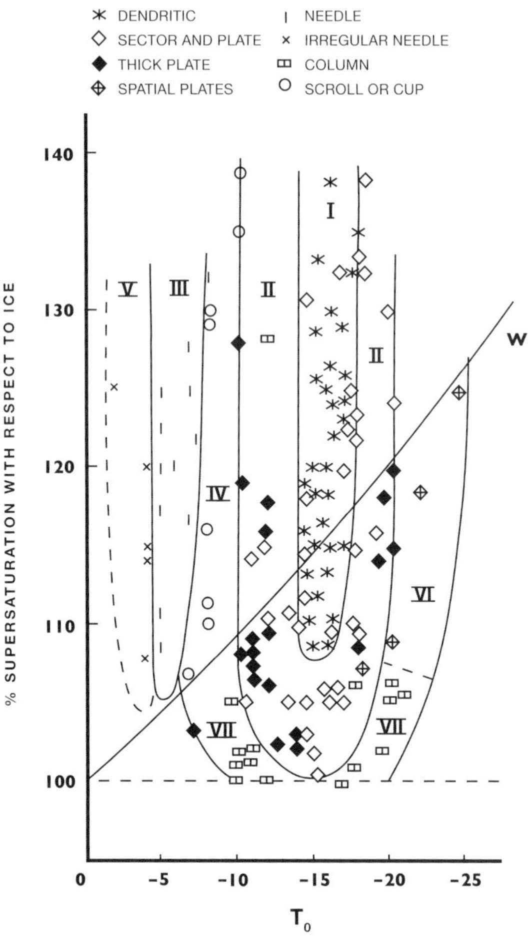

✳ DENDRITIC   I NEEDLE
◇ SECTOR AND PLATE   × IRREGULAR NEEDLE
◆ THICK PLATE   ⊡ COLUMN
⊕ SPATIAL PLATES   O SCROLL OR CUP

FIGURE 4.5 Nakaya's diagram showing the consistent relationship between cloud conditions and ice crystal form. (Nakaya, 1954, courtesy of L. R. Dexter and K. Birkeland.)

snow or snow that has been on the ground a short period of time (a few hours to days depending on temperature), which is referred to as *new snow*.

The Seasonal Snowcover and Old Snow

Upon reaching the ground, snowflakes quickly lose their original shapes as they become packed together and undergo metamorphism (Seligman 1936; Bader et al. 1939; de Quervain 1963; Colbeck 1983). Snow, then, displays continual change during formation, falling, and accumulation on the ground, until it eventually melts. Snow may form in the atmosphere at any latitude but, in order to maintain its identity, it must fall to the Earth in an area with temperatures sufficiently low to prevent it from melting. Most snow melts within a few days or months of the time it falls (referred to as the *seasonal snowcover*), but snow can remain year round depending on the amount received and on climatic conditions. Polar areas receive very little snow, owing to their extremely low temperatures, but what does fall is preserved with great efficiency. On the other hand, snow may persist even in areas where temperatures are above freezing if sufficient amounts fall. The snowline in the Himalaya extends much lower on the southern side than on the northern side, because the greater precipitation received on the southern side more than compensates for the effects of higher temperature. A similar situation exists in the tropics, where snow often reaches lower elevations in tropical mountains during summer (the period of high sun) than in winter. The increased precipitation and cloudiness in summer overrule the effect of the higher sun angle. Heavy snowpacks are found most commonly in middle-latitude and subpolar mountains, regions of relatively high precipitation and low temperatures. Even after the snow has disappeared from the surrounding lowlands in these areas, vast amounts may remain in the higher elevations.

The build-up of a snowcover (also called *old snow*) is in many ways analogous to the formation of a sedimentary rock. Snow accumulates as a sediment, with each layer reflecting the nature of its origin. Newly fallen snow has very low density, somewhat like fluffed goose down, with large amounts of air between the crystals. But with more accumulation, snow becomes compressed and settling takes place. Also, a related series of changes takes place over time at the crystal level, referred to as *snow metamorphism* (just as in geology where the metamorphic rock class represents a changed form deriving from other preexisting rock types affected by increased heat and pressure). The exact behavior and characteristic of old snow depends on its temperature structure, moisture content, internal pressures, and the age of each layer in the snowpack (Sommerfeld and LaChapelle 1970; LaChapelle and Armstrong 1977; Sturm and Benson 1997; Barry and Gan 2011). Snowpack metamorphism can take place by three fundamental processes: two that are largely two-phase, vapor-driven processes (i.e., without significant melting); and one that is a three-phase,

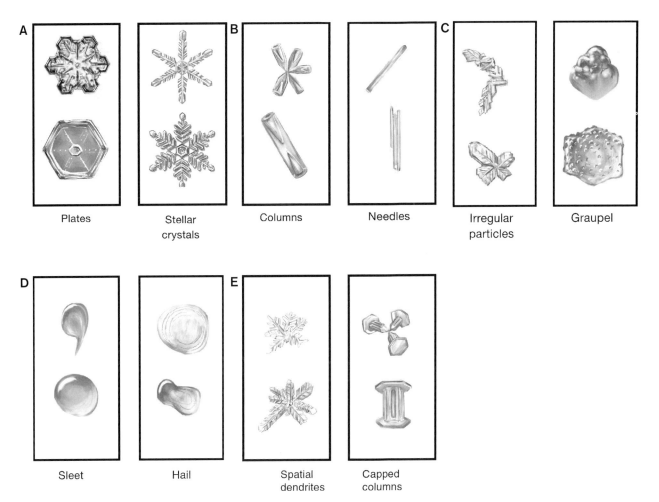

A — Plates — Stellar crystals

B — Columns — Needles

C — Irregular particles — Graupel

D — Sleet — Hail

E — Spatial dendrites — Capped columns

FIGURE 4.6 (A–E) Various classification schemes for new snow include eight to ten of the basic crystal types shown above. The newest ICSI classification for solid precipitation considers those crystals shown in panel E (spatial dendrites and capped columns) to be subtypes of more basic forms. (Images courtesy of L. R. Dexter and K. Birkeland.)

liquid-driven process (i.e., melting is now significant and liquid water is in the pore space to some degree).

EQUILIBRIUM METAMORPHISM

The first process is *equilibrium* metamorphism (referred to in older literature as equi-temperature, ET, or destructive metamorphism) (Fig. 4.7). This process, commonly called "rounding," occurs when the snowpack is subfreezing (i.e., not melting) and free of large vapor pressure and temperature variations. When these conditions are met, grain geometry (crystal shape) and pressure contact between adjacent grains controls the metamorphism. Points of grains are locations of higher vapor pressure, while grain declivities are locations of lower vapor pressure. A vapor flow is set up that transfers mass, molecule by molecule, from the tips of the grains to the branch junctions, leading, in time, to a spherical form often referred to as *rounded grains* or *rounds*. Where grains are in contact in these conditions, *sintering* (i.e., bonding) can take place through both vapor diffusion and solid diffusion, forming continuous ice

"necks" connecting adjacent grains and hence producing a mechanically strong snowpack (Colbeck 1982, 1983).

KINETIC METAMORPHISM

The second process is the *kinetic* metamorphic process (referred to in older literature as temperature gradient, TG, or constructive metamorphism) (Fig. 4.8). In this process, commonly called "faceting," the snowpack is also subfreezing (i.e., not melting) but, unlike equilibrium metamorphism, this process is dominated by large vapor pressure and temperature variations across sections of the snowpack (usually in a vertical direction). When temperature gradients become greater than approximately 10°C per m (5.5° F/ft), depending on the layer temperature, snow density, and other factors, vapor diffusion can occur (Akitaya 1974). An example of such conditions can be found in a shallow snowpack with a warm ground interface below and a cold air interface above. Water vapor flowing through the pores between the individual grains via this mechanism leads to metamorphism. Grain bodies serve as areas of

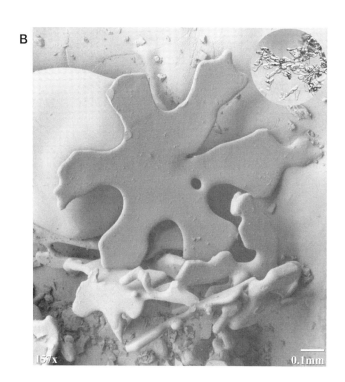

FIGURE 4.7 (A) Equilibrium metamorphism diagram showing the crystal change over time in the presence of relatively small temperature gradients (Image courtesy of L.R. Dexter and K. Birkeland.) and (B) a scanning electron microscope image of a sample crystal. (Image courtesy of E. Erbe.)

vapor deposition (i.e., the change of state from a gas directly to a solid), while the grain contacts receive little deposition. As a result, grains can become very large, with angular and stepped edges growing into the direction of the vapor flow (Sturm and Benson 1997). These growth forms are often referred to as *angles* or *facets* and can become completely three-dimensional *cup crystals* if sufficient space is available. It is interesting to note that these kinetic crystals are relatively strong in compressive strength (top to bottom loading), but very weak in shear strength (sideways loading). The rate of grain growth overpowers the sintering (bonding) effect, resulting in larger grains with fewer bonds per unit volume and a correspondingly weaker layer (Colbeck 1982, 1983). Several subtypes of this process occur, depending on the location and source of the vapor and temperature gradients (i.e., rates of temperature change). Steep temperature gradients near the ground (a common condition in cold mountains with low snowfall) can lead to weak zones lower in the snowpack, called *depth hoar* (McClung and Schaerer 2006: 57); temperature gradients driven by a variety of sources in the upper snowpack result in at least three types of *near-surface faceting* (Birkeland 1998), including *radiation*

recrystallization (LaChapelle and Armstrong 1976). Temperature gradients immediately above the surface can lead to *surface hoar* formation (Fig. 4.9). Surface hoar forms due to atmospheric rather than snowpack processes, and its formation and persistence can be complicated by factors such as proximity to trees or exposure to sun and wind (Lutz and Birkeland 2011). In all cases, faceting forms weak layers of varying thickness and location within the snowpack, a key ingredient for many avalanches.

MELT–FREEZE METAMORPHISM

The third type of metamorphic process is *melt–freeze* metamorphism (also referred to as MF metamorphism) (Fig. 4.10). This occurs where the melting point has been reached somewhere in the snowpack. This could be just a surface layer during a sunny period or it could include the entire snowpack when *isothermal* conditions (melting throughout) are reached in the spring. This process is more complicated than the first two, as it involves all three phases of water. Liquid water fills the intergranular pore space to some extent. During the melt phase, large grains

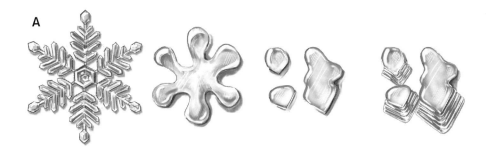

A

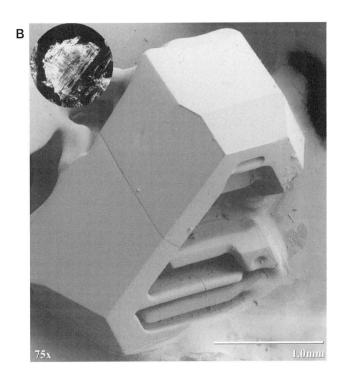

B

FIGURE 4.8 (A) Kinetic metamorphism diagram showing the crystal change over time in the presence of moderate temperature gradients. When temperature gradients are extremely large, crystals can progress directly from stellars at the left to more advanced faceted crystals shown on the right. (Image courtesy of L. R. Dexter and K. Birkeland.) (B) A scanning electron microscope image of a sample faceted crystal. (Image courtesy of E. Erbe.)

grow at the expense of smaller grains because of small but significant shape-related temperature differences (Colbeck 1982, 1983). The result is that large polygranular units form over time, often referred to as *corn snow*. In the warm part of the day, the snow may be mechanically weak because of the melting of intergranular bonds, whereas in the cold part of the evening, the snow may be very strong because of refreezing of the liquid water, especially near the surface where radiant energy exchange is pronounced. The process of repeated freezing and thawing causes increased densification and consolidation and is responsible for the formation of *firn* or *névé*, which is dense snow at least one year old. The snow may now be as much as 15 times denser than when it first fell, and is well on its way to becoming glacial ice (de Quervain 1963: 378).

THE INTERNATIONAL CLASSIFICATION FOR SEASONAL SNOW ON THE GROUND

A comprehensive snow classification system exists for all types of seasonal snow (including new snow, described previously): the *International Classification for Seasonal Snow on the Ground (ICSSG)* (Fierz et al. 2009). The ICSSG is fairly involved but, at the coarsest level, it consists of nine fundamental snow and ice types, based mainly on grain shape:

1. Precipitation particles (identical to the eight ICSI classes)
2. Machine-made snow (typically made for skiing)
3. Decomposing and fragmented precipitation particles (including windblown new snow)
4. Rounded grains (equilibrium metamorphism)
5. Faceted crystals (kinetic metamorphism)
6. Depth hoar crystals (advanced kinetic metamorphism)
7. Surface hoar (deposition of kinetic forms onto the snow surface)
8. Melt forms (melt–freeze metamorphism)
9. Ice formations (horizontal ice layers and vertical ice columns from piping)

This system is the standard used by most workers in snow-related endeavors around the world.

FIGURE 4.9 Surface hoar, a form of kinetic crystal growth by direct deposition of water vapor onto a cold snow surface. These crystals formed in southwestern Montana (K. Birkeland).

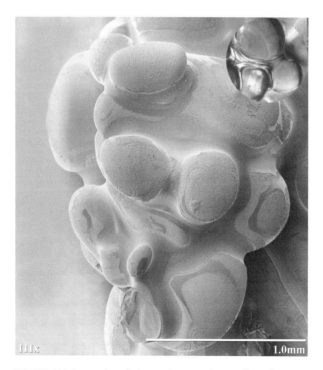

FIGURE 4.10 A scanning electron microscope image of a melt–refreeze grain. (Image courtesy of E. Erbe.)

The Mountain Snowpack as a Water Resource

The implications of mountain snow for human existence are discussed in Chapters 10 and 12, but the importance of meltwater cannot be overstressed. Numerous estimates indicate that 66 to 80 percent of all water resources used in the western United States originates as snowfall (e.g., Flaschka et al. 1987). The Pacific Northwest of the United States is largely dependent upon hydroelectric power from streams that head in the Cascade and Rocky Mountains, and California's bountiful farm production is derived largely from meltwater from the Sierra Nevada. In fact, it is safe to say the economy of the entire western United States is dependent upon meltwater from mountains (Serreze et al. 1999). Unfortunately, it appears that, as the climate warms, the average annual snowpack accumulation in the Western United States is on the decline (Mote et al. 2005).

From the point of view of snow metamorphism, snow-melt runoff is an extension of the melt–freeze process. The melting snowpack cannot deliver water to the river system until the available pore space is filled to field capacity with liquid water. This means that there is a lag time from the onset of melt until the snowpack fills up, becomes *ripe*, and can transfer water to the stream channel. During the ripening process, water can flow horizontally along dense layers; or, in cases where there are significant differences in grain size, between adjacent layers in the snowpack; or vertically through conduits referred to as *pipes*, leading to a complicated internal "plumbing system" during the snowmelt runoff period (Dunne and Leopold 1978; Barry and Gan 2011).

FORECASTING SNOWMELT-DERIVED WATER RESOURCES

In the western United States, the *Cooperative Federal Snow Survey* under the lead of the Natural Resources Conservation Service is charged with taking measurements and providing monthly reports on the status of the snowpack in different regions. This has become a vital operation in water-supply forecasting (Davis 1965; U.S. Department of Agriculture 1972; Palmer 1988; Pagano et al. 2004). Measurements are taken by two different techniques. The traditional technique was developed in the early 1900s by Dr. Frank Church, a professor of Romance languages at the University of Nevada at Reno, for forecasting runoff down the Truckee River. His technique involved simply shoving a length of pipe through the snowpack to the ground to capture a known volume of snow. The snow volume, reduced to its liquid content, is called the *snow water equivalent* (*SWE*). This basic technique is still in use today at many *snow survey courses* located throughout the country, where *federal sampler* tubes are used to take samples manually at several points along the snow course transect. However, this sampling method is increasingly being replaced or supplemented by a unique automated system called *SNOTEL* (SNOpack TELemetry) that has been implemented at over 600 sites across the western United States (Serreze et al. 1999). SNOTEL uses a large rubber or metal bladder filled with antifreeze. As snow accumulates on the bladder, a pressure transducer calibrated in inches of water senses the load. These data are sent to receiving stations using a solar-powered radio transmission system where signals are bounced off the ionized trails of burning meteors (called *meteor-burst* transmission).

SNOW AND SNOWMELT RUNOFF AUGMENTATION

Considerable research and effort has gone into developing methods of increasing and retaining the snowpack. They include installing fences in alpine grasslands, planting more trees, and experimental methods of timber cutting that alternate cut and standing patches of trees to preserve the snow from ablation (Martinelli 1967, 1975; Leaf 1975; Jarrell and Schmidt 1990). Efforts toward artificial stimulation of precipitation (*cloud seeding*) have largely been focused on increasing the snowfall in mountains (Weisbecker 1974; Steinhoff and Ives 1976; Bruintjes 1999). Cloud-seeding studies have produced contradictory results, and there are concerns about "sky water rights" and increased mountain hazards. Often, people living downwind of seeding projects feel that they are being deprived of some of "their water"; others are concerned that increased precipitation would lead to increased mountain hazards, such as snow avalanches and flooding. It should be pointed out that not all snow in the pack becomes stream runoff. A number of possible losses can and do occur as a result of soil infiltration, sublimation from the snow surface, sublimation from snow in trees, and evaporation from the melting snow surface (Avery et al. 1992; Niu and Yang 2004). Debate continues about the significance of such losses, but in some areas of the western United States, basin efficiencies are on the order of only 30–40 percent. It should be apparent that warm summers lead to high evaporation rates. This in turn can severely limit the effectiveness of summer rainfall as a water resource, unless it is torrential enough to fill stream channels (often dry in the summer) with flowing water that can be collected in reservoirs. Another interesting effect on the rate and timing of snowmelt is the possible addition of airborne dust carried in the winds, often from arid-land sources hundreds or thousands of kilometers away. Studies conducted in the San Juan Mountains of Colorado indicate that a series of dust deposition events (usually occurring during the late winter or spring) can decrease the snowpack albedo there and thus speed up the melt season by one full month (Painter et al. 2007).

"Permanent" Snow and the Snowline

Many areas of the globe are covered by snow and ice year round. Latitude plays a dominant role in the distribution of permanent snow and ice. However, high-altitude mountains can completely overpower the effect of latitude and provide a permanent abode for snow and ice even at the equator. The zone between seasonal snow that melts every summer and the permanent snow that does not melt is represented by the *snowline*. This zone has fundamental implications for environment and process. The varying disposition of the snowline in time and space has resulted in different interpretations of its significance, and has caused considerable confusion in the literature, with terms such as *climatic snowline, annual snowline, orographic snowline, temporary snowline, transient snowline,* and *regional snowline* (Charlesworth 1957; Flint 1971; Østrem 1964a, 1973, 1974; Seltzer 1994). Use of the term "snowline" without an accompanying explicit definition is fairly meaningless. To appreciate the problem, consider the following conditions: At one extreme is the delineation between a snow-covered and a snow-free area at any time of the year. Obviously, this snowline varies from day to day and will be lowest in the winter, reaching sea level in middle latitudes, and highest in summer. There is also a snowline establishing the lower limits of persistent snow in winter, a matter of great importance for the location of ski resorts and road maintenance (Elsasser 2001).

Our primary concern, however, is the location of the snowline after maximum melting in summer, since this is the level that establishes the glacial zone and largely limits the distribution of most plants and animals. The position of this line is likewise highly variable and difficult to delineate. For example, avalanches may transport large masses of snow to valley bottoms where, if shaded, they may persist for several years. Similarly, some mountain glaciers occupy sheltered topographic sites and receive greater accumulations from drifting snow and avalanches than do the surrounding slopes. Glaciers also can experience less melting because of their "shadow climate" and the natural cooling effect of the larger ice mass. As a result, the snowline is generally lower on glaciers than in the areas between them. In mountains without glaciers, or on slopes between glaciers, the snowline is commonly represented by small patches of perennial snow, where distribution is largely controlled by slope orientation and local topographic sites (Alford 1980; Brozovic et al. 1997; Mitchell and Montgomery 2005).

The disparity among the various snow limits, and the difficulty of establishing their exact locations, has led to the use of several indirect methods of approximation. One is to use the elevation where the average temperatures are 0°C (32°F) or less during the warmest month of the year. Since this is determined primarily through the use of radiosondes and weather balloons, a snowline can be established even where there are no mountains. The resulting snowline, although only theoretical, is useful for purposes of generalization. This is particularly true when investigating temperatures during the glacial age. For example, if a glacier exists today at 2,000 m (6,600 ft) that at one time extended to 1,000 m (3,300 ft), the difference in elevation can be converted to temperature (through use of the vertical lapse rate) to get an approximate idea of the temperature necessary to produce the lower snowline. There is some evidence that temperatures during the

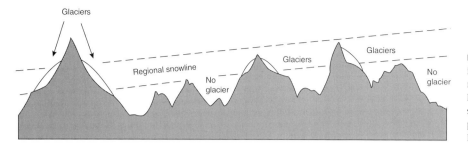

FIGURE 4.11 Method of approximating the regional snowline. The regional snowline occupies the zone lying between the highest peaks not supporting glaciers and the lowest peaks that do support glaciers. (After Flint 1971: 64 and Østrem 1974: 230.)

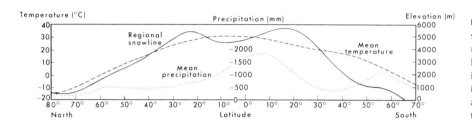

FIGURE 4.12 Generalized altitude of snowline on a north–south basis. The reason for a slightly depressed snowline elevation in the tropics is the increased precipitation and cloudiness in these latitudes. Mean temperature and mean precipitation are also illustrated. (After Charlesworth 1957: 9.)

Pleistocene glacials were about 4–7°C (7–13°F) lower than they are today (Wright 1961; Flint 1971: 72; Andrews 1975: 5; Porter 2001).

A more useful approach is to establish a zone or band about 200 m (660 ft) thick to represent the *regional snowline,* or the *glaciation level* as it is known, since it represents the minimum elevation in any given region where a glacier may form (Østrem 1964a, 1974; Porter 1977, 2001). The location of this zone is based on the difference in elevation between the lowest peak in an area bearing small glaciers and the highest peak in the same area without a glacier (but with slopes gentle enough to retain snow). For example, if one mountain is 2,000 m (6,600 ft) high but has no glacier even though its slopes are gentle enough to accommodate one, and another mountain 2,200 m (7,300 ft) high does have a glacier, the local glaciation level and the regional snowline lie between these two elevations (Østrem 1974: 230–233; Fig. 4.11).

The regional snowline is lowest in the polar regions, where it may occur at sea level, and highest in the subtropics, where it occurs between 5,000 and 6,500 m (16,500–21,500 ft). This is not a straight-line relationship, of course, owing to the interplay of temperature and precipitation. The highest snowlines are found between 6,000 and 6,500 m (19,800–21,500 ft) in the arid Puna de Atacama of the Andes (25°S latitude) and the Tibetan Highlands (32°N latitude). The greater precipitation and cloudiness experienced in the tropics depress the snowline, while areas under the influence of the subtropical high at 20–30°N and S latitude receive less precipitation and fewer clouds, resulting in a higher snowline even though temperatures are lower (Fig. 4.12). At any given latitude, the snowline is generally lowest in areas of heavy precipitation (e.g., coastal mountains) and highest in areas of low precipitation (e.g., continental mountains). Accordingly, there is a tendency for snowlines to rise in elevation toward the west in the tropics

and toward the east in middle latitudes, in accordance with the prevailing winds. The middle-latitude situation is illustrated by the snowline in the western United States, which rises from 1,800 m (6,000 ft) in the Olympic Mountains, Washington, at 48°N latitude, to 3,000 m (10,000 ft) in Glacier National Park, Montana, in the Rockies, 800 km (480 mi) to the east (Flint 1971: 66). A similar tendency for the snowline to rise from west to east is found in the mountains of Scandinavia, the Andes of southern Chile, and the Southern Alps of New Zealand (Østrem 1964a; Porter 1975).

Other Occurrences of Frozen Water in Mountains

RIME ICE OR HOARFROST

Rime ice, sometimes called *hoarfrost,* forms by contact freezing of supercooled water droplets and direct deposition of water vapor onto various nucleating objects in the surrounding environment (Graham 2009). These rime icing events are most often accompanied by high-velocity winds. Nucleating objects can be natural (trees, rocks, falling snowflakes, an old snow surface, or even entire mountain peaks) or human-made (aircraft wings, buildings, ski lift towers, communications towers, fence posts) (Fig. 4.13). Rime loading on human structures can become so great that a structure may collapse.

Rime ice often takes on a blade-like form (*rime feathers*) that builds outward from the collecting object into the oncoming wind. Surprisingly, rime may provide the majority of winter water accumulation in some areas. Polar mountains, for example, receive so little direct precipitation that the contribution of rime and hoarfrost is often greater than that of snow. In some very rare instances, rime accumulations have been shown to release abruptly from their anchorage and "avalanche" in curious rime flow events (Kokenakais and Dexter 1998).

FIGURE 4.13 Mt. Washington summit observatory covered in rime. (Photo courtesy Mount Washington Observatory.)

FREEZING OF LAKES AND PONDS

Another mystery of nature that people often take for granted is the fact that ice floats in its own liquid! Of commonly found compounds, only ammonia shares this trait with water. Lakes, ponds, and other relatively quiet bodies of water in high mountain environments often freeze over in the winter, but only very small lakes and ponds freeze completely solid. As the lake water near the surface cools with the approach of winter, its density increases and the cooled water sinks. When the temperature of the lake water cools below +4°C (39°F), the colder water becomes less dense, as the molecules begin to take on the expanded volume associated with the crystal lattice of ice. During this process, the lake water will completely overturn (i.e., exchange bottom water with top water). The near-freezing water, with its abundance of freezing nuclei, now floats to the surface and the crystal lattices merge to form of a thin sheet of *skim ice*. With increased thickening, the so-called "C" crystal axis of the ice becomes oriented vertically, producing a dark ice called *black ice* (or *candle ice* because of the pronounced vertical ice columns observed during melt-out). The ice thickens rapidly at first, but the rate slows down because of a self-insulating feedback loop. A second layer in the lake ice pack forms when snow falls on the black ice and depresses it isostatically into the water. Cracks form in the black ice that allow lake water to flood the overlying snow, producing a frothy layer of *white ice* (Gray and Male 1981) (Fig. 4.14). While this process may seem merely an interesting curiosity, it has far-reaching consequences for aquatic life. For example, if water behaved like most other substances (i.e., sinking as it solidifies), freezing of lakes would be far more extensive. Ice would sink to the bottom, exposing the liquid water to further surface cooling in a feedback cycle that would rapidly freeze even fairly deep lakes solid, leaving the hapless fish stranded on the ice surface! (Marchand 1996; Prowse and Beltaos 2002; Barry and Gan 2011).

FREEZING OF RIVERS AND STREAMS

The freezing of streams and rivers follows a different path. The turbulent water is thought to splash small droplets up into the cold air to initiate the freezing of seed crystals. As these seed crystals fall back into the flowing water, they serve as centers for further freezing. Water that freezes onto the seed crystals produces small, disk-shaped grains that collect into a mass of oatmeal-like mush called *frazil ice*. This can become a nuisance to human works (like inlet gratings for power plants) by clogging openings and freezing onto structures in the river. In addition to frazil ice, clear-water streams often cool at the bottom by radiation loss, producing another location for enhanced freezing directly on the channel bed. Ice that forms in this fashion is *anchor ice* (Fig. 4.15). Through these processes, streams freeze up by progressively being choked with frazil ice and anchor ice (Marchand 1996; Prowse and Beltaos 2002; Barry and Gan 2011).

FREEZING IN ROCK REGOLITH AND SOIL

Water that freezes in the interstices (i.e., pores, cracks, and other voids) of rock and soil can exert tremendous pressure during freezing. The pressure is great enough to lead to the splitting of apparently solid rock. This process is especially effective in seasons or environments where the diurnal temperature swings across the freezing point, allowing for

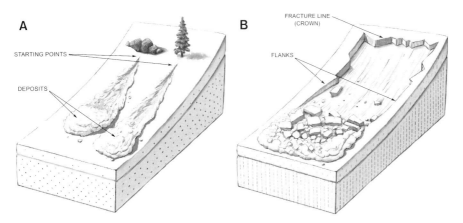

FIGURE 4.18 Typical forms displayed by (A) loose-snow avalanches and (B) slab avalanches. (Image courtesy of K. Birkeland.)

Martinelli 1976: 68; McClung and Schaerer 2006; Barry and Gan 2011).

Dangerous slab avalanches occur less frequently than loose-snow avalanches. Slab avalanches originate in all types of snow, from old to newly fallen and from dry to wet. The chief distinguishing characteristic is that the snow breaks away with enough internal cohesion to act as a single unit until it disaggregates during its journey downslope. The zone of release, or *starting zone,* is marked by fracture lines that are perpendicular to the slope and extend to a well-defined basal-fracture plane (Figs. 4.18b, 4.20). The size of a slab avalanche depends on many factors, but it is often confined to a specific area on the slope because of the nature of the terrain. However, during times of extreme instability, whole mountainsides may become involved, with the fractures racing along for several kilometers, releasing snow in numerous slide paths. Though it has been assumed that the entire mass of a slab avalanche is set in motion at once, research shows that the mass of the slide often increases in a downhill direction, as the avalanche erodes and entrains snow in the path (Sovilla et al. 2001). Avalanches reach their maximum velocity quickly, so their destructive power is significant near the point of origin (Atwater 1954: 27; McClung and Schaerer 2006). The exact behavior, of course, depends on the nature of the snow and several other factors. If the snow is dry, a *powder-snow avalanche* may develop. These move as much in the air as on the ground, and their turbulent motion may create a dense dust cloud of ice crystals, which behave like a body of heavy gas preceding the rapidly sliding snow. Such windblasts may achieve a velocity of 320 km (200 mi) per hour and can cause damage well beyond the normal avalanche zone (Seligman 1936; LaChapelle 1966). On the other hand, wet-snow avalanches tend to slide at slower speeds with no particular dust cloud, but their impressive mass can still cause great damage.

FIGURE 4.19 Loose snow avalanches (sluffs) above Snowbasin Ski area in Utah, U.S.A. Such avalanches are typically small and harmless, often occurring during or shortly after storms. Note the large overhanging cornice on the ridgeline above. (Photo by K. Birkeland.)

Factors Influencing Avalanche Occurrence

Snow is a highly variable material. It occurs under different environmental conditions, displays vastly different mechanical characteristics as its temperature and microstructure change, and is susceptible to constant modification once on the ground (McClung and Schaerer 2006). Modern avalanche forecasters use both their practical experience and their scientific knowledge of snow and avalanches to predict the probable occurrence of avalanches in time and space (LaChapelle 1980; McClung 2002). Though many of the tools used for avalanche forecasting have improved (e.g., remote weather systems, geographic information systems [GIS], and some computer models), most avalanche forecasters still use techniques pioneered over 50 years ago (Seligman 1936; Atwater 1954; Perla and Martinelli 1976; McClung and Schaerer 2006). Ultimately, there are three main ingredients needed for avalanching: (1) favored terrain, (2) unstable snow (which is a product of both the weather and the existing snowpack), and (3) a trigger.

FIGURE 4.20 An avalanche forecaster inspects the crown face of a large slab avalanche in Glacier National Park, Montana. Note the sharp crown fracture and the base over which the snow moved. (Photo by K. Birkeland.)

FIGURE 4.21 An aerial view of avalanche paths immediately north of Hebgen Lake in southwest Montana. The complex starting zone, well-defined track, and a runout zone that crosses the highway are shown for one of the paths. (Photo by K. Birkeland.)

TERRAIN

Avalanches usually occur repeatedly in the same places (referred to as *avalanche paths*) because of weather and terrain relationships. The area where an avalanche initiates is the *starting zone,* the area through which the avalanche runs is the *track,* and the area where deceleration and deposition take place is the *runout zone* (Fig. 4.21). In timbered areas, the path and its component parts are usually easy to identify because of the damage to, and clearing of, vegetation (Tremper 2008). The most important terrain factor for avalanches is the slope angle of the starting zone, with favored angles between 30° and 45°; the majority of

avalanches originate on slopes from 36° to 39° (Fig. 4.22). In snow climates with moist snowfall, slope angles for avalanche starting zones may be much steeper than 45°. Avalanches on slopes steeper than 55° occur in some mountain ranges but, typically, continuous *sluffing* cleans the steeper slopes. Slopes less than about 30° are generally not steep enough for avalanche initiation, because snow that fractures cannot overcome the residual friction and slide down the slope, although avalanches in motion can flow down slopes shallower than 10° for quite a distance. The momentum of large avalanches can even carry snow up the opposite canyon wall. Overall, the probability of

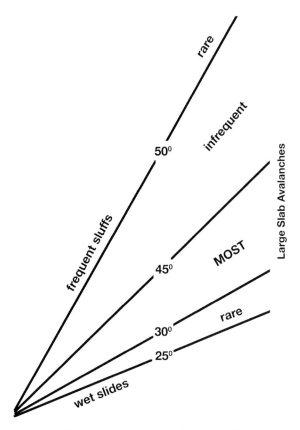

FIGURE 4.22 Characteristic slope angles for snow avalanches. (L. R. Dexter and K. Birkeland.)

avalanches increases with steeper slopes up to a certain point and then decreases.

Many other terrain features favor or inhibit avalanches. For example, a convex slope will be slightly more prone to avalanches than a concave slope, as the outward bend of a convex slope puts tensional stress on its snowcover, while a concavity of slope slightly strengthens the snow's cohesion through compression. More important than the slope shape is its orientation with respect to both the wind and the sun. Leeward slopes are typically more dangerous, since the wind can quickly load large quantities of snow onto the slope. They are also often overhung by *cornices,* which can fall on the slope and act as triggers (Figs. 4.19, 4.23). Exposure to the sun greatly influences the behavior of snow. North-facing slopes receive little sun during the winter (in the northern hemisphere). This typically results in cold conditions where instabilities may persist for longer periods. On south-facing slopes (in the northern hemisphere), the sun strikes the surfaces at a higher angle, causing kinetic and/or melt–refreeze metamorphism. South-facing slopes can be sites of instability because of near-surface faceting (Birkeland 1998) and wet-snow avalanches (McClung and Schaerer 2006).

Other important terrain variables include surface roughness and groundcover. A slope strewn with large boulders

is not as susceptible to avalanching as a smooth surface, at least until the snow covers the boulders. On the other hand, a smooth, grassy slope provides no major surface inequalities to be filled by the snow and offers little resistance to sliding. Densely forested slopes (>1,000 conifer trees per hectare [>1,000 trees per 2.5 acres] on steeper slopes) generally offer good protection from avalanche initiation (McClung and Schearer 2006). However, avalanches may start above the timbered zone and destroy strips of the forest in their path (see Fig. 4.21), and many mountain forests have been cut or destroyed in recent centuries (Aulitzky 1967). Once this happens, these zones become more vulnerable to avalanching and it is difficult for the forest to regenerate, since trees in the path of the avalanching are continually damaged or killed (Frutiger 1964; Schaerer 1972; Martinelli 1974; McClung and Schaerer 2006).

WEATHER

Any weather factors that change the mechanical state of the snowpack may also quickly change the avalanche conditions. The three most important weather factors are new snow (or rain), wind, and changes in temperature. More than 80 percent of all large slides occur either during or shortly after storms. The more snow, the more weight added to the snowpack, and the greater the stress acting on any weak layers in the existing snowpack or on the interface between the new snow and the previously existing snowpack. The rate of snowfall, or the *snowfall intensity,* is critically important. If the snow falls slowly enough, the snowpack may be able to adjust to the new snow load, but rapid snowfall may quickly overload weak layers before they can adjust, thereby causing avalanches. Rain can also be an important factor, since it adds weight to the snowpack without any addition in strength; rain falling on a midwinter snowpack often rapidly initiates a large number of avalanches.

Wind is also a critically important weather factor for avalanches. Wind redistributes snow onto the lee sides of ridges, gullies, and other terrain features, where it may pile into thick wind drifts. Even relatively gentle winds (about 15 km/hr [9 mi/hr]) are sufficient to move low-density snow (Tremper 2008). Wind also breaks snowflakes into smaller particles, which then bond together quickly, forming cohesive *wind slabs.* Areas of wind deposition are areas where more stress has been added to the snowpack, thereby creating more unstable conditions.

Temperature is another important weather factor, affecting the mechanical strength of both falling and accumulated snow. The temperature at the time of snow deposition affects the snow crystal type, the density, the rate of settlement, and the general cohesion. After deposition, the temperature of the snow layer and of the air above is critical to the rate of settling, compaction, internal creep, and metamorphosis. In general, instabilities in the snowpack tend to persist longer when temperatures are cold, and stabilize more quickly when temperatures

FIGURE 4.23 Large snow cornices overhang the main ridge of southwest Montana's Bridger Range. The dominant wind direction is from right to left, and when the cornices break off they are effective avalanche triggers. (Photo by K. Birkeland.)

are moderate. However, high temperatures permit melting and the loss of cohesion among snow layers, resulting in wet-snow avalanches. In addition, changes in temperature may affect how easy it is to trigger avalanches, with warmer temperatures decreasing the stiffness of the snow and allowing stresses to penetrate farther into the snowpack, thereby facilitating triggering by skiers or snowmobilers (McClung and Schweizer 1999).

SNOWPACK

How the weather interacts with the existing snowpack determines whether the snow is capable of producing avalanches, and we will touch briefly on some of the major points regarding unstable snowpacks. Schweizer (1999) reviews some of the details, and Heierli and others (2008) provide a model describing the current thinking on the fracture mechanics behind avalanche release. Slab avalanches require three basic snowpack ingredients: a slab, a weak layer, and a bed surface. The slab is a relatively cohesive layer of snow that overlies the weak layer. Slab densities can be quite variable, ranging from 50 to 450 kg/m^3 (3 to 28 lbs/ft^3) (McClung and Schaerer 2006). Slabs may be composed of any type of snow, but new snow, equilibrium metamorphosed snow, and wind slabs form the most common slab layers. The weak layer is simply a less cohesive layer underlying the slab, commonly composed of faceted crystals formed by kinetic-growth metamorphism like depth hoar, surface hoar, or near-surface facets. In some cases, the weak layer may be no more than a weak interface between the slab and the underlying snow. Bed surfaces are not critical for slab avalanches; in some cases (i.e.,

with a depth-hoar avalanche) no bed surface is necessary. However, a hard bed surface, such as a frozen rain crust, may create particularly unstable conditions when a weak layer and a slab are deposited on top of it.

Slabs, weak layers, and bed surfaces often occur in the snowpack, but the snow is not always unstable. For unstable conditions, the stress on the weak layer must exceed its strength at some location. The gravitational force of the preexisting overlying slab, plus any added weight from new or windblown snow, causes the stress on the weak layer. When this exceeds the strength of the weak layer locally, cracks initiate within the weak layer. After growing to a critical size (on the order of the slab depth), the cracks rapidly propagate around the slope and, if the slope is steep enough, an avalanche will release. The mechanics are further complicated by the highly spatially variable nature of the snowcover (Schweizer et al. 2008; Birkeland et al. 2010).

TRIGGERS

Snow on a slope that is unstable enough to be triggered is called *conditionally stable*. This situation is particularly dangerous since adding a person on skis, a snowboard, or a snowmobile to the slope may result in an avalanche (Heierli et al. 2011). The most common triggers for natural avalanches include new or windblown snow, though falling cornices are also important in some areas. Rapidly changing air temperature has occasionally been blamed for avalanche release, but its importance has not been convincingly demonstrated. Explosive blasts can also be used to trigger avalanches as a mitigation measure (see below). Loud sounds have been implicated as avalanche triggers, and there was an ancient regulation in Switzerland against yodeling during the avalanche season (Allix 1924). However, research suggests this is highly improbable. Even enormous sonic booms are only capable of triggering avalanches in rare cases of extremely unstable snowpack (Martinelli 1972).

The Avalanche as a Hazard, Avalanche Victims, and Avalanche Rescue

Once humans become involved in the avalanche equation, we have a *hazard*. It is ironic that, despite our greater scientific understanding of avalanches and our considerable investment in their prediction and prevention, the number of accidents continues to increase, primarily because more and more people, especially recreationists, go to the mountains during the winter (Fig. 4.24). This is graphically illustrated by an analysis of avalanche accidents in the United States in a series of volumes called the *Snowy Torrents* (Williams 1975; Williams and Armstrong 1984; Logan and Atkins 1996), as well as the more recent Canadian volume *Avalanche Accidents in Canada* (Jamieson et al. 2010).

In the United States, avalanche fatalities have increased greatly over the last few decades, for a number of reasons. Fatalities increased sharply in the late 1970s as ski equipment improved and backcountry skiing became more popular. A second sharp increase in fatalities occurred in the 1990s as ski and snowmobile equipment improved, with the five-year average becoming nearly 30 deaths per year. Since the mid-1990s, snowmobilers as a group have overtaken skiers and climbers in leading avalanche fatalities (Fig. 4.25). Snowmobile technology has improved greatly since the 1990s, allowing riders to access more avalanche-prone terrain more quickly after storms (CAIC 2012).

Avalanche rescue can be viewed in terms of two distinctly different sets of responses and personnel involved. The first response is the immediate attempt at recovery by the victim's companions. Because of the likelihood of suffocation, this approach is typically the victim's best chance of survival. Recent research suggests that over 90 percent of fully buried avalanche victims survive the first 15 minutes, but survival probability drops rapidly to around 30 percent after 30 minutes (Tremper 2008). The second response category is organized as outside rescue, such as a ski patrol or search and rescue unit. Some responses here can mean 24 hours or longer in a backcountry situation, so survival in these cases is unlikely unless the victim is in a vehicle or structure. The use of cell phones to alert rescue teams and helicopters to transport them have dramatically shortened response times for organized rescues in some situations.

Live avalanche rescue is greatly enhanced by prior training in the use of an electronic avalanche beacon, probe poles, and shovel; these should be worn or carried by each

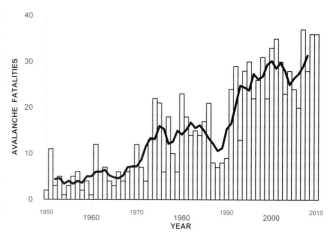

FIGURE 4.24 Annual United States avalanche fatalities (vertical bars), and a five-year running average (line), from 1950 to 2010. Fatalities increased in the 1970s and again in the 1990s as the numbers of people recreating in the backcountry increased. (Data courtesy Colorado Avalanche Information Center. Graph produced by L. R. Dexter from CAIC data.)

party member. Other new equipment like *Avalungs*, which allow a person to breath under the snow if they are not buried so deeply that they cannot expand their chest, and avalanche airbags, which help to keep victims on or near the snow surface, can enhance the chances of surviving an avalanche. In addition, each member of the party should realize that they are the victim's best hope and should not go for help until all other on-site efforts have been exhausted or unless assured help is very close (within minutes). Trained dogs are also very effective at locating victims (Tremper 2008).

Avalanche Forecasting and Mitigation

FORECASTING AVALANCHES

The safest way to deal with avalanches is to avoid them, and this is possible only by avoiding all snow-covered avalanche terrain. This will never occur so long as people wish to live in, play in, and travel through mountains in the winter. Therefore, reducing avalanche accidents relies on avoiding avalanche terrain during times of unstable snowpack conditions, and those times can be best determined through avalanche forecasts. Given the increasing number of avalanche fatalities, there is a need for improved forecast methods and broader forecast area coverage. Scientifically based avalanche forecasting originated in Europe in the early 1900s and has been practiced in the United States since the late 1940s (Atwater 1954; LaChapelle 1980; McClung 2002). Modern avalanche forecasting is a sophisticated yet inexact endeavor. Most current avalanche forecasters bring an extensive knowledge of mountain meteorology, snow mechanics, and terrain analysis to bear on the forecast. They use meteorological data from remote sites, snowpit

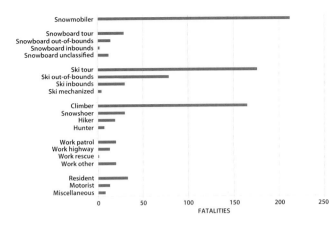

FIGURE 4.25 Avalanche fatalities in the United States by activity from 1950 to 2010, with snowmobilers, skiers, and climbers leading the list. The numbers of snowmobilers involved in avalanches has increased dramatically since 1990 as new snowmobile technology has allowed more riders to easily access avalanche terrain. (Data courtesy Colorado Avalanche Information Center. Graph produced by L. R. Dexter from CAIC data.)

and fracture-line profile analysis from field sites, the results of explosive tests from nearby ski areas or heli-ski operations, the results of computer simulations for weather and sometimes snowpack behavior and, most importantly, a great deal of personal experience to derive the forecast for the day. The bottom line is that modern avalanche forecasters still use so-called conventional techniques relying on an analysis of terrain with respect to many of the contributory factors defined by Atwater (1954) more than a half century ago. Currently, regional forecast centers exist in the United States (Alaska, California, Colorado, Idaho, Montana, Wyoming, Utah, the Pacific Northwest, and Mount Washington in New Hampshire) as well as Canada, New Zealand, mountainous European countries, and India. These centers offer public access to a mountain weather forecast and avalanche danger assessment on their websites, with North American forecasts available at www.avalanche.org and www.avalanche.ca. Canada and the United States recently updated the avalanche danger scale, emphasizing the concept of risk and providing more specific travel advice to the public. This scale is also used in New Zealand (Statham et al. 2010).

EXTREME (Black): Natural and human-triggered avalanches are certain, large to very large avalanches are expected in many areas, people should avoid all avalanche terrain.

HIGH (Red): Natural avalanches are likely and human-triggered avalanches are very likely. Large avalanches are expected in many areas and very large avalanches are expected in specific areas. These are very dangerous avalanche conditions and travel in avalanche terrain is not recommended.

CONSIDERABLE (Orange): Natural avalanches are possible and human-triggered avalanches are likely. People can expect small avalanches in many areas or large avalanches in specific area or very large avalanches in isolated areas. These are dangerous avalanche conditions where careful snowpack evaluation, cautious route finding, and conservative decision making are essential.

MODERATE (Yellow): Natural avalanches are unlikely, but human-triggered avalanches are still possible. People can expect small avalanches in specific areas or large avalanches in isolated areas. In these conditions, there are heightened avalanche conditions on specific terrain features, and people must evaluate the snow and terrain carefully, indentifying the features of concern.

LOW (Green): Natural and human-triggered avalanches are unlikely. People can expect small avalanches in isolated areas or extreme terrain. There are generally safe avalanche conditions, though there may be unstable snow on isolated terrain features.

MITIGATING AVALANCHES

The attempt to directly mitigate avalanches has been practiced in the Alps for centuries, but emerged as a relatively new endeavor in North America following World War II. There are two basic approaches to the problem: passive and active mitigation. Passive mitigation measures are relatively effective, but they can be expensive and require continual maintenance. Therefore, they are most appropriate in areas where permanent structures are threatened by avalanches. Active mitigation, such as triggering avalanches, is much less expensive but must be applied repeatedly. This technique is appropriate for areas where avalanches can be triggered when people are not in the area, such as ski runs and mountain highways.

Passive mitigation through terrain modification consists of placing structures such as walls, pylons, dams, and wedges of various designs either in the snow accumulation zone or immediately above the area to be protected (McClung and Schaerer 2006). The strategy in the snow accumulation zone is to break up the solid mass of the snow into smaller units, to anchor the snow base, and to create terraces so that there is less effective slope for each snow unit (LaChapelle 1968: 1024; Fig. 4.26). In the runout zone, the structures consist of barricades, walls, and wedges to dam or divert the avalanche (Fig. 4.27). Roofs or sheds are frequently constructed over highways and railway lines along avalanche paths (Fig. 4.28). An interesting technique is the use of alternately spaced earthen mounds (Fig. 4.29). These apparently break up and slow the avalanche by dividing it into cross-currents that dissipate its kinetic energy (LaChapelle 1966: 96).

The other major approach to avalanche mitigation is active modification of the snow itself. The oldest, and perhaps still the most successful, method of this type is the artificial triggering of avalanches, typically with explosives. This is generally done by using charges placed directly on the slope or through artillery fire (Martinelli 1972; Perla 1978; McClung and Schaerer 2006). The use of explosives allows the release

FIGURE 4.26 Avalanche fences in the snow accumulation zone above Wengen, Switzerland. Such structures tend to retain the snow and stabilize the slopes. (Photo by L. R. Dexter.)

FIGURE 4.27 A wedge built into the back of a church near Davos, Switzerland, splits avalanche flow around the building, thereby protecting it. (Photo by K. Birkeland.)

of avalanches from a safe distance, and allows avalanche workers to trigger avalanches when no people are present. The traditional artillery weapons are recoilless rifles and howitzers, since they have good range and accuracy. These weapons are extremely useful for avalanche work in difficult-to-access areas; they are in use across North America at ski areas and are especially important for keeping many mountain highways open. Due to the cost and complexity of using traditional military artillery, some of these pieces are being replaced by lighter-weight and cheaper methods, such as the gas-launched *Avalauncher*. Another innovation that is gaining acceptance in North America is the use of fixed, reuseable explosive devices like the European *GazEx* units. However, no alternative has been developed that provides the accuracy, reliability, and flexibility of military artillery for difficult-to-access starting zones.

FIGURE 4.28 Approaching an avalanche shed protecting a highway in the Italian Alps below Simplon Pass. Snow is allowed to cover the structures and avalanches slide harmlessly over the top. Structures like this are very common throughout the Alps. (Photo by L. R. Dexter.)

FIGURE 4.29 Avalanche diversion mounds like these near Chamonix, France, are often placed in runout zones to dissipate the energy of a flowing avalanche. (Photo by K. Birkeland.)

Another method of snow stabilization is simply to pack the snow down, often referred to as *boot packing*. This is used at some ski resorts, where skiers and tracked vehicles are constantly packing the newly fallen snow (Sahn 2010). Packing the snow increases its strength and may help to inhibit weak-layer formation. In addition, the increased variability introduced by boot packing increases the spatial variability of the snowcover, and modeling suggests this may also be helpful for slope stabilization (Kronholm and Birkeland 2005). Research is constantly being done to discover new ways of mitigating avalanches. However, the current mitigation and forecasting techniques have clearly reduced the potential hazards in many areas, allowing their winter use. Mitigating avalanches is expensive, and it will

never be possible to protect all people from all avalanches, especially the dispersed backcountry skiers, snowboarders, and snowmobilers who play in steep, avalanche-prone terrain. In the long run, the best defense is carefully locating facilities to avoid avalanche terrain, and carefully timing activities to avoid times of high avalanche danger.

Glaciers

A *glacier* is a mass of relatively slow-moving ice created by the accumulation of snow. The transformation of snow into ice is basically a continuation of the processes of snow metamorphism already discussed. These processes are accomplished by sublimation, melting, refreezing, and compaction of the ice grains. Sublimation, melting, and refreezing are most important when the snow is still near the surface in the *active layer;* compaction becomes more important after the snow has been buried under successive annual accumulations (Martini et al. 2001; Benn and Evans 2010; Barry and Gan 2011). First, newly fallen snow turns into pea-sized melt-freeze polycrystals of ice at the end of the season (corn snow). The corn snow then becomes *firn* (also called *névé*) as it survives from one year to the next. As the snow is compressed further, the air spaces between the particles are diminished and eventually close off to become bubbles. Once this stage is reached, the snow has become *glacial ice.* The difference between firn and glacial ice is not always clearly marked, but they can usually be differentiated by the color and density of the material. If there are air spaces between the ice crystals and the ice has a whitish color when viewed in mass, it is firn. If the material has a massive structure with no air spaces between the ice crystals, and a vitreous appearance reflecting and transmitting a blue or greenish color (because of the absorption of red wavelengths), it is glacial ice and has attained densities between 0.700 and 0.914 (Seligman 1936: 118; Paterson 1994; Zwally and Li 2002).

Types of Glaciers

The full classification of glaciers, which includes both *alpine* (mountain) and *continental* types, is quite involved, and we will consider only the most basic forms of purely mountain glaciers here (Martini et al. 2001; Benn and Evans 2010). Mountain or alpine glaciers range from small *cirque glaciers* occupying isolated depressions on mountain slopes (Fig. 4.30) to major *ice caps* or *icefields* covering all but the highest peaks.

Cirque glaciers are found at almost all latitudes, while ice-cap glaciers are typically found in subpolar and polar areas. Intermediate between these is the *valley glacier,* which heads in an accumulation basin (the cirque) and extends downvalley for some distance (e.g., the main glaciers shown in Figs. 4.32 and 4.38). Where the ice is sufficient to flow through the valley and accumulate at the base of the mountains, it may spread out upon reaching the flats to form a spatulate tongue. This is a *piedmont glacier,* good examples being the Malaspina Glacier, Alaska, and Skeiðarárjökull, Iceland.

The various forms of alpine glaciers result from both topography and climate. A glacier cannot develop if the slopes are too steep, since the snow cannot accumulate, even if climatic conditions are favorable. At the opposite extreme, it is unlikely that a glacier would develop on an exposed level upland of limited size, because of wind and sun exposure. Topography can be viewed as the initial mold into which the snow and ice must fit, while climate determines at what level and to what extent glaciers develop in any given topographic situation. In the simplest terms, all that is required for a glacier to form is for more snow to fall

FIGURE 4.30 Several small cirque glaciers along a northeast-facing ridge in the central Sierra Nevada, California. The photo was taken in late September 1972, and the snowline (firn line) shows up between the bright white tone of firn and the darker gray tone of glacial ice. The lobate deposits represent very recent morainal material; the bare rock further downslope owes its exposure to strong ice scouring in the past when glacial ice coverage was more extensive. (Austin Post, U.S. Geological Survey.)

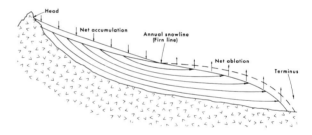

FIGURE 4.31 Longitudinal section of a typical valley glacier showing areas of accumulation and ablation separated by the annual snowline (or equilibrium line). Long arrows within the glacier represent flow streamlines. (After Sharp 1960: 9 and Flint 1971: 36.)

than melts. This may be accomplished by combinations of various environmental factors. Consider the differences in energy flux and temperature versus precipitation regimes in mountains at various latitudes. Midlatitude mountains receive heavy amounts of snow, but summers are relatively warm, resulting in quick melting and relatively rapid hydrologic turnover within the system. By contrast, polar mountains receive so little precipitation that the contribution of rime and hoarfrost is often greater than that of snow. At the same time, there is little or no melting. *Calving-off* of icebergs when glaciers move into the sea is the principal method of depletion. At the other extreme, tropical mountains often display a curious situation: The lower part of the glacier receives more precipitation than the higher part (owing to the zone of maximum precipitation), and melting

FIGURE 4.32 Small valley glacier located in the North Cascades, Washington. The photo was taken near the end of the summer (9 September 1966) and the firn line is evident midway up the glacier. The very lightest-toned snow had fallen within the previous few days. (Austin Post, U.S. Geological Survey.)

may take place every day of the year rather than just during summer. Consequently, tropical glaciers are typically quite short. There are also major differences in environmental conditions through time.

Glacial Climatic Response and Mass Balance

Today's glaciers are only a vestige of what existed during the height of the *Pleistocene epoch;* nevertheless, active mountain glaciers still occur in all latitudes. The Pleistocene is the most recent of several "ice ages" affecting the Earth; it represents approximately 2.6 million years of major fluctuation in environmental conditions which "ended" about 12,000 years ago (Riccardi 2009). Early researchers found that four major ice advances had occurred in the Pleistocene (Martini et al. 2001). More recent studies at finer temporal resolution have identified many more advances and retreats during that epoch (Benn and Evans 2010; Barry and Gan 2011). During glacial advances, continental glaciers developed in the high latitudes, then advanced equator-ward into the middle latitudes of the continents, while mountain glaciers advanced downslope from highlands and spread into the surrounding lowlands. It is generally felt that each major glacial advance

coincided with a period of lower temperatures, but the exact requirements for glacial growth are complex and may vary for different regions. Glaciers are created not simply by the lowering of temperature but by the interplay of different climatic factors (Ageta and Higuchi 1984; Bitz and Battisti 1999; Kaser and Osmaston 2002, Benn and Evans 2010; Barry and Gan 2011). Nevertheless, temperature remains the crucial factor, and there is evidence that temperatures were several degrees lower during the height of the ice ages. Once formed, a glacier responds to and reflects changing climatic conditions.

A glacier's "state of health" from a climatic point of view can be determined by analysis of its *mass balance* (Ahlmann 1948, Paterson 1994). Whether a glacier grows, retreats, or maintains itself depends on its mass balance, or budget. This is determined by total snow *accumulation* as opposed to what is lost through *ablation* (i.e., melting, calving, evaporation, and sublimation). A simple indicator of glacial status on a year-to-year basis is the location of the annual snowline (as discussed earlier in this chapter), or *firn limit,* which represents the maximum extent of summer melting (Fig. 4.31). Since firn is snow at least one year old, the firn limit is the zone dividing one year's snow from that of the previous year (or, in some cases, fresh snow from

FIGURE 4.33 Muir Glacier photographed in 1941. (Photo by W. Field. Courtesy National Snow and Ice Data Center.)

FIGURE 4.34 Muir Glacier photographed in 2004. (Photo by B. Molina. Courtesy National Snow and Ice Data Center.)

glacial ice). The firn limit on a glacier is generally quite distinct near the end of the summer, and can be identified by field examination or from aerial photographs (Fig. 4.32).

A similar but more sophisticated approach to the study of glacial mass balance is the use of the *equilibrium line* (or *equilibrium line altitude [ELA]*). Since the ELA is calculated from measurements of snow density, water equivalent, ablation loss, and other internal qualities, it does not always coincide with the firn limit. The ELA marks the zone on the glacier where the mass of the glacier stays approximately the same during the year. The area above the equilibrium line receives an excess of winter snow, resulting in increased mass. The area below the equilibrium line loses more to ablation than is gained by accumulation, resulting in decreased mass. If the mass gain above the line equals the mass loss below the line, the glacier is in a (rare) steady-state condition. If the mass gain above the equilibrium line exceeds the mass loss below the line, the result will be mass transfer to lower levels by glacial thickening of the ice and advance of the toe, whereas if mass loss exceeds accumulation, the glacier will shrink. Positive or negative trends will only become apparent over a period of several years (Posamentier 1977; Braithwaite 2002; Dyurgerov 2002; Barry and Gan 2011). In all cases, mass transfer across the equilibrium line occurs whether the glacier is retreating or advancing, except at the extreme of final decay, when the entire ice mass may become stagnant.

There are several problems involved in conducting formal glacial mass balance studies, as discussed by Meier

(1962). In spite of these problems, numerous mass balance observations have been conducted in many mountain and subpolar locations (Østrem and Brugman 1991; Braithwaite 2002; Dyurgerov 2002; Kaser et al. 2003; Hubbard and Glasser 2005). The National Snow and Ice Data Center (NSIDC) reports, ". . . about 70 percent of the observations come from the mountains of Europe, North America and the former Soviet Union. Mass balance on more than 280 glaciers has been measured at one time or another since 1946" (NSIDC 2008). The most striking fact about glacial mass balance behavior in the last century has been a widespread glacial retreat (Figs. 4.33, 4.34, and 4.35). There have been short cooling periods with glacial advance, as in the 1920s (Hoinkes 1968) and from the 1940s through the 1960s (Meier 1965: 803), but since the late 1900s and early 2000s, global temperatures have continued to rise, and most glaciers continue to shrink (Charlesworth 1957; Flint 1971; Leggett 1990; Dyurgerov and Meier 1997a, 1997b; Haeberli et al. 1998; Ding et al. 2006; Barry and Gan 2011). The NSIDC report continues, ". . . we only have a continuous record from about 40 glaciers since the early 1960s. These results indicate that, in most regions of the world, glaciers are shrinking in mass. For the period 1961–2003 'small' glaciers lost approximately 7 meters in thickness, or the equivalent of more than 4,000 cubic kilometers of water. The Global Glacier Mass Balance graph [Fig. 4.35] contains data for average global mass balance for each year from 1961 to 2003 as well as the plot of the cumulative change in mass balance, expressed in cubic kilometers of water, for this period" (NSIDC 2008). Especially alarming is the rate of loss of tropical glaciers (Thompson 2001; Thompson et al. 2003).

Climatic variations and glacial fluctuations such as this are the norm rather than the exception over long periods of time. It was formerly thought that the major ice age advances (stadials) had each lasted about 100,000 years, separated by somewhat longer interglacials (interstadials) with more stable warmer and drier climates. However, it is now believed that there were more frequent stadial and interstadial periods, each lasting only 10,000 to 30,000 years (Emiliani 1972; Woillard 1978). The last major stadial ended about 10,000–12,000 years ago, and we are now in an interglacial period. Shorter-term climatic fluctuations have continued to occur, superimposed on these larger trends. There is abundant evidence for glacial oscillations on time scales of hundreds to thousands of years (Benn and Evans 2010; Barry and Gan 2011) and, in some cases, even decades (Nolan 2003). For example, the final melting of the continental ice was followed by a distinctly warm and dry period, known as the Thermal Maximum or hypsithermal, which lasted from 4,000 to 10,000 years ago (Deevey and Flint 1957; Yoshino 2005). The next major change was a widespread readvance of mountain glaciers 2,000 to 4,000 years ago (Denton and Karlen 1973). Subsequent climatic fluctuations, most notably a warming trend about 1,200 to 800 years ago, were followed by a period of glacial

advance known as the Little Ice Age that occurred between the sixteenth and nineteenth centuries (Grove 1988, 2001; Matthews and Briffa 2005; Solomina et al. 2008). Glacial advance during this period did considerable damage to farmland and villages in the Alps and the mountains of Norway. In Iceland, the situation was compounded by a series of major volcanic eruptions, so that total evacuation of the entire population was seriously considered by the Danish king (Ives 2007). Dutch Masters paintings show people skating on the often frozen canals of Holland (Grove 1972; Messerli et al. 1978), and there are reports of ice parties on the Thames River in England (Ives 2007).

The period of modern glacial retreat witnessed during the twentieth and twenty-first centuries apparently reflects warming and amelioration of conditions following the Little Ice Age. While these temporal generalizations apply to most high mountains on the broad scale, recent evidence reinforces the idea that mountain glaciation is often asynchronous in different mountain ranges, even though they may be relatively close neighbors (Gillespie and Molnar 1995; Benn and Owen 1998; Thompson et al. 2005). It is, of course, too soon to know where this will end. Is it just another small deviation from the norm, or are we in fact near the end of the interglacial period and on the verge of another ice age? If the Milankovitch cycles (long-term variations in the Earth's orbit) are responsible for ice age/interglacial variations over the last few million years, we should expect to return to glacial conditions over tens of thousands of years, in spite of the current concern over global warming (Imbrie et al. 1992, 1993; Raymo and Nisancioglu 2003; Jouzel et al. 2007; Bintanja and van de Wal 2008).

Glacier Thermodynamics and Hydrology

Glaciers can be classified based on thermal conditions at the surface and at the base of the ice. Polar glaciers (or cold-based glaciers) remain well frozen throughout, are frozen to the rock of their beds, and move very slowly. Temperate glaciers (or warm-based glaciers) remain at the pressure melting point, are warm based, except for the winter freezing of a thin upper layer in many instances, and hence are not frozen to their bed. Temperate glaciers are, therefore, able to slide over their bed and move more rapidly. Polythermal glaciers are the intermediate condition; these are composed of both warm-based and cold-based sections. Polythermal glaciers can be further subdivided into letter designations—a through f—depending on the relative distribution of cold and warm ice within (Benn and Evans 2010). While this classification carries important implications for the behavior of glaciers in different climates, it is simplistic. Often, a single glacier that flows through a large altitudinal range may display multiple thermal conditions, and it is important to avoid lumping entire glaciers into a single thermal classification (Boulton 1972; Sugden 1977; Denton and Hughes 1981; Robin 1976; Blatter and Hutter 1991; Paterson 1994; Barry and Gan 2011). The portions of glaciers at the pressure melting point can

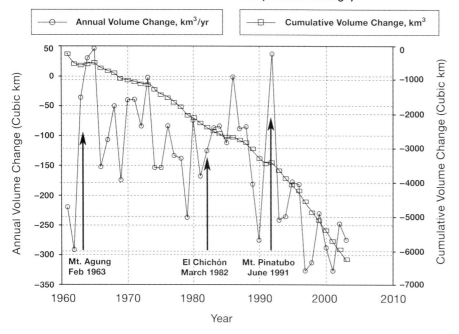

Global Glacier Mass Balance (Volume Change)

FIGURE 4.35 A graph showing glacier retreat based on many mass balance studies. (National Snow and Ice Data Center data. Redrawn courtesy of L. R. Dexter and K. Birkeland.)

cause the development of complex internal meltwater routing systems that take surface meltwater and feed it through a network of *englacial* and *basal* tunnels (Ilken and Bindschadler 1986). Ultimately, a large volume of water emerges as an outlet stream at the *snout* of the glacier. Such glacier rivers are augmented in summer by rainfall that percolates throughout the glacier. Since the flow of surface water into these glacial plumbing systems is strongly dependent upon temperature and radiation, the discharge from the outlet stream exhibits considerable diurnal variation. Maximum melt occurs well after the daily maximum temperature, and minimum melt occurs after sunrise, following the typical diurnal temperature swing (Gerrard 1990). Given the lag time in response through the internal channels of the ice, similar discharge fluctuations propagate downstream with a significant time lag. Thus, incautious hikers who wade glacier rivers in the early morning may be faced with a serious problem when attempting to retrace their steps late in the day.

Glacial Movement

The first recorded observation that glaciers may flow can be attributed to the Icelander Sveinn Pálsson in 1794. In that year he climbed Öræfajökull and, looking down onto one of its outlet glaciers, observed the prominent arcuate bands that have since come to be known as *ogives* (see below). This prompted him to the remarkable observation for the time "that glacier ice, without actually melting, has some kind of fluidity, like several resins" (translated by Williams and Sigurðsson, 2004: 68).

Glacial movement is determined by the thickness of the ice, its temperature, the steepness of its surface slope, the condition at the bed (frozen or thawed), and the

configuration of the underlying and confining topography. When the thickness of the ice exceeds about 60 m (197 ft), internal deformation can occur and the glacier can move (Sharp 1988). In general, the fastest movement occurs in the center of the glacier and decreases toward the margins and frontal zone. Movement is also fastest at the surface and decreases with depth, except where rapid sliding over the bed becomes a significant component of total flow. On a longitudinal basis, movement is greatest near the center at the equilibrium line and least at the head and terminus. The area above the equilibrium line is the zone of accumulation; the area below is the zone of ablation (Fig. 4.31). Therefore, if a glacier is to maintain its form and profile, transfer of mass must be greatest in this zone (Paterson 1994). Movement in the accumulation area is generally greatest in winter, because of increased snow load, while movement in the ablation zone is greatest in the summer, because temperatures are higher and there is more meltwater for lubrication (Ilken and Bindschadler 1986). Ives and King (1955), working on Morsárjökull in southeast Iceland, recorded some of the earliest observations that indicated a significant increase in rate of glacier flow following the onset of heavy rain.

Velocities are highly variable, ranging from a few centimeters to several meters per day. In steep reaches of the glacier, particularly where *icefalls* occur, velocities may be much higher. The highest velocities occur during the so-called *surges* of glaciers, when speeds exceeding 100 m (330 ft) per day may occur for short periods (Sharp 1988). Surging mechanisms are divided into two models: the *hydrological switch model,* where surging is controlled by basal hydrology, and the *thermal switch model,* where surging is controlled by basal temperature (Murray et al. 2003). This still little-understood phenomenon has received increasing attention within the last several decades; it is not fully understood, for instance, why some glaciers surge while neighboring glaciers do not (Meier 1969).

Glaciers are thought to move by one or two basic mechanisms (internal deformation and basal sliding), depending on whether the portion of the glacier being considered is frozen to its bed (cold based) or at the pressure melting point (warm based). Several theories have been developed over the years to describe the internal deformation of ice. It is currently believed that ice behaves as a *viscoplastic* or *pseudoplastic* polycrystalline solid where internal deformation occurs because of flow or creep, as in the creep of metals. This idea has been mathematically formalized in *Glen's Flow Law* (Glen 1955; Paterson 1994; Hooke 2005). Ice, of course, is much weaker than most crystalline solids. It deforms easily through the action of gravity, producing shear stress on its mass and causing *intragranular yielding*, in which the ice crystals yield to shear stress by gliding over one another along basal planes within the lattices of the ice crystals. The individual ice crystals should become internally elongated, but since no such crystal deformation is found in glaciers, a progressive recrystallization apparently accompanies the deformation (Sharp 1988: 46; Hooke 2005). The primary factors controlling the rate of internal deformation are the depth of the ice, the surface slope of the glacier, and ice temperature. The steepness of the bedrock slope beneath the ice is less important, since plastic flow may continue, even where there are bedrock depressions and obstacles (Paterson 1994; Barry and Gan 2011).

The other major mechanism involved in glacier movement is that of *basal sliding*, which involves the slippage of ice en masse over the rock surface at its base. The abrasions and striations left on bedrock across which glaciers have moved are evidence for this kind of movement. The processes involved are even less well understood than those of viscoplastic flow, since the base of a glacier is inaccessible to direct observation except in rare cases. The important controls on basal sliding are the temperature of the ice at the base and the presence of water to serve as a lubricant (Sharp 1988). Basal sliding does not generally occur in polar glaciers, since the ice is frozen to the underlying rock surface. In other regions, the temperature of the glacial ice is higher, and water may be present along the base. This is part of the explanation for why polar, or cold, glaciers move rather slowly, except where they calve into a lake or the sea.

Water may also be released when ice reaches the pressure melting point. This can happen when an obstacle is encountered during glacial movement: The ice is compressed on the upstream side of the obstruction, and the increased pressure causes melting. The meltwater then flows around the obstacle and refreezes to the downstream side where the pressure is less (referred to as *regelation*). The process is maintained by the *latent heat of fusion* (given off upon refreezing), which is transmitted by conduction from the freezing area to the melting area, where it helps maintain the melting. Regelation operates only on small

FIGURE 4.36 A crevasse on Collier Glacier, Three Sisters Wilderness, Oregon Cascades (downslope is to the left). The rocky debris has fallen onto the ice from a nearby projecting ridge. (Photo by L. W. Price.)

obstacles 1–2 m (3–7 ft) in length, however, because the heat cannot be effectively transmitted through larger features. On larger obstacles, the ice undergoes greater deformation and movement, probably due mainly to plastic flow, since the ice immediately next to the obstacle must travel farther and faster in order to keep up with the surrounding glacier mass. The larger the obstacle, the more rapid the ice deformation and movement near the bedrock interface (Weertman 1957, 1964; Lliboutry 1968, 1987, 1993). The processes involved in glacial flow are still a topic of active research (Martini et al. 2001; Hooke 2005).

Structures within Glacial Ice

Glaciers contain a number of interesting features resulting from the transformation of snow to ice and from downslope movement. Most of these are beyond our present scope, but three of them—crevasses, ogives, and moraines—require mention. A *crevasse* is a crack in the ice that may range up to 15 m (50 ft) in width, 35 m (115 ft) in depth, and several tens to hundreds of meters in length. Most are smaller than this, especially in temperate mountain glaciers, where the average crevasse is only 1–2 m (3–7 ft) wide and 5–10 m (16–33 ft) deep (Fig. 4.36). Crevasses are among the first structural

FIGURE 4.37 Ogives in East Twin Glacier, Alaska. (Austin Post photo courtesy of the Geophysical Institute GeoData Center, University of Alaska, Fairbanks.)

features to appear on a glacier and may develop anywhere from the head to the terminus. Crevasse formation is primarily a response to tensional stress, so their distribution, size, and arrangement provide useful information on the flow behavior of the ice (Sharp 1988; Benn and Evans 2010). Crevasses occur most often where the middle and the sides of the glacier move at different rates, or where the ice curves around a bend, or where the slope steepens and the rate of movement increases (Fig. 4.33). Crevasses are most often transverse to the direction of flow, but they can be oriented in any direction. Radial crevasses often occur where a glacier extends beyond its valley walls to spread out as a piedmont lobe. They are also largely restricted to the surface, where the ice is more brittle and fractures easily; the greater pressure at depth results in closure by plastic flow. A special type of crevasse develops at the upslope end of the glacier where the ice pulls away from the rocky headwall. This is known as the *bergschrund*. Rock debris from the headwall and valley sides falls into the bergschrund and other crevasses and becomes incorporated into the glacier, often not to be seen again until it is released by glacial melting at the terminus. The presence of crevasses, therefore, increases the efficiency of rock transport. Crevasses also hasten ablation by increasing the glacier's surface area, by the pooling of meltwater, and by disaggregating the ice near the terminus. Crevasses pose great danger to travel across glaciers. This is particularly true after fresh snow has bridged the surface, hiding the underlying chasms from view.

Another interesting type of ice structure is the *ogive*. Ogives are arcuate structures in the ice bowing down-glacier (Fig. 4.37). Two types of ogives are recognized: *band ogives* (alternate light and dark coloration of the ice) and *wave ogives* (alternate crest and trough structures within the ice) (Benn and Evans 2010; Godsell et al. 2002). Ogives usually form downslope from an icefall. They are thought to represent the annual flow of ice through the icefall: the dark portion of the band from the summer, when debris falls or is blown onto the glacier surface; and the light portion from the winter, when the icefall is mainly snow-covered. Some of the earliest systematic observations on ogives were made on Morsárjökull, southeast Iceland, and in the Jotunheimen, Norway (King and Ives 1956; King and Lewis 1961).

One of the most conspicuous surface features of mountain glaciers is the linear accumulation of rocky debris oriented in the direction of flow. Known as *lateral* and *medial moraines,* these accumulations result from rocks that have fallen onto the ice, ablated from the edges of the ice, and accumulated from the debris input of tributary glaciers (Benn and Evans 2010). When a smaller ice stream joins a larger glacier, it usually carries with it a load of rocky debris along its edges (lateral moraine) that becomes incorporated into the ice as a vertical partition between the two ice masses. The material then becomes a medial moraine on the main glacier (Fig. 4.38). What we see is only the surface expression of the rock debris, which extends into the ice, frequently all the way to the bottom (except for material contributed by smaller ice streams that join at shallower levels) (Fig. 4.39). The presence of moraines on the ice alters the mass balance, since the rock material is dark in color and can absorb more of the sun's energy. On the other hand, if the rocky burden is thick enough, it may serve as an insulative cover and inhibit local melting of the underlying ice. This leads to more rapid melting of the ice on either side of the debris, leaving the moraine, often with an ice core, exposed as a higher ridge. As the ridge builds through differential melting, some of the rocky material may slide or tumble onto the ice; in this way the moraine is widened and the underlying ice is again exposed to melting. The moraines gradually widen toward the terminus, eventually ending up as a jumble of rock debris covering the terminus of the glacier (called an *ablation moraine*). If the glacier is retreating, the underlying ice may melt, leaving the rocks lying about in heaps. On the other hand, isolated masses of ice may be preserved indefinitely under the debris as ice-cored moraine. Such moraines can easily be confused with moraine-covered snow accumulations. Snow can accumulate along the margins of glaciers and be buried by morainal material falling off the glacier front. In time, the buried snow will metamorphose to ice. This ice can be distinguished from glacier ice because it is composed of very small crystals, in contrast to the large crystals of the latter. This understanding was pioneered by Gunnar Østrem (1962, 1964b) based on research in Lappland and Baffin Island. The glacier terminus is essentially the end of the journey for the larger rock material. The finer debris,

FIGURE 4.38 Moraines in the Barnard Glacier, Saint Elias Mountains, southeastern Alaska, 1984. Those at the very edges of the glacier are lateral moraines. Those in between are all medial moraines and represent the confluence of two lateral moraines somewhere up-ice, as can be seen in several places in the photo. (Austin Post photo courtesy of the Geophysical Institute GeoData Center, University of Alaska, Fairbanks.)

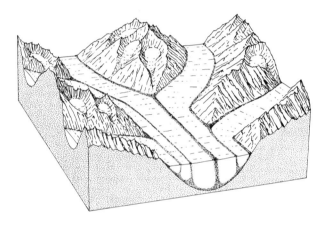

FIGURE 4.39 Idealized cross section of a valley glacier showing the relationship between lateral and medial moraines along with their subsurface extension into the glacier. Note that the moraines from the small tributary glacier on the right maintain themselves at the depth at which they join the main glacier. The photo in Figure 38 shows actual field example of this moraine–glacier relationship. (T. M. Oberlander, University of California.)

however, can be transported farther through the action of glacial melt streams and wind.

Glaciers as Landscape-Forming Tools

MECHANISMS OF GLACIAL EROSION

When a glacier moves over an area, the ice undergoes plastic deformation to fill every nook and cranny. Movement at the ice–rock interface results in modification of the underlying surface through glacial erosion and transport. The primary processes are abrasion, crushing, and plucking or quarrying. *Abrasion* is the scratching, gouging, and

grooving of the surface as the ice, carrying rock particles as tools, moves across it. Obviously, this is most effective when the rocks in the ice are harder than the surface over which they are passing. Pure ice or ice containing softer rocks is relatively ineffective at abrasion, although it may produce smoothed and shiny surfaces (*glacial polish*). A bedload of fine material will result in tiny scratches and smoothing of surfaces, while large embedded rocks can produce scratches several centimeters deep. Such *striations* are found in greatest abundance on gently inclined terrain where the ice was forced to ascend, since this places greater pressure on the glacial base. Striations provide good evidence for the direction of glacial movement; some caution should be used in their interpretation, however, since they can be caused by other processes, such as avalanches and mass movement, as well as lake- and sea-ice pushed ridges. *Crushing* is the pulverization of rock because of the glacial mass above. *Plucking* or *quarrying* is generally considered to be the most potent erosional tool of glaciers. Plucking involves regelation and the lifting and incorporation of ground surface rubble and bedrock segments into the moving ice. Plucking is aided in its work by crushing and frost weathering, which operates in front of the glacier, producing frost-shattered rock with many cracks and crevices. As the ice moves, it easily incorporates the loose material, and the ice undergoes plastic deformation around the larger rocks until they too are swept along with the mass. This debris becomes part of the glacier's bedload and serves as a tool for abrasion. Plucking also operates when ice reaches the pressure melting point on the upstream side of obstacles and the water moves downslope and refreezes (regelation) in cracks in the bedrock, creating a bond between the glacial ice and the rock; the continued movement of the ice plucks the individual segments from the bedrock. This process gives an asymmetric profile to the underlying obstacles: The *stoss* (upstream) side is smoothed and gentle, while the *lee* (downstream) side becomes steep and irregular, owing to the quarrying which has taken place. Such features (called *roche moutonées*) provide excellent evidence for the direction of glacial movement.

The landscape that extends above the glacial ice is a product of both frost shattering and glacial erosion (Russell 1933). Frost-shattered rocks eventually tumble onto the ice surface for further transport. Glacial erosion takes place constantly as well. A glacier can be thought of as a huge malleable mass completely smothering the surface and picking up loose rock and soil as it moves along. In this way, a new surface is continually being exposed to the erosive power of the ice. The load a glacier can carry is almost unlimited; a large glacier can easily transport rocks as big as a house.

MECHANISMS OF GLACIAL TRANSPORT

Rock material incorporated into the flow of glacial ice can be transported in one of three modes: *supraglacial* (on top of the ice surface), *englacial* (within the glacial ice), or *basal* (at the bottom of the glacier). The most important notion to keep in mind about glacially transported and deposited sediment is that ice can carry any size particle anywhere in its flow, including huge boulders right on the surface! In addition to material carried directly by the ice, sediment can be transported by meltwater through the complex plumbing system within the glacier. Sediment carried in this way is subject to the same hydraulics as sediment in rivers, and hence displays different characteristics from sediments laid down directly from the melting ice. The most pronounced difference is the sorted nature of the *glaciofluvial* sediments compared with the unsorted nature of the ice-laid deposits (Benn and Evans 2010).

Glaciers generally continuously bury surface rock material under deepening layers of snow and ice in the accumulation zone, and expose melted-out rock material in the ablation zone. All of this takes place while glacier ice is moving from the accumulation zone to the ablation zone. The net result is a set of curved *flow streamlines* that are nested from the surface at the ELA to the sole of the glacier at the head and toe of the glacier (Fig. 4.31; Sharp 1988; Benn and Evans 2010). This has the practical effect of taking rocks falling onto the glacier surface near the head on a long trip deep into the glacier at or near the sole of the ice before releasing them from their icy surroundings at the surface near the snout. On the other hand, rocks falling onto the surface of the glacier just above the ELA are taken on a short, shallow ride into the glacier before reemerging just downslope of the ELA. A dramatic demonstration of this effect can be found in the disappearance and subsequent discovery of the infamous missing airliner named *Stardust*. *Stardust* vanished without a trace in 1947 during a trans-Andean flight from Buenos Aires, Argentina, to Santiago, Chile. The disappearance was so immediate and complete that the loss of *Stardust* was attributed to a UFO abduction. In reality, the plane had crashed in poor weather at the head of the Tupangato Glacier and became quickly entombed by an avalanche triggered by the impact. Thus *Stardust* vanished and began a long slow ride through the bowels of the glacier following the flow streamlines, only to reemerge in 2000 to be discovered by climbers ascending the nearby peak (NOVA 2001).

MECHANISMS OF GLACIAL DEPOSITION

Using the transport mechanisms described above, glaciers may liberate their load (collectively called *drift*) in one of three ways: (a) by deposition directly from the melting ice (the deposit material is then referred to generically as glacial *till*); (b) by intermediate deposition from meltwater within the ice, then by melting of the ice (referred to as *glaciofluvial* deposits); or (c) by direct deposition from meltwater below the terminus of the ice (referred to as *outwash*).

The importance of knowing these depositional mechanisms is evident when trying to interpret various landforms found in previously glaciated terrain. Only by understanding these processes can anomalous features such as huge boulders of alien rock type littering a landscape (*glacial erratics*) or hundred-foot high sinuous mounds of sorted stream deposits running miles across a landscape (*eskers*) be explained (Benn and Evans 2010).

Glaciated Mountain Landscapes

The landscapes of glaciated mountains are among the most distinctive and striking on Earth. The features and forms created by ice sculpting are very different from those caused by running water, and glaciated mountains possess a ruggedness and grandeur seldom achieved in unglaciated mountains. For most of us, the visual image of high mountains is typified by glaciated landscapes, with their pyramidal peaks, jagged sawtooth ridges, amphitheater-like basins, and deep elongated valleys where occasional jewel-blue lakes sparkle amid surrounding meadows. It is a landscape largely inherited from the past, when the ice was much more extensive than now (see Fig. 4.30 for an example). In the western United States alone, there were over 75 separate high-altitude glacial areas (Fig. 4.40). Cirque or valley glaciers occupied most of these, but in some areas there were mountain ice caps. The largest examples are in the Yellowstone–Grand Teton–Wind River ranges, the Sierra Nevada, the Colorado Rockies, and the Cascades (Flint 1971: 471–474). Mountains farther north (i.e., the Canadian Rockies, the Coast Ranges, and the Alaska and Brooks Ranges) were almost totally inundated, while the Yukon River Valley remained ice free.

The most characteristic and dominant feature of mountain glaciation is erosion. Glacial erosion in mountains is facilitated by the channeling of ice into preexisting valleys which accentuate its depth and velocity. For this reason, glaciers erode deeper in mountain areas than the former ice sheets did in continental areas, with erosive depths often exceeding 600 m (2,000 ft) (Flint 1971: 114). There is a sharp contrast between the appearance of glaciated uplands and valleys. The ice on upper surfaces is thinner and prone to earlier melting than that in the valleys, where the ice is deeper and more sheltered. The higher surfaces are thus exposed to prolonged weathering. Typical features include sharp, angular ridges and peaks, and accumulations of frost-loosened rock. By contrast, the valleys (*glacial troughs*) are so smoothed and shaped by the ice that very few sharp or rugged features remain. An exception occurs where the entire upland surface has been overrun by ice so that both upland and valley are generally smoothed by the ice. The Scottish Highlands and the Presidential Range in New Hampshire are examples (Goldthwait 1970). Features of deposition, moraines and glaciofluvial debris are largely restricted to the lower elevations, and generally mark the point of maximum extent of the ice or places where the

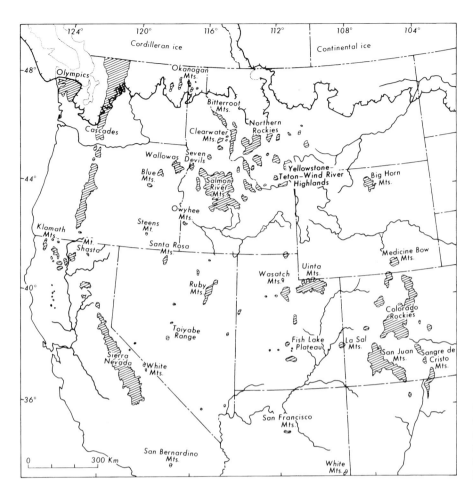

FIGURE 4.40 Generalized areas of mountain glaciation in the western coterminous United States. The southern extent of the Laurentide continental ice sheet is also shown. (Adapted from Flint 1971: 475.)

glacier remained for the longest periods, or where it re-advanced slightly as it receded.

FEATURES RESULTING FROM GLACIAL EROSION

The growth and decline of mountain glaciers leads to a pre-dictable pattern of landform development (Fig. 4.41).Upon initial accumulation, the snow and ice adapt to the preex-isting topography. If the snow accumulation is sufficient, the mountains may be totally submerged beneath glacial ice. When the landscape is submerged beneath a very thin mantle of ice, it is actually somewhat protected, since tem-peratures under the ice remain below the freezing level and surfaces are not subject to intense frost shattering. The rug-ged topography revealed in Figures 4.41b and c, however, is believed to develop under a partial ice cover or in a situa-tion where the higher areas are mantled by thin ice frozen to its bed; in the valleys, the base of the thicker ice remains at the pressure melting point and flows rapidly, causing considerable erosion (Sugden and John 1976; Sugden et al. 2005). Frost and mass-wasting processes attack the exposed surfaces while glaciers occupy the valleys and slope depressions to carve, deepen, and sculpt the topogra-phy into a distinctive landscape (Cotton 1942, 1958; Flint 1971; Embleton and King 1975; Phillips et al. 2006). The

dominant features of this type of landscape are *cirques, gla-cial troughs, horns, arêtes* (sawtooth ridges), *tarns,* and *pater-noster lakes* (strings of lakes in rocky basins), and *hanging valleys.*

"Few landforms have caught the imagination of geomor-phologists more than the glacial cirque (corrie)" (Sugden and John 1976; Benn and Evans 2010). A *cirque* is a semicir-cular, bowl-like depression carved into the side of a moun-tain where a small glacier has existed (see Figs. 4.30 and 4.42 for examples). Cirques are typically located at the heads of valleys, but may develop anywhere along a mountain slope. They vary in size from shallow basins a few meters in diameter to the huge excavations several kilometers deep and wide found from Antarctica to the Himalaya. A well-developed cirque usually contains a headwall, a basin, and a threshold. The *headwall* is the steep and smoothed bed-rock surface at the back of the cirque, extending concavely upward to the ridge. The *basin* is a circular or elongated depression at the base of the headwall and is frequently overdeepened as the result of glacial erosion. It rises to a *threshold,* a lip or slightly elevated rampart at the outlet end of the basin. The threshold, composed of either bedrock or depositional material, results from the decreasing gla-cial mass and rate of movement at the periphery, so that the intensity of erosion is less and deposition occurs. If the

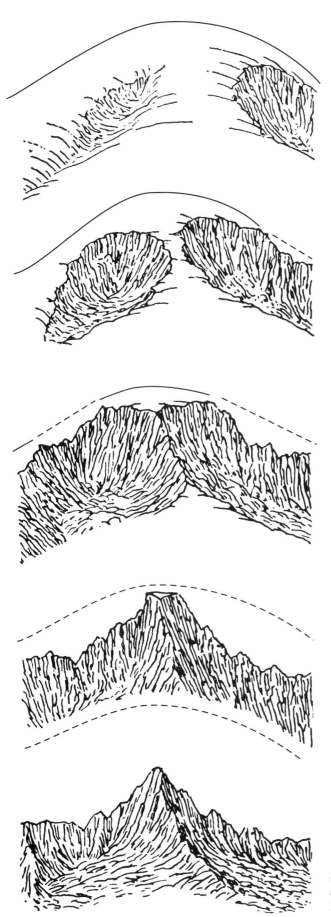

FIGURE 4.43 Sequence of events showing headward erosion by cirque glaciers to create steep sawtooth ridges (arêtes) and glacial horns. (From Davis 1911; Lobeck 1939; Cotton 1942)

erosion of cirque glaciers on all sides of a summit may result in a pyramidal peak called a *horn*. The Matterhorn in the Alps is the classic example, but such features are common in most glaciated mountains (Figs. 4.43, 4.41c, 4.30). An indication of their prominent position in glaciated mountains is that many languages have their own distinctive term: *tind* (Norwegian); *tindur* (Icelandic); *pike* (English Lake District).

Although cirque erosion is the dominant glacial process operating on upper slopes and depressions, larger glaciers may overflow the cirque basins to form valley glaciers. The ice commonly adapts to and modifies a preexisting drainage system, and the former stream channels are eventually transformed into *glacial troughs* (Figs. 4.41a, b, c). Most stream-cut valleys in mountain regions have roughly V-shaped cross-profiles, whereas a glaciated valley is typically U-shaped (see Fig. 4.42 for an example). Streams are limited to channel cutting along their beds, while other processes (especially *mass wasting*) erode the valley slopes and transport material to the stream. A glacier, on the other hand, occupies the entire valley, and its much greater mass and erosive capacity soon widen and deepen the valley into a semicircular or elliptical cross section with steep rock walls (Hooke 2005). The valley floor may be bare rock, or it may be back-filled with glacial meltwater deposits, resulting in flat valley bottoms, as seen in Yosemite Valley, California. In longitudinal profile, glacial valleys have a more irregular surface than stream valleys and often display a series of *steps* and *risers* (or *reigels*). Various origins have been postulated for the stepped nature of glacial valleys, including differential rates of erosion controlled by valley width, different rock types, more intensely fractured zones within the same rock type, greater erosion occurring at the base of deep crevasses, and association with places where tributary glaciers join the main stream (Thornbury 1969; Flint 1971; Embleton and King 1975; Sugden and John 1976; Anderson et al. 2006). Massive erosion and excavation of material by the ice deepens, widens, and straightens the former stream valley along its axis so that the lower reaches of tributary streams and their interfluves are cut off, leaving them truncated at some height above the main valley. After the glacier melts, the water of these streams cascades down as waterfalls over the trough sidewall. Such tributary valleys with floors higher than the floor of the trunk valley, known as *hanging valleys,* are a scenic feature of glaciated mountains (Fig. 4.41c).

Many glaciated mountain regions are situated close to oceans and experience predominantly onshore winds responsible for their characteristic heavy snowfall. Examples include the west coasts of Norway, Alaska and British Columbia, New Zealand's South Island, southern Chile, and parts of Greenland. These areas display one of the most spectacular of glacial landforms, the *fjord*. Fjords, above all, exemplify the glacial erosional phenomenon of *overdeepening,* that is, deep valley bottom erosion below *base level*. For river systems, ultimate base level is represented by sea level, and fluvial erosion can occur only slightly below this level. Fast-moving glaciers have the ability to erode far below sea level—more than 600 m below in Sognefjord, Norway. The counterparts of coastal fjords in glaciated mountains far from sea coasts are the great trough valleys such as the Lauterbrunnen Valley in the Swiss Bernese Oberland and Yosemite Valley in California's Sierra Nevada.

Because of the rough and irregular terrain left behind by glaciers, lakes are common in such landscapes (see Fig. 4.42 for examples). *Tarns* are lakes commonly found in cirques (also known as *cirque lakes*). Tarns are characteristically clear and blue, since the glacier has removed most of the loose debris, leaving a smoothed bedrock depression. Some tarns and other glacial water bodies may appear "milky" or opaque if they contain large amounts of fine glacier-derived rock powder in suspension (called *glacial flour*), especially if a glacier still exists in the cirque. Lakes frequently form in the depressions behind the treads downvalley from the cirque lakes. These lakes often occur in "chains" along glacial troughs, and are called *paternoster lakes* because of their resemblance to beads on a rosary (Fig. 4.42).

FEATURES RESULTING FROM GLACIAL DEPOSITION

Sooner or later, a glacier must put down the load of earth and rock that it has picked up or that has fallen onto its surface. The landforms created by glacial deposition, less spectacular than the features caused by glacial erosion, are nonetheless distinctive. Most glacial deposition takes place upon melting and retreat of the ice. Moraine material is deposited directly from the ice, while glaciofluvial material is deposited by meltwater streams. Moraines typically consist of large and small particles mixed in an unsorted matrix. They may occur along the sides of the glacier as *lateral moraines,* or around the end of the glacial tongue as a *terminal moraine* or, as it recedes, as *recessional moraines*. In other cases, the moraine may be less distinct, occurring as a jumble of rocky debris like the tailings from a deserted strip mine. Lateral and terminal moraines can be quite impressive (see Fig 4.38, for example), reaching heights of 100–300 m (330–1,000 ft) or more.

The larger rock debris can only be transported directly by the glacier or by *ice rafting* (blocks of ice floating in water), but the smaller material may be carried considerable distances by wind and glacial meltwater streams. The winds that blow off the glacier in summer are often very effective at picking up and transporting the finely ground rock particles produced by grinding and scraping during glacial transport (glacial flour). In some valleys where glaciers exist, the development of such winds is an almost daily occurrence during clear weather in summer. Ecologically, the deposition of this silt (called *loess*) is beneficial, as it expedites soil development and greatly improves local productivity. Several major agricultural regions have developed on loess soils (e.g., the Palouse region of eastern Washington State).

Glacial melt streams are the main mechanism for transport of the smaller material. The amount a stream can carry depends primarily upon the stream's velocity, which in turn depends, among other factors, upon the volume. Glacial streams, of course, display great fluctuations in flow between winter and summer as well as between day and night. Such volume fluctuations produce an irregular pattern of erosion and redeposition; during periods of high velocity, the stream erodes and carries a large load of material, only to drop it again as the water volume subsides and the velocity decreases (Price 1973). Glacial streams are characteristically choked with sediment, much of which is eventually deposited near the glacial terminus. Such deposits, called *valley train, glacial outwash plain,* or *sandur* (Church 1972; Sugden and John 1976; Martini et al. 2001; Benn and Evans 2010) create flat-floored valleys and may reach considerable depths and extend for several kilometers beyond the glacial terminus. An extreme example is Yosemite Valley of California, where seismic investigations reveal that over 600 m (2,000 ft) of deposits cover the original bedrock floor excavated by the glacier (Gutenberg et al. 1956). Glacial and glaciofluvial deposits are important ecologically because soil and vegetation develop much more rapidly on aggregate material than on bare rock. Such areas frequently become locally important agricultural regions. The annual contribution of glacial meltwater streams to the runoff of many watersheds amounts to millions of liters. On the negative side, glacial streams are commonly so choked with sediment that the water is not immediately usable by people. The sediments can be transported long distances, leading to increased deposition and infilling of the stream or lake into which they empty. The quality of the lake water is affected in several ways, the most obvious of which is that the normal crystal blue of the lake is transformed to a murky gray around the mouth of the stream (Bryan 1974a, 1974b).

Rock Glaciers

In periglacial environments, water can remain perennially frozen inside a matrix of loose talus and scree to form *rock glaciers*. Rock glaciers can flow and deform in much the same manner as ice glaciers, and are they often found as residual forms in the later stages of glacial retreat. The San Juan Mountains of southwestern Colorado serve as a classic example where many valleys formerly filled with glaciers now contain active rock glaciers. Rock glaciers are addressed more completely in Chapter 5.

Glacial Lake Outburst Floods

Glaciers can produce lakes on several scales, from small meltwater ponds sitting directly on the ice surface to very large lakes where major rivers are dammed by adjacent ice-sheet glaciers or by valley glaciers issuing from lateral tributaries. As an example of the latter, the Lowell Glacier

descends from the St. Elias Mountains as a tributary of the Alsek River, a major drainage in the southwestern Yukon Territory, Canada. The Lowell has surged at least five times in the past 3,000 years, extending its length beyond its valley mouth and thus blocking the flow of the Alsek. The resulting glacier-dammed lake (called Neoglacial Lake Alsek) has backed up as far as present-day Haines Junction, 60 km to the northeast. This lake formed as recently as 1850 (Clague and Rampton 1982).

A consequence of lakes forming behind ice dams is the likelihood of failure and subsequent flooding if the dam is overtopped, melts, or otherwise changes configuration, or if the terminal moraine itself fails. Such failures are termed *glacial lake outburst floods* (or *GLOFs*) (Post and Mayo 1971). Closely related failures are *jökulhlaups* (Björnsson 2002; Clarke 1982) from water bodies located under the ice and *marginal lake drainage* from water bodies along the edge of the ice. Regardless of the specific form, glacier outburst floods can be extremely hazardous for people living in the valleys below (Hewitt 1982). While GLOFs can occur in any state of climate change, the risk of such outbursts is on the rise along with global temperature (Ives et al. 2010).

One notable GLOF in recent history occurred in the spring and summer of 2002 when the Hubbard Glacier surged, damming the adjacent Russell Fjord where the two valleys converge on Yakutat Bay along the coast of the Alaskan panhandle. The Russell is normally a coastal estuary with ecosystems adapted to a mixture of salt and fresh waters. As the advancing Hubbard Glacier blocked the Russell's outlet, the proportion of fresh to sea water increased and the water level rose within the fjord, impacting numerous habitats. At one point, water from the Russell almost overtopped a mountain gap leading into the nearby Situk River, a prime salmon fishery. Plans were made to blast the toe of the glacier to drain the fjord and save the Situk. Nature intervened, and the Hubbard Glacier snout was overtopped on 14 August before the Situk could flood. The outpouring at the mouth of the Hubbard produced the second-largest observed GLOF in history (Motyka and Truffer 2007).

The magnitude of glacial outbursts found from prehistoric times is almost unbelievable. In fact, early researchers like J. Harlan Bretz and J. T. Pardee had trouble convincing other scientists of their day that floods of such magnitude could occur. Large marginal lakes had formed adjacent to the ends of the Laurentide and Cordilleran ice-sheet glaciers during the Pleistocene. The Great Lakes of North America are remnants of some of those lakes. To the west of the Great Lakes were other equally impressive glacially dammed lakes like Missoula, Spokane, and Columbia. In a complex series of giant GLOFs, Lake Missoula and others overtopped their glacial dams, sending mammoth floods down the course of the Columbia River and some of its tributaries in present-day eastern Washington. These megafloods eroded thousands of square miles into giant gorges, scour channels, ripples, and other flood features in what is today called *the channeled scablands* (Bretz 1923).

References

Ageta, Y., and Higuchi, K. 1984. Estimation of mass balance components of a summer accumulation type glacier in the Nepal Himalaya. *Geografiska Annaler* 66A: 249–255.

Ahlmann, H. W. 1948. *Glaciological Research on the North Atlantic Coasts*. Research Series 1. London: Royal Geographical Society.

Akitaya, E. 1974. Studies on depth hoar. In Proceedings of the Snow Mechanics Symposium April 1974, Grindewald, Switzerland. IAHS-AISH Publication 114: 42–48.

Akitaya, E. 1980. The orientation gradient: Regional variations of accumulation and ablation in alpine basins. In J. D. Ives, ed., *Geoecology of the Front Range: A Study of Alpine and Subalpine Environments* (pp. 214–223). Boulder, CO: Westview Press.

Alford, D. 1980. Spatial patterns of snow accumulation in the Alpine terrain. *Journal of Glaciology* 26: 517.

Allix, A. 1924. Avalanches. *Geographical Review* 14(4): 519–560.

Anderson, R. S., Molnar, P., and Kessler, M. A.. 2006. Features of glacial valley profiles simply explained. *Journal of Geophysical Research* 111: F01004, doi:10.1029/2005JF000344.

Andrews, J. T. 1975. *Glacial Systems: An Approach to Glaciers and their Environments*. North Scituate, MA: Duxbury Press,.

Atwater, M. M. 1954. Snow avalanches. *Scientific American* 190(1): 26–31.

Aulitzky, H. 1967. Significance of small climatic differences for the proper afforestation of highlands in Austria. In W. E. Sopper and H. W. Lull, eds., *International Symposium on Forest Hydrology* (pp. 639–653). Oxford, UK: Pergamon.

Avery, C. C., Dexter, L. R., Wier, R. R., Delinger, W. G., Tecle, A., and Becker, R. J. 1992. Where has all the snow gone? Snowpack sublimation in northern Arizona. Proceedings of the 61st Western Snow Conference, Jackson, WY (pp. 84–94).

Bader, H. P., Haefeli, R., Bucher, E., Neher, J., Eckel, O., Tharms, C., and Niggle, P. 1939. *Snow and Its Metamorphism*. U.S. Army Corps of Engineers Snow, Ice, and Permafrost Research Establishment Translation 14.

Barry, R. G., and Gan, T. Y. 2011. *The Global Cryosphere: Past, Present and Future*. Cambridge, UK: Cambridge University Press.

Benn, D. I., and Evans. D. J. A. 2010. *Glaciers and Glaciations*. 2nd ed. London: Hodder Education.

Benn, D. I., and Owen, L. A. 1998. The role of the South Asian monsoon and mid-latitude cooling in Himalayan glacial cycles: review and speculative discussion. *Journal of the Geological Society* 155(2): 353–363.

Bentley, W. A., and Johnson, W. J. 1931. *Snow Crystals*. New York: McGraw Hill.

Bintanja, R., and van de Wal, R. S. 2008. North American ice-sheet dynamics and the onset of 100,000 year glacial cycles. *Nature* 454: 869–871.

Birkeland, K. W. 1998. Terminology and predominant processes associated with the formation of weak layers of near-surface faceted crystals in the mountain snowpack. *Arctic and Alpine Research* 30(2): 193–199.

Birkeland, K.W., Hendrikx, J., and Clark, M. 2010. On optimal stability-test spacing for assessing snow avalanche conditions. *Journal of Glaciology* 56(199): 795–804.

Bitz, C. M., and Battisti, D. S. 1999. Interannual to decadal variability in climate and glacier mass balance in Washington, Western Canada, and Alaska. *Journal of Climate* 12: 3181–3196.

Björnsson, H. 2002. Subglacial lakes and jökulhlaups in Iceland. *Global and Planetary Change* 35: 255–271.

Blatter, H., and Hutter, K. 1991. Polythermal conditions in arctic glaciers. *Journal of Glaciology* 41: 333–343.

Boellstorff, J. 1978. North American Pleistocene stages reconsidered in light of probable Pliocene–Pleistocene continental glaciation. *Science* 202(4365): 305–307.

Boulton, G. S. 1972. The role of the thermal regime in glacial sedimentation. In R. J. Price and D. E. Sugden, eds., *Polar Geomorphology* (pp. 1–19). Institute of British Geographers, Special Publication 4.

Braitwaithe, R. J. 2002. Glacier mass balance: The first 50 years of international monitoring. *Progress in Physical Geography* 26: 76–95.

Bretz, J. H. 1923. The channeled scablands of the Columbia Plateau. *Journal of Geology* 31(8): 617–649.

Brozovic, N., Burbank, D. W., and Meigs, A. J. 1997. Climatic limits on landscape development in the northwestern Himalaya. *Science* 276: 571–574.

Bruintjes, R. T. 1999. A review of cloud seeding experiments to enhance precipitation and some new prospects. *Bulletin of the American Meteorological Society* 90(5): 805–820.

Bryan, M. L. 1974a. Water masses in Southern Kluane Lake. In V. C. Bushnell and M. G. Marcus, eds., *Icefields Ranges Research Project Scientific Results,* Vol. 4 (pp. 163–169). Washington, DC: American Geographical Society and Arctic Institute of North America.

Bryan, M. L. 1974b. Sublacustrine morphology and deposition, Kluane Lake. In V. C. Bushnell and M. G. Marcus, eds., *Icefields Ranges Research Project Scientific Sesults,* Vol. 4 (pp. 171–188). Washington, DC: American Geographical Society and Arctic Institute of North America.

CAIC. 2012. *Avalanche Accident Statistics (U.S.)*. Denver, CO: Avalanche Information Center. <http://avalanche.state.co.us/acc/acc_stats.php>.

Charlesworth, J. K. 1957. *The Quaternary Era*. London: Edward Arnold.

Church, M. 1972. *Baffin Island Sandurs: A Study of Arctic fluvial Processes*. Bulletin 216. Ottawa, ON: Geological Survey of Canada.

Clague, J. J., and Rampton, V. N. 1982. Neoglacial Lake Alsek. *Canadian Journal of Earth Sciences*. 19 (1): 94–117.

Clarke, G. K. C. 1982. Glacier outburst floods from Hazard Lake, Yukon Territory, and the problem of flood magnitude prediction. *Journal of Glaciology* 28(98): 3–21.

Colbeck, S. C. 1982. An overview of seasonal snow metamorphism. *Review of Geophysics* 20(1): 45–61, doi:10.1029/RG020i001p00045.

Colbeck, S. C. 1983. Snow particle morphology in the seasonal snowcover. *Bulletin of the American Meteorological Society* 64(6): 602–609.

Cotton, C. A. 1942. *Climatic Accidents in Landscape Making*. New York: John Wiley and Sons.

Cotton, C. A. 1958. *Geomorphology: An Introduction to the Study of Landforms*. 7th ed. London: Whitcombe and Tombs.

Davis, R. T. 1965. *Snow Surveys*. Agriculture Information Bulletin 302. Washington, DC: USDA Soil Conservation Service.

Davis, W. M. 1911. The Colorado Front Range: A study in physiographic presentations. *Annals of the Association of American Geographers* 1(1): 21–83.

Deevey, E. S., and Flint, R. F. 1957. Postglacial hypsithermal interval. *Science* 125: 182–184.

Denton, G. H., and Hughes, T. J. 1981. *The Last Great Ice Sheets*. New York: J. W. Wiley and Sons.

Denton, G. H., and Karlen, W. 1973. Holocene climate variations: Their pattern and possible cause. *Quaternary Research* 3(2): 155–205.

de Quervain, M. R. 1963. On the metamorphism of snow. In W. D. Kingery, ed., *Ice and Snow Properties, Processes, and Applications* (pp. 377–390). Cambridge, MA: MIT Press.

Derbyshire, E., and Evans, I. S. 1976. The climatic factor in cirque variations. In E. Derbyshire, ed., *Geomorphology and Climate* (pp. 447–494). London: John Wiley and Sons.

Ding, Y., Liu, S., Li, J., and Shangguan, D. 2006. The retreat of glaciers in response to recent climate warming in western China. *Annals of Glaciology* 43(1): 97–105.

Dunne, T., and Leopold, L. B. 1978. *Water in Environmental Planning*. San Francisco: W. H. Freeman and Co.

Dyurgerov, M. B. 2002. *Glacier Mass Balance and Regime: Data of Measurements and Analysis*. Institute of Arctic and Alpine Research Occasional Paper 55, Boulder: University of Colorado.

Dyurgerov, M. B., and Meier, M. F. 1997a. Mass balance of mountain and subpolar glaciers: A new global assessment for 1961–1990. *Arctic and Alpine Research* 29(4): 379–391.

Dyurgerov, M. B., and Meier, M. F. 1997b. Year-to-year fluctuation of global mass balance of small glaciers and their contribution to sea level changes. *Arctic and Alpine Research* 29(4): 392–401.

Elsasser, H. 2001. The vulnerability of the snow industry in the Swiss Alps. *Mountain Research and Development* 21(4): 335–339.

Embleton, C., and King, C. A. M. 1975. *Glacial Geomorphology*. New York: Halstead Press.

Emiliani, C. 1972. Quaternary hypsithermals. *Quaternary Research* 2(2): 270–273.

Federici, P. R., and Spagnolo, M. 2004. Morphometric analysis on the size, shape and areal distribution of glacial cirques in the Maritime Alps (Western French–Italian Alps). *Geografiska Annaler* 86A(3): 235–248.

Fierz, C., Armstrong, R. L., Durand, Y., Etchevers, P., Greene, E., McClung, D. M., Nishimura, K., Satyawali, P. K., and Sakrotov, S. A. 2009. *The International Classification for Seasonal Snow on the Ground*. IHP-VII Technical Documents in Hydrology 83, IACS Contribution 1. Paris: UNESCO.

Flaschka, I., Stockton, C. W., and Boggess, W. R. 1987. Climatic variation and surface water resources in the Great Basin Region. *Water Resources Bulletin* 23(1): 47–57.

Flint, R. F. 1971. *Glacial and quaternary Geology*. New York: John Wiley and Sons.

Frutiger, H. 1964. *Snow Avalanches along Colorado Mountain Highways*. Research Paper RM-7. Fort Collins, CO: USDA Forest Service.

Gerrard, A. J. 1990. *Mountain Environments*. Cambridge, MA: MIT Press.

Gillespie, A., and Molnar, P. 1995. Asynchronous maximum advances of mountain and continental glaciers. *Review of Geophysics* 33: 311–364.

Glen, J. W. 1955. The creep of polycrystalline ice. *Proceedings of the Royal Society Series A* 228 (1175): 519–538.

Godsell, B., Hambrey, M. J., and Glasser, N. F. 2002. Formation of band ogives and associated structures at Bas Glacier d'Arolla, Valais, Switzerland. *Journal of Glaciology* 51: 139–146.

Goldthwait, R. P. 1970. Mountain glaciers of the Presidential Range, New Hampshire. *Arctic and Alpine Research* 2(2): 85–102.

Gordon, J. E. 1977. Morphometry of cirques in the Kintail–Affric–Cannich area of northwest Scotland. *Geografiska Annaler* 59A: 177–194.

Graf, W. L. 1976. Cirques as glacier locations. *Arctic and Alpine Research* 8(1): 79–90.

Graham, R. 2009. Significant rime ice events in Utah's Wasatch Mountains: Impacts and conceptual models. 23rd Conference on Weather Analysis and Forecasting/19th Conference on Numerical Weather Prediction, 1–5 June , Omaha, NE.

Gray, D. M., and Male, D. H. 1981. *Handbook of Snow*. Toronto, ON: Pergamon Press.

Grove, J. M.1972. The incidence of landslides, avalanches, and floods in western Norway during the Little Ice Age. *Arctic and Alpine Research* 4(2): 131–139.

Grove, J. M. 1988. *The Little Ice Age*. New York: Routledge.

Grove, J. M. 2001. The initiation of "the Little Ice Age" in regions round the North Atlantic. *Climatic Change* 48: 53–82.

Gutenberg, B., Buwalda, J. P., and Sharp, R. P. 1956. Seismic explorations on the floor of Yosemite Valley. *Bulletin of the Geological Society of America* 67: 1051–1078.

Haeberli, W., Hoelzle, M., and Suter, S., eds. 1998. *Into the second century of world glacier monitoring: Prospects and strategies*. Studies and Reports in Hydrology: A Contribution to the IHP and the GEMS 56. World Glacier Monitoring Service.

Hall, K. 1996. Freeze–thaw weathering: the cold region "panacea." *Polar Geography* 19(2): 79–87.

Heierli, J., Birkeland, K. W., Simenhois, R., and Gumbsch, P. 2011. Anticrack model for skier triggering of slab avalanches. *Cold Regions Science and Technology* 65(3): 372–381.

Heierli, J., Gumbsch, P., and Zaiser. M. 2008. Anticrack nucleation as triggering mechanism for snow slab avalanches. *Science* 321: 240–243.

Hewitt, K. 1982. Natural dams and outburst floods of the Karakoram Himalaya. In J. Glen, ed., *Hydrological Aspects of Alpine and High Mountain Areas* (pp. 259–269). IAHS Publication No. 138.

Hobbs, P. V. 1974. *Ice Physics*. Oxford, UK: Oxford University Press.

Hoinkes, H. C. 1968. Glacial variation and weather. *Journal of Glaciology* 7(49): 3–19.

Hooke, R. L. 2005. *Principles of glacier mechanics*. Cambridge: Cambridge University Press.

Hubbard, B., and Glasser, N. 2005. *Field Techniques in Glaciology and Glacial Geomorphology*. Chichester, UK: Wiley.

Ilken, A., and Bindschalder, R. A. 1986. Combined measurements of subglacial water pressure and surface velocity of Findelengletscher, Switzerland: Conclusions about drainage systems and sliding mechanism. *Journal of Glaciology* 32: 101–119.

Imbrie, J., Berger, A., Boyle, E., Clemens, S., Duffy, A., Howard, W., Kukla, G., Kutzbach, J., Martinson, D., McIntyre, A., Mix, A., Molfino, B., Morley, J., Peterson, L., Pisias, N., Prell, W., Raymo, M., Shackleton, N., and Togweiller, J. 1993. On the structure and origin of major glaciation cycles 2: The 100,000 year cycle. *Paleoceanography* 8: 699–735.

Imbrie, J., Boyle, E., Clemens, S., Duffy, A., Howard, W., Kukla, G., Kutzbach, J., Martinson, D., McIntyre, A., Mix, A., Molfino, B., Morley, J., Peterson, L., Pisias, N., Prell, W., Raymo, M., Shackleton, N., and Togweiller, J.1992. On the structure and origin of major glaciation cycles 1: Linear responses to Milankovitch forcing. *Paleoceanography* 7: 701–738.

Ives, J. D. 2007. *Skaftafell in Iceland: A Thousand Years of Change*. Reykjavik: Ormstunga.

Ives, J. D., and King, C. A. M. 1955. Glaciological observations on Morsárjökull S.W. Vatnajökull, Part II: Regime of the glacier, present and past. *Journal of Glaciology* 17: 477–482.

Ives, J. D., Shrestha, R. B., and Mool, P. K. 2010. *Formation of Glacial Lakes in the Hindu Kush–Himalayas and GLOF Risk Assessment*. Katmandu: International Centre for Integrated Mountain Development.

Jamieson, B., Haegeli, P., and Gauthier, D. 2010. *Avalanche Accidents in Canada*, Vol. 5: *1996–2007*. Revelstoke, BC: Canadian Avalanche Association.

Jarrell, R. L., and Schmidt, R. A. 1990. Snow fencing near pit reservoirs to improve water supplies. Proceedings of the 58th Western Snow Conference, 17–19 April 1990, Sacramento, CA (pp. 156–159).

Jouzel, J., et al. 2007. Orbital and millennial Antarctic climate variability over the past 800,000 years. *Science* 317: 793–796.

Kariya, Y. 2005. Holocene landscape evolution of a nivation hollow on Gassan volcano, northern Japan. *Catena* 62(1): 57–76.

Kaser, G., Fountain, A., and Jansson, P. 2003. *A Manual for Monitoring the Mass Balance of Mountain Glaciers*. International Hydrological Program, Technical Developments in Hydrology 59. Paris: UNESCO.

Kaser, G., and Osmaston, H. 2002. *Tropical Glaciers*. Cambridge, UK: Cambridge University Press.

King, C. A. M., and Ives, J. D. 1956. Glaciological observations on some of the outlet glaciers of southwest Vatnajökull, Iceland, 1954, Part II: Ogives. *Journal of Glaciology* 2: 646–652.

King, C. A. M., and Lewis, V. W. 1961. A tentative theory of ogive formation. *Journal of Glaciology* 3: 913–939.

Knight, C. A. 1967. *The Freezing of Supercooled Liquids*. Princeton, NJ: Van Nostrand.

Kokenakais, A., and Dexter, L. R. 1998. The rime river. Proceedings, International Snow Science Workshop, Sun River, OR (pp. 544–550).

Kronholm, K., and Birkeland, K. W. 2005. Integrating spatial patterns into a snow avalanche cellular automata model. *Geophysical Research Letters* 32, L19504, doi:10.1029/2005GL024373.

LaChapelle, E. R. 1966. The control of snow avalanches. *Scientific Ameri*can 214: 92–101.

LaChapelle, E. R. 1980. The fundamental processes in conventional avalanche forecasting. *Journal of Glaciology* 26(94): 75–84.

LaChapelle, E. R., and R. L. Armstrong. 1976. Nature and causes of avalanches in the San Juan Mountains. In R. L. Armstrong and J. D. Ives, eds., *Avalanche Release and Snow Characteristics, San Juan Mountains, Colorado* (pp. 23–40). Occasional Paper 19. Boulder, CO: Institute of Arctic and Alpine Research.

LaChapelle, E. R., and R. L. Armstrong. 1977. *Temperature Patterns in an Alpine Snow Cover and Their Influence on Snow Metamorphism*. Technical Report February 1977. Boulder, CO: Institute of Arctic and Alpine Research.

Lamb, H. H. 1965. The early medieval warm epoch and its sequel. *Paleogeography, Paleoclimatology, Paleoecology* 1: 13–37.

Leaf, C. F. 1975. Watershed management in the Rocky Mountain subalpine zone: The status of our knowledge. Research Paper RM-137. Fort Collins, CO: USDA Forest Service.

Leggett, J., ed. 1990. *Global Warming: The Greenpeace Report*. Oxford, UK: Oxford University Press.

Lliboutry, L. 1968. General theory of subglacial cavitation and sliding of temperate glaciers. *Journal of Glaciology* 7: 67–95.

Lliboutry, L. 1987. Realistic, yet simple, bottom boundary conditions for glaciers and ice sheets. *Journal of Geophysical Research* 92(B9): 9101–9109, doi:10.1029/JB092iB09p09101.

Lliboutry, L. 1993. Internal melting and ice accretion at the bottom of temperate glaciers. *Journal of Glaciology* 39: 50–64.

Lobeck, A. K. 1939. *Geomorphology: An Introduction to the Study of Landscapes*. New York: McGraw-Hill.

Logan N., and Atkins, D. 1996. *Snowy Torrents 1980–86*. Special Publication 39. Denver, CO: Colorado Geological Survey.

Lutz, E. R., and Birkeland, K. W. 2011. Spatial patterns of surface hoar properties and incoming radiation on an inclined forest opening. *Journal of Glaciology* 57(202): 355–366.

Marchand, P. 1996. *Life in the Cold: An Introduction to Winter Ecology*. 3rd ed. Hanover, NH: University Press of New England.

Martinelli, M. 1967. Possibilities of snow management in alpine areas. In W. E. Sopper and H. W. Lull, eds., *International symposium on forest hydrology* (pp. 120–127). Oxford: Pergamon.

Martinelli, M. 1972. *Simulated Sonic Boom as an Avalanche Trigger*. Research Note RM-224. Fort Collins, CO: USDA Forest Service.

Martinelli, M. 1974. *Snow Avalanche Sites: Their Identification and Evaluation*. Agriculture Information Bulletin 360, Washington, DC: USDA Forest Service.

Martinelli, M. 1975. *Water Yield Improvement from Alpine Areas: The Status of Our Knowledge*. Research Paper RM-138. Fort Collins, CO: USDA Forest Service.

Martini, I. P., Brookfield, M. E., and Sadura, S. 2001. *Principles of Glacial Geomorphology and Geology*. Upper Saddle River, NJ: Prentice-Hall.

Matthews, J. A., and Briffa, K. 2005. The "Little Ice Age": Re-evaluation of an evolving concept. *Geografiska Annaler* 87A: 17–36.

McClung, D. 2002. The elements of applied avalanche forecasting, Parts I and II. *Natural Hazards* 25: 111–146.

McClung, D. and Schaerer, P. 2006. *The Avalanche Handbook*. Seattle: The Seattle Mountaineers.

McClung, D., and Schweizer, J. 1999. Skier triggering, snow temperatures and the stability index for dry slab avalanche initiation. *Journal of Glaciology* 45(150): 190–200.

Meier, M. F. 1962. Proposed definitions for glacier mass budget terms. *Journal of Glaciology* 4: 252–265.

Meier, M. F. 1965. Glaciers and climate. In H. E. Wright and D. G. Frey, eds., *The Quaternary of the United States* (pp. 547–566). Princeton, NJ: Princeton University Press.

Meier, M. F. 1969. Seminar on the causes and mechanics of glacier surges, St. Hilarie, Canada, September 10–11, 1968: A summary. *Canadian Journal of Earth Sciences* 6(4): 987–989.

Messerli, B., Messerli, P., Pfister, C., and Zumbuhl, H. J. 1978. Fluctuations of climate and glaciers in the Bernese Oberland, Switzerland, and their geoecological significance, 1600 to 1975. *Arctic and Alpine Research* 10(2): 247–260.

Mitchell, S. G., and Montgomery, D. R. 2005. Influence of a glacial buzzsaw on the height and morphology of the Cascade Range in central Washington State, USA. *Quaternary Research* 65: 96–107.

Mote, P. W., Hamlet, A. F., Clark, M. P., and Lettenmaier, D. P. 2005. Declining mountain snowpack in North America. *Bulletin of the American Meteorological Society* 86(1): 39–49.

Motyka, R. J., and Truffer, M. 2007. Hubbard Glacier, Alaska: 2002 closure and outburst of Russell Fjord and postflood conditions at Gilbert Point. *Journal of Geophysical Research* 112: F02004, doi:10.1029/2006JF000475.

Murray, T., Strozzi, T., Luckman, A., Jiskoot, H., and Chirstakos, P. 2003. Is there a single surge mechanism? Contrasts in dynamics between glacier surges in Svalbard and other regions. *Journal of Geophysical Research* 108(B5): 2273, doi:10.1029/2002JB001906.

Nakaya, U. 1954. *Snow Crystals: Natural and Artificial*. Cambridge.MA: Harvard University Press.

Nelson, J. T., and Baker, M. B. 1996. New theoretical framework for studies of vapor growth and sublimation of small ice crystals in the atmosphere. *Journal of Geophysical Research* 101(D3): 7033–7047, doi:10.1029/95JD03162.

Niu, G., and Yang, Z. 2004. Effects of vegetation canopy processes on snow surface energy and mass balances. *Journal of Geophysical Research* 109: D23111, doi:10.1029/2004JD004884.

Nolan, M. 2003. The "galloping glacier" trots: Decadal-scale speed oscillations within the quiescent phase. *Annals of Glaciology* 31(1): 7–13.

NOVA. 2001. *Vanished*. Washington, DC: Public Broadcasting System. <http://www.pbs.org/wgbh/nova/vanished/>.

NSIDC. 2008. *Mountain Glacier Fluctuations: Changes in Terminus Location and Mass Balance*. Boulder, CO: National Snow and Ice Data Center. <http://nsidc.org/cryosphere/sotc/glacier_balance.html>.

Østrem, G. 1962. Ice-cored moraines in the Kebnekajse area. *Biuletyn Peryglacyalny* 11: 271–278.

Østrem, G. 1964a. Ice-cored moraines in Scandinavia. *Geografiska Annaler* 64A: 228–337.

Østrem, G. 1964b. Ice crystals from an ice-cored moraine on Baffin Island. *Geographical Bulletin* 22: 72–79.

Østrem, G. 1973. The transient snowline and glacier mass balance in southern British Columbia and Alberta, Canada. *Geografiska Annaler* 55A(2): 93–106.

Østrem, G. 1974. Present alpine ice cover. In J. D. Ives and R. G. Barry, eds., *Arctic and Alpine Environments* (pp. 225–250). London: Methuen.

Østrem, G., and Brugman, M. 1991. *Mass Balance Measurements: a Manual for Field and Office Work*. Scientific Report 4. Saskatoon, SK: National Hydrological Research Institute.

Pagano, T., Garen, D., and Sorooshian, S. 2004. Evaluation of official western U. S. seasonal water supply outlooks, 1922–2002. *Journal of Hydrometeorology* 5: 896–909.

Painter, T. H., Barrett, A. P., Landry, C., Neff, J., Cassidy, M. P., Lawrence, C., McBride, K. E., and Farmer, G. L. 2007. Impact of disturbed desert soils on duration of mountain snowcover. *Geophysical Research Letters* 34: L12502, doi:10.1029/2007GL030284.

Palmer, P. L. 1988. The SCS snow survey water supply forecasting program: current operations and future directions. Proceedings, 57th Western Snow Conference, Kalispell, MT.

Paterson, W. S. B. 1994. *The Physics of Glaciers*. 3rd ed. Oxford, UK: Pergamon.

Perla, R. 1978. Artificial releases of avalanches in North America. *Arctic and Alpine Research* 10(2): 235–240.

Perla, R., and Martinelli, M. 1976. *Avalanche Handbook*. Handbook 489. Washington, DC: USDA Forest Service.

Phillips , W. M., Hall, A. M., Mottram, R., Fifeild, L. K., and Sugden, D. E. 2006. Cosmogenic ^{10}Be and ^{26}Al exposure ages of tors and erratics, Cairngorm Mountains, Scotland: Timescales for development of a classic landscape of selective linear glacial erosion. *Geomorphology* 73: 222–245.

Porter, S. C. 1975. Glaciation limit in New Zealand's Southern Alps. *Arctic and Alpine Research* 7(1): 33–38.

Porter, S. C. 1977. Present and past glaciation threshold in the Cascade Range, Washington, U.S.A.: Topographic and climatic controls and paleoclimatic implications. *Journal of Glaciology* 18(78): 101–115.

Porter, S. C. 1989. Some geological implications of average Quaternary glacial conditions. *Quaternary Research* 32(3): 245–261.

Porter, S. C. 2001. Snowline depression in the tropics during the last glaciation. *Quaternary Science Reviews* 20(10): 1067–1081, doi:10.1016/S0277-3791(00)00178-5.

Posamentier, H. W. 1977. A new climatic model for glacier behavior of the Austrian Alps. *Journal of Glaciology* 18(78): 57–65.

Post, A., and Mayo, L. R. 1971. Glacier dammed lakes and outburst floods in Alaska, accompanying Hydrological Investigations Atlas HA-455.Washington, DC: U.S. Geological Survey.

Price, L. W. 1981. *Mountains and Man: A Study of Process and Environment*. Berkeley: University of California Press.

Price, R. J. 1973. *Glacial and Fluvialglacial Landforms*. Edinburgh, UK: Oliver and Boyd.

Prowse, T. D., and Beltaos, S. 2002. Climatic control of river-ice hydrology: A review. *Hydrological Processes* 16(4): 805–822, doi:10.1002/hyp369.

Pruppacher, H. R., and Klett, J. D. 1997. *Microphysics of Clouds and Precipitation*. 2nd ed. Atmospheric and Oceanic Sciences Library 18. Dordrecht: Kluwer.

Raymo, M. E., and Nicancioglu, K. 2003. The 41 kyr world: Milankovitch's other unsolved mystery. *Paleooceanography* 18(1): 1011, doi:10.1029/2002PA000791.

Riccardi, A. C. 2009. IUGS ratified ICS Recommendation on redefinition of Pleistocene and formal definition of base of Quaternary [Letter]. International Union of Geological Sciences.

Robin, G. de Q. 1976. Is the basal ice of a temperate glacier at the pressure melting point? *Journal of Glaciology* 16: 259–271.

Russell, R. J. 1933. Alpine landforms of western United States. *Bulletin of the Geological Society of America* 44: 927–950.

Sahn, K. R. 2010. Avalanche reduction in the continental climate: How to implement an effective boot packing program. Proceedings of the 2010 International Snow Science Workshop, Squaw Valley, California (pp. 296–301).

Schaerer, P. A. 1972. Terrain and vegetation of snow avalanche sites of Rogers Pass, British Columbia. In O. Slaymaker and H. J. McPherson, eds., *Mountain Geomorphology* (pp. 215–222). Vancouver, BC: Tantalus Press.

Schweizer, J. 1999. Review of dry snow slab avalanche release. *Cold Regions Science and Technology* 30: 43–57.

Schweizer, J., Kronholm, K., Jamieson, J. B., and Birkeland, K. W. 2008. Review of spatial variability of snowpack properties and its importance for avalanche formation. *Cold Regions Science and Technology* 51: 253–272.

Selby, M. J. 1985. *Earth's Changing Surface: An Introduction to Geomorphology*. Oxford, UK: Clarendon Press.

Seligman, G. 1936. *Snow Structure and Ski Fields*. London: Macmillan.

Seltzer, G. O. 1994. Climatic interpretation of alpine snowline variations on millennial time scales. *Quaternary Research* 41(2): 154–159.

Serreze, M. C., Clark, M. P., Armstrong, R. L., McGinnis, D. A., and Pulwarty, R. S. 1999. Characteristics of the western United States snowpack from snowpack telemetry (SNOTEL) data. *Water Resources Research* 35(7): 2145–2160.

Sharp, R. P. 1960. *Glaciers*. Eugene: Oregon State System of Higher Education.

Sharp, R. P. 1988. *Living Ice: Understanding Glaciers and Glaciation*. Cambridge, UK: Cambridge University Press.

Solomina, O., Haeberli, W., Kull, C., and Wiles, G. 2008. Historical and Holocene glacier-climate variations: General concepts and overview. *Global and Planetary Change* 60: 1–9.

Sommerfeld, R., and LaChapelle, E. 1970. The classification of snow metamorphism. *Journal of Glaciology* 9(55): 3–17.

Sovilla, B., Sommavilla, F., and Tomaselli, A. 2001. Measurements of mass balance in dense snow avalanche events. *Annals of Glaciology* 32: 328–332.

Statham, G., Haegeli, P., Birkeland, K. W., Greene, E., Israelson, C., Tremper, B., Stethem, C., McMahon, B., White, B., and Kelly, J. 2010. The North American avalanche danger scale. Proceedings of the 2010 International Snow Science Workshop, Squaw Valley, CA (pp. 117–123).

Steinhoff, H. W., and Ives, J. D., eds. 1976. *Ecological Impacts of Snowpack Augmentation in the San Juan Mountains, Colorado*. Fort Collins: Colorado State University.

Sturm, M., and Benson, C. S. 1997. Vapor transport, grain growth and depth hoar development in the subarctic snow. *Journal of Glaciology* 43(143): 42–59.

Sugden, D. E. 1977. Reconstruction of the morphology, dynamics and thermal classification of the Laurentide Ice Sheet at its maximum. *Arctic and Alpine Research* 9: 27–47.

Sugden, D. E., Balco, G., Cowdrey, S. G., Stone, J. O., and Sass, L. C., III. 2005. Selective glacial erosion in the coastal mountains of Marie Byrd Land, Antarctica. *Geomorphology* 67: 317–334.

Sugden, D. E., and John, B. S. 1976. *Glaciers and landscapes*. London: Edward Arnold.

Tabazedeh, A., Djikaev, Y. S., and Reiss, H. 2002. Surface crystallization of supercooled water in clouds. *Proceedings of the National Academy of Sciences of the United States of America* 99(25): 15873–15878.

Thompson, L. G. 2001. Disappearing glaciers: Evidence of a rapidly changing Earth. Proceedings of the American Association of the Advancement of Science Meeting, 15–20 February 2001, San Francisco, CA.

Thompson, L. G., Davis M. E., Mosley-Thompson E., Lin P., Henderson K., and Mashiotta T. 2005. Tropical ice-core records: Evidence for asynchronous glaciations and Milankovitch timescales. *Journal of Quaternary Science* 20(7): 723–733.

Thompson, L. G., Mosley-Thompson, E., Davis, M., Lin P., Henderson K., and Mashiotta T. 2003. Tropical glacier and ice core evidence of climate change on annual to millennial time scales. *Climatic Change* 59(1–2): 137–155.

Thorn, C. E. 1979. Ground temperatures and surficial transport in colluvium during meltout: Colorado Front Range. *Arctic and Alpine Research* 11: 41–52.

Thorn, C. E., and Hall, K. 1980. Nivation, and arctic–alpine comparison and reappraisal. *Journal of Glaciology* 25: 109–124.

Thornbury, W. D. 1969. *Principles of Geomorphology*. New York: John Wiley and Sons.

Tremper, B. 2008. *Staying Alive in Avalanche Terrain*. Seattle: The Mountaineers.

Trenhaile, A. S. 1976. Cirque morphometry in the Canadian Cordillera. *Annals of the Association of American Geographers* 66(3): 451–462.

U. S. Department of Agriculture. 1968. *Snow Avalanches*. USDA Forest Service Handbook 194, Revised. Washington, DC: USDA Forest Service.

U. S. Department of Agriculture. 1972. *Snow Survey and Water Supply Forecasting*. National Engineering Handbook Section 22. Washington, DC: USDA Soil Conservation Service.

Weerteman, J. 1957. On the sliding of glaciers. *Journal of Glaciology* 3: 33–38.

Weerteman, J. 1964. The theory of glacial sliding. *Journal of Glaciology* 5: 287–303.

Weisbecker, L. W. 1974. *The Impacts of Snow Enhancement: Technology Assessment of Winter Orographic Snow Augmentation in the Upper Colorado River Basin*. Oklahoma City: University of Oklahoma Press.

Williams, K. 1975. *The Snowy Torrents: Avalanche Accidents in the United States, 1967–71*. General Technical Report RM-8. Fort Collins, CO: USDA Forest Service.

Williams, K., and Armstrong, B. 1984. *The Snowy Torrents: Avalanche Accidents in the United States 1972–79*. Jackson, WY: Teton Bookshop Publishing Company.

Williams, R. S., and Sigurðsson, O., trans. 2004. S. Pálsson (1796), *Icelandic Ice Mountains*. Reykjavik: Icelandic Literary Society.

Woillard, G. M. 1978. Grande Pile peat bog: A continuous pollen record for the last 140,000 years. *Quaternary Research* 9(1): 1–2.

Wood, S. E., Baker, M. B., and Calhoun, D. 2001. New model for the vapor growth of hexagonal ice crystals in the atmosphere. *Journal of Geophysical Research* 106(D5): 4845–4870.

Wright, H. E., Jr. 1961. Late Pleistocene climate of Europe: A review. *Geological Society of America Bulletin* 72(6): 933–983.

Yoshino, M. 2005. Climate change and ancient civilizations. In J. E. Oliver, ed., *Encyclopedia of World Climatology* (pp. 192–198). New York: Springer.

Zwally, H. J., and Li, J. 2002. Seasonal and interannual variations of firn densification and ice-sheet surface elevation at the Greenland summit. *Journal of Glaciology* 48: 199–207.

Mountain Landforms and Geomorphic Processes

JASON R. JANKE and LARRY W. PRICE

The mountain landscape is the product of both constructive and destructive processes. Mountains are created by forces originating from within the Earth, but they are soon modified and are eventually destroyed by external forces. Many mountain ranges have been created and destroyed throughout geologic time. The Alaska Range, Alps, Andes, Cascades, Himalaya, Rockies, and Sierra Nevada are all very young mountains and are still growing. Plate movement, which produces earthquakes and volcanic eruptions, may have destructive effects, but these processes are also fundamental to mountain construction. Present rates of *orogeny* in western North America exceed rates of erosion by about 7.5 m per 1,000 years (Schumm 1963). The Appalachian Mountains in eastern North America, created through collisions forming the super continent Pangaea, are considered old mountains. At one time, they may have been as high as the Alps today, but erosion has diminished their relief and elevation. In extreme cases, all evidence of relief and elevation may be erased, but the roots of the former mountain range give evidence of a more glorious past. The expansive, thinly soiled region of eastern Canada (the Canadian Shield) is an example of this.

While mountain ranges display different ages and degrees of development, all are relatively young compared to the vastness of geologic time. One reason for this may be the increased effectiveness of erosion at higher altitudes because of greater relief, steep slopes, unconsolidated material, and uplift. This provides much potential energy for erosion and redistribution of the eroded material, and greater precipitation totals increase the possibility of rock weathering. Mountains are "rapidly" worn away over the course of millions of years, and consequently, can never be very old in a geologic sense.

Relief has been shown to be an important control on the rate of *denudation* or removal of material through erosion and weathering that result in the lowering of the land (Hinderer 2001). Estimates from some of the world's largest river systems indicate that mechanical denudation ranges from 4 to more than 500 mm ka^{-1}, whereas rates of chemical denudation are lower, ranging from 1 to 30 mm ka^{-1} (Caine 2004). In the Alps, mean denudation rates have been about 6.2 cm per 100 years over the past 17,000 years (620 mm ka^{-1}). The exact rate of denudation varies in basins of different size, climate, vegetation regime, and rock type. For example, in the Wind River Range, denudation rates of 0.115 mm ka^{-1} have been reported, whereas in the nearby, yet geologically different, Front Range, rates of 0.1 mm ka^{-1} have been estimated (Ahnert 1970). Mean denudation in the Wasatch Mountains of Utah is consistent with longer-term exhumation, which suggests that denudation has been steady or decreasing over the past 5 million years (Stock et al. 2009). These lethargic rates illustrate the slowness of modern denudation in the Colorado Front Range (Caine 2001). Generally, as relief increases, perhaps through rejuvenation, denudation rates increase. This illustrates the importance of mountain systems as a source of sediment (Fig. 5.1) (Milliman and Syvitski 1992; Caine 2004).

The form, structure, and composition of mountains greatly affect the rate and type of geomorphic processes. Horizontally oriented rock, volcanic features, as well as folded, faulted, and domed structures with strata dipping in various directions, all present very different surfaces and starting points for erosional processes. Because water, soil, snow, and ice tend to follow the path of least resistance, the processes of erosion tend to reinforce and exaggerate existing slopes and structures.

Ollier and Pain (2000) provide an excellent discussion of how mountains originate (Fig. 5.2). The development of landscapes in different mountain regions depends on the rate of uplift and deformation, the nature of various rock

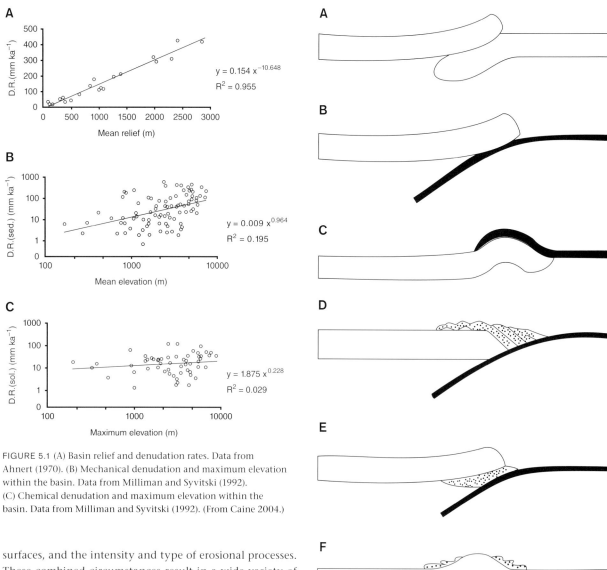

FIGURE 5.1 (A) Basin relief and denudation rates. Data from Ahnert (1970). (B) Mechanical denudation and maximum elevation within the basin. Data from Milliman and Syvitski (1992). (C) Chemical denudation and maximum elevation within the basin. Data from Milliman and Syvitski (1992). (From Caine 2004.)

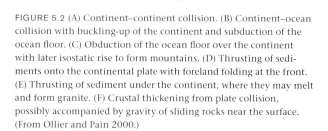

FIGURE 5.2 (A) Continent–continent collision. (B) Continent–ocean collision with buckling-up of the continent and subduction of the ocean floor. (C) Obduction of the ocean floor over the continent with later isostatic rise to form mountains. (D) Thrusting of sediments onto the continental plate with foreland folding at the front. (E) Thrusting of sediment under the continent, where they may melt and form granite. (F) Crustal thickening from plate collision, possibly accompanied by gravity of sliding rocks near the surface. (From Ollier and Pain 2000.)

surfaces, and the intensity and type of erosional processes. These combined circumstances result in a wide variety of conditions and products. For example, the structural modification of rock units often involves their compression in some areas and their stretching in others, producing alternating resistant and susceptible zones. Thus, in folded rock, anticlines are areas of extension and therefore are weak, whereas synclines are areas of compression and thus have greater strength. As a result, valleys often develop in anticlines, where stretching has made the strata more susceptible to breaching by erosion, and the elevated anticlines are worn away more rapidly than the synclines. This paradoxical situation results from the more aggressive and prolonged erosion of topographic promontories as well as the structural modifications associated with folding. The relationships are not the same everywhere; in some places, streams occupy synclinal valleys and anticlines form the ridges. Where folding is extremely rapid and the rock strata resistant, the structure may not be modified as much by erosion, so that it is expressed directly in the form of the land surface. Such features are the exception rather than the rule. Much depends upon the rate of uplift. Short-term

rates of uplift can be determined using high-resolution Global Positioning Satellite (GPS) measurements or satellite altimetry (Bishop et al. 2002). Long-term estimates can be derived by identifying corals, deltas, or other sea-level features (at known elevations) or radiometric dating of minerals at known depths. If an area has no uplift or deformation taking place, the major controls are the existing differences in bedrock, topography, and structure. However, if the area is tectonically active, a much more dynamic relationship exists between form and process because erosion is taking place simultaneously with uplift and deformation. Under these conditions, for example, streams may sometimes cut across the structural grain of the land with indifference to the underlying structural conditions.

The nature of the rock type itself has a major impact on landscape development. Massive and resistant crystalline rocks, such as granite or quartzite, have the potential to produce ridges, whereas landforms formed in weak rocks, such as shale, often form valleys. The exact outcomes depend on local circumstances, but the more resistant the rock, the more likely it is to produce peaks and ridges. Joints, fractures, and fault lines are zones of weakness and can experience more weathering. Streams exploit such features and develop preferentially along them, providing a good indicator of an area's structural framework.

While structure and rock type exert significant control on landform development, they are also largely innate and unique in their character and distribution, depending upon internal processes. External processes operating at the surface of the Earth, such as precipitation, temperature, or other climatic variables, display a very different distribution. Climate provides the basic framework into which all natural phenomena must fit. It is obvious that environmental conditions (temperature and precipitation) vary with climate type. Therefore, despite widespread regional differences in the character of mountains, many pervasive and overriding similarities suggest that similar geomorphic processes operate under comparable environmental conditions and give rise to similar types of landforms.

Landscape Development

As early as 1850, scientists began to question how flat erosional surfaces formed or why unconformities—breaks in the building structure—existed. The ideas of William Morris Davis coincide with these early attempts to reconstruct the denudational history of the Earth, the weathering and erosional processes that contribute to lowering land. Charles Darwin influenced the popular trend of thought during this period; as a result, Davis applied evolutionary ideas and terminology to describe landforms and the landscape. In 1889, Davis first presented his explanation of the erosion cycle, which explained landform development by investigating the rivers and valleys of Pennsylvania. He assumed that the landscape experienced an initial rapid tectonic uplift. Throughout the youthful, mature, and old stages, erosional forces reduced the landscape to its original form: a peneplain, a low surface with little relief. The cycle could then renew itself as more uplift or changes in climate occurred. Base level, the lowest point to which a river can flow, may have temporarily been met, but stream gradient can increase with uplift until a stream matures

and weathers relief. German geomorphologists Albrecht and Walther Penck offered an alternative view on landscape evolution. Rather than mountain summits consisting of peneplains from the past, they were simply upper limits to which mountains could grow, because uplift balanced denudation. The Pencks' ideas about landscape evolution were different from those of Davis, because they were rooted in crustal tectonic processes rather than strictly erosion.

Their ideas were simplistic, but their effect on mountain geomorphology was monumental. Each saw the evolutionary development of the landscape as a whole. This regionalized approach led to climatic geomorphology, according to which climatic regimes control process, and process controls landform development (Thorn 1992). There was, however, much discontent, since Davis's ideas were overgeneralized. Those seeking a firmer ground upon which to base their concepts branched off into detailed process studies, especially as the quantitative revolution unfolded.

During the mid-twentieth century, geomorphology became less concerned with evolution and more concerned with process, morphometry, and systems (Chorley 1965). In order to accomplish this and define the discipline, Hack (1960) and Chorley (1962) rejected the time dimension, thus examining the landscape within narrow temporal limits and justifying short-term process studies. Change was viewed as dynamic, moving about an average condition or causing a shift in the condition to a new or preexisting state. By viewing process at different temporal scales, one can pass from short-term *static* equilibrium states, to graded time over hundreds of years, to a *dynamic equilibrium* or metastable state in which thresholds are crossed, and finally to *cyclic* time over millions of years (Schumm 1977). These studies pursued a reductionist path. Unfortunately, researchers were unable to extend from a small spatial scale to a regional scale (Thorn 1992). Process could not always explain form at different spatial scales. This same dilemma exists today when trying to understand the formation of mountains. As a result, some feel that Davis's ideas will again become mainstream; geomorphology may be on the threshold of another golden era where cyclical approaches become the custom once again (Bishop 2007).

Another school of thought suggests that late Cenozoic uplift of mountains is a consequence of climate change (Molnar and England 1990). Based on the principle of *isostasy,* mean elevation should decrease by ΔT (amount of material eroded) divided by 6; thus the underlying rock (the crust and the Moho) should rise by $5\Delta T/6$. In other words, enhanced relief production by erosion can uplift mountains, while at the same time the mountains will experience a decrease in mean elevation, since material is being removed from between the peaks and high ridges. For example, glaciation will increase the depth of incision of the terrain and, concurrently, streams will increase erosion and deposition in the forelands, creating greater relief. Large sediment deposits in the Atlantic, Pacific, and Indian Oceans provide evidence of recent erosion and concurrent deposition. In the Gulf of Mexico, deposition rates were greater during Quaternary glaciation than during the 60 million years before glaciation. This suggests that glaciation was responsible for the late Cenozoic uplift of the Rocky Mountains. In the northern and central Apennines, Italy, hillslope erosion and river incision have balanced uplift for 1 My, but rates of exhumation have slowed since the emergence of the mountain chain, suggesting that some other process is operating (Cry and Granger 2008). In the southern Alps, both tectonics and uplift as a result of incision are present. Adams (1980) found that uplift and erosion rates are roughly equal. Isostatic forces must be active because if 1 km of material is removed, the landscape should lower by 1/6 km; thus tectonics are thickening the crust at a rate proportional to this. Plate convergence in this region has occurred for 10–15 million years, but deposits in the oceans are only 2.5 million years old, correlating well with recent glaciation, not tectonic events. Late Cenozoic uplift has been inferred for mountain regions across the world, yet the correlation with plate motions is small; only global climate change is regionally extensive enough to explain uplift. The combination of climate change, weathering, erosion, and isostatic rebound might even have created a system of positive feedbacks that allows glaciers to continue to grow, causing more erosion and therefore uplift.

Evidence for the link between erosion and climate also stems from a variety of other sources. Avouac and Burov (1996) hypothesized that removal of eroded material from mountains and its deposition in the foreland opposes the spreading of a crustal root, thus driving material toward the orogeny (inward flow). This idea could explain the formation of Colorado's Front Range during the Laramide orogeny. The Tien Shan and Himalaya both experience similar tectonic forces, but the Himalayas are shortening twice as fast (2 cm/yr compared to 1 cm/yr), which is attributed to accumulation of sediment in the forelands. The Tien Shan exists in an arid climate, but the Himalayas are in a monsoonal climate conductive to erosion, which suggests that climate may be the driving force of uplift.

Small and Anderson (1998) examined relief production for the Wind River (Wyoming), Beartooth (Montana), and Front Range (Colorado). Evidence suggests that erosion rates of summit flats in the Laramide Ranges are approximately 10 mm ka^{-1}, but erosion of valleys is about 100 mm ka^{-1} years. If summit erosion is slower than rock uplift driven by isostasy, then summit height will increase. Using *digital elevation models* (DEMs) and *geographic information system* (GIS) techniques, erosionally driven uplift was estimated to be 50–100 m. This rate of uplift is similar to summit erosion; thus summit erosion will be offset by uplift. Summit elevations have remained constant, but several hundred meters of relief have been produced. The onset of valley growth began some 2–3 million years ago, correlating well with the growth of ice sheets, suggesting that relief must be climatically driven.

Other studies have questioned the link between erosion (the process) and the growth of mountains (the form). While Small and Anderson (1995) showed that a significant component of uplift in California's Sierra Nevada was linked to the lithosphere's response to erosion coupled with deposition in the Central Valley, Brocklehurst and Whipple (2002) questioned these results after examining glaciated and nonglaciated basins in the Sierra Nevada. Relief was greatest in areas that experienced full glaciation. In fact, peaks were reduced at a rate greater than the isostatic uplift they would have induced. Relief was greatest where glaciers extended into other basins. This suggests that the isostatic response to incision is exaggerated in the Sierra Nevada. However, focused erosion at high elevations could contribute to a flexural response to uplift. Using cosmogenic nuclides to measure erosion in the Sierra Nevada, Riebe et al. (2001) showed that erosion rates vary by only a factor of 2.5 and are not correlated with climate across an eight-fold range in average annual precipitation and mean annual temperature conditions. These findings raise questions about the basic idea that changes in climate can produce more erosion, thus enhancing uplift.

Whipple et al. (1999) suggested that, although climate change may cause an increase in denudation, neither fluvial nor glacial erosion is significant enough to induce isostatic uplift. In a fluvial environment, both tributary and trunk relief would be reduced in a more erosive climate, which contradicts the idea that more erosion leads to greater relief. The transition to more erosive conditions will not only increase incision, but also increase landsliding concentrated in the upper part of the basin, thus leveling the landscape. Molnar and England (1990) thought that a transition to glacial erosion could create relief. However, Whipple et al. (1999) stated that this was not demonstrated empirically, and have challenged the claim that glacial erosion produces relief. In a glacial environment, they argue that glaciers only produce relief through valley widening, ice buttressing of rock slopes, and formation of hanging valleys. *Periglacial* processes, or cold climate processes without the presence of glacial ice, may further diminish relief, although mass wasting is often very slow. Aggradation of deposits at lower elevations will bury channels in alluvium, thus further reducing relief.

The mechanics of glacier erosion are insufficiently known. Glaciers are likely better transporters of material, thus providing further uncertainty as to an isostatic response of mountains (Caine 1986). Worldwide denudation rates in nonglacierized basins, not including some of the world's most active mountain belts, have indicated that no obvious relationship exists between precipitation or mean annual temperature and total denudation. In addition, topographic relief alone does not result in high rates of denudation. Denudation rates are highest in areas of rejuvenation, and the rates of weathering covary primarily with physical erosion and much less with temperature or precipitation (von Blanckenburg 2005).

Again, the link between process and resulting form is poorly developed and continues to puzzle geomorphologists. Spatial scale will continue to play an important role in future studies. As our ability to comprehend fluvial incision processes as well as glacial erosion and transport improve, so will our understanding of mountain growth (Tucker and Slingerland 1996; Hartshorn et al. 2002; Hovius et al. 2000). As more detailed observations become available, mountain geomorphologists will hopefully be able to "scale up" to answer such regionalized questions.

Hillslope Components

The basic components of the mountain landscape are the upland surfaces, the valley bottoms, and the slopes that connect them. Slopes are of the utmost importance, since they occupy the greatest area, provide a link between uplands and valley floors, and result in a high degree of energy transfer. Geomorphic processes are intensified in the presence of steep slopes and a mountain climate. While precipitation is often the major factor at intermediate altitudes, low temperatures become the dominant environmental feature at higher altitudes and give rise to glacial, nivational, and periglacial systems (Washburn 1980; Embleton and King 1975; Barsch and Caine 1984). The *glacial system* is one in which ice acts directly to shape the land. It generally occurs at the highest elevations (Benn and Evans 1998). *Nivation,* a special set of processes (and sometimes forms) that result from the presence of snow patches, is dominated by frost action and the downslope movement of earth material by gravity (mass wasting). The snow patch is not doing the work; rather, freezing and running water intensify erosion and transport, thus often hollowing a hillslope. Once begun, this is a self-perpetuating process: The larger the depression, the more snow that will accumulate, and the longer it will take to melt in summer, thereby increasing its erosive effect. Nivation processes are best developed at the snowline. The *periglacial system* is characterized by nonglacial, cold climatic conditions as well as frost action and mass wasting processes.

Since the processes that operate within these systems are often not mutually exclusive, others have devised broader classifications that focus on material fluxes. Barsch and Caine (1984) divided mountains into (1) a glacial system, (2) a coarse-grained debris system, (3) a fine-grained debris system, and (4) the geochemical system. Active glaciers in the *glacial system* have the greatest erosion potential (Embleton and King 1975; Barsch and Caine 1984). In the *coarse-grained system,* sediment is transferred from the cliffs to depositional features through rockfalls, landslides, or avalanches. Mass-movement landforms typically are present in the coarse debris system. The *fine-grained sediment system* is an open system that has both internal and external inputs fed from erosion of local soils or eolian (wind) deposition of dust. The *geochemical system* involves solution weathering. Bedrock and surficial geology as well

as nivation and fluvial processes all play important roles in the geochemical system (Darmody et al. 2000; Sueker et al. 2001). Other coarse-grained periglacial landforms, such as rock glaciers, have the potential to alter stream geochemistry by storing and releasing solutes (Williams et al. 2007).

Other classifications of alpine hillslopes have also been formulated. Caine (1974) categorized hillslopes in the Rocky Mountains into (1) interfluves, (2) free-face, (3) talus, (4) talus foot, (5) valley floor, and (6) stream channel components (Fig. 5.3). Interfluves are convex surfaces where slow mass wasting and soil creep dominate. The free-face consists of steep bedrock cliffs, where rapidly moving rocks are free to fall and bounce. The talus component experiences less rapid movement than the free-face since it accumulates alluvial and avalanche material. At the talus foot, landforms such as rock glaciers can be found. The valley floor also experiences mass wasting and overland flow, with some areas having relatively rapid movement. In Colorado, most valley floors are inactive, and it is rare that clastics ever reach them. In the stream channel, there is also insignificant input from upper slopes.

The Colorado Front Range has been intensively studied, but still provides a problem when linking geomorphic systems. Uncertainty is created when the systems are examined at the basin scale. Consider the hillslope categories mentioned above. At Niwot Ridge, lowering is at a rate less than 0.01 mm/yr, but cliff retreat is about 0.76 mm/yr. This suggests that Indian Peaks are relatively inactive and shows that zones of higher activity are surrounded by inactive zones with poorly developed linkages within basins. The interfluves, valley sides, and stream network should be viewed as independent entities with respect to coarse debris movement. There should be more coarse debris throughout the system, but this is not the case, because the system does not cascade all the way down (Gerrard 1990). At present, the hillslope sediment budgets are dominated by internal transfer of coarse debris within the free-face and talus components. Even where fine sediment is involved in movement, it is not transported out of the slope system, except possibly by high winds (Caine 1986). Mountain landscapes that were glaciated during the Pleistocene often have contemporary denudation rates that are an order of magnitude lower than the global maxima. Sediment storage dominates in a postglacial landscape with poor connections between fluvial systems and surrounding slopes, reflecting the extreme difference between valley and ridge lowering (Small et al. 1997; Anderson 2002). However, it is difficult to measure processes at the basin scale. Often these measurements are obtained for a few localized areas and then generalized for the entire basin. Remote sensing and GIS show promise in upscaling local measurements to regional assessments based on classification techniques.

The mountain landscape is characterized by instability and variability. Rock-strewn surfaces resulting from rapid,

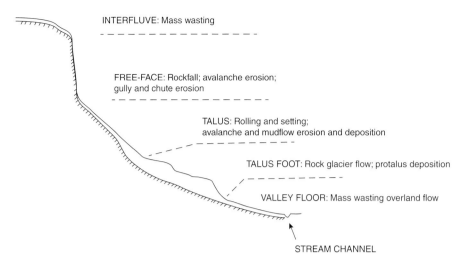

INTERFLUVE: Mass wasting

FREE-FACE: Rockfall; avalanche erosion;
gully and chute erosion

TALUS: Rolling and setting;
avalanche and mudflow erosion and deposition

TALUS FOOT: Rock glacier flow; protalus deposition

VALLEY FLOOR: Mass wasting overland flow

STREAM CHANNEL

FIGURE 5.3 Components of an alpine hillslope in the Rocky Mountains. Other extensively glaciated mountain systems would have additional horns, cirques, and greater relief at high elevations. (Adapted from Caine 1974.)

short-lived physical weathering are prevalent; exposed soil is continually being transported downslope. This is commonly observed in the rushing waters of alpine streams and in the movement of entire slopes. If you have ever spent much time above the timberline in mountains, you know that a very common sound is that of falling rocks. Large-scale features such as mudflows, landslides, and avalanches can reach catastrophic dimensions and do more geomorphic work in a matter of minutes than day-to-day processes accomplish in centuries. Such spectacular phenomena epitomize the inherent instability of the mountain system (Caine 1974). These catastrophic events begin, however, with the weakening and breakdown of bedrock.

Weathering

Weathering is the alteration and reduction of rock into finer particles, whereas erosion involves transporting the weathered particles. The processes of rock breakdown are normally divided into chemical and physical (mechanical) weathering. Recently, other biological weathering mechanisms, including biofilms such as fungi, algae, cyanobacteria, and heterotrophic bacteria, have been recognized for their potential to chemically and physically change rock (Gorbushina 2007). Some have suggested that weathering should be viewed from a critical zone framework in which unweathered rock and water are delivered to a reactor, enhancing the potential for physical, chemical, and biological weathering (Anderson et al. 2007).

Physical weathering refers to the mechanical forces that pry rocks apart or make their surfaces disintegrate. Minerals remain unchanged through physical weathering processes such as frost action, unloading, and salt crystal growth. *Chemical weathering* is the decomposition of rock by chemical alteration of the original minerals. The internal structure of minerals is destroyed, and new minerals are created. Chemical reactions tend to round the edges of rocks, producing pitted and grooved surfaces.

Physical Weathering

Frost action has typically been thought of as the main physical weathering process in mountains and cold climates. Other than frost action, little attention has been paid to other forms of weathering. The fundamental process of rock breakdown under the general category of frost action is *frost wedging,* the mechanical prying apart or shattering of rocks upon freezing. When water freezes, it expands by about 9 percent, producing pressure (Washburn 1980). For example, when liquid inside a capped bottle freezes, the expansive force can cause the bottle to break or the cap to be forced out! When water contained within rock freezes, it expands and breaks the rock apart. However, the force needed to break apart rock is much greater than that needed to break apart a bottle! The pressure exerted by freezing water is 2,100 kg/cm^2 at $-22°$C. Pressure also decreases at lower temperatures as the ice begins to crack. Under ideal conditions, ice must be in a closed system; if air bubbles were present, pressure would be reduced (French 1996). Unfortunately, rock does not meet these conditions; therefore, the simple volumetric expansion of water to ice cannot sufficiently explain frost shattering of rock. There may be other processes that enhance weathering by frost action. Taber (1929, 1930) found that ice crystals grow as additional water is attracted to the freezing plane through molecular cohesion. In this respect, water is attracted to the freezing surface from the surrounding area and accumulates (Taber 1929). As this occurs, the additional volume may help wedge rocks apart. The most important of several factors controlling the efficiency of these processes are the availability of water (height of the water table and degree of saturation), the composition of the rock (tensile strength, surface area, porosity, and hydraulic conductivity), and temperature (the rate, extent, and intensity of cooling) (Matsuoka 1991).

Some question the role of frost action in rock break-down, and instead support other processes such as *hydration shattering* (White 1976). In this process, molecular water is adsorbed on the surface of silicate minerals, so that pressure is exerted on surrounding rock surfaces. According to White (1976), a force of over a t cm^{-2} can be generated, greatly enhancing weathering. Numerous other types of weathering such as wetting and drying, rock fatigue, thermal shock, and salt crystal growth, as well as chemical and biological processes, could likely enhance frost wedging (Williams and Robinson 1991; French 1996; Hall 1998). For this reason, some prefer the term *cryogenic weathering,* which refers to mechanical–chemical process of rock breakdown (French 1996). Frost action remains difficult to corroborate in the field; what mountain geomorphologists need more than anything else is a method to determine the processes responsible for the weathering of rock fragments (Thorn 1992).

Frost weathering in mountains can be observed in the moist zone immediately below late-lying snow patches (Thorn 1978). The effects can also be seen at the moist bases of notched cliffs and rock walls (Gardner 1973). Porous and multijointed rocks that allow water to infiltrate usually experience more weathering than impervious, dense rocks. Sedimentary rocks are generally more susceptible to frost shattering than hard crystalline rocks, although much depends on their individual characteristics. Crystalline rocks with deep cracks may be more susceptible to frost action than compact sedimentary rocks. Frost wedging is most effective on rocks with reticulate hairline fractures, where moisture is allowed to penetrate but does not drain away, as is often the case with large cracks (Tricart 1969). Coarse and angular debris is the typical product of frost wedging. The exact size and shape of the by-products depend upon the type of rock and the intensity of the frost action. Rocks with stratification planes, such as slate or schist, typically break into flat slabs, whereas massive rocks, such as limestone or granite, shatter more randomly.

Once the bedrock has been broken, the effectiveness of frost weathering increases, regardless of the size of rock fragmentation. More surface area is exposed, the material diminishes in size, and its water-holding ability increases. The ultimate size to which material can be reduced by frost wedging is thought to be that of silt, although a small amount of clay may result from additional chemical weathering (Washburn 1980). Although the absence of clays in most mountain environments may be interpreted as the dominance of physical weathering over chemical weathering, it is likely that high winds remove most clay-sized particles.

Physical forces have been thought to account for the greater share of rock disintegration compared to chemical weathering. Rocks are subjected to rapid and unequal heating and cooling as well as to pressures exerted by ice-crystal growth. *Insolation,* or radiation from the sun, is intense at high altitudes, creating rapid and extreme temperature differences. In theory, each heating and cooling cycle causes unequal expansion between the surface and the interior of the rock, so that it may eventually become weakened and break apart. Other lines of evidence, however, suggest that heating and cooling alone is not sufficient to break rocks apart. Temperature fluctuations often do not penetrate deeper than 5 cm and would likely not break rock apart (Caine 1974). Laboratory experiments involving rapid heating and cooling cycles on rocks in dry air have also failed to produce any noticeable weakening of rocks, but when water is added, a definite weakening is observed. It was originally thought that number of freeze–thaw cycles were important for rock weathering, when in fact, the duration, intensity, and fluctuation in temperatures below 0°C are more important parameters of rock break-up (Barsch 1993). Fahey (1973) observed 238 diurnal freeze–thaws over 22 months at 2,600 m, but only 89 cycles at 3,750 m. On Baffin Island, there is little relationship between the air temperature outside and at the bottom of the bergschrund (30 m), the highest crevasse on a glacier. The sheer coldness of a frigid alpine environment creates stress which may help weather rock (Gerrard 1990). Therefore, it is likely that both physical and chemical processes are involved, each complementing and reinforcing the other to maximize weathering effects.

One result of these combined processes is *granular disintegration:* Grains are loosened around the periphery of a rock and crumble away. It is possible to brush the surface of such rocks and have the particles crumble into your hands! Granular disintegration is most effective in rocks with large crystals, particularly in rocks composed of light and dark minerals, which enhances differential heating and cooling (Washburn 1980). A related process is *exfoliation:* The rocks spall, or peel off, in concentric scales or layers, much like the skin of an onion, creating rounded features. Exfoliation may result from the chemical decay of minerals as well as from the volumetric expansion and contraction of the rock surface through heating and cooling. A similar process called *unloading* is an important weathering factor in mountains that have been heavily glaciated or are tectonically active and have had their overburden removed (Gerrard 1990).

Chemical Weathering

Very little research has been done on the role of chemical weathering in mountains. This is not only because of the apparent dominance of frost action, which tends to mask the more subtle effects of chemical action, but also because of the difficulty in measuring and studying chemical processes. Chemical weathering requires the compounds to be in solution and a certain amount of heat to be supplied for a chemical reaction to occur. Therefore, Peltier (1950) and others thought it was simply too cold and arid when water is frozen in deserts, polar regions, and most mountaintops for chemical weathering to occur (Fig. 5.4). Mountains in the humid tropics or near oceans generally show

FIGURE 5.4 Chemical surface weathering of a rock in the Sangre de Cristo Mountains of Colorado. Note the hollowed depressions on the face of the rock. (Photo by J. R. Janke.)

more active chemical weathering than mountains in continental and arid regions. There is also evidence of some chemical weathering even in arctic and continental mountain climates (Washburn 1980). Certain types of chemical processes such as *carbonation* (the alteration of minerals that contain basic oxides to carbonate salts by action of carbonic acid) may actually be accelerated in cold climates because of the greater solubility and concentration of carbon dioxide (CO_2) at low temperatures (Caine 1974). Furthermore, snow meltwaters are acidic; therefore, chemical weathering must play an important role in the alpine environment (Gerrard 1990).

Although reduced temperatures hinder chemical weathering, glacial activity can enhance chemical weathering by providing fine sediment that is easily weathered, as well as abundant water to enhance solution (Sharp et al. 1995). Silica fluxes and mean silica concentrations increase with water discharge in glacial catchments. As erosion increases, mineral surface area increases. With abundant meltwater, chemical weathering becomes more common (Anderson 2005). Anderson et al. (1997) could not differentiate between rates of chemical weathering in glacial and nonglacial catchments. In the granitic and metamorphic mountains of the southern and central Rockies, rates of chemical denudation are generally low, equivalent to 1 mm ka^{-1} (Caine 2001). In Europe, rates of solute denudation are only about 2 mm ka^{-1} (Walling and Webb 1986). Given the close correspondence with mechanical denudation, chemical weathering is likely more significant than previously recognized. Periglacial chemical weathering is generally slower compared to tropical areas; however, chemical weathering is often a dominant agent of mass removal (Dixon and Thorn 2005). These chemical weathering processes likely play an important role in mountains, though they are often overlooked (Thorn et al. 2006).

Rock surfaces often present evidence of chemical weathering in their altered and weakened exteriors. Chemical weathering may also be ascertained by measuring the quantity of dissolved minerals being transported by streams. The presence of a *weathering rind,* the chemically altered zone at the rock surface, can provide an estimate of the amount of chemical weathering. The depth of weathering provides an index of the age of the surface and the rate of the process under varying conditions. Although it penetrates only a few millimeters, the weathered zone can easily be identified when the rock is broken apart. The weathering rind commonly has a slightly reddish-brown color because of the oxidation of iron and manganese silicates. The exterior is strengthened at the expense of the interior, so that once the surface is breached, the underlying rock deteriorates at a rate faster than the protective covering.

Much of the effect of chemical weathering is not immediately visible, since it involves the removal of minerals through solution. When precipitation infiltrates rock, the water soon becomes loaded with dissolved salts that are subsequently transported away. A classic example of this is the solubility of limestone, which has created extensive caverns in various parts of the world. This process is called *hydrolysis;* it involves not merely absorption of water, as in a sponge (hydration), but a specific chemical change that produces a new mineral. For example, the process begins when water absorbs carbon dioxide (CO_2) to form a weak solution of carbonic acid (H_2CO_3). This then reacts with the calcium carbonate ($CaCO_3$) in limestone and produces a soluble salt, calcium bicarbonate, $Ca(HCO_3)_2$. Calcium bicarbonate is easily dissolved and transported in solution. Other minerals are less susceptible to chemical reactions than the calcite in limestone, but all undergo some reaction with water. Consequently, the quantity of dissolved solids in a mountain stream provides a good index of the rate of chemical weathering. However, some care must be taken to account for the quantity of particulate matter dissolved from the atmosphere and from the decomposition of organic material. Mountains located in marine environments and those containing highly soluble rocks display the most rapid rates of dissolution. In the limestone regions of the Alps, the solution rate is about 0.1 mm/yr (Caine 1974). In dolomite (limestone containing magnesium) areas of the White Mountains in eastern California, the rate is only about 0.02 mm/yr, since the climate is much drier and the dolomite is less soluble (Marchand 1971).

Rock types show considerably different rates of chemical weathering. In the Wind River Range, Wyoming, the solution removal from granitic rocks on the southwest side is about 7 t $km^{-2} yr^{-1}$, while sedimentary rocks on the northeast side dissolve at a rate of 19 t $km^{-2} yr^{-1}$ (Hembree and Rainwater 1961). Similarly, in the Sangre de Cristo Range, New Mexico, average solute contents of stream waters draining quartzite, granite, and sandstone are in the proportion of 2:5:20, an index of the relative solubility of each of these rocks in this area (Miller 1961). Similarly, Bluth and Kump (1994) suggested that granitic rocks tend to have lower rates of chemical weathering compared to sedimentary rocks.

Other forms of chemical weathering also deserve mention. *Dissolution* is a process whereby rock material turns directly into solution, like salt in water. It occurs because water is one of the most effective and universal solvents. Because of the polarity of the water molecule, practically all minerals are soluble to some extent in water. *Oxidation* is the combination of oxygen with a mineral to form a new mineral with a higher oxidation state (ionic charge). Of the elements that have variable charges, iron is the most important on Earth. After oxidation, rocks show rusted or reddish colored surfaces.

Frost-related Features and Processes

Frost also plays an important role in the unconsolidated deposits that result from weathering. Frost heaving, frost thrusting, and needle-ice growth give rise to distinctive geomorphic features. Since frost processes often increase as vegetation is destroyed, a landscape can become more vulnerable to disturbances such as overgrazing or shrub removal, so that it may take a very long time for disturbed surfaces to be colonized or stabilized by vegetation. The effectiveness of frost action is determined by its intensity and duration. Both of these factors are reflected in the presence of frozen ground.

Seasonally Frozen Ground

Seasonally frozen ground freezes and thaws every year. Winter freezing penetrates the surface most deeply in the subarctic and much less in lower latitudes. The zone of freezing and thawing in mountains migrates seasonally. The vertical mobility and extent of this belt are largely determined by latitude and altitude. The zone of freeze and thaw on a tropical mountain occupies a relatively narrow vertical extent and remains stationary. If the mountain is high enough, there may be an area near the top where thawing seldom occurs. At middle elevations, a transitional zone occurs where seasonal freezing and thawing take place. Freezing seldom takes place at lower elevations. In the middle latitudes, however, there may be a zone of continually low temperature at the top, but freezing and thawing extend to the lowlands in winter. The zone of freeze and thaw is narrowest in summer and widest in winter. It is compressed as the zone moves up the mountains in the spring, and is extended as it moves down in the fall; the width of this belt may also vary significantly with aspect.

Although the intensity and duration of low temperatures are the major factors controlling the depth of frost penetration, site conditions are also very important. Soil composition, moisture presence, vegetation, snowcover, and snow depth all greatly affect the rate and depth of freezing (Fahey 1973, 1974). For example, snow is an excellent insulator. Mountain surfaces covered with deep snow are relatively protected from freeze and thaw as well as from deep frost penetration. In the Cascade Mountains of Washington, frost penetration is relatively unimportant because the landscape is covered by a blanket of heavy snow in winter. In continental mountains, where there is less snowfall and temperatures are colder, the depth of freezing increases. This can pose special problems for human settlements, as special care must be taken when building on seasonally frozen ground.

Permafrost

Permafrost is defined as soil, bedrock, or any other material that has remained below 0°C continuously for two or more years (Ives 1974; Williams and Smith 1989; French 1996). This definition is based purely on thermal conditions and, in the strict sense of the word, a glacier is considered a form of permafrost, since its temperatures are continually at or below freezing. Usually, however, the term permafrost is used to describe the thermal state of the ground. Permafrost usually is superimposed by an *active layer,* a layer that experiences seasonal freezing and thawing. The thickness of the active layer depends on many variables, the most important of which are grain size, moisture content, and temperature.

Approximately 25 percent of the Earth's surface is underlain by arctic, alpine, and high-plateau permafrost (French 1996), including almost 80 percent of Alaska. Arctic permafrost has been traditionally grouped into continuous and discontinuous zones, although Ives (1974) argued for retention of the *sporadic* category. *Continuous* permafrost exists in the coldest areas; it is typically found in regions closer to the poles and in continental interiors, such as the central Keewatin Region, Northwest Territories, Canada. In Canada, the southern limit of continuous permafrost coincides with the −6 to −8°C mean annual air temperature (MAAT) isotherm, and discontinuous permafrost ends at the −1°C MAAT isotherm. *Discontinuous* permafrost often contains a more extensive network of *taliks,* or unfrozen sections linked to some local variable such as the presence of a former lake. Mountain permafrost can be categorized into continuous permafrost, found at the highest elevations, but below the firn line; discontinuous permafrost; and sporadic permafrost (Ives and Fahey 1971; Harris 1988; French 1996). Mountain permafrost is much more complex than arctic permafrost. Slope, aspect, elevation, vegetation cover, soil type, and snow distribution all play important roles in the formation of mountain permafrost. There is little precise information about permafrost distribution in the contiguous United States (Ives and Fahey 1971; Péwé 1983), though 100,000 km² of mountain permafrost could exist, with the lower limit extending as low as 2,500 m in Washington and as high as 3,500 m in Arizona.

The presence of permafrost may adversely affect engineering projects such as mining, road building, well drilling, and the installation of structures such as ski lifts, power lines, or cell phone communication towers. Special measures must be taken to maintain the thermal balance of frozen ground. For instance, in order to minimize heave, roads are often built on a well-drained, coarse layer that is underlain with an insulating material such as peat. If these conditions are not met, the road will quickly buckle under the force of periglacial processes. If ground ice is present, it may pose further problems because of differential melting and settling. For example, in the Colorado Rockies, buildings installed near the summits of Pikes Peak and Mount Evans at 4,000 m have experienced considerable settling because of melting ground ice (Ives 1974).

The thermal relationships involved in the origin and maintenance of mountain permafrost are complex. Ideally, permafrost should develop in any area where the MAAT is at or below −1°C. Local site conditions, however, cause considerable variation. Heat will flow from warm to cold areas. Snowfall can hinder or help develop mountain permafrost. A winter snowfall of 65–70 cm is sufficient to prevent the development of permafrost (French 1996). A heavy snowfall in the fall or early winter will inhibit frost penetration, whereas a low snowfall will do the reverse. If snow persists into the summer or if summer snowfall occurs, melting will be delayed, enhancing the possibility of permafrost presence. Wind or avalanche activity can redistribute snow in springtime, thus preserving cold temperatures at the base of slopes and protecting the permafrost from intense summer insolation (Harris and Corte 1992). Vegetation has a complex effect on mountain permafrost. Vegetation can act as an insulator if it is thick. However, trees can shade the ground from solar radiation and intercept winter snowfall. As a result, winter cold can

penetrate the ground more deeply, and summer radiation is restricted. Soil conditions can also affect permafrost formation. Permafrost is thickest beneath well-drained, bare soil or rock, and thinnest in areas of poorly drained soils. The presence of large surface boulders may enhance the Balch effect, in which cold air displaces warm air in a ventilation system between boulders (Harris and Corte 1992). Forest fires can also influence permafrost formation. If a fire passes rapidly through an area and only trees burn to the ground, the permafrost may not be affected; however, burning usually increases the thickness of the active layer. The spatial variability of these factors makes it difficult to predict permafrost presence.

In the continental United States, mountain permafrost is most easily found in the Rocky Mountains (Ives 1974). It is usually restricted to elevations above 3,500 m in Colorado but is lower to the north (in Montana and Wyoming) (Pierce 1961). At the Continental Divide between Alberta and British Columbia, permafrost is common above 2,600 m (Scotter 1975). On the west coast, isolated occurrences of permafrost occur in the White Mountains of California and the Sierra Nevada (Retzer 1965). Permafrost has been observed in the Coast Range of southern British Columbia at 1,800 m (Mathews 1955).

In the Alps, isolated occurrences of permafrost have been reported at elevations as low as 2,300 m; by 2,700 m, conditions become more favorable (Barsch 1969). Permafrost is also reported from the high ranges of the Caucasus, northern Urals, Pamirs, Tien Shan, Karakoram, and Himalayas (Gorbunov 1978). Permafrost exists at 3,969 m on Mauna Kea in Hawaii (Woodcock 1974). In the southern hemisphere, permafrost occurs on some of the higher peaks, especially in the central Andes and the Southern Alps of New Zealand (Schrott 1991; Kirkbride and Brazier 1995).

PERMAFROST DISTRIBUTION MODELING

The extent of permafrost has been investigated using various field techniques, including excavations, bottom temperature of winter snow (BTS), and geophysical methods (Gerrard 1990; King et al. 1992). These field methods, however, are often expensive, time consuming, restricted in spatial extent, and do not depict the spatial variability of permafrost occurrence. Modeling techniques using GIS and DEM variables have been used to estimate the distribution of mountain permafrost (Etzelmüller et al. 2001a, 2001b; Heginbottom 2002). Automated mapping of alpine permafrost began with the creation of the PERMAKART and PERMAMAP empirical models for parts of the Alps (Keller 1992; Hoelzle et al. 1993). PERMAKART utilizes three variables: (1) mean annual ground temperature (MAGT), derived from mean annual air temperature (MAAT); (2) potential solar radiation; and (3) thickness of snow cover. PERMAMAP is based on the relationship among BTS measurements, MAAT, and potential direct solar radiation (Hoelzle et al. 1993).

Similar models have been formulated using other physiographic, climatological, and biologic variables (Imhof 1996; Frauenfelder 1997; Gruber and Hoelzle 2001). PERMAMOD joins topoclimatic information with biogeographical features, such as cold water, absence of marmot burrows, perennial snow patches, and sites of rock glaciers (Frauenfelder 1997). Stocker et al. (2002) developed PERMEBAL, which simulates snow cover persistence and ground temperatures of snow-free points, based on meteorological and site-specific data. Guglielmin et al. (2003) designed the PERMACLIM model, which calculates ground temperatures for DEM points by including data from a climatic database and snow thermal characteristics. Janke (2005c, 2005d) used rock glacier topographic information (elevation, slope, and aspect) combined with a land cover weighting procedure to model distribution in the Front Range of Colorado. Despite the variety of models, uncertainty still exists as to which is most useful and effective for accurately representing the distribution of permafrost (Frauenfelder et al. 1998; Lugon and Delaloye 2001).

Modeling results need additional verification in the field. For example, several studies suggest that the D-1 site (3,739 m with a MAAT of −3.5°C) along Niwot ridge, Colorado, contains permafrost (Ives 1974; Ives and Fahey 1971; Janke 2005c). Modeling results suggest a 63 percent probability of permafrost occurrence at D-1 (Janke 2005c). However, through an electric resistivity tomography survey, Leopold et al. (2010) found no evidence of ice from the surface to a depth of 10 m near D-1. Permafrost occurrence, however, is defined by freezing temperature for at least two consecutive years; thus, it can be dry or ice-free. Boreholes lined with temperature data loggers would provide additional direct measurement of possible permafrost occurrence.

PERMAFROST AND CLIMATE CHANGE

Since the definition of permafrost is based on thermal criteria, scientists are naturally concerned with the implications of global warming. *General circulation models* (GCMs) usually take into account solar radiation, greenhouse gases, or boundary conditions to predict future climates. Predicting climate change in mountain regions is always difficult because of the need for high temporal and spatial resolution, which is often lost in GCMs (Barry 1994; Beniston 1994). In fact, the Alps are not even perceived by some models, nor is the physical influence of mountains (gravity-wave drag). Nested models, in which the GCM focuses on a detailed region, provide a method for improving prediction in the mountains. However, the feedback mechanisms of what appears to be a chaotic system still create a great deal of uncertainty.

Results from some GCMs provide insight as to potential future changes in climate. Barry (1992) stated that most GCMs based on a doubling of CO_2 by 2030 to 2050 suggest a 3° ± 1.5°C increase in mean global surface temperatures. Warming is amplified two to three times at the North Pole

because of greater land area in the northern hemisphere and positive feedbacks such as reduction of snow and sea ice extent. CO_2 levels have already risen from about 280 ppm (preindustrial) to 379 ppm in 2005. Other estimates are also available based on a doubling of CO_2: For the Rocky Mountains, Beniston (1994) showed a 2–4°C increase in temperature with no increase in precipitation and, for the Alps, a 2–4°C increase in temperature with drier conditions. However, another regional climate model of the Alps showed a 1.5°C increase in temperature in winter with more snow, and a 4°C increase in summer with less precipitation. According to data from SNOTEL sites across Colorado, annual median air temperatures have increased 0.7°C per decade from 1986 to 2007 (Clow 2010). In the Loch Vale watershed, temperatures have increased 1.3°C per decade from 1983 to 2007. Over the same period, the D-1 station on Niwot Ridge has warmed by 1.0°C per decade (Clow 2010). The time period analyzed is an important consideration: Short-term cooling anomalies superimposed on a longer warming trend may cause misinterpretation (Pepin and Losleben 2002).

According to the Intergovernmental Panel on Climate Change's 4th Assessment Report (Solomon et al. 2007), the next two decades will experience a warming of about 0.2°C per decade. Even if the concentrations of all greenhouse gases and aerosols had been kept constant at year 2000 levels, a further warming of about 0.1°C per decade would be expected. Beyond this time frame, temperature projections will depend on specific emissions scenarios. There is great uncertainty and variability as to feedbacks that may occur and their impacts on ecosystems. In particular, mountain climatic processes must be better understood in order to predict their effects on changing mountain environments (Beniston 2000).

Several sources of evidence describe the recent effects of climate change on permafrost. At a depth of 20 m on the Tibetan Plateau, temperatures have risen by 0.2–0.3°C over the last two decades. Lachenbruch and Marshall (1986) have shown a 2°C increase in the upper 2 m of permafrost during the last several decades to a century in Alaska. Permafrost in the Alps warmed at 0.1°C yr^{-1} because of warming in the 1980s. The response is slow because heat takes time to travel to great depths, but it is departing from the trend of natural variability (Haeberli 1994). Rock glaciers and glaciers in the Alps have shown accelerated rates of degradation (Haeberli 1994). Permafrost has responded by a thickening of the active layer, thaw settlement, and the alteration of the temperature profiles within the permafrost. At Gruben rock glacier, rates of surface subsidence accelerated by a factor of 2 to 3 in the warm 1980s to 1990s compared to the 1970s. Borehole temperatures on Murtèl I rock glacier have also shown an increase in temperature, a trend that was more consistent at depth as the influence of a seasonal signal was less. Recent measurements indicate that the ground has warmed by 0.2°C over the last 10 years in the Rocky Mountain Front Range, and a substantial rate of increase has occurred since 2000 (Caine 2010). Increased stream discharge in the late summer and fall is the result of increased melting of ice previously stored within permafrost (Caine 2010). The current rate of degradation seems to be evolving beyond the limit of natural Holocene variability.

Future warming will likely have diverse effects on the existence of permafrost. With increasing temperatures, permafrost may become a "historical curiosity" in which such landforms and processes no longer exist (Barsch 1993). Permafrost zones and processes may shift, and increased melting of permafrost will thicken active layers (Price and Barry 1997). Thermokarst could dominate in some regions where ground ice is being depleted, making development of roads, pipelines, or airfields more difficult. Development on permafrost will become difficult with future warming because most builders try to keep permafrost intact. In a warmer climate, developers must address differential heaving, sinking of portions of the surface, destruction of bridges by spring floods, burial of pavement by slumping or mass movements. Future development on permafrost will have to use new technology or techniques for construction (Ritter et al. 2002).

Climate change will also affect other microclimate variables related to permafrost aggradation or degradation (Williams and Smith 1989). A series of complex feedbacks will determine permafrost survival. Greater rates of evaporation will cause cooling of the ground. An increase in summer rainfall, or decreasing winter snowfall, may offset some permafrost warming compared to the magnitude of air temperature increase (Wu and Zhang 2008). Thermokarst from melting permafrost may act as snow traps in winter, thus further warming the soil. With more CO_2 available, vegetation could have higher productivity. Wetter (but not necessarily warmer) soils could enhance carbon and nitrogen production, which would provide an additional greenhouse gas contribution (Baumann et al. 2009). When warmer temperatures occur at higher elevations, vegetation will succeed or encroach upon alpine tundra where permafrost commonly occurs (Beniston 2000). Changes in vegetation patterns will also influence local microclimatic variables. Areas with no vegetation show the greatest annual range of temperature. As trees encroach, the annual amplitude will be less, which, depending upon the temperature regime, could cause permafrost to aggrade or degrade. Trees will also act as snow accumulators, which is far less important than their shading and cooling effects. Areas with trees will be cooler than tundra in the summer, but warmer in the winter because of snow fencing and trapping of snow (Williams and Smith 1989).

The timing, depth, and duration of snowpack will have an impact on permafrost existence. The date of spring snowpack melt has come earlier over the past 54 years in Australia (Green and Pickering 2009). Clow (2010) showed that MAAT at high elevations in the Colorado Front Range has increased by about 1.0°C per decade from 1983 to 2007, and that the timing of snowmelt is two to three weeks earlier. The mountain west of North America, especially the Cascade

Mountains and northern Colorado, has experienced declines in spring snowpack, despite increases in winter precipitation (Mote et al. 2005). An increase in winter snowpack has the potential to warm ground through insulation, and earlier spring snowmelt exposes the ground to warmer summer temperatures. Each is detrimental for permafrost occurrence.

Predicting future change will be improved through the use of geospatial models, such as ALPINE3D, which measures alpine snow processes (Lehning et al. 2006). Geospatial tools combined with field measurements can help investigate permafrost development or destruction under a variety of scenarios.

Warming of permafrost will also initiate other hazards. Groundwater seepage causes some rockfalls; therefore, if groundwater pressure rises with increased precipitation induced by a wetter climate, this could trigger rockfalls. As permafrost degrades, there will be less cohesion in the soil, and thus stable slopes may become active. Recent warming has accelerated the processes associated with recent deglaciation: glacial avalanches, landslides, slope instability, and outburst floods (Evans and Clague 1994). Sheet slides and skin flows could increase as a result of moisture oversaturation and high pore water pressure associated with melting permafrost or early snowmelt. Rock glaciers may become unstable and create hazards as debris is no longer consolidated by ice (Barsch 1993). Kääb and Vollmer (2000) used computer-aided aerial photogrammetry to map disasters and hazard potentials in a section of the Swiss Alps. Glacier lake outbursts in 1968 and 1970 from the Gruben area caused floods up to 400,000 m^3 and debris flows moving material at 15 m^3s^{-1}, leading to heavy damage in the village of Saas Balen. Creeping permafrost or rock glaciers can eventually move into steep channels, creating a dam that could block a river or cross a hiking path.

Because of the fragility of mountain environments, it is almost impossible for systems to return to their initial state after disturbance (Beniston 2000). The same conclusion can be drawn with respect to permafrost. The *sensitivity*, or the range of stress a system can experience without damage; the *vulnerability*, or the extent to which it may be damaged by stress; and the *adaptability*, or the degree to which it may adjust to stress, are factors that must be examined in detail.

Frost Heave and Thrust

When ice lenses (horizontal accumulations of ice crystals) build beneath the surface, they cause the ground to expand in the direction of the ice-crystal growth. The most common direction is vertical (*heave*), toward the soil surface, since this is the source of the cold. Expansion may also take place laterally (*thrust*) because of the variations in conductivity of heterogeneous materials (Washburn 1980). Frost heave and thrust are the major causes of stirring and disruption of the soil. As the soil freezes, water is attracted to the freezing plane, where it accumulates in the form of ice lenses that push the soil upward.

FIGURE 5.5 Freshly upheaved stones on a gentle slope at 3,400 m near Deluge Lake in the Gore Range near Vail, Colorado. (Photo by J. R. Janke.)

Frost heave and thrust are often evidenced by the upheaval and ejection of rocks from depth. Primary frost heave occurs at the frost line and is the result of ice expansion. Secondary frost heave occurs as ice continues to expand over time because of the attraction of water and concentration of ice in soil layers. Rocks that are freshly upheaved have newly exposed unweathered surfaces that are lichen-free. Heave is made evident by disrupted vegetation and soil surfaces. In some cases, rocks may be heaved 1–2 m above the surface and stand like lonely tombstones amid the tundra (Price 1970) (Fig. 5.5).

The principal processes involved in the movement of stones through frost heave and thrust fall into two groups: *frost pull* and *frost push*. Frost pull operates when the soil freezes and expands upward, pulling rocks with it. Upon thawing, the rocks do not return to their original positions, as soil fills the voids beneath the rock. Thus, the stones migrate toward the surface by increments with each freeze and thaw cycle. Frost push results from ice lenses growing underneath stones and pushing rocks upward. Because rocks have greater conductivity than soil, heat (or cold) can pass through them more quickly than through the soil. Therefore, ice lenses can accumulate at their bases and cause differential heaving. As with the frost pull mechanism, finer material then seeps into the cavity underneath and prevents the full return of the rock to its initial position (Washburn 1980). Ample water and a stable freezing front are critical for the development of ice lenses that produce heave (Smith and Patterson 1989). Otherwise, pore ice will form if the freezing front moves rapidly or water supply is cut off. Segregated ice is more likely to develop in fine-grained soils, whereas pore ice will develop in coarse-grained soils and will not produce

heaving. Under natural conditions, frost pull and frost push operate simultaneously and are difficult to separate, but each contributes to rocks moving toward the surface.

It was long believed that much of the frost-stirring evident at the surface in mountain environments was caused by frequent freeze–thaw cycles. However, Fahey (1973) and Thorn (1979) proved that diurnal (day to night) freeze and thaw generally penetrates no more than about 10 cm, even on bare and exposed surfaces. Frost heave and thrust depend on relatively deep freezing, so their operation is essentially limited to once a year, on the annual freeze–thaw cycle. The prolonged and intense cold of the alpine winter is more responsible for the major features of frost heave. Needle ice, however, operates at the surface in response to diurnal freeze and thaw.

Needle Ice

Needle ice consists of small individual columns or filaments of ice 1–3 cm high, projecting from the soil surface (Fig. 5.6). Each ice needle originates directly from the soil, and is usually capped by a thin ice layer on which dirt and small rocks can settle. The needles may be densely packed like brush bristles or in scattered columns, depending on variations in soil moisture and the extent of freezing. Needle ice typically forms at night and melts during the day. If it does not melt during the day, and new growth accumulates the following night, a tiered or storied effect may develop. Rarely, three to four layers may develop, each representing a separate needle-ice event.

Needle ice requires a calm, clear night with freezing temperatures and a fine soil with high moisture content. Under these conditions, the ice begins to segregate and grow as

FIGURE 5.6 A stone polygon with needle ice development. (Photo by J. D. Vitek.)

the water migrates to the freezing plane (Outcalt 1971). The needles grow from their bases and push upward at a rate that may reach several centimeters per hour. One requirement for good needle ice growth is a fine soil that can hold ample water and will allow rapid migration of the water to the freezing plane. A soil that is too coarse may not hold enough moisture, while tight clays may inhibit the migration of water. A silty loam seems to be the ideal texture.

Patterned Ground

Patterning of rocks, soil, and vegetation into various geometric forms is a common feature of high mountain landscapes (Fig. 5.7) These patterns fall into three basic categories: polygons, circles, and lines or stripes. These "surface markings" or "structure soils" as they have been called, range in size from tiny features measured in centimeters to large-scale forms several meters across. Because of their striking geometric arrangement and curious nature, they have attracted a great deal of attention (Troll 1958; Washburn 1956; Gleason et al. 1986; Odegard et al. 1988; Warburton and Caine 1999; Haugland and Owen 2005; Haugland 2006). The exact mechanism of their origin is still controversial. Washburn (1956) listed 19 different theories that have been proposed to account for their development! One problem is that similar patterns can be caused by different processes. The polygon, for example, is one of the most ubiquitous forms in nature. It can be created by thermal contraction upon freezing, by the drying and cracking of soil in a mud puddle or dry lakebed, or when molten lava solidifies. It is not surprising, then, that the process of freezing and thawing also creates polygons.

A useful contribution to the study of patterned ground was the classification devised by A. L. Washburn in 1956. It is based on only two criteria: geometric form and the presence or absence of sorting (the segregation of rocks and fines). The basic types of patterned ground are sorted and nonsorted circles, polygons, nets, steps, and stripes. A pattern is sorted if particle size varies from one part of the feature to another. A sorted circle, for example, has finer material in the center and larger particles around the perimeter. Frost sorting depends on (1) the amount of available moisture (best in saturated soils), (2) the rate of freezing (best with slow migration), (3) the particle size distribution of the soil (fine-grained is best), and (4) the orientation of the freezing plane. A nonsorted circle displays no difference in particle size; its form is usually determined by the bordering vegetation. Similarly, sorted polygons contain rocks along the pattern lines with finer material between them, while nonsorted polygons are the result of a cracking matrix at the surface with no difference in the size of material. In Finnish Lapland, remote sensing data helped classify nonsorted patterned ground, which is more common at lower elevations where the ground moisture and vegetation are abundant, and sorted patterned ground, which occurs at high elevations with steeper slopes and sparse vegetation (Hjort and Luoto 2006). Sorted stripes (stone stripes) are marked by alternating zones of larger and finer particles, while nonsorted stripes may be identified by ridges and furrows or by alternating bare and vegetated strips. Although both types of patterned ground present recognizable and conspicuous patterns, the sorted variety is the most intriguing, both because it has a more striking appearance and because the segregation of rock sizes suggests a more complex origin. Moreover, the sorting is both vertical and horizontal. The larger particles that compose the borders characteristically accumulate in a wedge shape. The cross section of the wedge has the widest part near the soil surface and tapers downward. In addition, the largest rocks are near the surface, progressively getting smaller with depth.

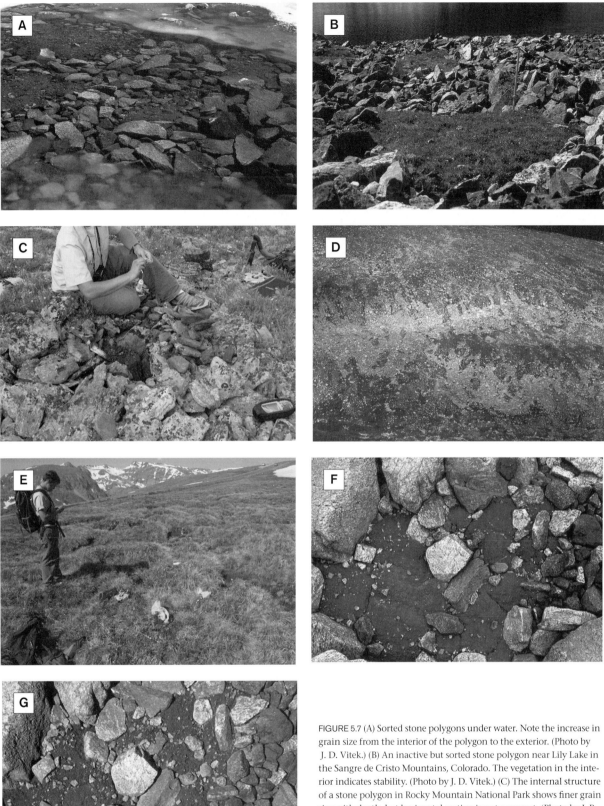

FIGURE 5.7 (A) Sorted stone polygons under water. Note the increase in grain size from the interior of the polygon to the exterior. (Photo by J. D. Vitek.) (B) An inactive but sorted stone polygon near Lily Lake in the Sangre de Cristo Mountains, Colorado. The vegetation in the interior indicates stability. (Photo by J. D. Vitek.) (C) The internal structure of a stone polygon in Rocky Mountain National Park shows finer grain size with depth, but horizontal sorting is not apparent. (Photo by J. R. Janke.) (D) Elongated stone stripes in the Sangre de Cristo Mountains, Colorado. (Photo by J. D. Vitek.) (E) Regularly spaced earth hummocks in the Colorado Rockies. (Photo by J. R. Janke.) (F) An active stone polygon from August 1978. (Photo by J. D. Vitek.) (G) The same stone polygon in July 1987. Note the displacements that have occurred. (Photo by J. D. Vitek.)

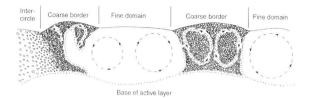

Base of active layer

FIGURE 5.8 Schematic model for free convection of soil within the active layer. (From Hallet et al. 1988.)

ORIGIN OF PATTERNED GROUND

Patterned ground can be created by a number of different processes in a wide variety of environments and materials (Washburn 1980). Patterned ground features have been identified on Mars and may indicate possible sources of ice or may aid in climatological reconstruction (Mangold 2005). The initiating process in polygonal patterns is generally cracking of the surface because of desiccation or frost cracking, while circular patterns are probably the result of frost heave. Sorting of materials is caused by various processes, but in cold climates it is mainly because of frost action.

Frost heave, frost thrust, and needle-ice growth all contribute to the segregation of rocks and fines. The precise mechanisms are still unknown, but the sorting can be explained in a general way. The surface of the ground typically consists of a heterogeneous mixture of coarse and fine particles. The fines hold more water than the coarse material and will expand more when they freeze. The fines also cohere and contract more when they thaw, compared to coarse material. With each expansion, particles move outward from the freezing nucleus; when they settle back, they fail to return completely to their original positions. The fines tend to congregate in these cycles of expansion and contraction, leaving the coarser materials in the intervening areas. The process continues until the centers of fine material begin to impinge on each other, with the larger particles forming the perimeters of the polygons or circles. In summary, vertical heaving selectively moves coarse particles up, and lateral sorting moves fine particles away from the advancing (top or sides) freezing front. Mechanical sorting occurs as frost mounds force coarser rocks to roll to the base of a mound. In another model based on convectional movement, a rock can migrate upward through denser overlying soil or rock, driven by buoyancy forces developed seasonally in the active layer (Gleason et al. 1986; Anderson 1988; Hallet and Waddington 1992; Hallet 1998; Harris 1998a, 1998b) (Fig. 5.8). Eventually, vegetation begins to establish at the edges of patterned ground features and gradually encroaches inward through the following successional

stages: (1) bryophytes and lichens, (2) grasses and sedges, and (3) woody shrubs (Haugland and Beatty 2005).

BLOCKFIELDS, BLOCKSLOPES, AND BLOCKSTREAMS

When bedrock is shattered by weathering, the result is an extensive array of angular stones. The German term *felsenmeer,* meaning "sea of rocks," has been used to describe this phenomenon. Such blockfields, blockslopes, and blockstreams are common in alpine landscapes (White 1976; Fig. 5.9). Blockfields and blockslopes run parallel to the contour and are differentiated according to their slope. Blockfields have the gentlest slope, ranging from 3 to 12°; blockslopes are steeper, with slopes greater than 12° (White 1981). Blockfields occur on crests, divides, low saddles, cols, or nearly level rock benches; they are considered a subtype of a blockslope (White 1981). Rockstreams, however, run perpendicular to the contour and often occupy valleys. They exhibit movement and are similar to stone stripes except for their larger size and because they may occur individually (Caine 1968).

Although blockfields have various origins, frost action was generally thought to be the primary formation factor (Washburn 1980). However, blocks may be produced from azonal weathering processes such as chemical weathering and thermal stress fatigue (Boelhouwers 1999). Bedrock fractures may be a prerequisite to augment chemical weathering, rather than an outcome of frost action.

The dimensions of blockfields vary depending on the rock type, present and past climate, and local geomorphic history. In mid- and high-latitude mountains, blockfields may cover extensive areas, particularly on unglaciated upland surfaces (Perov 1969). In tropical mountains, blockfields are much less impressive, appearing as local rubble-strewn surfaces with rocks usually only a few centimeters in diameter.

Most blockfields and blockslopes are sedentary features that develop in place and are not the result of rockfall from upslope. Others, however, may be capable of movement. In a classic study, Caine (1968) found that blockslopes in Tasmania moved when the entire deposit contained an abundant matrix of soil or interstitial ice, which has since been removed by groundwater or surface water flow. The open texture that exists now is not capable of producing movement. Considerable chemical weathering could be required to produce the fine-debris matrix (Rea et al. 1996). Solifluction of fine material could have produced movement, which helps explains terrace and lobe development. Caine (1968) proposed a three-layer system that could produce movement. At the surface, blocks with an open work texture (3 m) exist; in an intermediate layer (10–30 cm), there are muds and blocks; and finally, at the base is a basal layer on bedrock consisting of silty sands. Geophysical methods (ground-penetrating radar, seismic refraction, and DC resistivity) provide a method to investigate the internal structure of blockfields, talus slopes, landslides, or rock glaciers, but to avoid misinterpretation, they should not be used solely (Schrott and Sass 2008).

FIGURE 5.9 Extensive blockslope located in the Indian Peaks Wilderness, Front Range, Colorado. (Photo by J. R. Janke.)

Blockfields may be actively forming in some areas. Freshly shattered surfaces and the absence of soil and vegetation indicate recent activity. Active blockfields are relatively unstable, with unbalanced rocks. If you have ever crossed a blockfield, you know that it is safest to walk over the most heavily lichen-covered areas, since they exhibit greater stability, except when the rock surfaces are wet, when the lichens can be extremely slippery! Inactive or relict blockfields may provide evidence of a past climate (Potter and Moss 1968). During the Pleistocene, snowline and glaciers extended to lower elevations, as did colder climatic conditions; thus, the presence of inactive periglacial features like patterned ground and blockfields can help establish the extent of past environmental conditions. In some mountain areas, processes similar to that which formed these relict features at lower elevations may now operate at higher elevations. If the high-elevation surfaces were once covered by ice and protected from deep frost penetration, blockfields there have probably formed since deglaciation. Where peaks and ridges were not covered with ice, but stood like islands (*nunataks*) amid the ice, they would have been susceptible to frost shattering and other weathering processes for much longer. Such areas offer critical information about past environments, but their identification and interpretation are difficult and controversial (Ives 1966; Dahl 1966). For instance, Andre et al. (2008) suggested that the internal structure of stone runs (blockstreams, stone stripes, etc.) in the Falkland Islands implies that they are not of periglacial origin, but formed under subtropical or temperate climates.

Mass Wasting

Mass wasting is the downslope movement of material because of gravity. Many different processes are included, from slow *creep* and *solifluction,* to more rapid mudflows and

slumping, to spectacular rockfalls, debris avalanches, and landslides. Mass wasting occurs with the greatest intensity in periglacial regions because of surface movement through frost creep; high moisture content, which lubricates the soil and produces high pore water pressures; and because glaciers have oversteepened slopes (French 1996). At high altitudes, frost action is the chief agent of rock disintegration, and mass wasting is the chief agent of transport. With the exception of glacial sculpture, frost action and mass wasting account for most of the characteristic features of high mountain landscapes. Movement of coarse debris on alpine hillslopes is largely due to mass wasting (Caine 1986). Slow mass wasting accounts for only about 15 percent of the geomorphic work done in most mountains; its importance decreases with increasing basin size (Barsch and Caine 1984).

Forms of mass wasting have long been difficult to delineate. A mass of rock breaking loose from a mountain may slide, fall, and flow during various segments of its journey, resulting in very different forms. Processes and forms are transitional, and so are their identification and classification. The literature is filled with different names applied to similar features. Without becoming entangled in such problems, we will discuss several of the more distinct processes.

Creep

Creep, the slow downslope movement of surface material, is usually only detectable through long-term observations. It is a process common to all environments and can be caused by wetting and drying, heating and cooling, freezing and thawing, disturbance of the soil by organisms, or simply the effect of shear stress on slopes. The rate of creep is so slow in most lowlands that its effect can only be seen over decades or centuries. In mountains, however, the effect of these processes is greatly intensified. This is especially true of frost creep. Frost creep produces a ratchet-like motion in which a soil, during a freeze–thaw cycle, is displaced normal to its surface. The displaced soil is then moved downslope by gravity and settles parallel to the hillslope (Benedict 1970). Soil creep reveals a convex-upward velocity profile near the surface (Roering 2004), and there may be a retrograde motion (upslope) because of cohesion and attraction of fine-grained particles (Benedict 1976). Frost creep is primarily controlled by the frequency of freeze–thaw cycles, the angle of the slope, the moisture available for heave, the frequency of variation around 0°C, and the texture and frost susceptibility of the soil (Troll 1958; Benedict 1970, 1976; French 1996; Matsuoka 2005). The potential for frost creep is greatest in autumn when soils are saturated, and at snow-free sites within the discontinuous permafrost zone where seasonal freezing is deep (Benedict 1976). Two-sided freezing, or freezing from the permafrost upward, is important for inducing creep during the thaw period. Water can percolate through the soil, reach the freezing front created from the permafrost, freeze, and cause a summer heave (Lewkowicz 1988). On the crest slopes, frost creep occurs diurnally in areas that lack vegetation and have a thin debris mantle; seasonal frost heave can induce deeper movement (Matsuoka 2005).

Although frost creep is a clearly distinguishable process, its measurement and isolation are difficult because other processes are also involved, particularly when the soil is saturated and flowage occurs (Benedict 1970, 1976). On drier slopes, the effects of frost creep can often be seen, resulting in small steps and terraces. Below timberline, the downslope curvature of tree trunks may be good evidence of creep, although heavy snowfall, among other factors, may also be important (Phipps 1974).

Solifluction

Solifluction is derived from the Latin words *solum* (soil) and *fluere* (to flow) (Andersson 1906). It is best developed in cold climates where permafrost or a subsurface frozen layer prevents the downward percolation of water. Under these conditions, the soil becomes saturated, cohesion among soil particles is reduced, and the viscous mass begins to deform downslope in a porridge or concrete-like fashion (Williams 1957; Barsch 1993). Some have used the term *gelifluction* to describe such a process; others have referred to the process as solifluction when permafrost is not involved (French 1996). In some cases, solifluction can do more work to the alpine hillslope than rockfall and avalanche processes (Smith 1992). Solifluction and creep processes combine to produce forms such as turf and stone banked lobes, benches, and terraces (Benedict 1970). Solifluction lobes are one of the slower forms of mass wasting, with average rates of movement from 0.5 to 5.0 cm/yr (Benedict 1970, 1976; Fig. 5.10). In the Canadian Rockies, rates of solifluction averaged 0.47 cm/yr between 1980 and 1990 (Smith 1992). The greatest downslope movement occurs near the ground surface, whereas the greatest differential movement arises in an intermediate

FIGURE 5.10 Solifluction lobes on Niwot Ridge, Colorado. The lobes, denoted by darker vegetation from greater soil moisture, create a stepping, porridge-like appearance on the hillslope. (Photo by J. R. Janke.)

layer 20–50 cm below the ground surface. There is very little movement below the active layer (Price 1991). Solifluction, frost action, frost creep, and other weathering processes act to reduce overall slope and create *altiplanation* (*cryoplanation*) terraces (Péwé 1970; Reger and Péwé 1976).

The essential ingredient for solifluction is water originating from melting snow or ground ice. The affected area is often saturated for several weeks in the spring and early summer. Other important factors governing solifluction include soil texture, slope gradient, surrounding geology, and vegetation. Slopes of only 2°–5° are sufficient to induce solifluction in the Arctic, where there is ample permafrost and saturated soil, but in most mountains it occurs on slope gradients of 5°–20° (though some have reported flow on slopes as low as 1°; Ritter et al. 2002). On steeper slopes, water will be lost through overland flow. In general, the finer the soil, the more likely solifluction is to occur because of the greater water-holding ability, frost susceptibility, and potential for flowage. At a site in Norway, Harris et al. (2008) found that gelifluction is an important component of small, near-surface mass movements. Summer rainfall events had the potential to increase pore pressure, but did not initiate soil movement. The spatial and temporal variation in movement was influenced by snow distribution.

Vegetation can play a role in solifluction by increasing the moisture content through reducing runoff and decreasing evaporation, although solifluction lobes are best developed in areas with sparse vegetation or tundra grasses. Vegetation also acts as a binding agent, giving the downslope movement definition and form. Solifluction lobes and terraces often resemble huge soil tongues moving downslope. They often coalesce and form crenulated, lobate banks along the slope. Such features form a striking micro relief and have important ecological implications (Price 1971a, 1971b).

In many midlatitude mountains, solifluction features may be inactive, with well-vegetated, stabilized lobes. Little sign of movement exists and erosion may be occurring. Elsewhere, they may display evidence of reactivation. In any case, an understanding of their disposition can provide insight into past and present environmental conditions (Benedict 1966, 1976). Real-time monitoring techniques to detect the thermal status, hydraulic condition, phase changes, soil volume strain, and soil shear strain of solifluction can provide high temporal resolution measurements to improve our understanding of solifluction dynamics (Harris et al. 2007).

Mudflows

Mudflows consist of water-saturated heterogeneous material confined to a definite channel that flows quickly downslope; they are a major geomorphic process in mountains. They should not be confused with mudslides, such as those that occur in California, which involve the massive failure of large sections of slopes. Mudflows have a much greater speed of movement (up to several meters per second) than solifluction. The name mudflow is actually a misnomer, because the material is composed largely of rocks, resembling an aggregate of fresh concrete. However, mud provides the flow matrix and transporting medium. In mountainous environments, there are many other types of flows, all characterized by their high moisture content and sudden movement. Flows of volcanic origin are called *lahars* and have been observed after several volcanic eruptions, including those that occurred at Mount Saint Helens. Earthflows are typically less fluid than mudflows, contain more earth material, and exhibit slumping that produces a step-like terrain. A special term, *debris flow,* has been used to describe a watery type of flow that commonly occurs after forest fires in the western United States. Without having observed the actual event, it is often difficult to distinguish between different types of flow. However, common to all flows are high moisture content and rapid movement.

The conditions most favorable for mudflows include (1) abundant water to saturate the mass of mud and rocks, (2) a lack of stabilizing vegetation, (3) unconsolidated material with adequate fines to act as a lubricant, and (4) moderately steep slopes (Harris et al. 2001). They occur most often in spring and summer, when there is ample snowmelt and heavy rain during thunderstorms. Under these conditions, masses of unconsolidated material may become saturated, and the material may collapse and flow downslope where an unstable situation already exists.

The typical point of origin for a mudflow is at a high elevation on the slope where there is a source of moisture, such as a confluence of runoff or below snow patches. Mudflows also occur where gradient increases abruptly at a break in the slope. In the Ruby Mountains, Yukon Territory, solifluction lobes on gentle slopes above a glacial trough slowly migrate onto the steeper slope. Eventually, they become unable to maintain themselves and collapse, moving downslope as mudflows (Price 1969). Whatever the local

FIGURE 5.11 Mudflow deposits at the base of a rocky slope in the St. Elias Mountains, Yukon Territory. The lighter colored material in the center represents the most recent flow. Observe the mudflow channel and levees on the narrow part of the feature. (Photo by L. W. Price.)

situation, once a mudflow channel has been established, it is likely that future flows will follow the same path.

Mudflows are so sporadic and unpredictable that very few people have witnessed them. Saturated debris moves in a rapidly advancing snout of rocks, mud, and water. In its upper reaches, the mudflow is primarily an agent of erosion, scouring and removing debris in its path. This process frequently produces a steep-walled, canyon-like channel up to several meters wide and deep, lined with mud but otherwise free of debris. As the mudflow progresses downslope, it may slow or stop temporarily when the slope gradient decreases, or when the snout and other sections become too dry. The stationary sections serve as dams, pooling water behind them until the mass once again becomes saturated or the water breaks through the dam and material begins moving again. The movement of a mudflow, therefore, is often in a series of stops and starts, progressing downslope. Along the edge of the mudflow channel, particularly in the middle and lower reaches of the slope, debris is piled on either side like natural levees along a stream. At the base, the material spreads laterally in lobate fashion to form a debris fan (Fig. 5.11).

Mudflows may occasionally move several kilometers onto adjacent lowland and may transport boulders weighing several tons.

Mudflows have caused considerable destruction of life and property. Mudflows initiated at the peaks of inactive volcanoes were believed to contribute to the tragic mudslide at Guinsaugon, Philippines, in 2006. In the Andes of Peru, earthquakes and the melting of glacial ice during the past century have resulted in numerous major mudflows. Glacial meltwater collects in lakes behind moraine dams; if there is an earthquake or the dam is breached, a huge mudflow may result. A long history of such catastrophic mudflows can be interpreted in the geomorphology of the base of the Andes (Lliboutry et al. 1977).

Debris flows contain material ranging in size from clays to boulders, and may contain woody debris, such as logs and tree stumps. Available material is an important control on debris flow initiation; if material has been removed from a recent flow, a flood is more likely to occur (Glade 2005). The character of the surface cover and response to intense rainfall, rather than contributing area or slope, are more important for debris flow formation (Godt and Coe 2007). Evidence from a debris flow in the Italian Alps indicates a maximum flow depth of 7 to 8 m and a peak discharge between 350 and 400 m^3 per second, with a fluid-mud and grain flow behavior (Sosio et al. 2007).

Slumping

Slumping is the slippage of material moving as a unit or as several subsidiary units along a concave surface of rupture (Fig. 5.12). Slumping begins with a rapid movement at depth. The movement may evolve into a flow or slide. Slumping typically takes place along a zone of weakness where the area downslope has been disturbed and support has been removed. Consequently, this process is important along road cuts and where lakes or streams have undercut their banks. The rate of movement by slumping is rapid enough to be observable, but is not as rapid or destructive as mudflows or landslides.

Rockfalls

Rockfalls simply involve the rapid falling of rock downslope (Sharpe 1938). Rocks may fall directly from a cliff or headwall, or they may tumble downward and initiate other movement (Fig. 5.13). In many mountain areas, rockfall is frequent and is thought to be of considerable geomorphic significance (Gardner 1973; Luckman 1976). It is a localized and sporadic process, occurring most frequently in spring and autumn. During spring, rockfalls occur with greater frequency as the ice-cemented bond that holds cliffs together is lost. In autumn, rockfall occurs in a process similar to that of fracturing a pipe. Pressures increase because of the volumetric expansion of ice, but also because water cannot escape readily (Williams and Smith 1989). The increased

FIGURE 5.12 Slumping taking place on a disturbed slope in the Andes, near Santiago, Chile. Several homes were destroyed beneath the slump. (Photo by J. R. Janke.)

pressures lower the freezing point, and as the temperature cools, the pressure rises. Eventually, the pressure becomes so great that rupture occurs, initiating a rockfall event.

Rockfall is most intense above the treeline in steep terrain where frost action operates to loosen rocks from surrounding bedrock. Under these conditions, the slightest disturbance may dislodge rocks and send them plummeting. The triggering agent may be blowing wind, running water, melting snow, disturbance from animals or people, or the contraction and expansion of rock with diurnal or seasonal heating and cooling. The sound of falling rock is very common in mountains. Tumbling rocks sound like thunder as the dislodged rock strikes solid cliffs or bedrock. Sounds are often amplified, bouncing off nearby cliffs. Rockfalls are an eloquent testimony to the inherent instability of the alpine environment.

Gardner (1970) investigated the importance of rockfall as a geomorphic agent by simply listening to and recording rockfall events in a small area of the Rocky Mountains of Alberta, Canada. Over three summers, he listened for 842 hours, recording 563 rockfall events (0.7 per hour). The greatest number occurred at mid-afternoon, when the temperature was warmest, but a second period of high frequency was observed during the initial daily warming and thawing of the surface. In general, rockfalls were most frequent in the highest and steepest terrain on northeast- and east-facing slopes, where frost processes were more active (Gardner 1970).

Rockfall rates can also be used to determine cliff retreat rates in small watersheds. Caine (1986) examined the Green Lakes, Eldorado Lake basin, and Williams Fork in the Colorado Rockies to understand movement of coarse debris. The cliffs of the Green Lakes valley lose about 10 m^3 of rock each year, a mean cliff retreat rate of 0.02 mm yr^{-1}. The rockfall record at Eldorado basin gave a rate of 0.012 mm yr^{-1}. At the Williams Fork site, the rate is about

FIGURE 5.13 Evidence of rockfall events occurring in the Gore Range, Colorado. (Photo by J. R. Janke.)

0.3 mm yr^{-1}, an order of magnitude higher. The coarse debris systems in the upper Rhine basin (Jäckli 1956) and in Karkevagge (Rapp 1960) show much higher levels of geomorphic work by rockfall, talus shift, debris flow, and rock glacier flow, suggesting that the Colorado alpine environment is stable. In the Swiss Pre-alps, dendrochronology was used to identify 301 rockfall events between 1724 and 2002, showing that rockfall activity has increased over the last century, presumably because of warmer temperatures, not annual or seasonal precipitation totals, and peaks during the early spring when trees are dormant (Perret et al. 2006).

The geomorphic significance of rockfall lies in its ability to rapidly and powerfully transport material. The largest recorded rockfall and associated slide involved over 10 km^3 of mass along a fault plane in central Nepal (Heuberger et al. 1984). Falling rocks move at high speeds and can cause considerable damage on impact, whether they strike another rock, a tree, or simply the ground. Of course, falling rocks are very dangerous for people; this is why protective structures are often built next to mountain highways passing through rugged terrain, or large sections of netting are draped over critical sections of steep walls. Most of the time, however, highway travelers must settle for a sign warning about "Falling Rock."

Geospatial technology has been used to better understand rockfall characteristics and hazards. Abellan et al. (2006) used terrestrial laser scanning to determine rockfall trajectories and velocities as well as calculate the geometry and volume of the source area. Krautblatter and Dikau (2007) suggested that the complexity of rockfall modeling could be reduced by separating a hillslope into the stages of back weathering, filling and depletion of storage on the rock face, and finally rockfall supply onto the talus slope.

A GIS software program, RockFall Analyst, allows computation of 3D trajectories, spatial frequency, flying/bouncing height, and kinetic energy of falling rocks (Lan et al. 2007).

Landslides and Debris Avalanches

Because of media sensationalism, many types of mass movements are included in the general term "landslide." According to geomorphologists, the term refers to movements where there is a distinct zone of weakness that separates the slide material from stable underlying material. Sudden, rapid movement of a cohesive mass not saturated with moisture occurs above this boundary. *Translational* landslides occur when a displaced surface moves parallel to the slope angle, and *rotational* landslides slip and rotate along a concave stable surface. Debris avalanches involve the catastrophic falling and tumbling of rock, debris, snow, ice, and soil from melting or release of ice. The distinction is difficult to make in practice, however, since sliding, falling, and flowing are generally all involved to various degrees. For our purposes, they will be considered jointly, and simply called landslides, recognizing that the movement is not restricted to sliding alone.

Landslides are the most spectacular form of mass movement (Fig. 5.14). Steep slopes, great local relief, and sufficient room for cascading rock to accelerate provide ideal conditions for their formation. Most major landslides occur in the 5 percent of the steepest land surfaces on Earth where relief is close to the proposed upper strength limit of the rock; in other areas that lack relief, failures are attributed to soft rocks, low-angle discontinuities, high rates of fluvial incision, tectonic uplift, and slope loading (Korup et al. 2007).

Some landslides reach velocities of over 100 m s^{-1}, moving horizontally for several kilometers, and even ascend

FIGURE 5.14 Debris from a landslide in the Andes. Note the road cutting across the toe of the landform. (Photo by J. R. Janke.)

nearby slopes. In the Pamir Mountains of Tajikistan, a massive landslide blocked the Bartang River, causing a 60 km lake to form (Alford et al. 2000). If the dam breaks, nearly 5 million people could be affected, but with proper lake monitoring, the potential for a disaster could be averted. Although media propaganda exaggerated the risk since the likelihood of dam failure was so slight, this illustrates the potential effects of landslides on human settlements. When examining landslides, natural and artificial effects, as well as the time scales associated with response and recovery, must also be examined (Chang and Slaymaker 2002). Although a landslide may last only a few minutes, it disturbs and disrupts other systems, such as fluvial processes at the foot of a slide, which have a longer and potentially greater impact on the system (Hewitt et al. 2008).

Among the better-known mountain landslides are the great slide in 1881 at Elm, Switzerland, and the ancient Saidmarreh landslide in the Zagros Mountains of southwestern Iran. The latter was apparently the largest in the world: The side of a mountain broke loose, descended 1,500 m, traveled horizontally some 14 km, and eventually ascended 500 m over an intervening obstacle! The fallen rock covered 274 km^2 with a thickness of over 100 m. The material ranged in size from dust particles to huge blocks 18 m in diameter (Watson and Wright 1969).

In North America, the most famous slides have been the Turtle Mountain landslide that destroyed the small mining town of Frank, Alberta, in 1903; the Gros Ventre slide, Wyoming, in 1925; and the Sherman landslide, which came to rest atop the Sherman Glacier during the Alaska earthquake of 1964 (Cruden 1976; Fig. 5.15). An ancient slide that temporarily blocked the Columbia River midway through the Cascades and gave rise to the "Bridge of the Gods" of the Native American legend is now the site of Bonneville Dam (Waters 1973).

The most destructive landslide of the twentieth century took place in the Peruvian Andes during a great earthquake in 1970. Over 50,000 people were killed, 18,000 of them buried in the landslide that originated on Huascarán, a volcanic peak with an elevation of 6,768 m, about 350 km north of Lima. The earthquake caused a huge mass of overhanging snow and ice to break loose from the summit and fall 1,000 m, until it crashed into the mountain and pulverized. The impact dislodged unconsolidated slope material and caused massive slope failure. Frictional heat created from the collision also caused melting, so that vast amounts of water were available to saturate and lubricate the mass (Clapperton and Hamilton 1971; Browning 1973). The slide swiftly descended the mountain, going from an elevation of 5,500 m to 2,500 m in less than 3 minutes, traveling at speeds of up to 480 km h^{-1}. The survival of delicate moraine ridges and vegetation in its path suggests that the slide rode on a cushion of air for parts of its journey. The mass caromed back and forth from one side of the valley to the other in its descent, like a great sloshing liquid. Before the slide came to rest on the far flanks of the valley, it had traveled more than 16 km and had destroyed two villages in its path, Yungay and Ranrahirca. Ridges as high as 140 m were overridden, and blocks up to 6 m in diameter were scattered about like pebbles. The slide was preceded by a turbulent blast of air that demolished buildings even before the rock debris struck, and a dense dust cloud hung over the area for three days (Clapperton and Hamilton 1971).

These landslides were exceptionally large and involved millions of cubic meters of material, but most landslides are smaller and more frequent. Their effectiveness in transporting material downslope and in changing the face of the landscape is tremendous. It is difficult to spend much time in mountains without seeing landslide scars and deposits. Like many other processes in mountains, the landslide is a low-frequency, high-energy event capable of accomplishing more geomorphic work in a few seconds than day-to-day processes accomplish in centuries. It is important to

FIGURE 5.15 Landslide on the Sherman Glacier, Alaska, that was released during the great Alaska earthquake in 1964. Rock debris has since been transported to the terminus. (Photo by Austin Post, 27 March 1964, U.S. Geological Survey.)

realize, however, that although the landslide is a sudden and intense event, it depends on the less spectacular processes which have been slowly preparing the way for its eventual release.

Specific causes of landslides are usually divided into two groups: (1) internal condition of the rocks and (2) external factors affecting slopes. The first group includes such factors as weak rock formations, steeply dipping rock with bedding planes, joints, fault zones, and steep slopes; the second group includes climatic factors, erosion, and other types of disturbances, such as earthquakes (Howe 1909). It is difficult to point to any single factor as being responsible for a landslide. The Turtle Mountain slide at Frank, Alberta, is a good example of the interaction of several slide-causing factors. The mountain stood 940 m above the valley and was composed of massive limestones overthrust across softer sandstones and shales. This framework formed a natural zone of weakness, further exaggerated by coal mining in an interlying seam, which decreased the cohesiveness of the material. Finally, several earthquakes had disrupted the area over the previous two years. The precise trigger that caused the mountain to collapse is unknown, but this combination of factors surely led to its eventual release. Over 30,000,000 m³ of rock buried the town and railroad and killed 70 people. However, only part of the mountain fell. The northern shoulder still stands as an ever-present

threat to the citizens of the rebuilt town. The mountain may stand for centuries with no further problems, but may only require a slight trigger to release another tremendous landslide.

With statistics, remote sensing, and GIS, landslide hazard analysis has been improved. Guzzetti et al. (2005) used a temporal sequence of aerial photographs, statistics, and morphological, lithological, and land-use data to predict the location, frequency, and size of landslides. Ayalew and Yamagishi (2005) used a logistic regression to produce a landslide susceptibility map of central Japan and found that lithology, bedrock, slope, lineaments, aspect, elevation, and road networks played major roles in occurrence and distribution. Logistics regression with stepwise backward procedures produces lower error rates and best generalizations for landslide prediction (Brenning 2005). Electrical resistivity field measurements can help discern discontinuities between landslide materials and underlying bedrock, to help understand areas that may reactivate (Lapenna et al. 2005). Airborne laser altimetry is effective at differentiating landslide components and activity (Glenn et al. 2006).

Features of Mass Wasting

Many surface forms result from the aforementioned mass wasting processes. Some occur as specific features, such as stone stripes or solifluction lobes, while other less distinct features are simply identified as various sorts of deposits such as mudflow or landslide deposits. Two features deserve special mention because of their distinctive character and importance in the alpine landscape: talus and rock glaciers.

Talus

Talus is an accumulation of rock debris of various sizes transported from a mountain valley wall by gravity, rainwash, snowmelt, or avalanching snow (White 1981). It results primarily from rocks breaking off and falling until they come to rest to form a ramp or rock apron (Fig. 5.16). Talus accumulations are best developed above treeline, where there is little vegetation and frost action causes rapid breakdown and displacement of rocks (Rapp 1960). Most active talus accumulations are bare and rocky with very little fine material visible, but where mudflows, avalanches, and landslides feed into to the talus, fine material may be present (Luckman 1978). Because the larger rocks have more kinetic energy and therefore travel farther, they will be located at the bottom of the slope (Rapp 1960). The downslope sorting will not be perfect, however, since the rock size, shape, height of fall, nature of the surface, and interference from other obstacles will affect this arrangement (Gardner 1973).

White (1981) defined three different types of talus forms that occur in the alpine environment. *Rockfall talus* is a collection of angular rocks of all sizes that form below cliffs

FIGURE 5.16 Accumulation of alluvial talus in Rocky Mountain National Park, Colorado. (Photo by J. R. Janke.)

or steep rocky slopes. These are typically derived from falling, rolling, bouncing, or sliding to the base feature, and have a slope of 35–45°. Rockfall talus is often referred to as a rockfall cone, scree slope, talus cone, or talus slope. *Alluvial talus* forms from an accumulation of rocks of any size or shape that are carried by rainwash or snow through a gully or couloir to rest against a valley wall. Their slope is usually 35–38° with a concave up profile. *Avalanche talus,* an assemblage of angular rocks of any size, is derived from avalanched snow mixed with rock debris carried from cliffs or steep rocky slopes. Some talus forms may resemble a *protalus rampart,* or ridge of angular block created by rocks falling from cliffs that slide across a former snowbank or firn field. Protalus ramparts mark the downslope edge of old snowbanks.

The status of a talus slope is determined by the supply of material, its movement within the talus, and its removal. The rate of talus accumulation provides a gross index of the rate of weathering and denudation. Movement within the talus may also be substantial because of rockfall, snow avalanches, mudflows, running water, creep, and the removal of material from below by stream action. The average rate of surface movement ranges from about 5 to 20 cm yr^{-1} in the Canadian Rockies, although measurements of such features are circumstantial; individual rocks may move great distances, while others remain motionless (Gardner 1973). Rates of talus shift ranging from zero, from 6 to 111 cm yr^{-1} in the Canadian Rockies (Gardner 1973), and up to 22 cm yr^{-1} in northern Sweden (Rapp 1960) have been reported. There is also considerable regional variability because of the local environment, slope orientation, gradient, and rock type.

In some areas, talus is now inactive. Evidence for this includes lack of new material from above, dense concentration of lichens, weathering rinds on rock surfaces, infilling of rock voids by fine material, and encroachment by vegetation. Inactive talus can observed at lower elevations in some mountainous areas that were active in the past during colder climatic conditions. Although talus formation is best developed in cold climatic regimes, it may also form in other environments, so its interpretation as evidence of a cold climate must be made with caution (Hack 1960).

Rock Glaciers

Rock glaciers are an important component of high mountain systems, often serving as a visible indicator of mountain permafrost (Fig. 5.17; Barsch et al. 1979; Barsch 1996; Haeberli et al. 1999, 2006; Haeberli 2000). They consist of unconsolidated but frozen, ice-supersaturated debris that creeps or flows downslope at a rate of 1–100 cm yr^{-1} and exhibit a variety of forms, most typically tongue shaped and lobate. Their surface topography is quite variable, but some can display a sequence of transverse and longitudinal furrows, as well as a steep front slope that rests near the angle of repose (Wahrhaftig and Cox 1959; Benedict 1973; Haeberli 1985; Martin and Whalley 1987; Vitek and Giardino 1987; Barsch 1996). Humans have used rock glaciers as a source for construction material, a backdrop for residential areas, dam abutments, drill sites, shaft and tunnel portals, and a water source for urban areas (Burger et al. 1999; Giardino and Vick 1987).

A rock glacier's internal structure is thought to be a three-tiered system, with a top layer of rock fragments covering a second ice-cemented or ice-cored interior that overlies rock deposited and overridden by the top layers (Humlum 2000). Rock glaciers' periglacial or glacial origins are commonly debated, relating to whether or not

FIGURE 5.17 California rock glacier in the Sangre de Cristo Mountains, Colorado. (Photo by J. D. Vitek.)

rock glaciers have an ice-cemented (periglacial) or ice-cored (glacial) internal structure. Supporters of the periglacial model believe that cemented ice in the form of interstitial ice (pore ice) or segregated ice (ice lenses) produces creep (Wahrhaftig and Cox 1959; Haeberli 1985; Barsch 1996). Glacially derived or ice-cored rock glaciers form when debris from a rockfall covers a glacier, or when a glacier experiences excessive ablation during a stagnant period, allowing moraine or rock debris to melt out and occupy the surface (Wahrhaftig and Cox 1959; Outcalt and Benedict 1965; Potter 1972; Whalley 1974; Benedict 1973; White 1976; Ackert 1998; Potter et al. 1998). Abundant evidence supports each model, creating controversy (Whalley and Martin 1992; Clark et al. 1998). Some researchers have accepted an intermediate viewpoint, advocating both models in certain circumstances. Rock glaciers are sometimes considered part of a landscape continuum, a cycle describing the transition among glaciers, rock glaciers, and slope deposits (Corte 1987; Johnson 1987; Giardino and Vitek 1988). Followers of this viewpoint accept that both glacial and periglacial geomorphic processes can operate on any rock glacier, leading to questions about the usefulness of an active, inactive, and relict classification.

From their research in the Alaska Range, Wahrhaftig and Cox (1959) provided a framework for form classification. Lobate rock glaciers have single or multiple lobes which have a greater width than length, whereas tongue-shaped rock glaciers are longer, extending downslope from a cirque. Spatulate rock glaciers resemble tongue-shaped rock glaciers, but display an abrupt widening beyond lateral topographic constraints. Subsequent studies used classifications based upon topographical or geographical descriptors such as valley wall, valley floor, protalus, debris, talus, or glacial (Outcalt and Benedict 1965; Linder and Marks 1985; Barsch 1988; Humlum 1998). A single classification scheme has not gained widespread acceptance; however, a subset of Wahrhaftig and Cox's (1959) classification most commonly appears in the literature: (1) tongue-shaped and (2) lobate.

Rock glaciers that are moving exhibit pronounced ridge and furrow complexes, resembling a viscous flow substance such as lava (Vitek and Giardino 1988). Transverse ridges and furrows form perpendicular to the direction of movement; they originate from overthrusting of internal shear planes, differential movement of distinct layers, or changes in debris supply (Ives 1940; Wahrhaftig and Cox 1959; Potter 1972; Haeberli 1985; Barsch 1987; White 1987). Longitudinal ridges and furrows form parallel to the principal direction of movement and result from extensional flow or resistance to flow, or are remnants of lateral moraines (Barsch 1987; Calkin et al. 1987; Ackert 1998). Rock glaciers with greater flow rates have a steep front slope near the angle of repose, but near the head or rooting zone there is a gradual transition from the source of debris input to the rock glacier (Barsch 1996). Because of thermokarst or ice ablation, the surface morphology can also appear mottled, with spoon-shaped depressions (Washburn 1980). Remote sensing and terrain parameters show promise when mapping regional landforms such as rock glaciers over large mountainous areas (Brenning et al. 2007; Janke 2001).

Rock glaciers play an important role in the elevational distribution of landforms in glacial or periglacial environments (Caine 1984). Traditionally, rock glaciers are thought to exist mainly in low-precipitation, continental climates where frost weathering is dominant and temperatures are cool enough to maintain ice (Haeberli 1983). In the Alps, Haeberli (1983) restricted rock glaciers to dry continental climates at altitudes below the equilibrium line of glaciers, but above the lower permafrost limit. In Greenland, Humlum (1998) found that the locations of rock glaciers and glaciers are driven by topoclimates (elevation, slope, and aspect) and talus production rates, not regional climates. In the Colorado Front Range, Janke (2007) examined topoclimatic variables for rock glaciers and temperate glaciers using a GIS, finding that tongue-shaped rock glaciers are found at higher elevations and on slopes with more northerly aspects, compared to lobate rock glaciers. Active, inactive, and relict forms showed typical elevation and aspect gradients. Active rock glaciers are found at the highest elevations and most northerly aspects. Inactive rock glaciers are found at lower elevations on all aspects, with a tendency to face northeast; fossil rock glaciers occur at the lowest elevations on all aspects. Glaciers are mostly smaller, found at higher elevations, and restricted to more northern and northeastern slopes. Because these remaining small ice masses are mostly attached to steep cirque walls, their slope is usually steeper than the blocky accumulations of rock glaciers. Rock glaciers are larger and more abundant; as many are believed to contain ice, they could provide a small source of water in a future warmer, drier climate. Janke (2007) also found that active tongue-shaped rock glaciers have elevations and aspects similar to glaciers, but active lobate forms have different elevations and aspects compared to glaciers. This suggests that most tongue-shaped Front Range rock glaciers are glacially derived.

Present climates affect active and inactive rock glaciers by preserving or ablating an internal ice structure (Martin and Whalley 1987). Rock glaciers are therefore considered good indicators of climate change (Barsch 1988; Haeberli 1990). A rock glacier's response to climate is smoothed since surface debris serves as an insulator (Barsch 1996). Unlike glaciers that are sensitive to extreme fluctuations on a shorter time scale, a strong climatic signal must exist to produce change in a rock glacier system. Degradation of rock glaciers is generally measured from slumping surface morphology, frontal activity, amount of internal ice, variation in downslope movement, or temperature of frozen material (Francou et al. 1999).

In the Front Range of Colorado, temperature records from the 1960s did not indicate long-term cooling or warming trends, although most sites have shown a consistent warming recently (1997–2004) (Losleben 2004). Precipitation measurements show a slight increase in total precipitation; however, considerable variation occurs from year to year. Rock glacier velocities have remained consistent, showing no major increase in flow velocity over 40 years (Janke 2005a, 2005b). In the Alps, accelerated warming was pronounced in the 1980s and 1990s, and rock glaciers responded accordingly. Gruben rock glacier showed decreasing horizontal velocities and an increase in surface subsidence by a factor of 2 to 3 times over this period. Borehole temperatures on the Murtèl I rock glacier have shown a consistent increase in temperature with depth during the past few decades. Some rock glaciers in the Alps, such as the Hochebenkar and the Reichenkar rock glaciers, have shown increasing velocity since about 1990, the result of warmer temperatures, which have accelerated melting of permafrost, causing warmer ice to deform more quickly. In the San Juan Mountains of Colorado, some rock glaciers appear to have become reactivated, overriding forests in their paths. This may possibly be another effect of global warming: Ice that was previously too cold to flow effectively has now warmed and can deform like a plastic. Rock glaciers with ground temperatures close to 0°C flow faster compared to colder rock glaciers (Kääb et al. 2007; Janke and Frauenfelder 2008). Continued monitoring of rock glaciers in other parts of the world is necessary to help better understand global climate change, as well as monitor and predict potential hazards associated with melting ice.

Fluvial Processes and Landforms

Since ancient times, mountains have been recognized as a source of water. Mountains are the origin for most of the world's major rivers, and this water is more important to human populations now than ever before, as discussed in Chapter 12. Irrigation, hydroelectric power production, navigation, recreation, and domestic consumption are but a few of the uses of mountain streams and reservoirs. The Danube, which rises in the Alps and mountains of Eastern Europe, the great Indo-Gangetic rivers that originate in the Himalayas, or the Nile, which rises in the mountains of East Africa and Ethiopia, are of significant importance to the areas through which they flow. The hydrologic center of the continental United States is in the Rocky Mountains near Yellowstone National Park, where three great rivers (the Snake, the Colorado, and the Missouri) have their sources. The downstream components of each of these are vital to the regions through which they flow.

Running water is an important denudational agent in mountains. Material is continuously transported downslope by rivers. Precipitation is greater in most mountains, and this water has a high potential energy, owing to the presence of steep slopes and high local relief. In addition, bare areas, with abundant surface material and little protective vegetation, make mountain landscapes vulnerable to erosion by running water (Dingwall 1972). Much of the sediment load of streams is probably obtained from slopes adjacent to small tributaries near the headwaters of drainage basins. In the Amazon Basin, for example, it is estimated that 85 percent of the total stream load is provided by just 12 percent of the drainage area, which is concentrated in tributaries in the Andes (Bloom 1998). Streams also reach far beyond the mountains, serving as linkages to the lowlands, and ultimately, to the sea. Mountains may be worn down primarily by frost action and mass wasting, but if streams did not transport at least some of the material away, the valleys would eventually be buried by the weathered material. The fact that this does not happen is evidence for the importance of streams.

One would think that areas that have glaciers would more likely have rivers with increased sediment loads, given glacial erosion and the amount of available water. It is unclear, however, how glaciers link erosion and increased sediment load to rivers (Harbor and Warburton 1993). Fluvial specific sediment yield often does not correlate well with glacial cover (Desloges and Gilbert 1998). The degree of glacial activity, therefore, may be a more important variable. Local geology and surface deposits have important influences on the delivery of fine-grained sediment to mountain streams. Dedkov and Moszherin (1992) found that specific sediment yields vary from 100 t km^{-2} a^{-1} for igneous rocks to as high as 1,800 t km^{-2} a^{-1} for loess. Specific sediment yield is positively correlated with drainage area. Most often, a positive correlation occurs in regions with vegetated slopes, where erosion is minimal compared to channel erosion. For basins that have little established vegetation and high surface erosion, however, specific sediment yield is inversely related to increasing basin size.

Mountain watersheds have great erosional potential. The upper Indus and Kosi rivers draining the Himalaya show a regional denudational rate of 1.0 mm yr^{-1} (Hewitt 1972). Other mountain ranges may also approach this rate. The Alps are estimated to be eroding at between 0.4 and 1.0 mm yr^{-1} (Clark and Jäger 1969). The Canadian Rockies and the mountains of Alaska are being denuded at a rate of up to 0.6 mm yr^{-1} (McPherson 1971a; Slaymaker 1974; Slaymaker and McPherson 1977), while the rate in the

Rocky Mountains of Colorado and Wyoming is somewhat lower: 0.1 mm yr^{-1} (Caine 1974).

Characteristics of Mountain Streams

Mountain streams are similar to other streams in most respects, but they do have special attributes. Water flows down steep slopes through highly varied terrain with great local relief; the moisture supply is highly variable, depending on rainfall and melting snow and ice, and the debris delivered to the streams is often too large to be transported effectively. The physical behavior of water in mountains, however, is no different from that in any other natural environment. When the gradient or volume increases, so do the velocity and the ability of the stream to transport material. Since all of these factors change rapidly in mountains on spatial and temporal scales, the system frequently exhibits pulsation. Velocity varies greatly between steep slopes and flat valleys, between interconnecting pools or lakes, within the diurnal sequence, and with the sudden addition of rain from thunderstorms.

Smaller streams in high mountains are ephemeral, flowing only at certain times of the year (Leopold et al. 1964). Perennial streams are found at lower elevations where there is a greater drainage network. Discharge also increases downstream, making these regions more susceptible to erosion, although this is counteracted somewhat by well-developed vegetation.

The typical mountain stream is steepest in its upper reaches and gradually flattens downstream to form a concave longitudinal profile. Rocky ledges and other abrupt changes in the local relief called nickpoints create waterfalls. In semiarid mountains, the channels of ephemeral streams often consist of a series of steps composed of larger rocks interspersed with sand patches. Apparently, this is a response to intermittently heavy runoff and flooding which first scour and then refill the bed. Downstream, channel width increases and the bed materials tend to become finer (McPherson 1971b).

High mountain streams typically have small discharges and low velocities despite their steep slopes. They are consequently able to transport little coarse-grained material. The characteristic clearness of the water in high mountain streams, even at full-bank flow, is evidence of this. The beds of mountain streams, originating from present or past weathering, consist of gravel or boulders. The water flows around and between these boulders, which are really more a part of the channel than of the bedload. Maximum incision occurs when sediment supply is moderate since sediment promotes abrasion but also limits the extent of bedrock exposure (Sklar and Dietrich 2001). Glacial meltwater streams are an exception to the typical clear rushing brooks, as they have a continuous supply of fine sediment, called glacial flour. It is hard to believe that the "pure" glacial meltwater contained in some bottled water was once filled with suspended sediment!

A common characteristic of most lowland streams is the presence of alternating deep and shallow areas. Typically, deep areas exist on one side of the stream, with a gravel bar or shallow area on the opposite side. These bars tend to alternate from one side of the channel to the other. Smooth-water pools often form over the deeper areas and riffles (or rapids) over the bars, features well known to trout anglers. There is evidence, however, that many high mountain streams, especially those under semiarid regimes, lack pools and riffles, apparently because of the dominance of coarse bed material. The typically small discharges are unable to move the larger material. In the Rockies, streams do exhibit pools and riffles, but their spacing is much more variable than in lowland streams (Leopold et al. 1964). Step-pool sequences are typical in mountain streams that contain gravel bed with gradients steeper than 2–3 percent. The steps are areas of high gradient and increased velocity because of accumulations of wood, bedrock, cobbles, or boulders. Pools have fine-grained bed material with low gradient and slow flow. Floodplains and natural levees are less common in high mountains than in lowlands, since streams tend to be confined to bedrock channels and do not overflow their banks. There may be occasional valley flats, but these are often formed by other factors, such as moraine or beaver dams, landslides, and avalanches, rather than being strictly a result of stream action. Mountain streams with bedrock channels display little meandering except in stretches of low-gradient mountain meadows. Increased stream complexity in mountain regions increases the storage potential compared to urban and agricultural streams (Gooseff et al. 2007). Sediment sinks are important controls on mountain drainage. For instance, 50 percent recovery occurred within 2–4 km of a sink, but full recovery was not reached within 20 km downstream (Arp et al. 2007).

Discharge Regime

Mountain streams show periodicity on a daily as well as a seasonal basis. Snowmelt is reduced at night and increases during the day. Consequently, streams supplied by meltwater carry their greatest volume during the afternoon and early evening, and their lowest volume during the early morning. Hydrographs typically display a high daily fluctuation in flow during the summer. Snow stays in the high country until temperatures rise in the spring. Many high mountain streams freeze, and the flow decreases to a trickle, except in tropical or midlatitude coastal mountains where temperatures are not as cold and rainfall or wet snow is common throughout the winter. In the summer, however, water stored in the form of snow is released by melting (Fig. 5.18). In fact, 90 percent of mountain stream variability in the Green Lakes Valley in Colorado is due to snowmelt (Caine 1996). This may result in near-flood conditions for short periods in the spring; rain falling on snow in the summer could also induce flooding

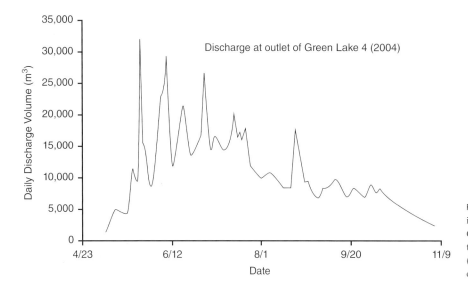

FIGURE 5.18 A hydrograph obtained in 2004 for a stream originating from Green Lake 4. Note the seasonality of the signal due to mainly snowmelt. (Graph by J. R. Janke using data obtained from Caine.)

(Rothlisberger and Lang 1987). In some years, great damage is done to bridges and highways in the lower stream channels. Discharge data from the Rocky Mountains show that winter flows have increased, spring runoff and peak flows have occurred earlier, and summer and autumn flows have been reduced. This may impose a drought stress on floodplain forests (Rood et al. 2008).

Where alpine glaciers exist, the normal spring runoff is augmented by glacial melting. In summer, mountains cannot store much water, so thunderstorm rainfall surges quickly through the drainage network. In the San Juan Mountains of southern Colorado, erosional rates are three to four times greater in summer than in winter because of the intensity of summer thunderstorms (Caine 1974). Exceptionally heavy precipitation can set off flash floods, increasing the potential for erosion or catastrophic *jökulhlaups,* Icelandic for outburst floods. Volcanic activity can also melt overlying ice and cause flooding. These are major environmental hazards to settlements in and near mountains. For example, Salt Lake City, Utah, where over a million people live in a narrow belt along the Wasatch Front, is susceptible to flash flooding (Marsell 1972). Sparsely vegetated slopes allow rapid runoff, collecting in the major drainage ways. The streams quickly pick up velocity and become charged with mud and rocky debris. Upon reaching the base of the mountains, the streams overflow their banks, or the water and debris disperse in alluvial fans, causing considerable damage. With increasing tourism and recreation in mountains, human developments may be at severe risk. Although development in these regions seems foolhardy, paradoxically, the people most affected are often those living in established areas at the base of mountains. In 1976, about 145 people died in the Big Thompson Creek flood when a moist air mass pushed up the Colorado Front Range and stalled. The impervious canyon walls funneled the precipitation downstream. The flood destroyed 418 homes, damaged another 138, removed 152 businesses, and

caused more than $40 million in damage. The late Gilbert White, a well-known hazards expert, warned about flooding in the city of Boulder, at the base of the Front Range. In fact, residents are reminded of the danger of fast and widespread flooding of creeks on the first of each month as the siren warning system is tested. The construction of paved streets and roads in canyons provides impermeable channels for flash flood waters to move swiftly downslope. Whereas large amounts of water previously infiltrated the rocky ground, lessening the potential for damage in lowlands, much more of the water is now directed to the foot of slopes. Intelligent and effective land-use planning is required to mitigate such problems.

There has been disagreement about when streams accomplish the most work. It has long been believed that major floods and catastrophic events can accomplish more erosion in a short time than day-to-day flow can accomplish in centuries. A number of other studies indicate that much of the work of streams is accomplished in only a few days of a decade or century, rather than on a more regularly spaced annual basis (Rapp 1974). However, measurements of several streams indicate that most erosion is accomplished by normal flow and by infrequent floods rather than by the rarer catastrophic floods (Leopold et al. 1964). Infrequent but larger events in mountains appear to have more validity compared to lowlands. Mountain streams, which contain a bedload of cobbles and boulders, need catastrophic events to move such material (Nanson 1974). It is during these events that the truly awesome potential of running water is demonstrated. In the future, as glaciers continue to recede, the frequency of catastrophic events may increase, as discussed in Chapter 4.

Drainage Features: Braided Channels and Alluvial Fans

A number of drainage features are well developed in mountains. These include braided channels, alluvial fans, and

FIGURE 5.19 A braided river channel (foreground) extending from Exit Glaciers (background) in Kenai Fjords National Park, Alaska. Most of the bed material is too coarse for the stream to carry, so it simply works its way through the material, dividing and reuniting. (Photo by J. R. Janke.)

asymmetric valleys. The basic drainage patterns, when viewed from the air or on a map, are also diagnostic of geomorphic processes and reflect the presence of certain types of rocks, structures, and past geologic events.

Braided channels have a low sinuosity and are wide, shallow, and straight, with numerous strands of water that are divided by coarse-grained bars. The water strands frequently change their location and numbers because of pulsating flow. Total channel width is large compared to channel depth, and the gradient is generally steeper than that of meandering rivers (Fig. 5.19). A typical misconception found in introductory physical geography or geology textbooks is that braided streams have too much sediment and that, therefore, aggradation occurs. In reality, this is not the case, as there are several factors associated with braiding. The most important variable is the erodibility of the banks. If banks are easily eroded and have low cohesion, widespread bank erosion occurs and a braided pattern emerges. Abundant bed material, especially coarse-sized bedload, is also required, as well as rapid, frequent variations in discharge that make it difficult for vegetation to establish and stabilize banks. If there is too much coarse bedload material for the stream to transport in a single deep channel, bars or miniature islands begin to develop. This raises the stream bed, increasing the slope and downstream velocity. At the same time, the channel is made shallower, which increases the flow velocity along the streambed and allows a larger amount of bedload to be carried by the same discharge. Braided sections of streams display gradients several times steeper than sections with single channels. Consequently, the shallow but relatively steep channels of braided streams facilitate the transport of

coarse particles along the stream bottom. Braiding, therefore, can be viewed as a natural response by streams to an increased bedload consisting of material too large for transport by a single channel.

Braided channels reach their highest development in mountain regions where runoff is rapid and periodic. Many glacial streams display extensive braiding because of the abundant debris contributed to the stream. Braiding is also found in intermediate-sized streams that are fed by smaller streams originating on steeper slopes. The small but turbulent streams often transport coarser material to the valley bottom than the intermediate stream can carry in a single normal channel. Nonbraided channels may also become braided because of human disturbance such as logging, agriculture, or mining in areas with steep slopes.

Braided channels typically have wide valleys without confining walls. In many cases, the braided stream has inherited a valley greatly widened and deepened by glaciers. Braided channels are composed of poorly sorted material, varying from fine silt, sand, and gravel to cobbles and boulders. Mid-channel bars composed of this loosely consolidated material are subject to continuous erosion as the current impinges upon their flanks. When discharge is high, erosion takes place on the upstream ends of the bars, with deposition occurring at their downstream ends. Thus, the bars continually change shape and migrate downstream. The channels themselves display much greater variation, since they are subject to rapid change during periods of high runoff. The White River draining the Emmons Glacier on Mount Rainier, Washington, has moved laterally more than 100 m in eight days (Fahnestock 1963). The Kosi River, which rises near Mount Everest in

FIGURE 5.20 An alluvial fan in the St. Elias Mountains, Yukon Territory. The fan has been deposited onto the floodplain of the Slim's River, indicating a relatively young age. Note that the left side of the feature is forested, while the right side is largely bare, owing to recent activity in the area. The stream now appears to be moving to the left and encroaching on the forest. Alluvial fans serve as good indicators of the amount of erosion and other geomorphic activities taking place in mountains. (Photo by L. W. Price.)

Nepal, has migrated westward a distance of 112 km over the last 200 years. The river has also shifted 19 km in a single year (Leopold et al. 1964). In 2008, the Kosi broke through its embankments and reverted to its old course, causing serious flooding and many deaths in Bihar State, India.

Alluvial fans are cone- or fan-shaped deposits occurring at the mouths of mountain valleys or canyons. Water running down the valley loses velocity, and therefore drops some of its load, since the gradient at the apex of the fan is reduced. Moreover, the water is no longer confined to a steep-walled valley, so it may spread laterally. Its energy is diffused, and a braided pattern may develop in the middle of the deposited sediment. The resulting debris accumulation builds up in the form of a concave-upward fan or an outward slope to a semicircular perimeter (Fig. 5.20). The larger material is dropped first because of the lower gradient and resulting loss of velocity; smaller-sized material is transported farther (McPherson and Hirst 1972). Alluvial fans are essentially composed of the portion of the bedload which mountain streams have been unable to transport farther. In general, large upland watershed and source regions produce larger alluvial fans. In Death Valley, California, high-resolution airborne laser swath mapping revealed that alluvial fan roughness becomes smoother until a threshold is reached, and incision dominates on the older surfaces (Frankel and Dolan 2007).

Alluvial fans are key areas for agriculture and settlement in mountainous regions because of their relatively smooth surfaces, productive soils, and availability of water. Streams are sometimes dammed in the mountains and the water diverted for use. In other areas, as alluvial fans provide excellent aquifers and are good sources of groundwater, wells are drilled into the fan.

Drainage Patterns

The most common drainage pattern on Earth is the *dendritic* pattern, which resembles the branches of a tree. Headwater tributaries or branches feed into the mainstream of a trunk. The dendritic stream pattern develops under homogeneous surface conditions and is generally accepted as evidence for lack of significant structural control. In mountains, however, where stream flow is more strongly controlled by slope and structure, other types of stream patterns develop. The presence of any given type of stream pattern may aid in understanding the geomorphic history of the landscape. For example, a dome or volcanic peak will frequently display a *radial* stream pattern, in which the streams flow outward from a central peak. If the mountain mass is domed upward but with a series of encircling ridges, as in the Black Hills of South Dakota, an *annular* drainage pattern may prevail. Streams follow the circular outcrops of less resistant rocks until they find a path through the ridges of resistant rock. In folded mountains with parallel ridges and valleys, such as the Appalachians or Jura, the common pattern is for a main stream to occupy the center of the valley with many short tributary streams joining it at right angles, forming a *trellis* pattern. Faulting and jointing greatly affect the direction of stream flow. Many parts of the Adirondack Mountains in New York display a *rectangular* pattern because the streams are forced to make right-angle bends in response to the regional joint system. Other types of stream patterns could be mentioned, but these are the principal types for mountains.

The development of stream valleys running transverse to landforms is a curious and enigmatic phenomenon, since this opposes the basic law of nature that water flows downhill. Many different situations can bring about such features. In one possibility, the lowest outlets of the stream exist across the ridges, so the water simply backs up until it spills to low ground and erodes a channel. In another, a stream erodes headward until it cuts across the apparent obstacle. Two other possibilities are called *antecedence* and *superimposition* (Thornbury 1969).

An *antecedent* stream is one that existed before the mountains were formed; when uplift took place, the stream kept pace in downcutting and maintained its channel through the mountains (Fig. 5.21). Antecedent streams have created some spectacular valleys. The Brahmaputra and Indus Rivers originate on the Tibetan Plateau and flow across the highest mountains on Earth to reach the sea. In the United States, the Columbia River cuts through the Cascades, a major barrier to the Pacific Ocean. *Superimposed* streams develop where the former topography has been buried by sediments, lava, or glacial deposits. New streams form on this surface, establish themselves, and continue to erode through the underlying topography. Later, when

FIGURE 5.21 East–west asymmetric valley in a subarctic alpine environment (Ruby Range, Yukon Territory). The gentle slope on the left is south-facing. This valley was recently glaciated; unconsolidated material has accumulated largely since the melting of ice. Solifluction is the primary process operating on the south-facing slope and accounts for the displacement of the lakes to the south side of the valley. The major processes operating on the bare and rocky north-facing slope are frost wedging, nivation, rockfall, and mudflow. The upper surface of this region is relatively gentle, owing to cryoplanation. (Photo by L. W. Price.)

the overlying softer material is stripped away, the streams appear to have cut transversely through the topography; in reality, the streams had no choice, since their pathways were established under different conditions. The Hudson and Susquehanna rivers are among the better-known superimposed streams.

When streams do cut across mountains, they create water gaps. These are valuable to people, since they provide easy passage across mountain barriers. The Columbia Gorge through the Cascade Mountains, for example, contains roads and railways. Many valleys have been left without a stream, either because the stream could not keep up with the rapid uplift or because water was diverted to some other part of the drainage system. Such features, known as wind gaps, are also important as mountain passes. One of these is the celebrated Cumberland Gap in the southern Appalachians, through which early settlers passed on their way to the interior.

Eolian (Wind) Processes

Elevated, exposed peaks and ridges in middle latitudes are among the windiest of environments. The geomorphic effect of wind in mountains is still not completely understood, however. The persistence of high winds on exposed slopes, with relatively little snow and vegetation to bind the soil, suggest considerable potential for erosion. However, the effects of wind must be placed in their proper perspective. For example, wind was formerly believed to be the primary agent of erosion in many desert areas, but it is now known that running water, even though infrequent,

is the major factor in landscape sculpture in most deserts. A similar condition exists in mountains. Wind becomes more important with increasing elevation and exposure, but its role as a denudational agent is still less important than the other processes we have discussed.

Wind Erosion

Wind erodes by picking up small particles, transporting them, and using them to create a sandblast effect. The size of the particles that can be picked up depends on the wind velocity and the shape and density of the particles. The greatest work is accomplished during fierce gusts under turbulent conditions (Fig. 5.22). During high winds, sand grains can be lifted up to 1–2 m and carried for much longer distances. Silt- and clay-sized material can be lifted far into the atmosphere and carried for great distances. Wind can also push and slide rocks across ice-glazed surfaces (Schumm 1956).

Several features result from wind erosion in mountains. Among the most distinctive are ventifacts—stones that have been polished and faceted by the abrasion of wind-carried particles. Ventifacts commonly have one or more flat faces or facets divided by sharp angles. On larger stones that remain relatively stationary for long periods, one side may show considerable abrasion and cutting by the wind, providing evidence of the prevailing wind direction (Rudberg 1968). Under certain windy conditions, snow, which increases in hardness with lower temperatures, can abrade rock surfaces (Fristrup 1953).

Wind has been deemed responsible for the cavernously weathered, fluted, honeycombed, and deeply pitted rock walls of some mountainous areas. Related features are weathering pits, circular depressions occurring on more or less horizontal rock surfaces. These are common in high mountains; although they are principally formed by chemical and physical weathering, wind removes the fine debris from the depressions and exposes fresh surfaces.

Wind Deposition

What the wind picks up, it must put down. In some areas, sand may accumulate, forming dunes. The finer material, predominantly of silt size, may be carried much farther, resulting in deep deposits and fertile loess soils. The mountain region itself is an area of loss rather than gain, but some of the fine material may accumulate in moist and sheltered sites, such as meadows, and produce some good soils within the mountains. Analysis of soil surface horizons on alpine glacial moraines in Colorado indicates deposition of loess and dust from eolian sources (Muhs and Benedict 2006).

Perhaps the single most important geomorphic aspect of wind in mountains is its role in the distribution of snow (Thorn 1978). Ridges and exposed slopes are typically blown free of snow, while lee slopes receive increased amounts.

FIGURE 5.22 A dust devil moving along the lateral moraine of Collier Glacier in the Three Sisters Wilderness Area, Oregon Cascades. A large rock (lower left of dust) has been dislodged and is shown in mid-air. (Photo by L. W. Price.)

FIGURE 5.23 Isabelle glacier in the Colorado Front Range. Blowing snow helps feed and maintain what is left of this glacier. (Photo by J. R. Janke.)

Snow depth is a function of small-scale topography where wind speed is reduced (Dadic et al. 2010). In some cases, blowing snow will even create and maintain small perennial snow patches or cirque glaciers. Many of the glaciers of the Colorado Front Range are of this type (Fig. 5.23).

A positive feedback exists because snow patches tend to quarry into the slope and establish larger areas for catching new snow. Wind and snowdrift thus influence the relative rates of frost action, surface runoff, and mass wasting from slope to slope. In turn, these processes have marked effects on the distribution of soil, vegetation, drainage patterns, and slope asymmetry (French 1972).

Wind is an effective transport agent that is capable of delivering particulates from great distances. Organic contaminants accumulate in mountain regions because of diurnal winds and high precipitation, but they may be reduced in a spring pulse during snowmelt (Daly and Wania 2005). Dust has increased by 500 percent above Holocene averages, which is linked to increased settlement of the western United States and to livestock grazing. The dust flux has resulted in a fivefold increase of K, Mg, CA, N, and P in alpine regions, which could impact soils, vegetation, and ecosystem health (Neff et al. 2008).

Outlook

Mountain systems are changing rapidly and are susceptible to a variety of environmental hazards (Messerli and Ives 1997; Ives 2004; UNEP 2003). Glacial retreat and mountain permafrost degradation threaten the nature and stability of mountains (Harris et al. 2001). People create more geomorphic risk by exposing themselves to danger by settling previously uninhabitable areas and by drastically altering the alpine landscape (Hewitt 2004). For example, through deforestation, people have gathered timber for fuel and building material but weakened the stability of mountain slopes. Soil erosion and degradation as well as the frequency of mass movements will all increase because of deforestation (Sandor and Nash 1995; Beniston 2000). However, intelligent land management, such as terracing, can protect mountain soils from erosion (Quine et al. 1992). Although deforestation was blamed for devastating flooding of Gangetic India and Bangladesh, the true cause was a combination of simultaneous high discharge on large rivers, heavy rainfall, high groundwater tables, and so on (Hofer and Messerli 2006; see Chapter 11, this volume). Solid scientific evidence must be provided to avoid potential hoaxes such as those that occurred in Asia.

GIS, remote sensing, and computer-enhanced field techniques and data sets should continue to be developed and explored since they can help assess geomorphological hazards. Web-based GIS structures that utilize remotely sensed data to monitor natural hazards could be used to improve risk assessment in mountainous areas (Metternicht et al. 2005). Not only will high-resolution DEMs integrated with GPS data, remotely sensed data obtained from new sensors, digital landcover, soils, or other geologic information will help predict the timing and location of hazards; these data sources will also improve our understanding of mountain processes at previously unattainable spatial and temporal scales (Butler and Walsh 1998; Walsh et al. 1998).

References

Abellan, A., Vilaplana, J. M., and Martinez, J. 2006. Application of a long-range terrestrial laser scanner to a detailed rockfall study at Vall de Nuria (Eastern Pyrenees, Spain). *Engineering Geology* 3–4: 136–148.

Ackert, R. P. 1998. A rock glacier/debris-covered glacier system at Galena Creek, Absaroka Mountains, Wyoming. *Geografiska Annaler, Series A: Physical Geography* 80A: 267–276.

Adams, J. 1980. Contemporary uplift and erosion of the Southern Alps, New Zealand. *Geological Society of America Bulletin* 91: 1–114.

Ahnert, F. 1970. Functional relationships between denudation, relief and uplift in large, mid-latitude drainage basins. *American Journal of Science* 268: 243–263.

Alford, D., Cunha, S., and Ives, J. 2000. Lake Sarez, Pamir Mountains, Tajikistan: Mountain hazards and development assistance. *Mountain Research and Development* 20: 20–23.

Anderson, R. S. 2002. Modeling the tor-dotted crests, bedrock edges, and parabolic profiles of high alpine surfaces of the Wind River Range, Wyoming *Geomorphology* 46: 35–58.

Anderson, R. S., Smith, S. J. and Koehler, P. A. 1997. Distribution of sites and radiocarbon dates in the Sierra Nevada: Implications for paleoecological prospecting *Radiocarbon* 39: 121–137.

Anderson, S. P. 1988. The upfreezing process: Experiments with a single clast. *Geological Society of America Bulletin* 100: 609–621.

Anderson, S. P. 2005. Glaciers show direct linkage between erosion rate and chemical weathering fluxes. *Geomorphology* 1–2: 147–157.

Anderson, S. P., von Blanckenburg, F., and White, A. F. 2007. Physical and chemical controls on the critical zone. *Elements* 5: 315–319.

Andersson, J. G. 1906. Solifluction, a component of subaerial denudation. *Journal of Geology* 14: 91–112.

Andre, M. F., Hall, K., Bertran, P., and Arocena, J. 2008. Stone runs in the Falkland Islands: Periglacial or tropical? *Geomorphology* 3–4: 524–543.

Arp, C. D., Schmidt, J. C., Baker, M. A., and Myers, A. K. 2007. Stream geomorphology in a mountain lake district: Hydraulic geometry, sediment sources and sinks, and downstream lake effects. *Earth Surface Processes and Landforms* 4: 525–543.

Avouac, J. P., and Burov, E. B. 1996. Erosion as a driving mechanism of intracontinental mountain growth. *Journal of Geophysical Research—Solid Earth* 101: 17747–17769.

Ayalew, L., and Yamagishi, H. 2005. The application of GIS-based logistic regression for landslide susceptibility mapping in the Kakuda–Yahiko Mountains, central Japan. *Geomorphology* 1–2: 15–31.

Barry, R. G. 1992. *Mountain Weather and Climate*. 2nd ed. New York: Routledge.

Barry, R. G. 1994. Past and potential future changes in mountain environments: A review. In M. Beniston, ed., *Mountain Environments in Changing Climates* (pp. 3–33). New York: Routledge.

Barsch, D. 1969. Permafrost in the upper subnival step of the Alps. *Geographica Helvetica* 24: 10–12.

Barsch, D. 1987. The problem of the ice-cored rock glacier. In J. R. Giardino, J. F. Shroder, and J. D. Vitek, eds., *Rock Glaciers* (pp. 45–53). London: Allen & Unwin.

Barsch, D. 1988. Rockglaciers. In M. J. Clark, eds., *Advances in Periglacial Geomorphology* (pp. 69–90). New York: John Wiley.

Barsch, D. 1993. Periglacial geomorphology in the 21st century. *Geomorphology* 7: 141–163.

Barsch, D. 1996. *Rockglaciers: Indicators for the Present and Former Geoecology in High Mountain Environments*. Berlin: Springer-Verlag.

Barsch, D., and Caine, N. 1984. The nature of mountain geomorphology. *Mountain Research and Development* 4: 287–298.

Barsch, D., Fierz, H., and Haeberli, W. 1979. Shallow core drilling and bore-hole measurements in the permafrost of an active rock glacier near the Grubengletscher, Wallis, Swiss Alps. *Arctic and Alpine Research* 11: 215–228.

Baumann, F., He, J. S., Schmidt, K., Kuhn, P., and Scholten, T. 2009. Pedogenesis, permafrost, and soil moisture as controlling factors for soil nitrogen and carbon contents across the Tibetan Plateau. *Global Change Biology* 12: 3001–3017.

Benedict, J. B. 1966. Radiocarbon dates from a stonebanked terrace in the Colorado Rocky Mountains, U.S.A. *Geografiska Annaler, Series A: Physical Geography* 48A: 24–31.

Benedict, J. B. 1970. Downslope soil movement in a Colorado alpine region: Rates, processes, and climatic significance. *Arctic and Alpine Research* 2: 165–226.

Benedict, J. B. 1973. Origin of rock glaciers. *Journal of Glaciology* 12: 520–522.

Benedict, J. B. 1976. Frost creep and gelifluction features: A review. *Quaternary Research* 6: 55–76.

Beniston, M. 1994. Climate scenarios for mountain regions: An overview of possible approaches. M. In Beniston, ed., *Mountain Environments in Changing Climates* (pp. 136–152). New York: Routledge.

Beniston, M. 2000. *Environmental Change in Mountains and Uplands*. New York: Arnold, Oxford University Press.

Benn, D. I., and Evans, D. J. A. 1998. *Glaciers and Glaciation*. New York: Arnold, Wiley.

Bishop, M. P., Shroder, J. F., Bonk, R., and Olsenholler, J. 2002. Geomorphic change in high mountains: A western Himalayan perspective. *Global and Planetary Change* 32: 311–329.

Bishop, P. 2007. Long-term landscape evolution: linking tectonics and surface processes. *Earth Surface Processes and Landforms* 3: 329–365.

Bloom, A. L. 1998. *Geomorphology : A Systematic Analysis of Late Cenozoic Landforms*. 3rd ed. Englewood Cliffs, NJ: Prentice-Hall.

Bluth, G. J. S., and Kump, L. R. 1994. Lithologic and climatologic controls of river chemistry. *Geochimica et Cosmochimica Acta* 58: 2341–2359.

Boelhouwers, J. C. 1999. Relict periglacial slope deposits in the Hex River Mountains, South Africa: Observations and palaeoenvironmental implications. *Geomorphology* 30: 245–258.

Brenning, A. 2005. Spatial prediction models for landslide hazards: Review, comparison and evaluation. *Natural Hazards and Earth System Sciences* 6: 853–862.

Brenning, A., Grasser, M., and Friend, D. A. 2007. Statistical estimation and generalized additive modeling of rock glacier distribution in the San Juan Mountains, Colorado, United States. *Journal of Geophysical Research-Earth Surface* 112(F02S15): 1–10.

Brocklehurst, S. H., and Whipple, K. X. 2002. Glacial erosion and relief production in the eastern Sierra Nevada, California. *Geomorphology* 42: 1–24.

Browning, J. M. 1973. Catastrophic rock slides, Mount Huascaran, northcentral Peru, May 31, 1970. *American Association of Petroleum Geology Bulletin* 57: 1335–1341.

Burger, K. C., Degenhardt, J. J. and Giardino, J. R. 1999. Engineering geomorphology of rock glaciers. *Geomorphology* 31: 93–132.

Butler, D. R., and Walsh, S. J. 1998. The application of remote sensing and geographic information systems in the study of geomorphology: An introduction. *Geomorphology* 21: 179–181.

Caine, T. N. 1968. *The Blockfields of Northeastern Tasmania.* Canberra: Australian National University.

Caine, T. N. 1974. The Geomorphic Processes of the Alpine Environment. In J. D. Ives and r. Barry, eds., *Arctic and Alpine Environments* (pp. 721–740). London: Methuen.

Caine, T. N. 1984. Elevational contrasts in contemporary geomorphic activity in the Colorado Front Range. *Studia Geomorphologica Carpatho-Balcanica* 18: 5–30.

Caine, T. N. 1986. Sediment movement and storage on alpine slopes in the Colorado Rocky Mountains. In A. D. Abrahams, ed., *Hillslope Processes* (pp. 115–137). Winchester, MA: Allen & Unwin.

Caine, T. N. 1996. Streamflow patterns in the alpine environment of North Boulder Creek, Colorado Front Range. *Zeitschrift fur Geomorphologie* 104: 27–42.

Caine, T. N. 2001. Geomorphic systems of Green Lakes Valley. In W. D. Bowman and T. R. Seastedt, eds., *Structure and Function of an Alpine Ecosystem, Niwot Ridge, Colorado* (pp. 45–74). New York: Oxford University Press, Inc.

Caine, T. N. 2004. Mechanical and chemical denudation in mountain systems. In P. Owens and O. Slaymaker, eds., *Mountain Geomorphology* (pp. 132–152). London: Edward Arnold.

Caine, T. N. 2010. Recent hydrologic change in a Colorado alpine basin: An indicator of permafrost thaw? *Annals of Glaciology* 51:130–134.

Calkin, P. E., Haworth, L. A., and Ellis, J. M. 1987. Rock glaciers of central Brooks Range, Alaska, U.S.A. In J. R. Giardino, J. F. Shroder, and J. D. Vitek, eds., *Rock Glaciers* (pp. 65–82). London: Allen & Unwin.

Chang, J.-C., and Slaymaker, O. 2002. Frequency and spatial distribution of landslides in a mountainous drainage basin: Western foothills, Taiwan. *Catena* 46: 285–307.

Chorley, R. 1962. Geomorphology and general systems theory. U.S. Geological Survey Professional Paper 500B.

Chorley, R. 1965. A re-evaluation of the geomorphic system of W M Davis. In R. Chorley and R. Haggett, eds., *Frontiers in Geographic Teaching* (pp. 21–40). London: Methuen.

Clapperton, C. M., and Hamilton, P. 1971. Peru beneath its external threat. *Geographical Magazine* 43: 632–639.

Clark, D. H., Steig, E. J., Potter, N. and Gillespie, A. R. 1998. Genetic variability of rock glaciers. *Geografiska Annaler, Series A: Physical Geography* 80A: 175–192.

Clark, S. P., Jr., and Jäger, E. 1969. Denudation rate in the Alps from geochronologic and heat flow data. *American Journal of Science* 267: 1143–1160.

Clow, D. W. 2010. Changes in the timing of snowmelt and streamflow in Colorado: A response to recent warming. *Journal of Climate* 23: 2293–2306.

Corte, A. E. 1987. Rock glacier taxonomy. In J. R. Giardino, J. F. Shroder, and J. D. Vitek, eds., *Rock Glaciers* (pp. 27–40). Boston: Allen & Unwin.

Cruden, D. M. 1976. Major rock slides in the Rockies. *Canadian Geotechnical Journal* 13: 8–20.

Cyr, A. J., and Granger, D. E. 2008. Dynamic equilibrium among erosion, river incision, and coastal uplift in the northern and central Apennines, Italy. *Geology* 2: 103–106.

Dadic, R., Mott, R., Lehning, M., and Burlando, P. 2010. Wind influence on snow depth distribution and accumulation over glaciers. *Journal of Geophysical Research—Earth Surface* 115: 1–8.

Dahl, R. 1966. Blockfields and other weathering forms in the Narvik Mountains. *Geografiska Annaler, Series A: Physical Geography* 48A: 224–227.

Daly, G. L., and Wania, F. 2005. Organic contaminants in mountains. *Environmental Science & Technology* 2: 385–398.

Darmody, R. G., Thorn, C. E., Harder, R. L., Schlyter, J. P. L., and Dixon, J. C. 2000. Weathering implications of water chemistry in an arctic-alpine environment, northern Sweden. *Geomorphology* 34: 89–100.

Dedkov, A. P., and Moszherin, V. T. 1992. Erosion and sediment yield in mountain areas of the world. *International Association of Hydrological Sciences Publication* 209: 29–36.

Desloges, J. R., and Gilbert, R. 1998. Sedimentation in Chilko Lake: A record of the geomorphic environment of the eastern Coast Mountains of British Columbia, Canada. *Geomorphology* 25: 75–91.

Dingwall, P. R. 1972. Erosion by overland flow on an alpine debris slope. In O. Slaymaker and H. J. McPherson, eds., *Mountain Geomorphology* (pp. 113–120). Vancouver, BC: Tantalus Research.

Dixon, J. C., and Thorn, C. E. 2005. Chemical weathering and landscape development in mid-latitude alpine environments. *Geomorphology* 1–2: 127–145.

Embleton, C., and King, C. 1975. *Periglacial Geomorphology.* New York: John Wiley and Sons.

Etzelmüller, B., Hoelzle, M., Heggem, E. S. F., Isaksen, K., Mittaz, C., Vonder Mühll, D., Ødegård, R. S., Haeberli, W., and Sollid, J. 2001a. Mapping and modelling the occurrence and distribution of mountain permafrost. *Norwegian Journal of Geography* 55: 186–194.

Etzelmüller, B., Odegard, R. S., Berthling, I. and Sollid, J. L. 2001b. Terrain parameters and remote sensing data in the analysis of permafrost distribution and periglacial processes: Principles and examples from southern Norway. *Permafrost Periglacial Processes* 12: 79–92.

Evans, S. G., and Clague, J. J. 1994. Recent climatic change and catastrophic geomorphic processes in mountain environments. *Geomorphology* 10: 107–128.

Fahey, B. D. 1973. An analysis of diurnal freeze–thaw and frost heave cycles in the Indian Peaks region of the Colorado Front Range. *Arctic and Alpine Research* 5: 269–281.

Fahey, B. D. 1974. Seasonal frost heave and frost penetration measurements in the Indian Peaks region of the Colorado Front Range. *Arctic and Alpine Research* 6: 63–70.

Fahnestock, R. K. 1963. Morphology and hydrology of a glacial stream, White River, Mt. Rainier, Washington. U.S. Geological Survey Professional Paper 422A.

Francou, B., Fabre, D., Pouyaud, B., Jomelli, V., and Arnaud, Y. 1999. Symptoms of degradation in a tropical rock glacier. *Permafrost Periglacial Processes* 10: 91–100.

Frankel, K. L., and Dolan, J. F. 2007. Characterizing arid region alluvial fan surface roughness with airborne laser swath mapping digital topographic data. *Journal of Geophysical Research—Earth Surface* 112(F02025): 1–14.

Frauenfelder, R. 1997. Permafrostuntersuchungen mit GIS-Eine Studie im Fletschhorngebiet. Zurich: University of Zurich.

Frauenfelder, R., B. Allgower, W. Haeberli, and Hoelzle, M. 1998. Permafrost investigations with GIS-A case study in the Fletschhorn area, Wallis, Swiss Alps. Proceedings of the 7th International Permafrost, Nordicana, Yellowknife, Canada (pp. 551–556).

French, H. M. 1996. *The Periglacial Environment.* Essex, UK: Addison Wesley Longman.

French, H. M. 1972. The role of wind in periglacial environments, with special reference to northwest Banks Island, western Canadian Arctic. In W. P. Adams and F. M. Helleiner, eds., *International Geography* (pp. 82–84). Toronto: University of Toronto Press.

Fristrup, B. 1953. Wind erosion within the arctic deserts. *Geografisk Tidsskrift* 52: 51–65.

Gardner, J. 1970. Rockfall: A geomorphic process in high mountain terrain. *Albertan Geographer* 6: 15–21.

Gardner, J. 1973. The nature of talus shift on alpine talus slopes: An example from the Canadian Rocky Mountains. In B. D. Fahey and R. D. Thompson, eds., *Research in Polar and Alpine Geomorphology* (pp. 95–105). Ontario: 3rd Guelph Symposium on Geomorphology.

Gerrard, A. J. 1990. *Mountain Environments: An Examination of the Physical Geography of Mountains.* Cambridge: MIT Press.

Giardino, J., and J. Vitek. 1988. The significance of rock glaciers in the glacial–periglacial landscape continuum. *Journal of Quarternary Science* 3: 97–103.

Giardino, J. R., and Vick, S. G. 1987. Geologic engineering aspects of rock glaciers. In J. R. Giardino, F. Shroder, and J. D. Vitek, eds., *Rock Glaciers* (pp. 265–288). Boston: Allen & Unwin.

Glade, T. 2005. Linking debris-flow hazard assessments with geomorphology. *Geomorphology* 1–4: 189–213.

Gleason, K. J., Krantz, W. B., Caine, N., George, J. H., and Gunn, R. D.. 1986. Geometrical aspects of sorted patterned ground in recurrently frozen soil. *Science* 232: 219–220.

Glenn, N. F., Streutker, D. R., Chadwick, D. J., Thackray, G. D., and Dorsch, S. J. 2006. Analysis of LiDAR-derived topographic information for characterizing and differentiating landslide morphology and activity. *Geomorphology* 1–2: 131–148.

Godt, J. W., and Coe, J. A. 2007. Alpine debris flows triggered by a 28 July 1999 thunderstorm in the central Front Range, Colorado. *Geomorphology* 1–2: 80–97.

Gooseff, M. N., Hall, R. O., and Tank, J. L. 2007. Relating transient storage to channel complexity in streams of varying land use in Jackson Hole, Wyoming. *Water Resources Research* 43(W01417): 1–10.

Gorbunov, A. P. 1978. Permafrost investigations in high mountain regions. *Arctic and Alpine Research* 10: 283–294.

Gorbushina, A. A. 2007. Life on the rocks. *Environmental Microbiology* 7: 1613–1631.

Green, K., and Pickering, C. M. 2009. The decline of snowpatches in the Snowy Mountains of Australia: Importance of climate warming, variable snow, and wind. *Arctic Antarctic and Alpine Research* 2: 212–218.

Gruber, S., and Hoelzle, M. 2001. Statistical modelling of mountain permafrost distribution: Local calibration and incorporation of remotely sensed data. *Permafrost Periglacial Processes* 12: 69–77.

Guglielmin, M., Aldighieri, B., and Testa, B. 2003. PERMACLIM: A model for the distribution of mountain permafrost, based on climatic observations. *Geomorphology* 51: 245–257.

Guzzetti, F., Reichenbach, P., Cardinali, M., Galli, M., and Ardizzone, F. 2005. Probabilistic landslide hazard assessment at the basin scale. *Geomorphology* 1–4: 272–299.

Hack, J. T. 1960. Origin of talus and scree in northern Virginia. *Geological Society of America Bulletin* 71: 1877–1878.

Haeberli, W. 1983. Permafrost–glacier relationships in the Swiss Alps: Today and in the past. Proceedings of the 4th International Conference on Permafrost (pp. 415–420). Washington, DC: National Academy Press.

Haeberli, W. 1985. Creep of mountain permafrost: Internal structure and flow of alpine rock glaciers. *Mitteilungen der Versuchsanstalt für Wasserbau, Hydrologie und Glaziologie* 77: 1–142.

Haeberli, W. 1990. Glacier and permafrost signals of 20th-century warming. *Annals of Glaciology* 14: 99–101.

Haeberli, W. 1994. Accelerated glacier and permafrost changes in the Alps. In M. Beniston, ed., *Mountain Environments and Changing Climates* (pp. 91–107). New York: Routledge.

Haeberli, W. 2000. Modern research perspectives relating to permafrost creep and rock glaciers: A discussion. *Permafrost Periglacial Processes* 11: 290–293.

Haeberli, W., Frauenfelder, R., Hoelzle, M., and Maisch, M. 1999. On rates and acceleration trends of global glacier mass changes. *Geografiska Annaler, Series A: Physical Geography* 81A: 585–591.

Haeberli, W., Hallet, B., Arenson, L., Elconin, R., Humlun, O., Kaab, A., Kaufmann, V., Ladanyi, B., Matsuoka, N., Springman, S., and Vonder Mühll, D. 2006. Permafrost creep and rock glacier dynamics. *Permafrost and Periglacial Processes* 3: 189–214.

Hall, K. 1998. Rock temperatures and implications for cold region weathering, II: New data from Rothera, Adelaide Island, Antarctica. *Permafrost Periglacial Processes* 9: 47–55.

Hallet, B. 1998. Measurement of soil motion in sorted circles, western Spitsbergen. Permafrost: Seventh International Conference, Yellowknife, NWT, Canada, *Proceedings* (pp. 415–420.

Hallet, B., Anderson, S., Stubbs, C. W., and Gregory, E. 1988. Surface soil displacements in sorted circles, western Spitsbergen. Permafrost: Fifth International Conference, Trondheim, Norway, *Proceedings* (pp. 770–775).

Hallet, B., and Waddington, E. D. 1992. Buoyancy forces induced by freeze–thaw in the active layer: Implications for diapirism and soil circulation. In J. C. Dixon and A. D. Abrahams, eds., *Periglacial geomorphology* (pp. 251–280). London: Allen & Unwin.

Harbor, J., and Warburton, J. 1993. Relative rates of glacial and nonglacial erosion in alpine environments. *Arctic and Alpine Research* 25: 1–7.

Harris, C., Kern-Luetschg, M., Smith, F., and Isaksen, K. 2008. Solifluction processes in an area of seasonal ground freezing, Dovrefjell, Norway. *Permafrost and Periglacial Processes* 1: 31–47.

Harris, C., Luetschg, M., Davies, M. C. R., Smith, F., Christiansen, H. H., and Isaksen, K. 2007. Field instrumentation for real-time monitoring of periglacial solifluction. *Permafrost and Periglacial Processes* 1: 105–114.

Harris, C., Rea, B., and Davies, M. 2001. Scaled modelling of mass movement process on thawing slopes. *Permafrost Periglacial Processes* 12: 125–135.

Harris, S. A. 1988. The alpine periglacial zone. In M. J. Clark, ed., *Advances in Periglacial Geomorphology* (pp. 369–413). New York: John Wiley.

Harris, S. A. 1998a. Non-sorted circles on Plateau Mountain, S.W. Alberta, Canada. Permafrost: Seventh International Conference, Yellowknife, NWT, Canada, *Proceedings* (pp. 441–448).

Harris, S. A. 1998b. A genetic classification of the palsa-like mounds in western Canada. *Biuletyn Peryglacjalny* 37: 115–129.

Harris, S. A., and Corte, A. E. 1992. Interactions and relations between mountain permafrost, glaciers, snow, and water. *Permafrost Periglacial Processes* 5: 103–110.

Hartshorn, K., Hovius, N., Dade, W. B., and Slingerland, R. L. 2002. Climate-driven bedrock incision in an active mountain belt. *Science* 297: 2036–2038.

Haugland, J. E. 2006. Short-term periglacial processes, vegetation succession, and soil development within sorted patterned ground: Jotunheimen, Norway. *Arctic Antarctic and Alpine Research* 38: 82–89.

Haugland, J. E., and Beatty, S.W. 2005. Vegetation establishment, succession and microsite frost disturbance on glacier forelands within patterned ground chronosequences. *Journal of Biogeography* 1: 145–153.

Haugland, J. E., and Owen, B. S. 2005. Temporal and spatial variability of soil pH in patterned-ground chronosequences: Jotunheimen, Norway. *Physical Geography* 26: 299–312.

Heginbottom, J. A. 2002. Permafrost mapping: A review. *Progress in Physical Geography* 26: 623–642.

Hembree, C. H., and Rainwater, F. H. 1961. Chemical degradation on opposite flanks of the Wind River Range, Wyoming. U.S. Geological Survey Water Supply Paper 1543E.

Heuberger, H., Masch, L., Preuss, E., and Schrocker, A. 1984. Quaternary landslides and rock fusion in central Nepal and in the Tyrolean Alps. *Mountain Research and Development* 4: 345–362.

Hewitt, K. 1972. The mountain environment and geomorphic processes. In O. Slaymaker and H. J. McPherson, eds., *Mountain Geomorphology* (pp. 17–36). Vancouver, BC: Tantalus Research.

Hewitt, K. 2004. Geomorphic hazards in mountain environments In P. Owens and OSlaymaker, eds., *Mountain Geomorphology* (pp. 187–218). London: Edward Arnold.

Hewitt, K., Clague, J. J., and Orwin, J. F. 2008. Legacies of catastrophic rock slope failures in mountain landscapes. *Earth-Science Reviews* 1–2: 1–38.

Hinderer, M. 2001. Late Quaternary denudation of the Alps, valley and lake fillings and modern river loads. *Geodinamica Acta* 14: 231–263.

Hjort, J., and Luoto, M. 2006. Modelling patterned ground distribution in Finnish Lapland: An integration of topographical, ground and remote sensing information. *Geografiska Annaler, Series A: Physical Geography* 1: 19–29.

Hoelzle, M., Haeberli, W., and Keller, F. 1993. Application of BTS-measurements for modelling mountain permafrost distribution. Proceedings of the 6th International Conference on Permafrost, Beijing, China (pp. 272–277).

Hofer, T., and Messerli, B. 2006. *Floods in Bangladesh: History, dynamics, and rethinking the role for the Himalayas.* Tokyo: United Nations University Press.

Hovius, N., Stark, C. P., Chu, H. T., and Lin, J. C. 2000. Supply removal of sediment in a landslide dominated mountain belt: Central Range, Taiwan. *Journal of Geology* 108: 73–89.

Howe, E. 1909. Landslides in the San Juan Mountains, Colorado: Including a consideration of their causes and their classification. U.S. Geological Survey Professional Paper 67.

Humlum, O. 1998. The climatic significance of rock glaciers. *Permafrost Periglacial Processes* 9: 375–395.

Humlum, O. 2000. The geomorphic significance of rock glaciers: Estimates of rock glacier debris volume and headwall recession rates in west Greenland. *Geomorphology* 35: 41–67.

Imhof, M. 1996. Modelling and verification of the permafrost distribution in the Bernese Alps. *Permafrost Periglacial Processes* 7: 267–280.

Ives, J. D. 1966. Blockfields, associated weathering forms on mountain tops and the Nunatak hypothesis. *Geografiska Annaler, Series A: Physical Geography* 4A: 220–223.

Ives, J. D. 1974. Permafrost. In J. D. Ives and R. Barry, eds., *Arctic and Alpine Environments* (pp. 159–194). New York: Methuen; distributed in the U.S. by Harper & Row.

Ives, J. D. 2004. *Himalayan Perceptions: Environmental Change and the Well-being of Mountain Peoples.* New York: Routledge.

Ives, J. D., and Fahey, B. D. 1971. Permafrost occurrence in the Front Range, Colorado Rocky Mountains, U.S.A. *Journal of Glaciology* 58: 105–111.

Ives, R. L. 1940. Rock glaciers in the Colorado Front Range. *Geological Society of America Bulletin* 51: 1271–1294.

Jäckli, H. 1956. Gegenwartsgelogie des bundnerischen Rheingebietes: Ein Beitrag zur exogen Dynamik alpiner Gebirgslandschafter. *Beitrage zur Geologie der Schweiz, Geotechnical Series* 36: 126.

Jahn, A. 1967. Some features of mass movement on Spitsbergen slopes. *Geografiska Annaler, Series A: Physical Geography* 49A: 213–225.

Janke, J. R. 2001. Rock glacier mapping: a method utilizing enhanced TM data and GIS modeling techniques. *Geocarto International* 3: 5–15.

Janke, J. R. 2005a. Long-term flow measurements (1961–2002) of the Arapaho, Taylor, and Fair rock glaciers, Front Range, Colorado. *Physical Geography* 26: 313–336.

Janke, J. R. 2005b. Photogrammetric analysis of Front Range rock glacier flow rates. *Geografiska Annaler, Series A: Physical Geography* 87A: 515–526.

Janke, J. R. 2005c. The occurrence of alpine permafrost in the Front Range of Colorado. *Geomorphology* 67: 375–389.

Janke, J. R. 2005d. Modeling past and future alpine permafrost distribution in the Colorado Front Range. *Earth Surface Processes and Landforms* 30: 1495–1508.

Janke, J. R. 2007. Colorado Front Range rock glaciers: Distribution and topographic characteristics. *Arctic, Antarctic, and Alpine Research* 39: 74–83.

Janke, J., and Frauenfelder, R. 2008. The relationship between rock glacier and contributing area parameters in the Front Range of Colorado. *Journal of Quaternary Science* 2: 153–163.

Johnson, P. G. 1987. Rock glaciers: Glaciers debris systems or high magnitude low-frequency flows. In J. R. Giardino, J. F. Shroder, and J. D. Vitek, eds., *Rock Glaciers* (pp. 175–192). London: Allen & Unwin.

Kääb, A., Frauenfelder, R., and Roer, I. 2007. On the response of rockglacier creep to surface temperature increase. *Global and Planetary Change* 1–2: 172–187.

Kääb, A., and Vollmer, M. 2000. Surface geometry, thickness changes and flow fields on creeping mountain permafrost: Automatic extraction by digital image analysis. *Permafrost Periglacial Processes* 11: 315–326.

Keller, F. 1992. Automated mapping of mountain permafrost using the program PERMAKART within the geographical system ARC/INFO. *Permafrost Periglacial Processes* 3: 133–138.

King, L., Gorbunov, A., and Evin, M. 1992. Prospecting and mapping of mountain permafrost and associated phenomena. *Permafrost Periglacial Processes* 3: 73–81.

Kirkbride, M., and Brazier, V. 1995. On the sensitivity of Holocene talus-derived rock glaciers to climate change in the Ben Ohau Range, New Zealand. *Journal of Quaternary Science* 10: 353–365.

Korup, O., Clague, J. J., Hermanns, R. L., Hewitt, K., Strom, A. L., and Weidinger, J. T. 2007. Giant landslides, topography, and erosion. *Earth and Planetary Science Letters* 261: 578–589.

Krautblatter, M., and Dikau, R. 2007. Towards a uniform concept for the comparison and extrapolation of rockwall retreat and rockfall supply. *Geografiska Annaler, Series A: Physical Geography* 1: 21–40.

Lachenbruch, A., and Marshall, B. 1986. Changing climate; geothermal evidence from permafrost in the Alaskan Arctic. *Science* 234: 689–696.

Lan, H. X., Martin, C. D., and Lim, C. H. 2007. RockFall analyst: A GIS extension for three-dimensional and spatially distributed rockfall hazard modeling. *Computers & Geosciences* 2: 262–279.

Lapenna, V., Lorenzo, P., Perrone, A., Piscitelli, S., Rizzo, E., and Sdao, F. 2005. 2D electrical resistivity imaging of some complex landslides in the Lucanian Apennine chain, southern Italy. *Geophysics* 3: B11–B18.

Lehning, M., Volksch, I., Gustafsson, D., Nguyen, T. A., Stahli, M., and Zappa, M. 2006. ALPINE3D: A detailed model of mountain surface processes and its application to snow hydrology. *Hydrological Processes* 10: 2111–2128.

Leopold, L. B., Wolman, M. G., and Miller, J. P. 1964. *Fluvial Processes in Geomorphology*. San Francisco: W. H. Freeman.

Leopold, M., Voelkel, J., Dethier, D., Williams, M. W., and Caine, N. 2010. Mountain permafrost: A valid archive to study climate change? Examples from the Rocky Mountains Front Range of Colorado, USA. *Nova Acta Leopoldina* NF 384: 281–289.

Lewkowicz, A. G. 1988. Slope processes. In M. J. Clark, ed., *Advances in Periglacial Geomorphology* (pp. 325–368). New York: John Wiley.

Linder, L., and Marks, L. 1985. Types of debris slope accumulation and rock glaciers in South Spitsbergen. *Boreas* 14: 139–153.

Lliboutry, L., Morales Arnao, B., Pautre, A., and Schneider, B. 1977. Glaciological problems set by the control of dangerous takes in Cordillera Blanca, Peru, I: Historical failures of morainic dams, their causes and prevention. *Journal of Glaciology* 18: 239–254.

Losleben, M. 2004. D-1 (3743 m) climate station: Temperature and precipitation data (1952–2004), edited. <http://culter.colorado.edu/exec/.extracttoolA?d-1tdayv.ml>.

Luckman, B. H. 1976. Rockfalls and rockfall inventory data: Some observations from Surprise Valley, Jasper National Park, Canada. *Earth Surface Processes* 1: 287–298.

Luckman, B. H. 1978. Geomorphic work of snow avalanches in the Canadian Rocky Mountains. *Arctic and Alpine Research* 10: 261–276.

Lugon, R., and Delaloye, R. 2001. Modelling alpine permafrost distribution, Val de Réchy, Valais Alps (Switzerland). *Norwegian Journal of Geography* 55: 224–229.

Mangold, N. 2005. High latitude patterned grounds on Mars: Classification, distribution and climatic control. *Icarus* 2: 336–359.

Marchand, D. E. 1971. Rates and modes of denucleation, White Mountain, eastern California. *American Journal of Science* 270: 109–135.

Marsell, R. E. 1972. Cloudburst and snowmelt floods. In L. S. Hilpert, ed., *Environmental Geology of the Wasatch Front* (pp. 1–18). Salt Lake City: Utah Geological Association.

Martin, H., and Whalley, W. B. 1987. Rock glaciers, Part 1: Rock glacier morphology: Classification and distribution. *Progress in Physical Geography* 11: 260–282.

Mathews, W. H. 1955. Permafrost and its occurrence in the southern coast mountains of British Columbia. *Canadian Alpine Journal* 28: 94–98.

Matsuoka, N. 1991. A model of the rate of frost shattering: Application to field data from Japan, Svalbard, and Antarctica. *Permafrost Periglacial Processes* 2: 271–281.

Matsuoka, N. 2005. Temporal and spatial variations in periglacial soil movements on alpine crest slopes. *Earth Surface Processes and Landforms* 1: 41–58.

McPherson, H. J. 1971a. Dissolved, suspended and bedload movement patterns in Two O'Clock Creek, Rocky Mountains, Canada, summer 1969. *Journal of Hydrology* 12: 221–233.

McPherson, H. J. 1971b. Downstream changes in sediment character in a high energy mountain stream channel. *Arctic and Alpine Research* 3: 65–79.

McPherson, H. J., and Hirst, F. 1972. Sediment changes on two alluvial fans in the Canadian Rocky Mountains. In O. Slaymaker and H. J. McPherson, eds., *Mountain Geomorphology* (pp. 161–175). Vancouver, BC: Tantalus Research.

Messerli, B., and Ives, J. D., eds. 1997. *Mountains of the World: A Global Priority*. New York: Parthenon.

Metternicht, G., Hurni, L., and Gogu, R. 2005. Remote sensing of landslides: An analysis of the potential contribution to geo-spatial systems for hazard assessment in mountainous environments. *Remote Sensing of Environment* 2–3: 284–303.

Miller, J. P. 1961. Solutes in small streams draining single rock types, Sangre de Cristo Range, New Mexico. U.S. Geological Survey Water Supply Paper 1535F.

Milliman, J. D., and Syvitski, J. P. M. 1992. Geomorphic tectonic control of sediment discharge to the ocean: The importance of small mountainous rivers. *Journal of Geology* 100: 525–544.

Molnar, P., and England, P. 1990. Late Cenozoic uplift of mountain-ranges and global climate change: Chicken or egg? *Nature* 346: 29–34.

Mote, P. W., Hamlet, A. F., Clark, M. P., and Lettenmaier, D. P. 2005. Declining mountain snowpack in western North America. *Bulletin of the American Meteorological Society* 1: 39–49.

Muhs, D. R., and Benedict, J. B. 2006. Eolian additions to late Quaternary alpine soils, Indian Peaks Wilderness Area, Colorado Front Range. *Arctic Antarctic and Alpine Research* 1: 120–130.

Nanson, G. C. 1974. Bedload and suspended load transport in a small, steep, mountain stream. *American Journal of Science* 274: 471–486.

Neff, J. C., Ballantyne, A. P., Farmer, G. L., Mahowald, N. M., Conroy, J. L., Landry, C. C., Overpeck, J. T., Painter, T. H., Lawrence, C. R., and Reynolds, R. L. 2008. Increasing eolian dust deposition in the western United States linked to human activity. *Nature Geoscience* 3: 189–195.

Oberlander, T. 1965. *The Zagros Streams: A New Interpretation of Transverse Drainage in an Orogenic Zone*. Syracuse, NY: Syracuse University Press.

Odegard, R., Liestol, O., and Sollid, J. L. 1988. Periglacial forms related to terrain parameters in Jotunheimen, southern Norway. Permafrost: Fifth International Conference, Trondheim, Norway, *Proceedings* (pp. 59–61).

Ollier, C. D., and Pain, C. F. 2000. *The Origin of Mountains*. London: Routledge.

Outcalt, S. I. 1971. An algorithm for needle ice growth. *Water Resources Research* 7: 394–400.

Outcalt, S. I., and Benedict, J. B. 1965. Photo-interpretation of two types of rock glaciers in the Colorado Front Range, U.S.A. *Journal of Glaciology* 5: 849–856.

Peltier, L. C. 1950. The geographic cycle in periglacial regions as it is related to climatic geomorphology. *Annals of the Association of American Geographers* 40: 214–236.

Pepin, N., and Losleben, M. 2002. Climate change in the Colorado Rocky Mountains: Free air versus surface temperature trends. *International Journal of Climatology* 22: 311–329.

Perov, V. F. 1969. Block fields in the Khibiny Mountains. *Biuletyn Peryglacjalny* 19: 381–389.

Perret, S., Stoffel, M., and Kienholz, H. 2006. Spatial and temporal rockfall activity in a forest stand in the Swiss Prealps: A dendrogeomorphological case study. *Geomorphology* 1–4: 219–231.

Péwé, T. L. 1970. Altiplanation terraces of early Quaternary age near Fairbanks, Alaska. *Acta Geographica Lodziensia* 24: 357–363.

Péwé, T. L. 1983. Alpine permafrost in the contiguous United States. *Arctic and Alpine Research* 15: 145–156.

Phipps, R. L. 1974. The soil creep curved tree fallacy. *Journal of Research of the U.S. Geological Survey* 2: 371–378.

Pierce, W. G. 1961. Permafrost and thaw depressions in a peat deposit in the Beartooth Mountains, northwestern Wyoming. U.S. Geological Survey Professional Paper 424B: 154-156.

Potter, N. 1972. Ice-cored rock glacier, Galena Creek, northern Absaroka Mountains, Wyoming. *Geological Society of America Bulletin* 83: 3025–3068.

Potter, N. and Moss, J. H. 1968. Origin of the Blue Rocks block field and adjacent deposits, Berks County, Pennsylvania. *Geological Society of America Bulletin* 79: 255–262.

Potter, N., Steig, E. J., Clark, D. H., Speece, M. A., Clark, G. M., and Updike, A. B. 1998. Galena Creek rock glacier revisited: New observations on an old controversy. *Geografiska Annaler, Series A: Physical Geography* 80A: 251–265.

Price, L. W. 1969. The collapse of solifluction lobes as a factor in vegetating blockfields. *Arctic* 22: 395–402.

Price, L. W. 1970. Up-heaved blocks: A curious feature of instability in the tundra. *Proceedings of the Association of American Geographers* 2: 106–110.

Price, L. W. 1971a. Vegetation, microtopography, and depth of active layer on different exposures in subarctic alpine tundra. *Ecology* 52: 638–647.

Price, L. W. 1971b. Geomorphic effect of the arctic ground squirrel in an alpine environment. *Geografiska Annaler, Series A: Physical Geography* 53A: 100–106.

Price, L. W. 1991. Subsurface Movement on solifluction slopes in the Ruby Range, Yukon Territory, Canada: A 20-year study. *Arctic and Alpine Research* 23: 200–205.

Price, M. F., and Barry, R. G. 1997. Climate change. In B. Messerli and J. D. Ives, eds., *Mountains of the World : A Global Priority* (pp. 409–445). New York: Parthenon.

Quine, T. A., Walling, D. E., Zhang, X., and Wang, Y. 1992. Investigations of soil erosion on terraced fields near Yanting, Sichuan province, China, using caesium-137. In D. E. Walling, T. R. Davies, and B. Hasholt, eds., *Erosion, Debris Flows and Environment in Mountain Regions* (pp. 155–168). Wallingford, UK: IAHS Press.

Rapp, A. 1960. Recent development of mountain slopes in Karkevagge and surroundings, northern Scandinavia. *Geografiska Annaler, Series A: Physical Geography* 42A: 73–200.

Rapp, A. 1974. Slope erosion due to extreme rainfall, with examples from tropical and arctic mountains. In H. Poser, ed., *Geomorphologische Prozesse und Prozesskombinationen unter verschiedenen Klimabedingungen* (pp. 118–136). Gottingen: Abhandlung der Akademie der Wissenschaften.

Rea, B. R., Whalley, W. B., Rainey, M. M., and Gordon, J. E. 1996. Blockfields, old or new? Evidence and implications from some plateaus in northern Norway. *Geomorphology* 15: 109–121.

Reger, R. D., and Péwé, T. L. 1976. Cryoplanation terraces: Indicators of a permafrost environment. *Quaternary Research* 6: 99–110.

Retzer, J. L. 1965. Alpine soils of the Rocky Mountains. *Journal of Soil Science* 7: 22–32.

Riebe, C. S., Kirchner, J. W., Granger, D. E., and Finkel, R. C. 2001. Minimal climatic control on erosion rates in the Sierra Nevada, California. *Geology* 29: 447–450.

Ritter, D. F., Kochel, R. C., and Miller, J. R. 2002. *Process Geomorphology*. New York: McGraw-Hill.

Roering, J. J. 2004. Soil creep and convex-upward velocity profiles: Theoretical and experimental investigation of disturbance-driven sediment transport on hillslopes. *Earth Surface Processes and Landforms* 13: 1597–1612.

Rood, S. B., Pan, J., Gill, K. M., Franks, C. G., Samuelson, G. M., and Shepherd, A. 2008. Declining summer flows of Rocky Mountain rivers: Changing seasonal hydrology and probable impacts on floodplain forests. *Journal of Hydrology* 3–4: 397–410.

Röthlisberger, H., and Lang, H. 1987. Glacier hydrology. In A. M. Gurnell and M. J. Clark, eds., *Glaciofluvial Sediment Transfer* (pp. 207–284). New York: Wiley.

Rudberg, S. 1968. Wind erosion preparation of maps showing the direction of eroding winds. *Biuletyn Peryglacjalny* 17: 181–194.

Sandor, J. A., and Nash, N. S. 1995. Ancient agricultural soils in the Andes of Peru. *Soil Science Society of America Journal* 59: 170–179.

Schrott, L. 1991. Global solar radiation, soil temperature and permafrost in the central Andes, Argentina: A progress report. *Permafrost Periglacial Process* 2: 59–66.

Schrott, L., and Sass, O. 2008. Application of field geophysics in geomorphology: Advances and limitations exemplified by case studies. *Geomorphology* 1–2: 55–73.

Schumm, S. A. 1956. The movement of rocks by wind. *Journal of Sedimentary Petrology* 26: 284–286.

Schumm, S. A. 1963. The disparity between present rates of denudation and orogeny. U.S. Geological Survey Professional Paper 454H.

Schumm, S. A. 1977. *The Fluvial System*. New York: John Wiley and Sons.

Scotter, G. W. 1975. Permafrost profiles in the continental divide region of Alberta and British Columbia. *Arctic and Alpine Research* 7: 93–96.

Sharp, M., Tranter, M., Brown, G. H., and Skidmore, M. 1995. Rates of chemical denudation and CO_2 drawdown in a glacier-covered alpine catchment. *Geology* 23: 61–64.

Sharpe, C. 1938. *Landslides and Related Phenomena*. New York: Columbia University Press.

Sklar, L. S., and Dietrich, W. E. 2001. Sediment and rock strength controls on river incision into bedrock. *Geology* 29: 1087–1090.

Slaymaker, O. 1974. Alpine hydrology. In J. D. Ives and R. Barry, eds., *Arctic and Alpine Environments* (pp. 235–245). London: Methuen: 235–245.

Slaymaker, O., and McPherson, H. J. 1977. An overview of geomorphic processes in the Canadian cordilerra. *Zeitschrift für Geomorphologie* 21: 169–186.

Small, E. E., and Anderson, R. S. 1995. Geomorphologically driven Late Cenozoic rock uplift in the Sierra-Nevada, California. *Science* 270: 277–280.

Small, E. E., and Anderson, R. S. 1998. Pleistocene relief production in Laramide mountain ranges, western United States. *Geology* 26: 123–126.

Small, E. E., Anderson, R. S., Repka, J. L., and Finkel, R. 1997. Erosion rates of alpine bedrock summit surfaces deduced from *in situ* Be-10 and Al-26. *Earth and Planetary Science Letters* 150: 413–425.

Smith, D. J. 1992. Long-term rates of contemporary solifluction in the Canadian Rocky Mountains. In J. C. Dixon and A. D. Abrahams, eds., *Periglacial Geomorphology* (pp. 201–221). London: Allen & Unwin.

Smith, M. W., and Patterson, D. E. 1989. Detailed observations on the nature of frost heaving at a field scale. *Canadian Geotechnical Journal* 26: 306–312.

Solomon, S., Qin, D., Manning, M., Chen, Z., Marquis, M., Avery, K. B., Tignor, M., and Miller, H., eds. *Climate Change 2007: The Physical Science Basis*. Contribution of Working Group I to the Fourth Assessment Report of the Intergovernmental Panel on Climate Change. Cambridge, UK: Cambridge University Press.

Sosio, R., Crosta, G. B., and Frattini, P. 2007. Field observations, rheological testing and numerical modelling of a debris-flow event. *Earth Surface Processes and Landforms* 2: 290–306.

Stock, G. M., Frankel, K. L., Ehlers, T. A., Schaller, M., Briggs, S. M., and Finkel, R. C. 2009. Spatial and temporal variations in denudation of the Wasatch Mountains, Utah, USA. *Lithosphere* 1: 34–40.

Stocker, M., Hoelzle, M., and Haeberli, W. 2002. Modelling alpine permafrost distribution based on energy-balance data: A first step. *Permafrost Periglacial Processes* 13: 271–282.

Strahler, A. N. 1965. *Introduction to Physical Geography*. New York: John Wiley and Sons.

Sueker, J. K., Clow, D. W., Ryan, J. N., and Jarrett, R. D. 2001. Effect of basin physical characteristics on solute fluxes in nine alpine/subalpine basins, Colorado, USA. *Hydrological Processes* 15: 2749–2769.

Taber, S. 1929. Frost heaving. *Journal of Geology* 37: 428–461.

Taber, S. 1930. The mechanics of frost heaving. *Journal of Geology* 38: 303–317.

Thorn, C. E. 1978. The geomorphic role of snow. *Annals of the Association of American Geographers* 68: 414–425.

Thorn, C. E. 1979. Bedrock freeze–thaw weathering regime in an arctic environment, Colorado Front Range. *Earth Surface Processes* 4: 211–228.

Thorn, C. E. 1992. Periglacial geomorphology: What, where, when? In J. A. Dixon and A. D. Abrahams, eds., *Periglacial Geomorphology: The Binghamton Symposia in Geomorphology Internationals Series*, No. 22 (pp. 3–31). New York: John Wiley and Sons.

Thorn, C. E., Dixon, J. C., Darmody, R. G., and Allen, C. E. 2006. A 10-year record of the weathering rates of surficial pebbles in Karkevagge, Swedish Lapland. *Catena* 65: 272–278.

Thornbury, W. D. 1969. *Principles of Geomorphology*. New York: John Wiley and Sons.

Tricart, J. 1969. *Geomorphology of Cold Environments*. London: Macmillan.

Troll, C. 1958. Structure soils, solifluction, and frost climates of the Earth. *Transactions of H. E. Wright and Associates*. U.S. Army Snow, Ice and Permafrost Research, Established Transactions 43. Willamette, OR: Corps of Engineers.

Tucker, G. E., and Slingerland, R. 1996. Predicting sediment flux from fold and thrust belts. *Basin Research* 8: 329–349.

United Nations Environment Programme (UNEP). 2003. Managing fragile ecosystems: sustainable mountain

development, edited. <http://www.unep.org/Documents/
Documents.htm>.

Vitek, J. D., and Giardino, J. 1987. Rock glaciers: A review
of the knowledge base. In J. Giardino, J. Shroder Jr., and
J. Vitek, eds., *Rock Glaciers* (pp. 1–26). Boston: Allen and
Unwin.

von Blanckenburg, F. 2005. The control mechanisms of
erosion and weathering at basin scale from cosmogenic
nuclides in river sediment. *Earth and Planetary Science Letters*
3–4: 462–479.

Wahrhaftig, C., and Cox, A. 1959. Rock glaciers in the Alaska
Range. *Bulletin of the Geologic Society of America* 70: 383–436.

Walling, D. E., and Webb, B. W. 1986. Solutes in river systems.
In S. T. Trudgill, ed., *Solute Processes* (pp. 251–327).
New York: Wiley-Interscience.

Walsh, S. J., Butler, D. R., and Malanson, G. P. 1998. An
overview of scale, pattern, process relationships in
geomorphology: A remote sensing and GIS perspective.
Geomorphology 21: 183–205.

Warburton, J., and Caine, N. 1999. Sorted patterned ground in
the English Lake District. *Permafrost Periglacial Processes* 10:
193–197.

Washburn, A. L. 1956. Classification of patterned ground and
review of suggested origins. *Geological Society of America
Bulletin* 67: 823–866.

Washburn, A. L. 1980. *Geocryology: A Survey of Periglacial
Processes and Environments*. New York: John Wiley and Sons.

Waters, A. C. 1973. The Columbia River Gorge: Basalt
stratigraphy, ancient lava dams, and landslide dams.
In *Geologic Field Trips in Northern Oregon and Southern
Washington* (pp. 133–162). Portland: Oregon Oregon
Department of Geology and Mineral Industries Bulletin.

Watson, R. A., and Wright, H. E., Jr. 1969. The Saidmarreh
landslide, Iran. In S. A. Schumm and W. C. Bradley, eds.,
United States Contributions to Quaternary Research (pp. 115–140).
Special Paper 123. New York: Geological Society of America.

Whalley, W. B. 1974. Origin of rock glaciers. *Journal of
Glaciology* 13: 323–324.

Whalley, W. B., and Martin, H. E. 1992. Rock glaciers, II:
Models and mechanisms. *Progress in Physical Geography* 16:
127–186.

Whipple, K. X., Kirby, E., and Brocklehurst, S. H. 1999.
Geomorphic limits to climate-induced increases in
topographic relief. *Nature* 401: 39–43.

White, S. E. 1976. Rock glaciers and block fields, review and
new data. *Quaternary Research* 6: 77–97.

White, S. E. 1981. Alpine mass movement forms
(noncatastrophic): Classification, description, and
significance. *Arctic and Alpine Research* 13: 127–137.

White, S. E. 1987. Differential movement across transverse
ridges on Arapaho rock glacier, Colorado Front Range,
USA. In J. R. Giardino, F. Shroder, and J. D. Vitek, eds., *Rock
Glaciers* (pp. 145–150). London: Allen & Unwin.

Williams, M. W., Knauf, M., Cory, R., Caine, N., and Liu, F.
2007. Nitrate content and potential microbial signature of
rock glacier outflow, Colorado Front Range. *Earth Surface
Processes and Landforms* 7: 1032–1047.

Williams, P. J. 1957. Some investigations into solifluction
features in Norway. *Geographical Journal* 72: 42–58.

Williams, P. J., and Smith, M. W. 1989. *The Frozen Earth:
Fundamentals of Geocryology*. New York: Cambridge
University Press.

Williams, R. B., and Robinson, D. 1991. Frost weathering of
rocks in the presence of salt: A review. *Permafrost Periglacial
Processes* 2: 347–353.

Woodcock, A. H. 1974. Permafrost and climatology of a
Hawaii crater. *Arctic and Alpine Research* 6: 49–63.

Wu, Q. B., and Zhang, T. J. 2008. Recent permafrost warming
on the Qinghai–Tibetan plateau. *Journal of Geophysical
Research—Atmospheres* 113(D13108): 1–22.

CHAPTER SIX

Mountain Soils

LARRY W. PRICE and CAROL P. HARDEN

Soil is so common that most people assume they know what it is. The scientific definition of soil depends on who uses the term. To an engineer, soil is the unconsolidated material at the surface of the Earth, whereas to a biologist, soil is alive with living organisms that contribute to its physical and biological characteristics. For our purpose of examining the broad range of mountain soils on Earth, soil will be considered as the uppermost layer of the Earth's surface, in which organisms live, and which has physical, chemical, biological, and mineralogical properties that differ from those of the underlying parent material.

Soils form by the weathering and breakdown of rocks and other mineral materials, including volcanic ash and landslide deposits, in combination with the movement of water and the activity and decay of plants and animals. Soil consists of mineral particles, living and nonliving organic matter, the spaces (pores) between solid particles, and the gas and liquid in those spaces. Functionally, soil is the interface between the mineral earth, the atmosphere, water, and living things. It may look like "dirt," but it supports life on our planet and controls the storage and runoff of freshwater. A typical soil consists of distinct layers called *soil horizons,* which, taken together, form the *soil profile.* These layers are distinguished by differences in their color, texture, structure, organic matter, clay content, and pH (acidity). Horizons form slowly over time, as water percolating through the soil dissolves available chemicals and carries them downward. This process, called *leaching,* leads to the removal of water-soluble elements in the upper layers of soil and the enrichment of lower horizons where the leached elements are redeposited. Tiny solid particles may also be moved downward by percolating water until redeposition takes place. Thus, the upper layers continually lose material to the lower layers.

The uppermost soil layers ("A" horizons) contain the most organic material and are darkest in color, while the middle layers ("B" horizons) are primarily composed of mineral fragments. The lowest part of the soil profile ("C" horizon) consists of partially weathered parent material. Well-defined horizons are the product of soil development under long-persisting, undisturbed conditions; hence they are uncommon in mountain soils. At high altitude, cold conditions restrict the pace of biological activity and the movement and geochemical activity of water, so soils form slowly. Mountain soils are characteristically shallow, rocky, acidic, infertile, and immature. The diversity of mountain landscapes has the effect of creating a discontinuous and heterogeneous patchwork of microenvironments characterized by continually changing and contrasting site conditions. It is within this framework that mountain soils develop and, accordingly, reflect the nature of their origin.

Soil-forming Factors

The primary factors responsible for different kinds of soils are climate, biological factors, topography, parent material, and time (Jenny 1941). The same factors affect soil development in mountain as in nonmountain environments, but the intensity and relative importance of these factors differ between mountains and plains, and among mountain locations. No one factor controls soil formation; rather, all are interrelated.

Climate

The primary climatic factors affecting soil development are temperature, precipitation, and wind. Climate has significance far beyond itself, since it controls the distribution of vegetation, which, in turn, influences the kind of soil that develops in any given area. Temperature is an important control of the rates and types of weathering that cause rock to break down. Physical weathering, considered dominant in cold climates, results in a coarse soil texture, but there is increasing evidence for more chemical weathering in mountains than was formerly thought (Thorn et al. 2001, 2011; Hall et al. 2002). One study of basalt weathering in Hawaii (Brady et al. 1999) determined that weathering rates were proportional to rainfall and less strongly affected by differences in temperature.

The periodicity of temperature is one factor involved in weathering and soil formation; others include rock type, presence of organic acids, and moisture availability (Brady and Weil 2008). In tropical high mountains, the soil surface may freeze every night, but warmth during the day can provide ample heat for chemical weathering to produce a moderate amount of clay. By contrast, in midlatitude and polar mountains, held in the grasp of winter for much of the year, chemical weathering is more restricted. As discussed in Chapter 5, low temperatures and frost action are responsible for the formation of various types of patterned ground. The daily frost of high tropical mountains penetrates only a few centimeters below the surface, whereas the seasonal frost in middle and high latitudes penetrates deeper. Consequently, rocks in tropical mountains are more readily reduced to small aggregates, and the resulting patterns are of small dimension; patterned ground in mid- and high latitudes occurs on a broader scale and may contain larger rocks. Related slope processes operating in low-temperature mountain climatic regimes include frost creep, *solifluction,* and other types of mass wasting, also discussed in Chapter 5. These cause the movement and intermixing of surface layers, which disrupt the soil profile and the processes of soil formation.

Low temperatures limit biological activity. At low temperatures, less organic material is added to the soil, soil organic matter decomposes very slowly, and soil fauna are rare (Townsend et al. 1995; King et al. 2010). A soil temperature of 5°C (43°F) has been proposed as the temperature below which biological activity in the soil becomes very slow (Retzer 1974). At an alpine tundra research station at 3,750 m on Niwot Ridge in the Colorado Rockies, soil temperatures remained above 5°C for the numbers of days shown in Table 6.1. Site-specific soil temperatures vary with exposure, snow cover, and other microsite conditions, but the Niwot Ridge data demonstrate the limited time available for biological activity and the decrease of this activity period with soil depth.

Many factors control the distribution of moisture in mountains. The trend of precipitation increasing with altitude is well known, although, above a certain maximum altitude, precipitation may again decrease. This is especially evident in the tropics. While the intermediate slopes of Kilimanjaro and Mount Kenya contain luxuriant vegetation, the summit areas are desert-like (Hedberg 1996).

TABLE 6.1
Numbers of Days with Soil Temperatures above 5°C on Niwot Ridge, Colorado

Soil Depth	Days with Temperature >5°C (>41°F)
5 cm	110 days
15 cm	93 days
30 cm	45 days
60 cm	21 days

SOURCE: Marr et al. 1968.

Strong contrasts in precipitation between the leeward and windward sides of mountains and differences in the effects of solar intensity and the distribution of sunlight on slopes produce different rates of heating and evaporation. Snow remains in place longer than rain, but is highly susceptible to further transport by wind. Exposed ridges and slopes are often snow-free and dry, while lee slopes become loaded with snow and retain snow patches that provide meltwater well into the summer. The distribution and redistribution of snow affect soil chemistry (Liptzin and Seastedt 2009) and also influence the location of permafrost patches in midlatitude mountains (Ives 1973; Ives and Barry 1994).

The lack of moisture in arid landscapes reduces biological activity, organic material, and decomposition rates. Where water is not available to promote chemical weathering, physical weathering processes become relatively more important. Excess moisture, on the other hand, results in waterlogging, poor aeration, and increased soil acidity. The most productive soils develop in intermediate, well-watered, but adequately drained sites; but mountain sites, such as exposed slopes and ridges or poorly drained meadows and bogs, often have a scarcity or an excess of moisture.

Wind achieves major importance as a climatic factor by causing evaporative stress on vegetated and bare surfaces, either directly or indirectly through the redistribution of snow. Wind also erodes and removes fine material from exposed surfaces, especially where frost action has left the soil susceptible to transport. Wind erosion is frequently observed at unvegetated sites around glacial and stream deposits, where ample fine material is available.

The corollary to erosion is deposition. Although some fine particles are blown away from the mountains, some are deposited locally and contribute to soil development. Volcanic ash is an important wind-borne source of soil material. In addition, very fine particles from nearby lowlands may be transported by wind into the mountains. The local-scale distribution of wind-deposited material is similar to that of snow, and the combined presence of fine material and moisture is favorable to the establishment of plants. Wind-blown deposits are often alkaline and thus help counteract the natural acidity of alpine soils (Delmas et al. 1996). Many mountain soils develop from wind-borne

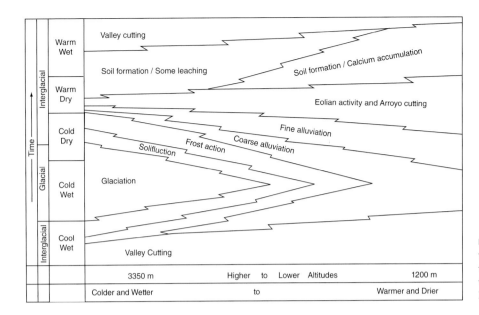

		Warm Wet	Valley cutting

FIGURE 6.1 Graphical portrayal of the combined effects of altitude, temperature, moisture, and time in the La Sal Mountains, Utah. (From Richmond 1962.)

deposits rather than through the breakdown of bedrock (Retzer 1965; Muhs and Benedict 2006), especially where deposits of volcanic ash produce deep soils, as in Iceland, Japan, and the northern Andes (Buytaert et al. 2005).

Climates change. During the past 900,000 years, glaciers in Europe and North America have advanced and retreated nearly 20 times. Midlatitude and polar mountains experienced the effects of these major climatic changes more acutely than lowlands, being glacier-covered even when the surrounding lowlands were not. Consequently, soil surfaces in mountains such as the Alps or Rockies only began to form after the ice melted, and thus are young (10,000–15,000 years). The climate has warmed since then, though marked fluctuations have occurred. Surfaces that escaped direct glaciation provide evidence of the soil-forming and geomorphic effects of these climatic events (Fig. 6.1). In some cases, older soils persist and retain the characteristics that evolved under very different climatic regimes. The Coast Range of northern Oregon contains ancient reddish lateritic soils (Oxisols, highly weathered lowland tropical soils) formed during the Miocene (20 million years ago). As the climate changed, its effect on soil development also changed, so that older soils served as parent materials for modern soils.

A soil can develop under one climatic regime in the lowlands and then be uplifted to higher elevations during mountain building. This probably happened in the Oregon Coast Range, where the major uplift occurred during the Pliocene (10 million years ago). Similarly, in the limestone Alps of northern Austria, tropical soils formed during the late Tertiary (20–30 million years ago) and were uplifted later. They now exist as soil relicts at an altitude of about 2,100 m (Kubiena 1970). In such cases, the complexity of environmental effects can be very great because the soils are exposed to a variety of environmental systems during uplift. Furthermore, uplift causes temperature and

moisture regimes to change. Uplift of the Peruvian Andes caused landscapes to pass from arid conditions in the lowlands through humid conditions at intermediate altitudes, and eventually into a cold, arid environment at the highest levels (Garner 1959). Such variable conditions have the potential to create polygenetic (multiple causes) soils and landscapes.

Biological Factors

The plants and animals affecting soil range from microorganisms to macrofauna, herbaceous plants, and forests. A handful of fertile soil is filled with living things, predominantly fungi and bacteria. Organic matter, including living organisms (biomass) and well-decomposed remains of living things (humus), usually makes up less than 10 percent of the volume of soil, but is very important in the physical and chemical processes of soil and provides nutrients for soil fauna. As they feed and respire, soil microbes release compounds, including carbon, stored within the soil organic matter. In this way, these unseen microorganisms help regulate the planet's geochemical cycles.

More than any other factor, vegetation gives the soil its distinctive character. In particular, vegetation controls the amount and kind of organic material added to soil. In grasslands, the aerial parts of plants die each year, adding organic matter from the top, while roots add organic matter below the surface. In forests, trees, even evergreens, lose their leaves. Certain evergreen leaves are tough, leathery, and inherently slow to decay, especially under the cool temperatures of high altitudes (Aerts 1995). Where people have replaced forest vegetation with pasture grass, the soil adjusts to differences in organic matter and in physical and chemical inputs. Deforestation followed by pasture establishment in the Alay Range in Kyrgystan caused soil phosphorus concentrations to increase (Turrion et al. 2000).

In a comparison of soil conditions under pasture and lower montane forest in northwestern Ecuador, Rhoades and Coleman (1999) found pasture soils to be wetter and denser than those under nearby mature or second-growth forest. In the Scottish uplands, afforestation with pines decreased soil pH (increased acidity) and reduced the turnover time of organic matter (Grieve 2001).

One of the sharpest soil boundaries is that between forest and grassland. In many areas, timberline has advanced or retreated, indicating either climatic change or some other modification, such as by fire, disease, or human intervention. Evidence in the fabric of the soils may allow researchers to reconstruct the former timberline (Molloy 1964; Reider et al. 1988; Carnelli et al. 2004). Soils above treeline may preserve profile characteristics (Earl-Goulet et al. 1998), stable carbon isotope ratios (Ambrose and Sikes 1991; Byers 2005), or wood charcoal (Di Pasquale et al. 2008; Talon 2010) that indicate the earlier presence of forest.

Much remains to be learned about the effects of animals and other organisms on mountain soils. Microfauna and microorganisms are less abundant at higher altitudes (Tolbert et al. 1977; Margesin et al. 2009), although some, including protozoa and nematodes, can thrive in moist, cool conditions. Certain groups of bacteria and fungi have adapted to extremely cold and water-scarce conditions, even on Mount Everest, through mechanisms that include the production of antioxidant enzymes and antifreeze proteins (Margesin 2012). They play an important role in decomposing organic matter and recycling nutrients. On the upper slopes of Mount Kinabalu, Borneo, and Mount Kosciusko, Australia, large earthworms and microorganisms incorporate surface litter into the soil (Costin et al. 1952). In the Luquillo Mountains of eastern Puerto Rico, the faunal species with the greatest biomass is thought to be earthworms. Termites transfer one ton of soil per hectare per year in the tropics, while ants are effective soil movers in tropical and temperate-zone soils (Bridges 1997).

Small burrowing animals are important to the alpine tundra. The pocket gopher (*Thomomys* spp.) is a common occupant of North American mountain meadows. Searching for plant roots, these little creatures create complex networks of shallow tunnels under the soil. The excavated material is taken above the surface and stuffed into similar tunnels in the overlying snow. The following spring, soil in these snow tunnels appears on the ground as interwoven soil casings. On Niwot Ridge, Colorado, researchers found pocket gopher mounds to contain less soil organic matter and significantly different amounts of total carbon, total nitrogen, and total phosphorus than their surroundings (Sherrod and Seastedt 2001). They concluded that erosion from pocket gopher mounds increased the redistribution of soil mass and soil nutrients.

Mixing and overturning of soil by burrowing animals increase its vulnerability to erosion (Butler 1995). This is particularly evident on slopes (Imeson 1976; Yoo et al. 2005). Price (1971) found material moved downslope by

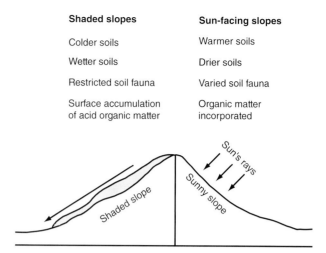

FIGURE 6.2 Differences between slopes of opposite aspects. In the northern hemisphere, south-facing slopes are sunnier. (From Bridges 1997: 26.)

the arctic ground squirrel (*Citellus undulatus*) in the Ruby Mountains in Canada's Yukon Territory to be around 0.36 tons per hectare per year. Years earlier, Grinnell (1923) had estimated that pocket gophers in Yosemite National Park, California, excavated 0.04 tons of soil per hectare per year. In comparison, earthworms have been estimated to move 20–300 tons per hectare per year (Goudie 1988).

Topography

Topography, particularly slope gradient, length, and aspect, affects soil formation through exposure to sun and wind, slope stability, and drainage. Exposure to the sun controls the energy available for biological activity and greatly affects physical and chemical processes (Fig. 6.2). For example, in the Springerville volcanic field, Arizona, soils on south-facing slopes receive more solar energy and have accelerated rates of weathering and soil development (Rech et al. 2001). Exposure to wind is also important: Soils formed on windy ridges differ from those in sheltered valleys. Wind enhances evaporation, redistributes snow and soil materials, and physically limits the growth of plants, especially when it carries abrasive mineral or ice fragments. Slope gradient and slope length strongly affect soil development. More soils on mountain slopes develop from transported materials than directly from the underlying bedrock (Parsons 1978). As slope gradient increases, soil depth decreases (Fig. 6.3; Ruhe and Walker 1968; Moore et al. 1993), primarily because gravity transfers soil downslope. Steeper upper slopes are more susceptible to erosion by water, while gentler slopes are more often areas of deposition. Moisture conditions also differ between upper and lower slopes. Soils on steeper slopes are typically well drained, whereas those on lower slopes and in valleys contain more moisture due to larger contributing areas, shallower water tables, and

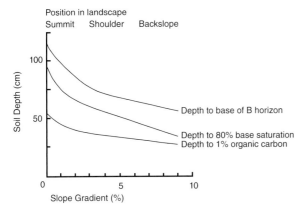

FIGURE 6.3 Relationships between slope gradient and several soil properties on slopes in Iowa provide a basis for understanding similar relationships on mountain slopes. (Data from Ruhe and Walker 1968; graph from Birkeland 1984, by permission of Oxford University Press.)

the greater likelihood of being shaded by surrounding slopes. These upslope–downslope relationships are used in wetness indices to predict soil moisture of different portions of slopes on digital elevation models (DEMs) (Woods et al. 1997).

Researchers have long recognized relationships between soil properties and topographic position, especially when other factors, like parent material and time, remain the same (Bushnell 1942). The term *catena* (chain) refers to the predictable patterns of soil types along a slope (*toposequence* is also used). Topographic differences also produce differences in other soil-forming factors. Mid-slope soils, for example, are less leached than soils on summits (Birkeland 1984). Soil properties change with elevation and exposure, but is this because of topography or changes in climate and vegetation? Topography is generally viewed as a relatively passive soil-forming factor, while climate and vegetation play active roles in soil development. In any event, topography is critical because gradient and drainage directly affect soil development.

Parent Material

The inorganic stuff from which soils are made affects soil characteristics. It may be solid bedrock or unconsolidated materials such as glacial deposits, talus, blockfields, mudflows, alluvial fans, windblown deposits, or even materials transported by people. Soils develop more rapidly on unconsolidated material than on solid bedrock, but much depends on the nature of the rock and the site conditions. Volcanic ash weathers very rapidly; its silicate minerals readily break down into clay. For this reason, volcanic soils tend to be highly productive. Within bedrock, soft and weak rocks (sedimentary) usually weather more rapidly than hard and resistant rocks (igneous or metamorphic). Fine-grained rocks, such as shale or sandstone, typically break into fine particles; coarse-grained rocks, such as granite, break into

coarse particles. Some rocks weather relatively rapidly into clays; others do not. Thus, the type of parent material affects soil texture, soil structure, moisture-holding ability, and base-exchange capacity (a measure of fertility).

Parent material also affects soil chemistry. Many igneous and metamorphic rocks are acidic, whereas sedimentary rocks tend to be basic. Some rocks undergo such strong reactions that they form distinctive soils in almost all climates. Limestone, for example, is famous for forming the *rendzina* (included in FAO leptosol group), a humus-rich carbonate soil that occurs in several mountain regions (Kubiena 1970). Other rock types with unusual chemical properties—including serpentine, gypsum, and calamite (an ore of zinc)—limit soil development by inhibiting plant growth and metabolism (Billings 1950; Brady et al. 2005). The effects of these rock types may be so strong that their distribution can be plotted by mapping the vegetation.

Parent material affects soil properties primarily in the early stages of soil development. Mountains are relatively young geologically; glacial advances and other erosional processes have exposed and created extensive new surfaces. For this reason, parent material plays a more important role in the distribution of soil types in mountains than in the surrounding lowlands.

Time

Time, like topography, is often considered a passive factor in soil formation because it makes no direct contribution to soil characteristics; nevertheless, soil profile development requires time. The older a soil, the more advanced soil formation will be. If a lava flow buries a mountain slope, soil formation begins anew and the former soil (*paleosol*) becomes part of the geologic record. If a mountain has been severely eroded by glaciers, its surfaces may have been scraped to bare rock. Where soil has been removed or buried, time starts at zero.

A number of studies have documented the relative rate of soil development on younger features such as glacial moraines and mudflows (Mahaney 1990; Morisada and Ohsumi 1993; Douglass and Bockheim 2006; Haugland and Burns 2006). Although the rate of soil formation is variable, evidence of the early stages of soil development usually appears within a century or so. In the moist tropics, the rate is greatly telescoped, particularly on volcanic parent material, so that a soil may begin to develop within decades; in mid-latitude and polar mountains, it may take thousands of years. The eruption of the Indonesian island volcano Krakatoa in 1883 deposited a new land surface on the remnant island, Rakata. Today, Rakata is covered with lush tropical vegetation and relatively well-developed soil. By contrast, in many areas in mid- and high-latitude mountains, several millennia of exposure to the elements have failed to result in a well-developed soil. One investigation of glacial deposits in the subalpine zone of the Indian Peaks region, Colorado Rockies, estimated that over 2,000 years were required for the soil to

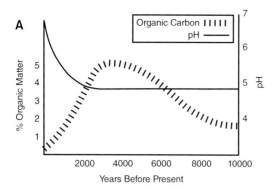

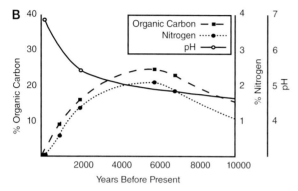

FIGURE 6.4 Generalized pattern of changes in pH (solid line) in soil over time. (A) Soil developed through time in the Indian Peaks regions, Colorado Rockies, showing percentage of organic matter as dashed line (Mahaney 1970). (B) Soil development on Mount Kenya, Kenya (Mahaney 1990), showing organic carbon as dashed line and nitrogen as dotted line.

develop distinctive horizons and achieve a steady pH (Retzer 1974). Soil development is limited by low moisture as well as by low temperatures: In maritime mountains, where moisture is plentiful, soil would develop faster than in arid mountains. On Mount Kenya, soil depth, degree of horizon development, surface color, and grade of soil structure are closely associated with the age of the Pleistocene and Holocene parent material deposits (Mahaney 1990). Soil horizons in the Afroalpine developed slowly: A "B" horizon formed within about 2,000 years (Fig. 6.4). Deposits near active glaciers or volcanoes provide excellent sites for studying the early stages of soil development (Douglass and Bockheim 2006). In Switzerland, Egli et al. (2001) used radiocarbon dating to determine the ages of six soils, ranging from 150 years to 10,000 years, and then studied chemical changes in the soil profiles over time. A useful aspect of the relationship of soil development to time is that, once rates of weathering and soil development have been determined in an area, the degree of soil development can be used to estimate the age of nearby surfaces.

Major Kinds of Mountain Soils

Mountain soils exhibit considerable local-scale diversity, and mountains occupy diverse positions with respect to Earth's environments. What is true of a moist, protected

meadow does not hold for an exposed ridge. Soil development slows as altitude increases in most humid midlatitude areas, but rates of soil development may increase with altitude in arid and tropical mountains. In midlatitude humid mountains, temperature is the primary limiting factor; in desert mountains, the increase in precipitation with altitude may more than compensate for decreasing temperature, and conditions for soil development may become more favorable than at lower altitudes (Whittaker et al. 1968; Messerli 1973). In the humid tropics, decreasing temperatures with altitude slow rates of decomposition and leaching, so more organic material accumulates at the surface and more nutrients are retained in higher-elevation soils (Coûteaux et al. 2002).

Adding to the problem of great diversity of mountain soils is the fragmented nature of our knowledge about them. Much of what we know comes from isolated research carried out in widely separated areas. Generally, less is known about desert and tropical mountain soils than about midlatitude humid mountain soils; and detailed soil maps are rare, even for midlatitude mountains. On most world soil maps, mountain areas are provisionally labeled "undifferentiated." But despite the heterogeneity of mountain soils and our lack of detailed information about them, their general nature under similar environmental systems can be predicted with a fair degree of assurance. National and international efforts to develop soil classification systems demonstrate common soil characteristics and relationships across the globe.

Soil Classification

A number of systems for identifying, classifying, and mapping soils have been implemented over the years, but none has achieved global acceptance. Many of the ideas about soil development originated in Russia. The Russian soil classification system was based on the fundamental assumption that soil development accompanies the development of vegetation as controlled by climate. Therefore, given enough time, regions with similar climates and vegetation should support similar soils, regardless of differences in parent material or topography. Vegetation and soil were thought to develop contemporaneously, each reflecting greater complexity through time until eventually achieving a dynamic equilibrium with climate. This stage of development was considered "climax" vegetation with "mature" soil profiles. No major changes were thought to take place beyond this point, and those that did occur were considered to be fluctuations within the context of a steady state, rather than directional. The similarity of soils under northern boreal forests across vast expanses of terrain supports the assumptions of the Russian system.

Today, we recognize that static systems rarely exist in nature because wind, human activity, fire, climate fluctuations, and other processes change environmental conditions. Nevertheless, we recognize that some soils represent

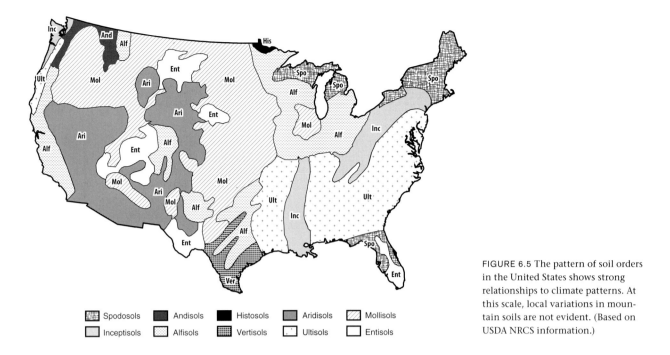

FIGURE 6.5 The pattern of soil orders in the United States shows strong relationships to climate patterns. At this scale, local variations in mountain soils are not evident. (Based on USDA NRCS information.)

Spodosols
Inceptisols
Andisols
Alfisols
Histosols
Vertisols
Aridisols
Ultisols
Mollisols
Entisols

long-term climatic and vegetative conditions. For example, "prairie soils" (Mollisols in the U.S. taxonomy, chernozems in Russia) exist in areas supporting prairie grasses, Alfisols and Ultisols in areas supporting broadleaf deciduous trees, and Spodosols (podzols in Russia) in the boreal forest. In mountain regions, soil mapping usually shows zones that correspond to the vegetation as it changes with increasing altitude. These are considered "mature" soil types created by the long-term presence of similar vegetation in the absence of climate change. Such *zonal* soils theoretically occupy typical sites with good drainage on gentle slopes. Soils at many locations within these broad regions do not match the zonal soil types, however, because of site-specific conditions. Soils that differ because of local factors such as drainage, rock type, and topography are called *intrazonal;* those on surfaces too young for full soil development are called *azonal* (Jenny 1941). In many mountainous landscapes, intrazonal and azonal soils occupy more territory than zonal soils.

Although still widely used, soil classification by vegetation cover has been criticized in recent years because the terminology is imprecise, the soil characteristics are difficult to quantify, and the system has a strong genetic bias that many specialists find limiting. As a result, a new system of soil classification, *Soil Taxonomy,* was developed in the United States and also used by scientists in 45 other countries (Brady and Weil 2008; Soil Survey Staff 2010). It is based entirely on soil properties that can be determined by human senses or by instrumentation.

The U.S. Soil Taxonomy classification has 12 soil orders, subdivided into suborders, great groups, subgroups, families, and series (Fig. 6.5, Table 6.2). Subdivisions are based on analyses of chemical and physical characteristics from the entire soil profile, rather than from surface samples. No soil order is specific to mountains; all can occur in mountain regions. The initial soil taxonomy contained ten orders; then *Andisols* (volcanic soils) were added in 1989 and *Gelisols* (soils with permafrost within 2 m of the surface) in 1996.

A major effort to standardize soil mapping across country boundaries was completed by the Food and Agriculture Organization (FAO) of the United Nations and revised in 1988 (FAO 2003). The FAO soil classification has 28 major soil groupings, based largely on environment. Because mountain environments consist of many microenvironments (e.g., well-drained slopes, poorly drained depressions, alluvial deposits), most FAO soil groups are present in mountain environments. Some soils, such as andosols (*Andisols* in the U.S. taxonomy) and podzols (*Spodosols*) are essentially the same in both the U.S. and FAO classification systems. Others do not have direct equivalents between the two systems because the structures of the classification systems are not the same (Table 6.3). Many excellent Internet resources are available on the broader topic of soil classification, including those of FAO and the U.S. Soil Taxonomy.

Humid Midlatitude Mountain Soils

In the humid mi-latitude mountains of North America, five soil orders predominate:

Entisols

Inceptisols

Mollisols

Spodosols

Histosols

Each of these is described briefly below.

TABLE 6.2
Soil Orders of the U.S. Soil Taxonomy

Name	Derivation	Characteristics
Alfisols	(invented syllable)	Typically develop under deciduous forest
Andisols	modified from *ando*	Developed on volcanic ejecta
Aridisols	*aridies* (L.) (dry)	Developed in arid and semiarid environments
Entisols	(invented syllable)	Little or no evidence of profile development
Gelisols	*gelare* (L.) (to freeze)	Evidence of frost churning; permafrost within 2 m of surface
Histosols	*histos* (Gk.) (tissue)	Peaty; more than 30% organic matter
Inceptisols	*inceptum* (L.) (beginning)	Early stages of profile development
Mollisols	*mollis* (L.) (soft)	Thick, dark, high base saturation
Oxisols	*oxide* (Fr.) (oxide)	Highly weathered; no clay horizon
Spodosols	*spodos* (Gk.) (wood ash)	Ashy horizon with Fe and Al oxides; acidic
Ultisols	*ultimus* (L.) (last)	Clay horizon; low base saturation
Vertisols	*verto* (L.) (turn)	Clays swell when wet; deep cracks when dry.

SOURCE: Soil Survey Staff 2010.

ENTISOLS

Entisols are thin, poorly developed soils found on surfaces such as steep slopes, exposed ridges, and fresh deposits. Owing to their shallow depth and exposure, they are usually dry and support a scanty cover of lichens, mosses, and cushion plants. Organic material may darken the surface, but there is little other evidence of horizons. Some Entisols may be simply accumulations of fine or rocky material rather than true soils. The ending "-ent" denotes an entisol. A "cryorth*ent*," for example, is an Entisol developed where the mean annual temperature at 50 cm depth is 0–8°C.

INCEPTISOLS

Inceptisols are soils in early stages of profile development or soils with diagnostic horizons that fail to meet the criteria for other orders (Soil Survey Staff 2010). They occur from the tropics to tundra regions, but do not include soils of arid environments. Inceptisols comprise most of the soils found above timberline in the Pyrenees, Caucasus, Rockies, and Alps (Retzer 1974). The ending "-ept" denotes an Inceptisol. The alpine-turf subgroup (cryumbrept) is well-drained and occurs on upper slopes and exposed areas. Turf stabilizes these surfaces. Once the turf is destroyed, however, the soil becomes highly susceptible to erosion (Bouma 1974; Byers 2005). Consequently, protection of turf is essential for the preservation of alpine landscapes.

Cryumbrepts may be relatively deep, extending to 30–80 cm. They show distinct horizon development and are weakly to strongly acidic, characteristics that represent some downward movement of water through the profile (Bäumler and Zech 1994). The finest particles occur at the surface, reflecting the greater weathering and biological activity near the soil/atmosphere interface (Fig. 6.6). *Eolian* (wind) deposition may also contribute to the finer surface texture. The "A" horizon is dark brown to black, with a high organic content, but nutrient status is fairly low.

MOLLISOLS

Mollisols are relatively deep subalpine and alpine grassland soils with distinct horizons and neutral to slightly alkaline pH. Compared to Inceptisols, Mollisols contain more exchangeable bases and, therefore, more plant nutrients. Mollisols characteristically have an organic-rich "A" horizon at least 25 cm thick. Mollisols develop under grass vegetation in environments that are neither humid nor dry. They occur in midlatitude prairie regions, and at high and low latitudes and high altitudes.

SPODOSOLS

Spodosols, or subalpine-forest soils, are moderately deep, well-drained, acidic soils with well-developed and distinct horizons. They form under a complete cover of coniferous forest on many types of parent material, predominantly on crystalline rocks. The primary soil-forming process involves leaching of the "A" horizon and translocation of bases and clays to the "B" horizon (Lundström et al. 2000). Rates of decomposition and incorporation of organic matter into the soil are slow, owing to the low temperatures and heavy snowpack of many high mountain forests. Under these conditions, a layer of litter

TABLE 6.3
Soil Classification Worldwide[a]

U.S. Soil Order	Canadian	FAO (Food and Agriculture Organization of the United Nations)[b]
Alfisols	Luvisolic, Solonetzic	Luvisols, Planosols
Andisols	—	Andosols
—	—	Anthrosols
Aridisols	—	Xerosols
Entisols	Regosolic, Brunisolic	Regosols, Arenosols, Fluvisols, Anthrosols
Gelisols	Cryosolic	—
Histosols	Organic	Histosols
Inceptisols	Brunisolic	Cambisols, Leptosols
Mollisols	Chernozemic	Chernozems, Kastanozems, Phaeozems
Oxisols	—	Ferralsols
Spodosols	Podzolic	Podzols
Ultisols	—	Acrisols, Alisols
Vertisols	Vertisolic	Vertisols
—	Gleyosolic	Gleysols

SOURCES: Soil Survey Staff 2010; Agriculture and Agri-Food Canada 2003; FAO 2003.

[a] Correspondence between classification systems is necessarily approximate because the systems differ.

[b] Some FAO soils without U.S. or Canadian equivalents (e.g., Calcisols) are not shown here.

and partially decomposed organic material accumulates above the mineral soil. Water moving down through the profile carries humic acid from the decomposing humus, creating an acidified profile in which oxides of iron and aluminum can be mobilized, dissolved, and transported downward. This set of processes, called *podzolization,* leaches the lower part of the "A" horizon, leaving a grayish, bleached, siliceous (silica-rich) zone (spodic horizon), the trademark of the true Spodosol. Oxides of iron and aluminum accumulate in the "B" horizon, giving it a reddish-yellow color. The translocation of clays and bases to the "B" horizon often results in a blocky or prismatic soil structure that may inhibit drainage. Spodosols are not very fertile.

At the upper forest limit, in the krummholz and tundra zone, Spodosols grade into alpine turf and/or alpine-meadow soils. Low temperatures at these altitudes inhibit rates of organic-matter breakdown and leaching. Below the subalpine forests and the zone of true Spodosols (podzols), in the direction of increasing temperatures and decreasing snowfall, better incorporation of organic material into the mineral soil produces a black or dark brown "A" horizon that grades into a lighter brown "B" horizon. Depending on the climate, forests may extend into the lowlands or give way to grasses and desert shrubs. As the vegetation changes, the pH and the type of organic matter and biotic processes change; thus, soil development also changes.

HISTOSOLS

Histosols, bog or peat soils, form where drainage is poor, typically in depressions and areas where water accumulates from seeps or springs. The predominant characteristic of Histosols is an abundance of mosses and partially decomposed plant remains at the surface; these vary in thickness from a few centimeters to over a meter. The mineral matter of the soil is generally derived from the erosion of upland slopes, and the transition from the mineral to the organic material is abrupt. The mineral soil is frequently mottled with orange and bluish gray, due to the lack of soil aeration. Bog soils are strongly acidic, with a pH of 4.0 to 5.0. By definition (Soil Survey Staff 2010), Histosols do not contain permafrost within a meter of the surface. If permafrost is present, it contributes to poor drainage and the accumulation of sedges and mosses. Bog soils are intrazonal because they develop under locally restrictive conditions. They are found in all mountain areas except deserts. Bog and peaty soils are particularly abundant in the oceanic mountains of Europe (Montanarella et al. 2006). With their high organic matter content, Histosols are important for carbon storage in the world's mountains.

Arid Mountain Soils

In arid regions, increasing altitude frequently presents the apparent paradox of deeper soils, more plant and animal species, greater biomass, and higher productivity than in

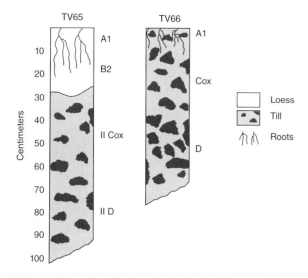

FIGURE 6.6 Two soil profiles from the foreland of the Lewis Glacier on Mount Kenya. The deeper profile on the left shows an Inceptisol, developed on late glacial deposits, while the profile on the right, with less vertical development, is an Entisol developed on younger neoglacial deposits. (From Mahaney 1990.).

the lowlands. As discussed in Chapter 3, this is because moisture increases with altitude. Forests often occupy intermediate elevations in semiarid mountains and display both upper and lower timberlines, the lower timberline due to lack of moisture. Many mountains in dry areas provide "islands" of greater moisture. Tropical examples include Kilimanjaro and Mount Kenya in East Africa and the west slope of the Andes in Peru, Bolivia, and northern Chile. Midlatitude examples include the Australian Alps, the Caucasus, various ranges of the Trans-Himalayas, the Rockies, Great Basin ranges, and the east slopes of the Cascades and Sierra Nevada. Such intermediate-altitude forests do not occur in the drier subpolar regions because temperatures are too low for forest growth.

In arid and semiarid regions, the increase of moisture with increasing altitude promotes soil development by supporting more complete plant cover and greater production of organic material and by intensifying the chemical and physical weathering processes. *Aridisols* are characteristically light in color and low in organic matter. Evaporation of soil moisture can deposit soluble salts in the upper horizons or at the surface. Arid soils are generally alkaline (Fig. 6.7), but higher-elevation soils, with more moisture, more organic matter, and more microorganisms, are darker in color and less alkaline. The nitrogen content and base-exchange capacity of soils also increase with increasing elevation in arid mountains (Hanawalt and Whittaker 1977).

Marked differences in exposure to the sun may occur and affect soil and vegetation on north- and south-facing slopes (Whittaker et al. 1968; Istanbulluoglu et al. 2008). The elevation of a soil zone may differ by tens of meters between slopes of different *aspects* (surfaces facing different directions) (Fig. 6.8). In hot, dry regions, the more favorable sites

for plants usually extend to higher elevations on poleward-facing slopes and shaded areas. Conditions for plant and soil development in arid mountains are optimal only in a narrow vertical zone between the low-temperature high altitudes and the arid low altitudes. If the topography within this optimal zone includes steep slopes and narrow ridges, the area favorable to soil development will be relatively small; if it consists of broad and gentle upland slopes, the area favorable to soil development will be considerably larger. Examples of broad, gently sloped areas situated within favorable climatic zones at higher altitudes include the Tibetan Plateau and the Altiplano of Bolivia and Peru. Broad uplands also occur on the gentle dip slopes of fault blocks in the Basin and Range mountains of the western United States.

The optimal zone for plant growth may occasionally be found in the subalpine zone, as in the Australian Alps. The maximum elevation of 2,220 m on Mount Kosciusko coincides with broad, gentle upland surfaces. The timberline is located at about 1,800 m. Lack of extensive glaciation, along with adequate moisture in the subalpine zone, has favored soil development to an unusual degree. In fact, Costin (1955) called the Australian Alps "soil mountains" to distinguish them from others, such as the European Alps, where upland surfaces are predominantly peat or rock.

Andisols

Andisols (andosols in the FAO system) develop in volcanic ash or other volcanoclastic parent material. They occur throughout the world on volcanoes and are especially prominent in the Pacific Rim Ring of Fire, including the Andes, the Cascade Range, southern Alaska, the Kamchatka Peninsula, Japan, and Hawai'i. They also occur in Iceland, along the African rift zone, and in other volcanic regions. Clay minerals associated with Andisols develop from microbial action and the alteration of minerals in place (Shoji et al. 1993). These soils are typically porous, due to the presence of noncrystalline clays and high concentrations of organic material (Chen et al. 1999). The clays bind readily to organic matter and to phosphorus, making it unavailable to plants. Agriculture on Andisols can be productive if phosphorus is added (Shoji et al. 1993). High water retention capacities make Andisols an important component of the water resource for people living in high inter-Andean valleys (Podwojewski et al. 2002). Hofstede (1995) found soil in an undisturbed grassland soil in the Colombian Andes to contain up to 2.5 grams of water per gram of dry soil.

Humid Tropical Mountain Soils

Mountains of the humid tropics characteristically rise from lowlands supporting dense tropical forests. As elevation increases, lowland forests become montane forests, which become shorter and less dense until trees give way to

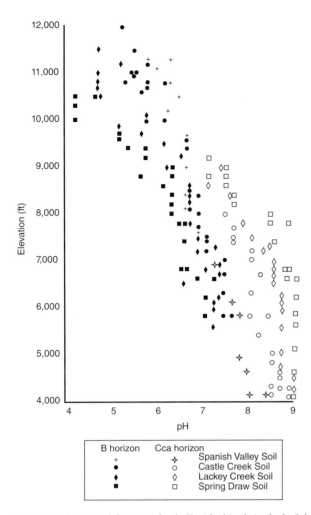

FIGURE 6.7 The general decrease of soil pH with altitude in the La Sal Mountains, Utah, a semiarid region. (From Richmond 1962: 89.)

shrubs, grasses, and herbaceous plants at the highest levels. Old soils of most tropical lowlands are notoriously nutrient poor; they exist in a hot and humid environment with excessive rainfall, rapid decomposition of organic material, and intensive leaching of nutrients. Whereas most soils serve as nutrient sources for plants, *Oxisols* (highly weathered lowland tropical soils) are so leached that the living biomass of the forest, rather than the soil, is the major nutrient reservoir. As long as the forest remains, so do the nutrient reserves, which are recycled through leaf fall, decomposition, leaching, and uptake by tree roots. If the forest is removed, then the source of nutrients is removed. For this reason, most attempts to grow annual crops on a large scale in the tropics have failed. The primary soil-forming process on old surfaces in the lowland humid tropics, called *laterization,* is the selective leaching of silica from the upper layers of the soil, leaving iron and aluminum, which impart red and yellow colors to the soil. Optimum conditions for laterization are high rainfall, high temperature, fluctuating water table, and the absence of soil organic material. Oxides of iron and aluminum are relatively insoluble in the absence of humic acids, so they are retained, while silica is soluble under these conditions and easily transported downward.

In humid tropical mountain soils, soil organic matter increases with increasing altitude (Buytaert et al. 2011). In the lower montane forests of New Guinea, at 2,500 m, the ratio of organic matter in the soil is 4:1, compared to 0.5:1 in adjacent lowland tropical forests (Edwards and Grubb 1977). At higher elevations, leaching decreases, the accumulated organic material provides a ready source of nutrients, and nutrients are depleted less rapidly. As a consequence, tropical mountain soils commonly have a higher nutrient content than lowland soils (but see Grubb 1977 for examples of lower nutrient content in upland tropical soils). Additional reasons for the greater nutrient content of humid tropical mountain soils at higher elevations are the lower rates of weathering that occur with cooler temperatures and the relative youthfulness of the mountain landforms. Many tropical mountains formed only during the late Tertiary period (i.e., about 70 million to 1 million years ago). In New Guinea, for example, soils old enough to be strongly weathered are restricted to elevations below 2,100 m. On the other hand, weathering is sufficiently rapid in the humid tropics so that Entisols, the least weathered soils, are relatively scarce except on the highest summits and ridges.

Although usually associated with coniferous forests of midlatitude and subpolar mountains, podzolization also operates at higher altitudes in the humid tropics (Schawe et al. 2007). This involves the selective leaching of iron and aluminum, leaving silica behind in the upper layers of the soil. As organic matter increases, creating humic acids, and temperatures decrease, podzolization prevails, resulting in increasing acidity and more distinct soil horizons with increasing altitude. These tendencies only hold true up to the timberline; beyond this point, nocturnal low temperatures limit the development of vegetation and soils (Hedberg 1996). In the tropics, as long as soils are fertile and temperatures favorable, annual crops are more successful at intermediate altitudes than in the lowlands. In fact, maize, common beans, and some types of squash were first domesticated and grown in neotropical highlands. Agriculturally suitable soils plus a moderate climate at intermediate altitudes often result in denser human populations in tropical highlands than in lowlands, as discussed in Chapters 10 and 11.

Other Mountain Soils

Four soil orders of the U.S. Soil Taxonomy system—Gelisols, Vertisols, Ultisols, and Alfisols—also occur on mountains as well as in nonmountainous regions. *Gelisols,* which are associated with permafrost environments, are strongly affected by changes in volume caused by the freezing and thawing of water. Permafrost presents a barrier to downward movement of water in Gelisols, and the disturbances of freezing and thawing may prevent the formation of diagnostic horizons.

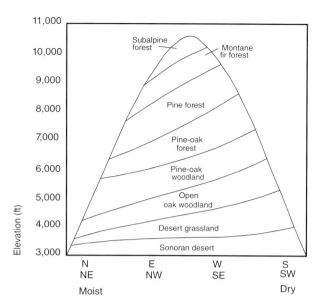

FIGURE 6.8 This diagram shows an idealized cross section of plant community patterns from north- to south-facing slopes in the desert environment of the Santa Catalina Mountains, Arizona. Soils would be expected to follow the same pattern if the vegetation zones remained the same over long periods of time. (Adapted from Whittaker et al. 1968: 441.)

Alfisols and Ultisols have well-defined horizons. *Alfisols*, which have developed in many mid- and low-latitude environments, cover about 13 percent of the land area of the United States, including parts of the northern Rockies (Brady and Weil 2008). *Ultisols* are more weathered than Alfisols, but less strongly weathered than Oxisols and less acidic than Spodosols. They are relatively old soils, formed in humid tropical and subtropical environments. The key diagnostic feature of *Vertisols* is the presence of clays that swell when wet and shrink, forming cracks, when dry. Vertisols typically occupy gentle slopes and lowland or valley sites with poor drainage; however, they are occasionally found on steep slopes (Soil Survey Staff 2010). Mountain locations of Vertisols include areas of Ecuador, Bolivia, and Ethiopia.

Recognition of the influence of human activities on soils prompted an international effort to develop a system for classifying anthropogenic (affected by human activities) soils—anthrosols (Bryant and Galbraith 2003; IUSS Working Group WRB 2007). Human activities in mountain regions, notably mining, terracing, agriculture, traffic, waste disposal, and settlement, alter the parent material, chemistry, fertility, texture, structure, and rate of development of horizons in mountain soils. Thus, the identification and classification of anthrosols is important for mountain regions.

Potential and Limitations of Mountain Soils

With the exception of intermediate elevations in arid and humid tropical mountains, soils become less developed and less productive as elevation increases. They also become thinner, coarser and rocky, acidic, unstable, infertile, and immature. Human use of soil for agriculture is constrained by characteristics of the soil at intermediate and high altitude. Other human uses of mountain environments, including grazing, trekking, and road construction, may be limited by mountain soils and can, in turn, alter soil characteristics.

Soil productivity is affected by the totality of processes operating at any location. If soils remain frozen or snow-covered for long periods, their operational level of physical, chemical, and biological processes is greatly reduced. Low temperatures during the growing season effectively limit the number and kinds of species that exist in a given environment, and lack of moisture sets similar limits. Certain characteristics help to explain why many mountain soils are relatively unproductive. Shallow soils restrict the penetration of roots, absorb less water, and do not insulate well against cold nighttime or winter temperatures. Coarse-textured mountain soils that lack well-developed structure have low base-exchange capacities and low water-holding abilities. Low nutrient levels may also limit plant growth in mountain soils. Luxuriant plant growth around animal burrows and bird roosts in alpine areas reflects the high nitrogen and phosphorus content of animal droppings at those sites, compared to the low values of those essential elements in the surrounding soil.

Recently, heightened interest in terrestrial carbon storage has brought global attention to the organic-matter dynamics and carbon storage of mountain soils. Concentrations of soil organic carbon increase with elevation, provided that enough vegetation is present to supply organic matter to the soil. This increase is attributed to the slower rates of decomposition in cooler temperatures (Leifeld et al. 2009; Budge et al. 2011). Although soil organic matter increases with altitude, this does not always connote an increase in fertility. In fact, the reverse may be true. A study in Taiwan attributed high humus (decomposed organic matter) contents of Andisols to aluminum toxicity and the lack of phosphorus (Chen et al. 1999). Land-use practices that reduce soil organic matter or cause soil erosion are significant because they reduce the amount of carbon stored in mountain soils.

Organic material is also important to mountain soils for its capacity to store and regulate the release of water. The importance of mountains in terms of their role in capturing, storing, and releasing the freshwater used by over half of the world's population is discussed further in Chapter 12. Mountains store water in snowbanks and glaciers, lakes and bogs, and soils. Most human uses of mountain regions compact the soil, reducing the available pore space and decreasing the ability of soil to absorb water. One study in the Colombian Andes found soils to contain less moisture following long-term grazing because soil compaction by grazing animals caused a higher proportion of rainfall to flow away as runoff (Hofstede 1995). Similarly, a 20-year record of sheep grazing in high-elevation grasslands of the Ecuadorian Andes demonstrated multiple impacts: an increase of bare land, a three- to fourfold reduction of water-retention capacity, and an increased tendency toward water-repellent soil surfaces (Podwojewski et al. 2002). The combination of these effects increases rainfall runoff and decreases the capacity of the soils to regulate the storage and release of water.

Soil erosion can have dramatic consequences for thin mountain soils. The stability of soils above timberline is largely due to the presence of a fairly complete cover of herbs and grasses, which form a tight matrix of organic material and interwoven, fine roots. If the vegetation is damaged, the soil becomes much more susceptible to erosion. Herein lies one of the greatest limiting factors to the use of mountain soils for agricultural purposes. In North America, mountain areas are rarely used for agriculture, while in other parts of the world (e.g., the Alps, Andes, and Himalaya), pressures of dense populations and other factors, such as the lack of fertile, low-relief agricultural lands and land ownership systems, have led to centuries of intensive agriculture, grazing, and other anthropogenic stresses on steep slopes, as discussed in more detail in Chapter 11.

Cultivation of mountain lands is limited not only by steep surfaces and thin soils, but also by climate. If the climate is adequate and the need is great, even the steepest slopes may be cultivated. If steep slopes are to be brought under cultivation, terracing is often employed to decrease soil erosion and water loss, but it is an expensive and high-maintenance endeavor. In most cases, terracing has been accomplished by organized cultures, which have continued to maintain the fields and terraces. Many of the world's mountainsides, notably those in the Philippines and Southeast Asia, have been extensively terraced for agriculture. Before the mid-1980s, many "Western experts" and institutions assumed that conversion of hillsides from forest to terraced agriculture was causing an environmental catastrophe in the Himalaya, an assumption that has been largely disproved (Ives and Messerli 1989). Terracing for rain-fed agriculture has been shown to decrease soil erosion rates in the Middle Hills of Nepal (Gardner and Gerrard 2003) and in Jamaica, El Salvador, and Taiwan (Sheng 1981). Where terraces are actively maintained and used for agriculture, the soil can become enriched in plant nutrients and biological activity. In the Colca Valley of Peru, after 1,500 years of cultivation, terraces, both those still cultivated and abandoned, have higher concentrations of phosphorus, organic carbon, and nitrogen, and higher levels of three soil enzymes than nearby uncultivated slopes (Sandor and Eash 1995). Terrace abandonment, however, is often associated with higher rates of soil erosion (Inbar and Llenera 2000) and a higher incidence of landsliding (Gerrard and Gardner 2002).

Without terraces, most mountain agriculture is less sustainable and more vulnerable to soil erosion. An intermediate situation is found in the Alps, which have a long history of settlement and agriculture. These landscapes have

FIGURE 6.9 The black páramo soil of the Ecuadorian Andes, here an andisol above a light-colored tephra layer in the province of Tungurahua, has a high water-holding capacity. (Photo by C. P. Harden.)

been preserved remarkably well, considering the length of time they have been used; nevertheless, the replacement of natural vegetation by crops has depleted the soil (Bouma 1974). In the northern Andes, a recent increase of greenhouse-based agriculture created a market for highly organic mountain soils (Fig. 6.9). The resulting excavation and removal of soil from sites in the high-elevation grasslands creates an anthropogenically accelerated process of soil loss in the uplands.

Most high mountain areas can withstand some grazing, but animals that graze in herds, especially sheep, goats, or yaks, can rapidly damage an area. Many examples of the effects of overgrazing in mountains exist. One classic area is in the Mediterranean region, where after several thousands of years of grazing, hills have little natural vegetation and soils are severely eroded. Even in the mountains of the western United States, after only about two centuries of grazing, evidence of its effects on mountain soils and vegetation is cause for concern (Milchunas 2006). Fortunately, steps have been and are being taken to counter this trend. In many higher-elevation areas of public land that were once freely grazed, grazing is now either regulated or not allowed.

Grazing and cultivation are not the only land uses that promote soil erosion in mountains. Stormwater and snowmelt runoff from roads and trails on mountain slopes can erode roads and trails and spill over onto and erode vulnerable surfaces downhill (Harden 1992, 2001) (Fig. 6.10). As highly compacted surfaces with very low infiltration capacities, roads and trails generate runoff sooner during a rainstorm and at lower rainfall intensities than agricultural fields or forested lands. The significant erosional effect of roads and trails is a serious concern in mountains where recreation is the dominant land use, and also where other human activities, such as grazing, agriculture, mining, or forestry, alter rainfall runoff patterns.

Mountain land use in the countries of Western Europe and North America is now regulated, but unrestricted use of marginal lands continues in many other areas of the world.

FIGURE 6.10 Roads and trails promote the erosive flow of rainfall runoff, which erodes the road or trail surface and also spills onto and erodes the surrounding slopes, Province of Pichincha, Ecuador. (Photo by C. P. Harden.)

Mountain soils have a limited capacity to support life, but the increasing pressure on mountain lands requires us to learn a great deal more about them and about the consequences of our actions on them. International attention to mountains and mountain soils is an important step toward better understanding and more sustainably managing the mountain soil resource.

References

Aerts, R. 1995. The advantages of being evergreen. *Tree* 10(10): 401–407.

Agriculture and Agri-Food Canada. 2003. Canadian Soil Information System (CanSIS). <http://sis.agr.gc.ca/cansis /taxa/index.html>.

Ambrose, S. H., and Sikes, N. E. 1991. Soil carbon isotope evidence for Holocene habitat change in the Kenya Rift-Valley. *Science* 253(5026):1402–1405.

Bäumler, R., and Zech, W. 1994. Soils of the high mountain region of Eastern Nepal: Classification, distribution and soil forming processes. *Catena* 22(2): 85–103.

Billings, W. D. 1950. Vegetation and plant growth as affected by chemically altered rocks in the western Great Basin. *Ecology* 31: 62–74.

Birkeland, P. 1984. *Soils and Geomorphology.* 2nd ed. New York: Oxford University Press.

Bouma, J. 1974. Soil dynamics in an alpine environment. *Soil Survey Horizons* 15(3): 3–7.

Brady, K., Kruckeberg, A., and Bradshaw, H., Jr. 2005. Evolutionary ecology of plant adaptation to serpentine soils. *Annual Review of Ecology Evolution and Systematics* 36: 243–266.

Brady, N., and Weil, R. 2008. *The Nature and Properties of Soils.* 14th ed. Upper Saddle River, NJ: Prentice-Hall.

Brady, P. V., Dorn, R. I., Brazel, A. J., Clark, J., Moore, R. B., Glidewell, T. 1999. Direct measurement of the combined effects of lichen, rainfall, and temperature on silicate weathering. *Geochemica et Cosmochimica Acta* 63(19–20): 3293–3300.

Bridges, E. M. 1997. *World Soils.* Cambridge, UK: Cambridge University Press.

Bryant, R. B., and Galbraith, J. M. 2003. Incorporating anthropogenic processes in soil classification. In H. Eswaran, T. Rice, R. Ahrens, and B.A. Stewart, eds., *Soil Classification, a Global Desk Reference* (pp. 57–65). Boca Raton: CRC Press.

Budge, K., Leifeld, J., Hiltbrunner, E., and Fuhrer, J. 2011. Alpine grassland soils contain large proportion of labile carbon but indicate long turnover times. *Biogeosciences* 8: 1911–1923, doi:10.5194/bg-8-1911-2011.

Bushnell, T. M. 1942. Some aspects of the soil catena concept. *Soil Science Society of America Proceedings* 7: 466–476.

Butler, D. R. 1995. *Zoogeomorphology:Animals as Geomorphic Agents.* Cambridge, UK: Cambridge University Press.

Buytaert, W., Cuesta-Camacho, F., and Tobón, C. 2011. Potential impacts of climate change on the environmental services of humid tropical alpine regions. *Global Ecology and Biogeography* 20: 19–33.

Buytaert, W., Sevink, J., De Leeuw, B., and Deckers, J. 2005. Clay mineralogy of the soils in the south Ecuadorian páramo region. *Geoderma* 127: 114–129.

Byers, A. 2005. Contemporary human impacts on alpine ecosystems in the Sagarmatha (Mt. Everest) National Park, Khumbu, Nepal. *Annals of the Association of American Geographers* 95(1): 112–140.

Carnelli, A., Theurillat, J.-P., Michel Thinon, M., Vadi, G., and Talon, B. 2004. Past uppermost tree limit in the Central European Alps (Switzerland) based on soil and soil charcoal. *The Holocene* 14(3): 393–405.

Chen, A., Asio, V., and Yi, D. 1999. Characteristics and genesis of volcanic soils along a toposequence under a subtropical climate in Taiwan. *Soil Science* 164(7): 510–525.

Costin, A. B. 1955. Alpine soils in Australia with reference to conditions in Europe and New Zealand. *Journal of Soil Science* 6: 35–50.

Costin, A. B., Hallsworth, E. G., and Woof, M. 1952. Studies in pedogenesis in New South Wales, III: The alpine humus soils. *Journal of Soil Science* 3: 197–218.

Coûteaux, M., Sarmiento, L., Bottner, P., Acevedo, D., and Thiéry, J. M. 2002. Decomposition of standard plant material along an altitudinal transect (65–3968 m) in the tropical Andes. *Soil Biology & Biochemistry* 34: 69–78.

Delmas, V., Jones, H., Tranters, M., and Delmas, R. 1996. The weathering of aeolian dusts in alpine snows. *Atmospheric Environment* 30(8): 1317–1325.

Di Pasquale, G., Marziano, M., Impagliazzo, S., Lubritto, Ca. De Natale, A., and Bader, M. 2008. The Holocene treeline in the northern Andes (Ecuador): First evidence from soil charcoal. *Palaeogeography, Palaeoclimatology, Palaeoecology* 259: 17–34.

Douglass, D., and Bockheim, J. 2006. Soil-forming rates and processes on Quaternary moraines near Lago Buenos Aires, Argentina. *Quaternary Research* 65: 293–307.

Earl-Goulet, J. R., Mahaney, W. C., Sanmugadas, K., Kalm, V., and Hancock, R. G. V. 1998. Middle-Holocene timberline fluctuation: Influence on the genesis of podzols (spodosols), Norra Storfjallet Massif, northern Sweden. *Holocene* 8(6): 705–718.

Edwards, P. J., and Grubb, P. J. 1977. Studies of mineral cycling in a montane rainforest in New Guinea, 1: The distribution of organic matter in the vegetation and soil. *Journal of Ecology* 65: 943–969.

Egli, M., Fitze, P., and Mirabella, A. 2001. Weathering and evolution of soils formed on granitic glacial deposits: Results from chronosequences of Swiss alpine environments. *Catena* 45: 19–47.

FAO. 2003. WRB (World Reference Base) map of world. Rome: Soil Resources, Food and Agriculture Organization of the United Nations. <http://www.fao.org/ag/agl/agll/wrb/soilres.stm>.

Gardner, R., and Gerrard, A. J. 2003. Runoff and soil erosion on cultivated rainfed terraces in the Middle Hills of Nepal. *Applied Geography* 23(1): 23–45.

Garner, H. F. 1959. Stratigraphic-sedimentary significance of contemporary climate and relief in four regions of the Andes mountains. *Bulletin of the Geological Society of America* 70: 1327–1368.

Gerrard, A. J., and Gardner, R. 2002. Relationships between landsliding and land use in the Likhu Khola Drainage Basin, Middle Hills, Nepal. *Mountain Research and Development* 22(1): 48–55.

Goudie, A. 1988. The geomorphological role of termites and earthworms in the tropics. In H. A. Viles, ed. *Biogeomorphology* (pp. 166–192). New York: Basil Blackwell.

Grieve, I. C. 2001. Human impacts on soil properties and their implications for the sensitivity of soil systems in Scotland. *Catena* 42(2–4): 361–374.

Grinnell, J. 1923. The burrowing rodents of California as agents in soil formation. *Journal of Mammalogy* 4(3): 137–149.

Grubb, P. J. 1977. Control of forest growth and distribution on wet tropical mountains. *Annual Review of Ecology and Systematics* 8: 83–107.

Hall, K., Thorn, C. E., Matsuoka, N., and Prick, A. 2002. Weathering in cold regions: Some thoughts and perspectives. *Progress in Physical Geography* 26(4): 577–603.

Hanawalt, R. B., and Whittaker, R. H. 1977. Altitudinal patterns of Na, K, Ca, and Mg in soils and plants in the San Jacinto Mountains, California. *Soil Science* 123(1): 25–36.

Harden, C. 1992. Incorporating roads and footpaths in watershed-scale hydrologic and soil erosion models. *Physical Geography* 13(40): 368–385.

Harden, C. 2001. Soil erosion and sustainable mountain development: Experiments, observations, and recommendations from the Ecuadorian Andes. *Mountain Research and Development* 21(1): 77–83.

Haugland, J., and Burns, S. 2006. Soils and geomorphology in the Oregon Cascades: A comparative study of Illinoian-aged and Wisconsin-aged moraines. *Physical Geography* 27(4): 363–377.

Hedberg, O. 1996. *Features of Afroalpine Plant Ecology.* Uppsala: Swedish Science Press.

Hofstede, R. 1995. The effects of grazing and burning on soil and plant nutrient concentrations in Colombian paramo grasslands. *Plant and Soil* 173(1): 111–132.

Imeson, A. C. 1976. Some effects of burrowing animals on slope processes in Luxembourg Ardennes. *Geografiska Annaler* 58A(1–2): 115–125.

Inbar, M., and Llerena, C. 2000. Erosion processes in high mountain agricultural terraces in Peru. *Mountain Research and Development* 20(1): 72–79.

Istanbulluoglu, E., Yetemen, O., Vivoni, E., Gutierrez-Jurado, H., and Bras, R. 2008. Eco-geomorphic implications of hillslope aspect: Inferences from analysis of landscape morphology in central New Mexico. *Geophysical Research Letters* 35(L14403), doi:10.1029/2008GL034477.

IUSS Working Group WRB. 2007. World Reference Base for Soil Resources 2006, first update 2007. World Soil Resources Reports 103. Rome: Food and Agriculture Organization of the United Nations.

Ives, J. D. 1973. Permafrost and ist relationship to other environmental parameters in a midlatitude, high-altitude setting, Front Range, Colorado Rocky Mountains. In North American Contribution, Second international Conference on Permafrost, 13–28 July 1973, Yakutsk, USSR (pp. 121–125). Washington, DC: National Academy of Sciences.

Ives, J. D., and Barry, R. G., eds. 1994. *Arctic and Alpine Environments.* New York: Harper & Row.

Ives, J. D., and Messerli, B. 1989. *The Himalayan Dilemma: Reconciling Development and Conservation.* London: Routledge.

Jenny, H. 1941. *Factors of Soil Formation.* New York: McGraw-Hill.

King, A. J., Freeman, K., McCormick, K., Lynch, R., Lozupone, C., Knight, R., and Schmidt, S. 2010. Biogeography and habitat modelling of high-alpine bacteria. *Nature Communications* 1: 53, doi:10.1038/ ncomms1055.

King, A. J., Karki, D., Nagy, L., Racoviteanu, A., and Schmidt, S. K. 2010. Microbial biomass and activity in high elevation (>5100 meters) soils from the Annapurna and Sagarmatha regions of the Nepalese Himalayas. *Himalayan Journal of Sciences* 6(8): 11–18.

Kubiena, W. L. 1970. *Micromorphological Features of Soil Geography.* New Brunswick, NJ: Rutgers University Press.

Leifeld, J., Zimmermann, M., Fuhrer, J., and Conen, F. 2009. Storage and turnover of carbon in grassland soils along an elevation gradient in the Swiss Alps. *Global Change Biology* 15: 668–679, doi: 10.1111/j.1365-2486.2008.01782.x.

Liptzin, D., and Seastedt, T. 2009. Patterns of snow, deposition, and soil nutrients at multiple spatial scales at a Rocky Mountain tree line ecotone. *Journal of Geophysical Research* 114, G04002, doi: 10.1029/2009JG000941.

Lundström, U. N., van Breemen, N., and Bain, D. 2000. The podzolization process: A review. *Geoderma* 94: 91–107.

Mahaney, W. 1970. Soil genesis on deposits of Neoglacial and late Pleistocene age in the Indian Peaks of the Colorado Front Range. Doctoral dissertation, University of Colorado.

Mahaney, W. 1990. *Ice on the Equator: Quaternary Geology of Mount Kenya, East Africa*. Sister Bay, WI: Wm. Caxton.

Margesin, R. 2012. Psychrophilic microorganisms in alpine soils. In C. Lütz, ed., *Plants in Alpine Regions* (pp. 187–198). Vienna: Springer-Verlag.

Margesin, R., Jud, M., Tscherko, D., and Schinner, F. 2009. Microbial communities and activities in alpine and subalpine soils. *FEMS Microbial Ecology* 67: 208–218.

Marr, J. W., Johnson, W., Osbum, W. S., and Knorr, O. A. 1968. Data on mountain environments, II: Front Range, Colorado, four climax regions, 1953–1958. University of Colorado Studies Series in Biology 28.

Messerli, B. 1973. Problems of vertical and horizontal arrangement in the high mountains of the extreme and zone (Central Sahara). *Arctic and Alpine Research* 5(3, Pt. 2): A139–148.

Milchunas, D. 2006. Responses of plant communities to grazing in the southwestern United States. General Technical Report RMRS-GTR-169. Fort Collins, CO: U.S. Department of Agriculture, Forest Service, Rocky Mountain Research Station.

Molloy, B. P. J. 1964. Soil genesis and plant succession in the subalpine and alpine zones of Torlesse Range, Canterbury, New Zealand, Part 2: Introduction and description. *New Zealand Journal of Botany* 1: 137–148.

Montanarella, L., Jones, R., and Hiederer, R. 2005. The distribution of peatland in Europe. In *Mires and Peatland*, Vol. 1, Article 01. <http://www.mires-and-peatland.net>.

Moore, I. D., Gessler, P., Nielsen, G., and Peterson, G. 1993. Soil attribute prediction using terrain analysis. *Soil Science Society of America Journal* 57(2): 443–452.

Morisada, K., and Ohsumi, Y. 1993, Soil development on the 1888 Bandai mudflow deposits in Japan. *Geoderma* 57(4): 443–458.

Muhs, D., and Benedict, J. 2006. Eolian additions to Late Quaternary alpine soils, Indian Peaks Wilderness Area, Colorado Front Range. *Arctic, Antarctic, and Alpine Research* 38(1): 120–130.

Parsons, R. B. 1978. Soil-geomorphology relations in mountains of Oregon. *Geoderma* 21: 25–39.

Podwojewski, P., Poulenard, J., Zambrana, T., and Hofstede, R. 2002. Overgrazing effects on vegetation cover and properties of volcanic ash soil in the paramo of Llanganua and La Esperanza (Tungurahua, Ecuador). *Soil Use and Management* 18(1): 45–55.

Price, L. W. 1971. Geomorphic effect of the arctic ground squirrel in an alpine environment. *Geografiska Annaler* 53A(2): 100–106.

Rech, J. A., Reeves, R. W., and Hendricks, D. M. 2001. The influence of slope aspect on soil weathering processes in the Springerville volcanic field, Arizona. *Catena* 43(1): 49–62.

Reider, R., Huckleberry, G., and Frison, G. 1988. Soil evidence for postglacial forest-grassland fluctuation in the Absaroka Mountains of Northwestern Wyoming, U.S.A. *Arctic and Alpine Research,* 20(2): 188–198.

Retzer, J. L. 1965. Alpine soils of the Rocky Mountains. *Journal of Soil Science* 7: 22–32.

Retzer, J. L. 1974. Alpine soils. In J. D. Ives and R. G. Barry, eds., *Arctic and Alpine Environments* (pp. 771–804). London: Methuen.

Rhoades, C. C., and Coleman, D. C. 1999. Nitrogen mineralization and nitrification following land conversion in montane Ecuador. *Soil Biology & Biochemistry* 31(10): 1347–1354.

Richmond, G. M. 1962. Quaternary stratigraphy of the La Sal Mountains, Utah. U.S. Geological Survey Professional Paper 324.

Ruhe, R. V., and Walker, P. H. 1968. Hillslope models and soil formation, I. Open systems. Transactions of the 9th International Congress of Soil Science, Adelaide, Australia, Vol. 4 (pp. 551–560).

Sandor, S. A., and Eash, N. S. 1995. Ancient agricultural soils in the Andes of southern Peru. *Soil Science Society of America Journal* 59: 170–179.

Schawe, M., Glatzel, S., and Gerold, G. 2007. Soil development along an altitudinal transect in a Bolivian tropical montane rainforest: Podzolization vs. hydromorphy. *Catena* 69: 83–90.

Sheng, T. 1981. The need for soil conservation structures for steep cultivated slopes in the humid tropics. In R. Lal and E. W. Russell, eds., *Tropical Agricultural Hydrology* (pp. 357–372). New York: John Wiley & Sons.

Sherrod, S. K., and Seastedt, T. R. 2001 Effects of the northern pocket gopher (*Thomomys talpoides*) on alpine soil characteristics, Niwot Ridge, CO. *Biogeochemistry* 55(2): 195–218.

Shoji, S., Dahlgren, R., and Nanzyo, M., 1993. Genesis of volcanic ash soils. In S. Shoji, M. Nanzyo, and R. Dahlgran, eds., *Volcanic Ash Soils* (pp. 37–71). Amsterdam: Elsevier.

Soil Survey Staff, 2010. *Keys to Soil Taxonomy.* 11th ed. Washington, DC: USDA Natural Resources Conservation Service.

Talon, B., 2010. Reconstruction of Holocene high-altitude vegetation cover in the French southern Alps: Evidence from soil charcoal. *The Holocene* 20(1): 35–44.

Thorn, C. E., Darmody, R. G., and Dixon, J. C. 2011. Rethinking weathering and pedogenesis in alpine periglacial regions: Some Scandinavian evidence. *Geological Society of London, Special Publications* 354: 183–193.

Thorn, C. E., Darmody, R. G., Dixon, J. C., and Schlyter, P. 2001. The chemical weathering regime of Karkevagge, arctic-alpine Sweden. *Geomorphology* 41(1): 37–52.

Tolbert, W. W., Tolbert, V. R., and Ambrose, R. E. 1977. Distribution, abundance, and biomass of Colorado alpine tundra arthropods. *Arctic and Alpine Research* 9(3): 221–234.

Townsend, A., Vitousek, P., and Trumbore, S. 1995. Soil organic matter dynamics along gradients in temperature and land use on the island of Hawaii. *Ecology* 76(3): 721–733.

Turrion, M. B., Glaser, B., Solomon, D., Ni, A., and Zech, W. 2000. Effects of deforestation on phosphorus pools in mountain soils of the Alay Range, Khyrgyzia. *Biology and Fertility of Soils* 31(2): 134–142.

Whittaker, R. H., Buol, S. W., Niering, W. A., and Havens, Y. H. 1968. A soil and vegetation pattern in the Santa Catalina Mountains, Arizona. *Soil Science* 105: 440–451.

Woods, R., Silvapalan, M., and Robinson, J. 1997. Modeling the spatial variability of subsurface runoff using a topographic index. *Water Resources Research* 33(5): 1061–1073.

Yoo, K., Amundson, R., Heimsath, A., and Dietrich, W. 2005. Process-based model linking pocket gopher (*Thomomys bottae*) activity to sediment transport and soil thickness. *Geology* 33(11): 917–920.

CHAPTER SEVEN

Mountain Vegetation

KEITH S. HADLEY, LARRY W. PRICE,
and GEORG GRABHERR

Mountain environments display some of the most striking examples of vegetation transition on Earth, including such well-known patterns as vegetation zonation, treeline, and elevation-induced decreases in plant stature and species diversity. These and other mountain-related vegetation patterns have long intrigued biogeographers and ecologists, and they continue to provide a contextual background for exploring fundamental biogeographic patterns such as vegetation boundaries, species diversity gradients, and the geographical history of plants. This knowledge has further contributed to the development of such important scientific concepts as natural selection, community succession, and environmental change.

The goal of this chapter is to summarize the ideas central to the understanding of mountain vegetation. Although much of this material focuses on the distribution and characteristics of mountain vegetation, it does so within the context of important biogeographical and ecological concepts. To achieve this objective, the content of the chapter is subdivided into several topics. The first section presents a brief overview of vegetation zones and a comparison of arctic and alpine environments, providing a comparative context for the remainder of the chapter. The next section examines plant distributions (*phytogeography*) and the history of mountain plant communities. The main body of the chapter is then presented in three major sections: (1) mountain forests, (2) timberline, and (3) alpine vegetation. The first of these sections examines the characteristics of mountain forests, primarily using examples from midlatitude and tropical regions. The second and third sections discuss the positioning, patterns, and causes of treelines and the characteristics of alpine vegetation. The chapter ends with a brief discussion of the historical and future human impacts on mountain vegetation.

Vegetation Zones and the Mountain Environment

Vegetation Zones and High-Elevation Climates

The concept of elevation zones is among the oldest and most useful characterizations of mountain environments (Fig. 7.1). This chapter adheres to this traditional presentation of vegetation patterns, but does so with a few caveats. The first of these is the recognition that many of the spatial patterns associated with vegetation are the result of local environmental differences and ecological processes present at the landscape, community, and individual plant scales (Gosz 1993). This requires that we recognize that local differences in species composition and the positioning of vegetation communities are the result of steep *environmental gradients* (changing in abiotic conditions over space) superimposed upon the discontinuous geologic substrates, soil types, geomorphic surfaces, topo- and microclimatic patterns, and other environmental factors (e.g., Whittaker 1953; Peterson et al. 1997; Ferreyra et al. 1998; Löffler and Pape 2008).

Second, we recognize that zonal vegetation patterns are dynamic and are continually shifting their locations, composition, and structure in response to environmental change. Third, we need to acknowledge that our current level of understanding is incomplete with regard to zonal transitions (*ecotones*) (van der Maarel, 1990), the divergent effects of climate on zonal boundaries in subtropical versus higher-latitude environments (e.g., Morales et al. 2004), and the role of human influences on the spatial organization of vegetation in transition zones (e.g., Sarmiento and Frolich 2002; Young and León 2007). Evolution, dispersal, and biotic interactions, such as competition, coexistence, predation, and mutualism, must further be accounted for

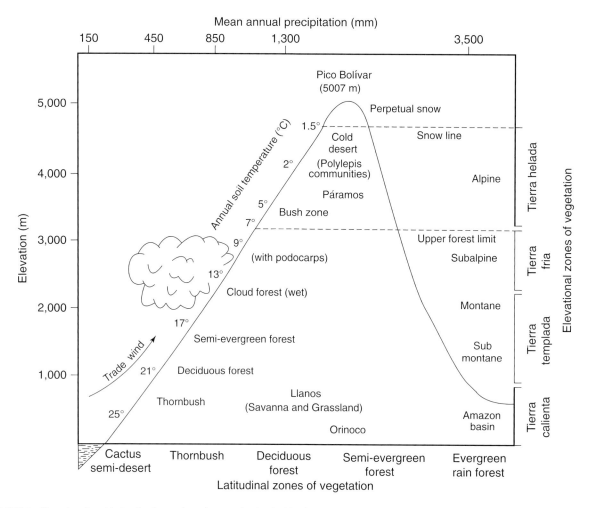

Mean annual precipitation (mm)

FIGURE 7.1 Elevational and latitudinal zonation of vegetation in the Northern Andes. (Huggett 1995.)

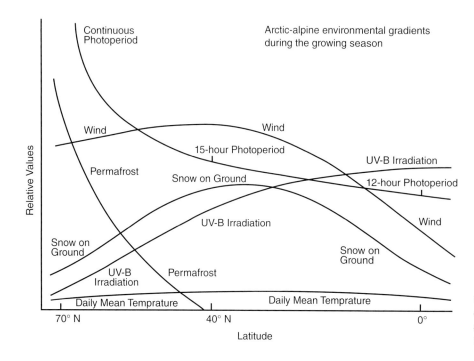

FIGURE 7.2 Latitudinal variation of environmental factors influencing arctic and alpine vegetation. (Billings 2000.)

in order to more thoroughly understand the characteristics of mountain vegetation (e.g., Callaway et al. 2002; Pierce et al. 2007; Young and León 2007).

Finally, we seek to clarify the common misconception that arctic and alpine environments are climatically similar (e.g., Löve 1970; Billings 1979; Walker et al. 1999). These environments share few climatic similarities beyond their low average annual temperatures (Fig. 7.2; Table 7.1). Consequently, beyond the general purposes of broad generalizations, environmental comparisons between arctic and mountain locations are too variable to be directly compared.

Phytogeography of Mountain Areas

Mountains have played a profound role in the evolution and distribution of terrestrial plants through their dual capacity as migration *corridors* and *barriers*. Mountains serve as corridors by extending the geographic range of environmental conditions into areas of dissimilar regional climate, allowing alpine and other mountain-adapted plants to move freely along mountain axes. For example, north–south-oriented cordilleras consisting of several mountain ranges, such as the Andes, Rockies, Cascades, and Sierra Nevada of California, play an important role in facilitating the interchange of species from different latitudes (Van der Hammen 1968; Troll 1968; Lauer 1973; Taylor 1977; Hadley 1987). Under cooler climate conditions, these corridors have also expanded longitudinally to include arctic tundra areas in Asia and North America (Major and Bamberg 1967a), the Altai Mountains (Weber 1965), and Changaishan area in China (Qian et al. 1999). Many Afroalpine plants are related to temperate-latitude

TABLE 7.1
Relative Differences between Alpine and Arctic Environments

Environmental Condition	Alpine	Arctic
Length of growing season	Longer	Shorter
Maximum radiation	High	Low
Daily Total radiation	Similar	(Similar)
UV Radiation	High	Low
Mean daily temperature	Similar	(Similar)
Daily temperature range	High	Low
Maximum temperature	High	Low
Minimum temperature	Higher to Similar	Lower
Diurnal temperature variation	High	Low
Surface wind speed—range	High	Low
Atmospheric vapor pressure gradient	High	Low
Leaf/atmosphere vapor pressure difference	High	Low
Precipitation—mean annual range	High to Low	Low
Precipitation—seasonal variation	High	Low
Mechanical soil stability	Low	High
Slope stability	Low	High
Soil carbon pool	Low	Higher
Cryogenic soil processes—summer	Greater	Lower
Cryogenic soil processes—winter	Lower	Greater
Soil permafrost	Rare	Common
Soil moisture	Low to High	High
Soil pH	Higher	Lower
Regional isolation of floras	Higher	Lower
Habitat diversity	Higher	Lower

NOTE: Modified after Körner (1995) and Billings (1973, 1979).

species despite the barriers represented by the Mediterranean Sea and the Saharan and Middle East deserts, and may have benefited from cooler and more moist conditions during the Pleistocene (Smith and Cleef 1988), similar to those suggested for other areas bordering desert regions (e.g., Major and Bamberg 1967b; Hadley 1987). In Australasia, the east–west orientation of the Himalaya has been less effective in hindering the exchange of mountain species because of its large geographic extent and the presence of the Hengduan and adjoining ranges in Malaysia and Indonesia (Van Steenis 1934, 1935, 1962, 1964; Raven 1973).

Although mountain environments may appear to be stable over centuries to millennia, biogeographic corridors often disappear or become discontinuous in response to long-term climate change. This results in *disjunct* (discontinuous) plant distributions through the development of migration *filters* that require organisms have long-distance *dispersal capacities* in order to be successful migrants (Lomolino et al. 2005). In some cases, however, unusual *conditions for dispersal,* such as high regional winds (e.g., Bonde 1969), or temporary corridors, such as periglacial bog complexes (Major and Bamberg 1967b), can facilitate the dissemination of reproductive plant parts (*propagules*) and the development of disjunct distributions.

While mountains may represent migration corridors for some species, they create migration *barriers* for others through the creation of unfavorable environmental conditions for lowland species (e.g., Janzen 1967). The selective nature of these conditions is enhanced by the dissimilar conditions that commonly develop on the windward and leeward sides of mountain ranges. These disparate conditions decrease the likelihood of successful plant colonization by presenting dispersing species with stressful environmental conditions as well as competition with locally adapted (*endemic*) species and subspecies (*ecotypes*).

Mountain ranges separated by expansive lowlands often serve as isolated *"habitat islands"* that share some of the biogeographic characteristics of oceanic islands (e.g., Brown 1971, MacArthur 1972; Carlquist 1974; see Chapter 8; Fig. 7.3). These "islands" often share a common regional history but differ in their species richness and composition. Several studies have confirmed the insular nature of these areas (e.g., Billings 1978; Riebesell 1982; Hadley 1987; Agakhanjanz and Breckle 1995), consistent with MacArthur and Wilson's *Equilibrium Theory of Island Biogeography* (1963, 1967), which predicts greater species numbers for larger islands and islands closer to "mainland" regions, in addition to genetic divergence. Patterns of increasing species richness with increasing area are well established for several regions, including the Great Basin and Rocky Mountains of the western United States (e.g., Billings 1978; Hadley 1987), Middle Asia (Agakhanjanz and Breckle 1995), and the mountains of Africa (Hedberg 1971, 1995; Burke 2005), but are less evident in other

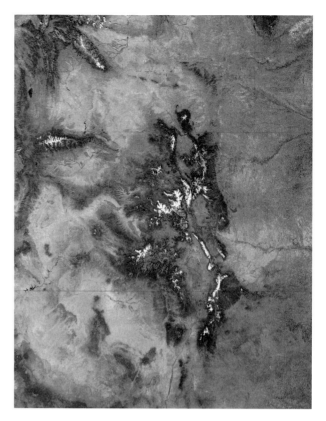

FIGURE 7.3 Alpine and forest areas of the southern Rocky Mountains, USA. Dark areas represent forest cover between upper treeline bounded by the snow-covered alpine zone, and lower treeline bounded by grassland and steppe vegetation. Note the forest and alpine connectivity (corridors) between mountain ranges and insular nature of others. (NASA Earth Observatory.)

regions such as Europe. In this case, the large, high-latitude mountain areas of northern Europe (e.g., Scandes) have fewer species than the smaller, lower-latitude mountain ranges of the Southern and Central Europe (e.g., Alps and Pyrenees), where environmental conditions, including calcareous soils, promote higher plant diversity (Virtanen et al. 2003).

The geographic isolation of mountain areas has several important evolutionary and floristic consequences. First, isolation, in conjunction with limited area, decreases the probability of genetic infusion from outside sources. Consequently, adaptations derived through natural selection tend to reflect specific, highly localized environmental conditions and the development of endemic species or subspecies (Körner 1995; DeChaine and Martin 2005). This process can be further magnified on mountainous oceanic islands where *genetic drift,* the random selection of genetic traits from a few colonizing individuals (*founder effect*) can accelerate the development of endemic taxa (Templeton 1980). In terrestrial environments, isolated, lower-latitude mountain floras tend to resemble adjacent, lower-elevation floras. The alpine flora of San Francisco Peaks in northern Arizona, for example, is almost exclusively composed of subalpine species (Schaack 1983) and has a low similarity to the rest of the Rocky Mountains in alpine species (Hadley 1987).

Plant Community Characteristics

Many of the characteristics of mountain plant communities change with increasing elevation, including a decrease in species richness and changes in plant stature, form, and structure (*physiognomy*). The most obvious visual change in mountain plants is the tendency toward smaller stature and less elaborate plants with increasing elevation. Other characteristics of high-elevation plants include lower annual growth rates and lower annual productivity. Exceptions to these general trends occur in some tropical cloud forests and desert mountains, where diversity and plant size increase along with increases in precipitation (Myers 1969; Pearson and Ralph 1978).

Mountain Forests

Mountain forests are generally categorized according to their zonal elevational distributions. Lower mountain forests within an elevational sequence are referred to as the submontane or montane zones, which are often divided into upper and lower areas. Higher-elevation forests comprise the subalpine zone, below the high-elevation, treeless alpine zone (Fig. 7.1). Zonal patterns in equatorial and tropical mountains diverge from this general trend in response to strong regional influences exerted by the Intertropical Convergence Zone (ITCZ), smaller continental landmasses, and monsoon conditions. Consequently, the number and distribution of elevational zones in tropical high mountains are far more complex than at higher latitudes. This calls for additional clarification and standardization of zonal nomenclature to improve comparisons of vegetation zones at different latitudes (e.g., Fickert and Richter 2004).

The trees of mountain forests are most easily categorized according to three distinct leaf growth forms (*habits*): needle-leaf conifers; broadleaf evergreen trees; and broadleaf deciduous trees. Needle-leaf conifers dominate most of the mid- to high-latitude regions of the northern hemisphere. Conifers are important montane forest trees in the southern hemisphere, but are generally subdominant to broadleaf evergreen trees. These once broadly distributed southern hemisphere conifers represent different genera with life forms (e.g., araucaria "pine" or monkey-puzzle tree [*Araucaria* spp.]) that are distinctly different from those that are dominant in the northern hemisphere, and are now most concentrated in wet environments (Hill and Brodribb 1999). Broadleaf evergreen trees tend to dominate in warm, humid regions with small temperature ranges, including both the tropics and the marine-influenced midlatitude regions of the southern hemisphere. In the northern hemisphere, deciduous broadleaf trees dominate midlatitude forests in humid continental climates with sharp seasonal contrasts

associated with subcontinental-scale atmospheric circulation patterns (Delcourt and Delcourt 2000).

Northern Hemisphere Mountain Forests

NORTHERN HEMISPHERE CONIFER FORESTS

Evergreen conifers dominate most higher-elevation forests in the northern hemisphere; they include several species closely related to those found in the boreal forests of North America and Eurasia. This similarity suggests a common ancestry for these forests and helps to explain the dominance of three tree genera, pine (*Pinus* spp.), spruce (*Picea* spp.), and fir (*Abies* spp.), which comprise 90 percent of the conifer species present in both regions (Eyre 1968). The shared *adaptive zones* (environmental regions where species coexist while exploiting the same resources) and current distribution of these genera also implicate the historical role of mountain climates as a critical factor leading to the southerly migration of these trees into the midlatitudes and subtropics (e.g., Perry et al. 1998).

Many needle-leaf conifers exhibit broad *ecological amplitudes* (environmental tolerance among species) developed over a long history of natural selection under diverse environmental conditions. Consequently, conifers dominate such diverse mountain forests as the cloudy, marine climate of the Pacific Northwest in the United States, the xeric slopes of the Atlas Mountains in North Africa, and the cold and windy slopes at the upper timberlines throughout the northern hemisphere. In all these environments, conifers have proven to be well suited to cold winters and short growing seasons.

Some of the success of conifers can be attributed to their compact and often conical crowns, which expose a comparatively small surface area. This has the advantage of shedding snow quickly and being less vulnerable to wind damage. The small, needle-shaped leaves of conifers provide further advantages, including low transpiration rates, low susceptibility to mechanical damage and wind abrasion, and high photosynthesis efficiency under diffuse light conditions. Moisture retention is critical during periods of high summer temperatures and the onset of freezing conditions in fall and winter, when the desiccation potential is high. The ability to use diffuse light is advantageous in both high-latitude regions, where incident radiation occurs at a low (oblique) angle, and in areas with cloudy, marine climates. In the latter case, the ability of conifers to photosynthesize under low light conditions extends their growing season well beyond that of competing deciduous broadleaf trees (Waring and Franklin 1979).

The evergreen habit of most needle-leaf conifers provides the additional advantage of early growing season photosynthesis by avoiding the delay caused by new leaf growth. This adaptation is critical in areas characterized by short growing seasons. Northern hemisphere conifers can also exhibit high resistance to frost, depending on their evolutionary history. For example, *Pinus cembra,* a European treeline species, can survive temperatures below −75°C, having evolved its frost hardiness in a colder, more continental climate region such as Siberia (Buchner and Neuner 2011). Paradoxically, an evergreen habit can also be maladaptive in areas of cold continental climates, such as portions of northern Siberia and Scandinavia. In these areas winter needle mortality favors the replacement of evergreen conifers by larch (*Larix* spp.), a deciduous conifer, or by broadleaf deciduous trees, such as birch (*Betula* spp.) or aspen (*Populus* spp.) (Troll 1973).

The low nutrient requirements of conifers are another important adaptation that allows them to occupy rocky substrates and unusual *edaphic* (soil–plant) conditions (Kruckeberg 2002). Conifers also return fewer nutrients to the soil through leaf fall and litter accumulation, resulting in relatively acid and infertile soils relative to those that develop under broadleaf deciduous forests (Perry 1994). These conditions, in conjunction with canopy shading, result in less diverse understory communities than those typically found in broadleaf deciduous forests.

BROADLEAF DECIDUOUS FORESTS

Broadleaf deciduous trees constitute the second common forest type found in northern hemisphere mountains. These forests occur at low to middle elevations in humid midlatitude regions and are best developed in Western Europe, eastern Asia, and the eastern United States. Although consisting of distinct species, the similar genera, appearance, and structures of these forests suggest former connections among these widely separated regions during the Tertiary period (ca. 65 to 1.8 million years before present) (Delcourt and Delcourt 2000). The dominant genera in these forests include oak (*Quercus* spp.), maple (*Acer* spp.), beech (*Fagus* spp.), elm (*Ulmus* spp.), hickory (*Carya* spp.), chestnut (*Castanea* spp.), ash (*Fraxinus* spp.), hornbeam (*Carpinus* spp.), and birch (*Betula* spp.).

Broadleaf deciduous forests in general differ from needle-leaf conifer forests by possessing multilayered canopies and higher species diversity (Delcourt and Delcourt 2000). European beech forests are an exception to this "rule," differing from their North American counterpart by virtue of their single-layered canopy and low species richness. These forests occur as a natural "monoculture," free of surface mosses resulting from leaf fall and burial.

Midlatitude deciduous trees are generally smaller than those found in tropical rainforests, but share their proclivity to grow larger and in more diverse communities with increasing moisture. Seasonal leaf loss in broadleaf deciduous forests can result from several factors, including low air temperatures and frost damage, low light levels and shorter photoperiods, seasonal moisture deficit, and freezing ground that limits the availability of water, thus inhibiting transpiration.

The broadleaf deciduous forests generally give way to conifers at higher elevations, but broadleaf deciduous trees (mainly birches) serve as the upper timberline species in

some areas (Troll 1973; Price 1978; Nagy and Grabherr 2009). Disturbances, including fire, windstorms, landslides, and avalanches, may also allow broadleaf deciduous trees to extend into higher elevations normally dominated by conifer forests. Where both conifers and deciduous trees are present, disturbance can result in a mosaic of intermingling communities of different ages and compositions.

Tropical and Subtropical Mountain Forests

Tropical mountain forests are characterized by high species diversity and the dominance of broadleaf evergreen trees. The high species diversity of these forests is enhanced by several factors, including a multilayered canopy and a nearly continuous cycle of reproduction under the near-constant thermic conditions. Although distinct thermic seasons are generally absent in the tropics below 15°–20° latitude, many tropical areas experience seasonal variations in the form of alternating wet and dry periods. Deciduous trees that shed their leaves during the dry season dominate mountain forests in areas with pronounced wet/dry seasons.

Similar to other mountain forests, the characteristics of tropical forests change markedly with elevation. These changes include gradual decreases in tree height, species richness, and structural complexity, as well as a shift from a three-storied to a two-, and then a single-storied forest with increasing elevation. Changes in tree characteristics with increasing elevation include a decrease in leaf size, loss of leaf drip tips, a decrease in the frequency of trunk *buttressing,* and thickening of tree bark. The abundance of *lianas* (hanging vines), and the profusion of elaborate flowering plants also decreases, while lichens and mosses increase on upper slopes (Richards 1966; Grubb 1977).

Temperate-latitude genera also occur in tropical forests above 1,000 m (3,300 ft), contributing to a far greater number of species than found in forests at higher latitudes. For example, the tropical montane forests of Mexico harbor over 40 species of pine (*Pinus* spp.) and 200 species of oak (*Quercus* spp.) (Perry et al. 1998). Comparable diversity trends are observed in Malaysia (Troll 1960). Similar to mid- and high-latitude areas, the number of woody species found in tropical forests declines with increasing elevation (Grabherr 2000).

Tropical forests generally show a well-developed zonation among the major plant communities (Fig. 7.1), with a typical progression that includes lowland tropical, submontane, montane, subalpine forest, and alpine grasses and shrubs. One of the more distinctive subzones is the *mossy forest, cloud forest,* or *elfin woodland* located at 1,000–3,000 m (3,300–10,000 ft) within the broader montane or subalpine zones. This subzone also occurs in subtropical climates where adiabatic cooling and higher precipitation are capped by the subtropical trade wind inversion, creating more moist conditions than at higher elevations (Leuschner 1996; Martin et al. 2007). These woodlands are a form of cloud forest adapted to low light levels and

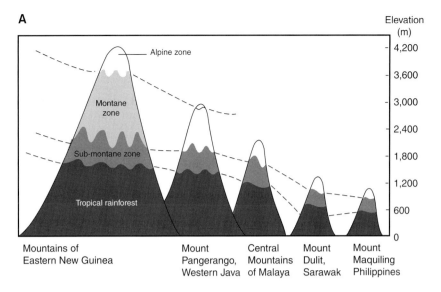

A

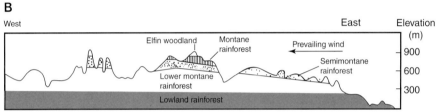

B

FIGURE 7.4 Treeline locations as a function of mountain mass and prevailing winds (A) in the humid tropics of Indo-Malaysia (Price 1981, adapted from Eyre 1968) and (B) a vegetation cross section of the Northern Range, Trinidad, West Indies (horizontal distance: 29 km (18 mi). (Price 1981, modified from Beard 1946.)

saturated atmospheric conditions. The trees in these forests and woodlands occur in a single layer of stunted and gnarled individuals about 6 m (20 ft) in height, covered with epiphytic mosses and liverworts; they are found at approximately the same elevations in Malaysia, Africa, and South America. Although species differ in these widely separated regions, the general appearance of the forests is similar (Grubb 1974). Areas above cloud forests often show a temporary increase in tree stature (lower latitudes) or shrubland or grassland vegetation types (subtropics) in response to higher light levels and drier conditions.

At high elevations, tropical forests become discontinuous and consist of patches interspersed with grasses. Lower temperatures become an increasingly important limiting factor for trees and, by 3,500–4,000 m (11,500–13,200 ft), grasses, shrubs, and bare ground become the dominant surface cover. The exact location of this boundary is often difficult to interpret because of the long history of human-set fires (e.g., Horn 1998; Young and Keating 2001) that have lowered timberlines well below their climatic limits (Smith 1975, 1977; Young and León 2007).

The position of tropical vegetation zones is not exclusively a response to changes in elevation, as shown by the shift of vegetation zones to lower elevations on smaller mountains relative to their larger neighbors (Fig. 7.4). Historically, this has been explained by the *"Massenerhebung"* or *"mountain-mass"* effect, based on the supposition that larger mountains exert a proportionally greater effect on their own climates. While this concept may hold some validity, it fails to explain why zones would shift under regionally constant climate conditions (Van Steenis 1961; Grubb 1971). A more recent explanation suggests that tropical vegetation zones are controlled by cloud cover, and that the lowering of elevational vegetation zones on smaller mountains results from lower cloud levels, higher soil moisture, and slower decomposition rate of organic matter. These conditions are thought to reduce plant nutrients, causing plant communities to adjust their location downward (Grubb 1971, 1974, 1977). Regardless of the cause of lower vegetation zones on smaller mountains, mountain zones on tropical islands are typically depressed in the direction of the prevailing trade winds (e.g., Leuschner 1996; Fig. 7.4). A similar pattern of forest-zone depression can be seen on windward and coastal mountains, compared to leeward or more continental mountains (e.g., Caccianiga et al. 2008).

Not all tropical mountains support well-developed forests (e.g., Wesche et al. 2000; Burke 2005). Examples include Kilimanjaro and Mount Kenya in East Africa, which rise steeply above dry plateaus covered by scattered forests and savanna grasses. A mossy forest occurs at approximately 2,400 m (8,000 ft) but fails to support the typical wet rainforest vegetation sequence above this level. Bamboo forests are a distinctive plant community on the mountains of East Africa, except on Kilimanjaro, and are important vegetation components on other tropical and midlatitude mountains (e.g., Veblen, 1982; Taylor 1992).

Central Himalayan forests share many of the angiosperm families and genera characteristic of tropical (e.g., *Shorea*) and temperate forests (*Quercus, Betula*), but have evolved several distinguishing traits in response to the climatatic influence of the Asian Monsoon and a complex climate history associated with the rapid tectonic rise of the region (Zobel and Singh 1997). Consequently, this region exhibits a high degree of endemism, and important traits include a high proportion of broadleaf tree species with multiyear leaf longevity, high concentrations of leaf nitrogen, and a range of flowering and late reproductive stages corresponding to different monsoonal phases (Zobel and Singh 1997). In general, these forests are highly productive, as measured by net primary productivity and *biomass* (the dry weight of organic matter per unit area), and well adapted to the variations in insolation, temperature, and precipitation characteristic of monsoonal climate conditions.

The most *xeric* (dry) mountain areas are located above tropical deserts. Many of these mountains, such as the Ahaggar and Tibesti in the Sahara Desert, exceed 3,000 m (10,000 ft) and support scattered *xerophytic* (drought-adapted) shrubs and trees above 1,500 m (5,000 ft), but lack forests (Messerli 1973). Several regions on the western slopes of the Andes are also treeless.

The Andes display a wide range of environmental conditions and illustrate how latitude, elevation, and moisture availability interact to shape montane forests (Veblen et al. 2007). Near the equator in northern Ecuador and Colombia, the rainforest ascends to similar elevations on both the lee and windward sides of the mountains, in response to convectional processes that operate on both slopes, creating similar precipitation patterns (Fig. 7.5). Within a few degrees to the north or south, the Andes come under the influence of the northeast and southeast trade winds, creating humid conditions on their east side and a pronounced rainshadow and desert conditions on the western slopes. This shift in air flow results in an asymmetrical distribution of vegetation zones across the range (Fig. 7.5) (Troll 1968).

Despite the general lack of moisture, trees attain their highest elevations in the dry subtropics. This paradoxical situation is possible because of the high surface heating and development of convection storms. Examples of these high-elevation forests include the pine forests (*Pinus hartwegii*) of central Mexico (20°N), at elevations up to 4,000 m (13,200 ft) (Lauer 1973), and the stunted *Polylepis* forest (*Polylepis tomentella*) in northern Chile and western Bolivia (18°S), where upper treeline elevations reach 4,600 to >4,800 m (15,100 to 15,800 ft) (Hoch and Körner 2005) and isolated trees occur at elevations exceeding 5,000 m (16,400 ft) (Navarro et al. 2005).

Northern versus Southern Hemisphere Mountain Forests

Mountain forests in the southern hemisphere are composed primarily of broadleaf evergreen trees and include more species of trees, shrubs, lianas, epiphytes, and herbs than those of the northern hemisphere. Tree ferns are common understory plants in many temperate southern hemisphere forests, even near the timberline. Although some species interchange has taken place between the northern and southern hemispheres, the floras on either side of the equator evolved independently and are distinct.

The humid tropics show little difference in species composition north and south of the equator. This pattern changes dramatically by 15° to 20°S latitude, where the seasonal contrasts become more strongly expressed, particularly at higher elevations. Montane forests extending from Bolivia to Mexico have similar genera, with *Weinmannia, Podocarpus,* and *Fuchsia* being widely distributed. Above 2,000 m (6,600 ft), however, forest composition north of the equator becomes distinctly boreal, while forests south of the equator are dominated by southern hemisphere species. This pattern is an artifact of geological history and plant evolution, rather than a result of current climatic differences. The situation changes rapidly beyond the Tropics of Cancer and Capricorn because of the contrast in the landmasses of the northern and southern hemispheres and a greater marine influence in the southern hemisphere.

The cool lowland temperate forests of Patagonia and New Zealand are similar to the upper-montane forests of tropical Malaysia, East Africa, and the Andes, suggesting that the southern hemisphere exhibits elevation and latitude zonation patterns similar to the northern hemisphere, and that many tree species and genera found in the southern hemisphere midlatitudes are also found in tropical mountains (Troll 1960). For example, both the equatorial mountain forests of New Guinea and the forests of New Zealand, 40° farther south, are co-dominated by the evergreen beech *Nothofagus* and conifers such as *Podocarpus* spp. and *Phyllocladus* spp. The major differences between these forests are the greater number of species in New Guinea and the predominance of pure *Nothofagus* stands in New Zealand (Wardle 1973a). The tendency toward single-species dominance at higher latitudes also occurs in Australia, where evergreen gums, especially the snow gum (*Eucalyptus niphophila*), dominate higher-elevation forests (Costin 1957, 1959). In southern Chile, the genus *Nothofagus*, including deciduous species in Patagonia, is again important (Veblen et al. 1996), as is *Araucaria* and the species *alerce* (*Fitzroya cupressoides*) and cordilleran cypress (*Austrocedrus chilensis*). Similar to the northern hemisphere, (1) there is a distinct decrease in timberline elevations with increasing latitude; (2) there is a marked decrease in species with distance from the equator; and (3) there are lower midlatitude treelines on the lee side of the southern Andes (Daniels and Veblen 2003).

Forest Succession and Disturbance

The dominance of a particular plant type or species depends on its relationship with the physical environment and other organisms. The prevalence of either conifers

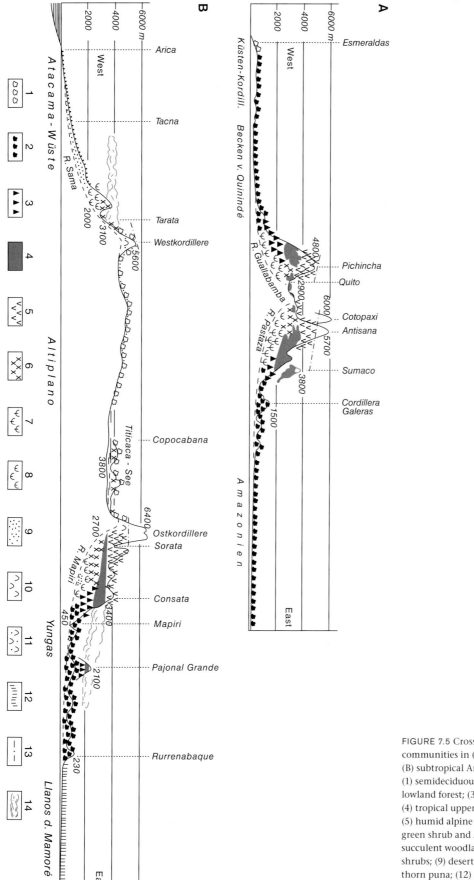

FIGURE 7.5 Cross section of major plant communities in (A) the equatorial Andes and (B) subtropical Andes at 16°S. Symbols represent: (1) semideciduous lowland forest; (2) tropical lowland forest; (3) tropical lower-montane forest; (4) tropical upper-montane forest (cloud forest); (5) humid alpine communities (páramos); (6) evergreen shrub and *Polylepis* woodland; (7) thorn and succulent woodlands; (8) thorn and succulent shrubs; (9) desert; (10) moist grass puna; (11) dry thorn puna; (12) moist lowland savanna; (13) snowline; (14) cloud belt. (From Troll 1968.)

or broadleaf deciduous trees in a given forest depends on which are better suited to the local conditions, including their relationship with each other. In marine and higher-latitude climates, deciduous tree species are typically less shade tolerant than conifers, and reproduce poorly under closed forest canopies. These conditions hasten the rate of species turnover and relegate broadleaf deciduous trees to early successional stages. In midlatitude deciduous forests, these successional trends are reversed, with pines and hemlock as the early successional shade-intolerant species. The presence of these successional patterns are, however, highly variable and may be locally dependent on environmental conditions and human activities.

Many western North American forests have experienced dramatic changes in their composition and structure over the past 100 years as a result of fire exclusion (e.g., Covington and Moore 1994). The outcome is an acceleration of succession toward more shade-tolerant species and a corresponding decrease in community and landscape diversity (Agee 1993). Fire exclusion has thus led to a dramatic reduction of habitat diversity and species richness for both plants and animals (e.g., Noss et al. 2006).

Lower fire frequencies combined with greater fire severities may also have long-term effects on forest stability by promoting certain life-history characteristics of fire-adapted tree species (e.g., Kauffman 1990). This is especially notable in areas where aspen stands contribute a large portion of the plant species diversity in subalpine and montane forests. Aspen (*Populus tremuloides*) is a broadly distributed pioneer species that is becoming less dominant in western North American forests as the result of fire suppression and its replacement by more shade-tolerant species, such as subalpine fir (*Abies lasiocarpa*) and Engelmann spruce (*Picea engelmannii*) (e.g., Peet 2000).

Mountain meadows are important components of many forest landscapes and are often related to forest succession and disturbance. The origin of meadows can be the result of environmental factors including poor drainage, shallow soils, low soil nutrients, snowpack, or disturbance (e.g., Jakubos and Romme 1993; Miller and Halpren 1998). Research on physical and biological processes, such as climate variability, fire occurrence, and fire exclusion (Taylor 1990; Rochefort and Peterson 1996; Hadley 1999), grazing (Miller and Halpren 1998), and competition among plants (e.g., Magee and Antos 1992), reveals that many mountain meadows in the western United States have experienced high rates of recent tree establishment and meadow closure (e.g., 0.1 to 0.5 m/year or 0.3–1.6 ft/year; Hadley and Savage 1996), resulting from episodes of favorable climate conditions for seedling establishment in conjunction with changes in management strategies around fire suppression and grazing allotments (e.g., Belsky and Blumenthal 1997).

Similar to the issue of declining aspen in the Rocky Mountains, resource managers are confronted with two difficult choices regarding the sustained presence of mountain meadows: (1) refraining from intervention that might allow tree encroachment to continue and decrease the size or eliminate many meadows; or (2) attempt to preserve meadows through active management, such as prescribed burning or clearing young trees and saplings (e.g., Franklin et al. 1971). Little is known about how active management might influence these meadows; in some cases, these strategies could increase the rate of seedling establishment by removing herbaceous plants that compete and inhibit seedling establishment (cf. Magee and Antos 1992).

In contrast to the predominantly natural meadows of western North America, mountain meadows in Europe are generally the result of human activity. These meadows are common landscape features of the Alps, Carpathians, and Caucasus, exhibiting high species richness, with as many as 60 species occurring in areas as small as 25 m^2 (269 ft^2). Initially maintained by hay harvests, many of these visually attractive meadows are now abandoned or preserved as "cultural treasures" financed by external funds. Natural mountain meadows below timberline are generally restricted to active avalanche paths.

Timberline

The transition between the subalpine forest and alpine tundra vegetation identifies the location of the *alpine ecotone*. Ecotones indicate changes in ecological conditions where differences in plant stature and form (*physiognomy*) become important adaptations to more extreme environmental conditions (alpine) or resource competition (subalpine forest) (van der Maarel 1990). The forest to alpine tundra ecotone is notable because of its global occurrence and sensitivity to changes in regional and global climate, especially temperature changes. Lower forest boundaries are also good indicators of climatic change in semi-arid mountain areas, where the lower treeline merges into grassland or steppe vegetation because of a lack of moisture (Weltzin and McPherson 2000). In both cases, the position and shape of local treeline can be highly variable, depending on the role of several factors related to climate variability, topography, soil type, disturbance, and the interaction among different species.

Biogeographers and ecologists have long been fascinated by the upper timberline, and their research has introduced a wide range of terms (Nagy and Grabherr 2009). *Timberline* refers to the overall transition from closed forest to open treeless tundra. *Forestline* represents the upper limit of contiguous forest. Abrupt forestlines coincide with timberline but may be indistinct where there is a gradual transition (*ecotone*) from forest to tundra. *Treeline* is the upper limit of erect arborescent growth, represented by scattered clumps of trees or isolated individuals above the forestline. Trees in this case are usually defined as being 2–4 m (6–13 ft) high to differentiate between treeline and the *scrubline* or *krummholz line*, the upper limit of stunted shrub-like trees

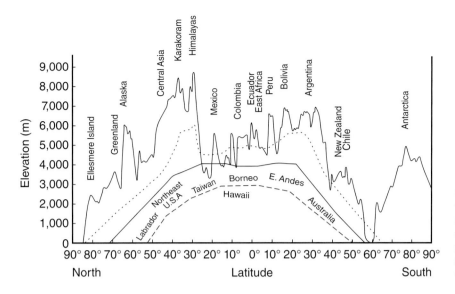

FIGURE 7.6 Global maximum elevation with modeled upper limit of the alpine zone (dotted line), approximate location of upper treeline (solid line), and the generalized temperature regimes supporting forest vegetation (dashed line) (Modified from Körner 2007.)

(Arno and Hammerly 1984). Krummholz is a German term meaning "crooked wood" and refers to stunted, contorted trees living in extreme and usually windy environments. Little (1979) further identifies a cushion krummholz zone located above the erect and often flagged trees of the krummholz zone. These smaller cushion krummholz often extend well into the alpine zone in areas protected by microtopographic ridges.

Characteristics of Timberline

In general, upper timberline elevations increase toward the equator (Fig. 7.6). The highest timberlines, however, occur in the subtropics between 20° and 30°S latitude in the Peruvian and Bolivian Andes and on the high plateau of Tibet (Miehe et al. 2007). Global timberline ranges from sea level in the Arctic to 3,500–4,000 m (11,500–13,200 ft) in the moist tropics, and more than 4,500 m (14,900 ft) at its extreme upper limits in the dry subtropics. Timberline elevation increases in the middle latitudes at an average of 110 m (360 ft) per degree of latitude until it reaches the subtropics at about 30°, where it peaks before descending near the equator (Holtmeier 2003; Körner 2007). This pattern differs on islands and in marine locations, where timberlines are lower relative to continental areas, in response to higher precipitation and cloud cover (e.g., Höllermann 1978; Fig. 7.4). In the United States, timberline varies from west to east, occurring at 1,800 m (6,000 ft) in the marine climate of the Olympic Mountains and Cascade Range, rising to >3,000 m (10,000 ft) in the dry continental climate of the central Rocky Mountains, and descending to 1,500 m (5,000 ft) in the humid continental climate of Mount Washington in New Hampshire. Similar to islands, timberlines often reach higher elevations on larger mountains than on smaller mountains because of the mountain-mass effect noted above. This relationship can be observed in all climatic regions, including the central

Alps and the Rocky Mountains, where trees grow at higher elevations than they do on marginal ranges (Brockmann-Jerosch 1919; Griggs 1938).

An important characteristic of global timberline is the dominance of evergreen tree species. These tree species are typically long-lived, have long reproductive life spans, and are capable of regenerating asexually (e.g., Lloyd 1997). Each of these traits is likely to contribute to the superior adaptation of conifers to the extreme environmental conditions that occur at upper timberline (Daubenmire 1954). Nonetheless, broadleaf deciduous species also form timberline in some midlatitude areas and the tropics, and can be found along with the needle-leaf conifers in the northern hemisphere.

Timberline Floristics

The vast majority of timberline trees of northern hemisphere mountains are pine (*Pinus*), spruce (*Picea*), and fir (*Abies*). These genera are represented by similar species throughout the northern hemisphere and occupy similar ecological niches, as illustrated by the ecology of the whitebark pine (*P. albicaulis*), a western North America timberline species, and its close relatives in Eurasia. Whitebark pine produces large, wingless seeds that rely on a bird, the Clark's nutcracker (*Nucifraga columbiana*), for seed dispersal through the harvesting and caching of seeds as a winter food source (Vander Wall and Balda 1977). Unused portions of these caches subsequently serve as an important seed source for seedling establishment. An identical relationship exists between pines and nutcrackers in Western Europe from the Alps to the Caucasus (e.g., between *P. cembra* and *Nucifraga caryocatactes*), and in the mountains of eastern Asia, between *P. pumila* and *N. caryocatactes* var. *japonicus* (Holtmeier 1973).

Other, less common genera of the pine family found at timberline include hemlocks and larches. The timberline hemlocks are represented by mountain hemlock (*Tsuga mertensiana*) in the Cascade and Olympic ranges of western North America. The deciduous larch (*Larix* spp.) is represented by several species in North America and Eurasia. Juniper (*Juniperus communis*) is also found in most northern hemisphere mountains, but usually grows as a scattered and low shrub (Wardle 1974).

Broadleaf deciduous trees form the timberline in some areas, especially in Scandinavia, eastern Asia, and parts of the Himalaya (Troll 1973). The dominant genus is birch, but other genera such as aspen, alder, and beech also occur. The upper treeline occurrence of broadleaf genera usually occurs in the absence of competing conifers. Broadleaf deciduous trees such as willow, birch, alder, and maple may also be present near timberline in areas experiencing slope instability and avalanches. These trees are well adapted for unstable slopes because of their ability to resprout from their root systems and because their flexible stems resist breakage by moving snow.

Timberline areas in tropical and southern hemisphere mountains are typically smaller but more floristically diverse than in the northern hemisphere (Wardle 1974). Timberline occurs along the axis of the Andes in South America, but is limited to the tropical mountains in Africa and exists as scattered enclaves in Australasia, with the exception of New Zealand. Broadleaf evergreen species, particularly *Nothofagus,* dominate timberline throughout the southern hemisphere. The timberline species of *Nothofagus* are generally evergreen, but can also occur in deciduous form in the drier and more continental areas of Patagonia, Tasmania, and New Zealand. Coniferous species of *Araucaria, Podocarpus, Libocedrus,* and *Papuacedrus* may also occur at timberline but are subdominant to *Nothofagus.* In Australia, upper timberline is dominated almost entirely by the snow gum (*Eucalyptus niphophila*) (Costin 1959).

Members of the heath family (Ericaceae) dominate the timberline vegetation of many tropical mountains, forming an "ericaceous belt" (Hedberg 1951; Troll 1968; Wesche et al. 2000). The ecological significance of Ericaceae in tropical mountains is unknown, but many species are fire resistant and able to reproduce quickly following fire and other types of disturbance (Sleumer 1965; Wesche et al. 2000). Ericaceous species typically grow as evergreen shrubs and trees with tough and leathery leaves. Dominant genera include *Vaccinium, Rododendron, Befaria, Pernettya, Erica,* and *Philippia.* In the tropical Andes, a rosaceous genus, *Polylepis,* grows as shrubby trees on steep rocky slopes above the cloud forest. *Polylepis tomentella* grows to >4,800 m (15,800 ft), the highest elevation of tree growth in the world (Hoch and Körner 2005).

In the high mountains of tropical Africa, ericaceous woodlands of *Erica* and *Philippia* grow to elevations of 3,500–4,000 m (11,500–13,200 ft). Tussock grasslands occur above this, along with giant *Senecios* and *Lobelias.* Equivalent plant forms grow in the high mountain grasslands of the tropical Andes (*Espeletia* spp.), as well as on the highest reaches of Hawai'i (*Argyroxiphiurn* spp.). The presence of similar life forms in these widely separated areas is generally interpreted as representing *convergent evolution* under similar conditions. Although they grow to heights of several meters and are tree-like, these plants are tall, columnar, massive, and almost unbranched herbs. The confusion as to whether to classify them as trees or shrubs has made the delimitation of timberline in these areas somewhat problematic (Hedberg 1951; Troll 1973).

Timberline Patterns

Timberline patterns vary considerably from the tropics to the high latitudes, reflecting the many different species and life forms, different environmental regimes, and variations in local conditions (e.g., Gieger and Leuschner 2004; Harsch and Bader 2011; Fischer et al. 2013). Nevertheless, all latitudes share the general timberline pattern of decreasing forest density and height and the development of a forest–alpine ecotone.

THE POSITIONING OF TREELINE

The location of upper treeline is sensitive to several environmental factors including latitude, continentality, and topography. This is illustrated by the topographic position of timberline in the tropics, where trees grow to their highest elevations in the valleys but reach their maximum elevations on ridges at higher latitudes (Troll 1968; Körner 2007; Fig. 7.7). These differences can be partly explained by the increasing depth and duration of snowcover with increasing latitude and its influence on the length of the growing season and regeneration success (e.g., Hättenschwiler and Smith 1999; Cuevas 2000; Wipf et al. 2009). Where snowpacks are deep, trees in northern hemisphere mountains can be found above the valley floors on the ridges or valley slopes. Aspect also influences seedling establishment patterns, inhibiting regeneration success on warmer and drier slopes exposed to higher levels of solar radiation (Elliott and Kipfmueller 2010).

Snow is comparatively unimportant in the tropics, making valleys more favorable habitat than exposed ridges. Under these diurnal climatic regimes, valleys experience smaller temperature ranges and a lower probability of frost than the ridges. Valleys also have deeper soils and are less exposed to direct sunlight, thereby increasing local moisture availability (Troll 1968). Mountain valley bottoms surrounded by high valley walls can provide an exception to this rule, supporting alpine vegetation rather than trees. The result is an inverted timberline caused by poorly drained soils, fire, cold air drainage (Barry 2008), and frost pockets (cf. Goldblum and Riggs 2002). Mountain meadows occurring below timberline in midlatitude mountains experience similar conditions. In these areas, however, greater snow accumulation and poorly drained soils are

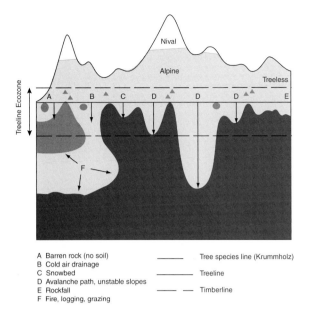

A Barren rock (no soil)
B Cold air drainage
C Snowbed
D Avalanche path, unstable slopes
E Rockfall
F Fire, logging, grazing

——— Tree species line (Krummholz)

——— Treeline

— — — Timberline

FIGURE 7.7 Hypothetical temperature-predicted treeline limits modified by local environmental conditions and disturbance. Treeline (solid line), timberline and tree species limits (dashed line), and ecotone breadth are indicated by horizontal lines. Causes and hypothetical extent of treeline depression are indicated by letters, with treeline depression indicated by arrows and light gray shading. Krummholtz are noted by gray triangles. Gray circles represent isolated populations of erect (non-krummholtz) trees. Dark gray indicates closed forest conditions. White areas represent the nival zone (areas of permanent snow). (Modified from Körner 2007).

probably more important than cold air drainage (Wardle 1974; Miller and Halpren 1998).

Another unusual form of local timberline is the *ribbon forest, glade,* or *"shimagare"* common in the central Rocky Mountains (Billings 1969, Miles and Singleton 1975) and other midlatitude mountains (e.g., Iwasa et al. 1991; Figure 7.8). These long, narrow features form as a consequence of downwind snow accumulation and subsequent periods of prolonged soil saturation, resulting in a short growing season that inhibits tree growth and seedling establishment (Billings 1969; Hättenschwiler and Smith 1999). A similar but more insular form of timberline occurs in regions of heavy snowfall such as the Pacific Northwest region of North America (Franklin and Dyrness 1988; Fig. 7.8).

Timberline can also be depressed in response to the underlying geology, grazing, forest clearing, lack of local seed sources, and fire. Such factors can result in the formation of grassy balds found in the southern Appalachians and Oregon Coast Range, and the parks found in the Rocky Mountains and Great Basin ranges in Nevada (e.g., Billings and Mark 1957; Gersmehl 1971, 1973; Lindsay and Bratton 1979). Fire plays a similar role in determining the location of treeline in arid and semiarid regions. In these mountain environments, forest succession is typically slow, because most timberline tree species are not fire-adapted. As a

result, trees growing near the previous treeline may require hundreds of years to reestablish (Peet 1981; Shankman 1984; Daly and Shankman 1985). In the absence of conifers, aspen can become both the dominant species and the treeline species, as it has in several of the Basin and Range mountains in Nevada and southeast Oregon (Faegri 1966; Price 1978).

TEMPORAL SHIFTS IN TREELINE

The elevational location of both global and regional treelines is strongly related to temperature and is sensitive to climatic variation. Both arctic and alpine treelines correspond closely with the position of the 10°C (50°F) isotherm for the warmest month (e.g., Körner 1998). The ability of mature trees to tolerate extended cool periods and survive periods of minor climatic fluctuations reinforces treeline stability (e.g., Lavoie and Payette 1996, Lloyd 1997).

While short-term (i.e., decades to centuries) treeline stability might be argued for some locations (e.g., Cullen et al. 2001), longer-term evidence of shifting treeline (e.g., macrofossil, pollen, charcoal, tree-ring data, and historical measurements) provides evidence of climate-induced changes in treeline position over periods of decades to millennia. These data document changing alpine treeline positions during the late Pleistocene in the central Alps (e.g., Tinner and Theurillat 2003), a mid-Holocene treeline advance in the Andes (Pasquale et al. 2008) and the western United States (e.g., Scuderi 1987), and post–Little Ice Age advance in Sweden (Kullman and Öberg 2009). LaMarche and Mooney's (1967) classic research on bristlecone pine (*P. longaeva*) in California and Nevada, for example, documented a 150-m (500 ft) upward advance of treeline during the Hypsithermal (ca. 7,000–4,000 years before present [YBP]) followed by an elevational depression of treeline during the Little Ice Age (ca. 2,800–120 YBP). Scuderi's (1987) research on foxtail pine (*P. balfouriana*) in California's Sierra Nevada further suggests a period of treeline depression occurring between 3,400 and 3,200 YBP, with a maximum depression of 20 m (80 ft) below current timberline. During the historical period, Kullman and Öberg (2009) report a 200-m (660 ft) maximum rise in Swedish treelines since the end of the Little Ice Age, based on repeated treeline location measurements along elevational transects. Other research has documented recent and decadal changes in treeline positioning and tree growth in response to climate change (e.g., Jacoby et al. 1996; Paulsen et al. 2000; Grace et al. 2002; Moiseev and Shiyatov 2003; Díaz-Varela et al. 2010).

Short-term climate variability can also lead to fluctuations in treeline positions. Warm temperatures and drought can lead to treeline depression in arid regions (Lloyd 1997; Lloyd and Graumlich 1997), but tree species abundance may serve as a more sensitive indicator of climate change than treeline position (Lloyd and Graumlich 1997). These latter results are consistent with those presented by

FIGURE 7.8 (A) Ribbon forests (photo by L. W. Price) and (B) "insular" treeline pattern in the Olympic Mountains, Washington (photo by K. S. Hadley).

Miller et al. (2004), who suggest directional, long-term growth responses of high-elevation conifers to long-term warming, but abrupt and reversible responses to the warm to cool phase shifts of the Pacific Decadal Oscillation (PDO). Daniels and Veblen (2004) report similar responses to short-term climate fluctuations in the southern Andes, where *Nothofagus* seedling establishment and growth respond more closely to El Niño–Southern Oscillation (ENSO) conditions rather than general atmospheric warming.

The idea that more than one climate variable controls treeline has also been shown using pollen evidence. Markgraf and Scott (1981), for example, suggest that the upper and lower treelines advanced in opposite directions between 10,000 and 4,000 YBP in central Colorado in response to warmer temperatures (upper treeline) and increased precipitation (lower treeline), associated with the development of strong monsoon conditions. Their idea of diverging treelines, and the importance of drought in treeline dynamics (Lloyd and Graumlich 1997) illustrate the complex relationship between treeline and environmental factors.

Causes of Timberline

Several theories have been proposed to explain the cause of upper treeline; all remain disputed. At any given site, timberline can be related to such factors as low temperatures, UV radiation, carbon dioxide concentrations, aridity, excessive snow, strong winds, poor or excessive drainage, lack of soil, recent disturbance by fire, disease, avalanches, volcanic eruptions (e.g., Lawrence 1938; Holtmeier 2003; Bader et al. 2007; Körner 2007; Harsch and Bader 2011), or by stature- or growth-related factors, such as plant competition and facilitation, deep root systems, or carbon balance (Stevens and Fox 1991; Callaway et al. 2002; Nagy and Grabherr 2009). At the global scale, however, the relatively consistent pattern of timberlines (Fig. 7.6) suggests that some overriding environmental or ecological factor transcends local conditions and accounts for the majority of tree behavior at timberline (Wardle 1971, 1974; Körner 1998; Körner and Paulsen 2004). While acknowledging that no single factor can explain local timberlines (e.g., Körner 2012), a review of some of the more important factors known to limit tree growth provides insight into the complexity of the problem and possible avenues to its solution.

SNOW

Snow has both a positive and a negative influence on timberline. Snow enhances tree survival at timberline by providing protection from low temperatures and high winds, and it provides moisture during the growing season. Excessive snow, however, can smother trees or cause physical damage by promoting avalanches or snow creep. Late-lying snow harbors molds and fungi that attack trees

(Stevens and Fox 1991; Körner 1998), reduces the length of the growing season, and influences local competition for resources with vascular plants and cryptograms (Moir et al. 1999). The depth of snowpack may also explain why trees grow preferentially along ridges rather than in valleys in most midlatitude mountains (Shaw 1909) and how ribbon forests develop (Billings 1969; Hättenschwiler and Smith 1999; Figs. 7.9, 7.10). Regardless of its ecological importance in controlling the position of local timberlines, snow is less significant in many tropical mountain areas, thus eliminating it as a controlling factor of timberline at the global scale. Even in midlatitude areas where snowpack strongly influences and is influenced by local conditions (Hiemstra et al. 2002), it is only one of several important factors operating at timberline.

WIND

Wind velocity generally increases with altitude and produces high levels of both mechanical and physiological stress for trees. Wind can cause significant mechanical damage by felling or causing the breakage of stems or limbs, particularly when they are coated with rime, ice, or snow. Wind also acts as an abrasive agent by transporting small mineral grains and snow and ice crystals. This results in bark and limb erosion and, when combined with branch breakage on the windward side, gives rise to flagged trees (Fig. 7.9a). Wind desiccation also shapes primary shoots, encouraging lateral branch growth, flagged branches, krummholz tree forms, and flat tree canopies (Fig. 7.9b) (Cairns 2001). Finally, winter wind damages young shoots and buds by removing their *cuticle,* an impermeable waxy surface (Hadley and Smith 1983), causing accelerated moisture loss when water is largely unavailable because of the freezing temperatures (e.g., Tranquillini 1979).

Seedling regeneration on wind-exposed sites is generally low, because timberline trees rarely produce viable seeds (e.g., Wardle 1968; Caldwell 1970). Seed supply can be further compromised by poor seed development in many cones. European spruce (*Picea abies*) in the Alps, for example, produce cones at lower-elevation sites every three to five years, but only once every six to eight years at higher elevations, and once every nine to eleven years at timberline (Tranquillini 1979). This behavior suggests that many tree seeds in alpine areas may be wind-transported from lower-elevation forests. Conversely, abundant seed rain can contribute little to the upward movement of treeline where seed dispersal is limited and seed viability is low (Cuevas 2000).

In addition to exceeding a minimum threshold in seed rain, seed dispersal, and seed viability, germination and seedling survival at treeline require suitable germination sites. Such "safe sites" include geomorphic features such as boulders and terrace risers that decrease surface wind velocities, increase winter snow cover and summer soil moisture, and provide additional heating through the irradiance of long-wave radiation (e.g., Resler 2006). Once established,

FIGURE 7.9 (A) Flag and (B) krumm-holz mat, Colorado Front Range. Note exposure of upper branches and snow accumulation within the krummholz mat relative to the surrounding surface. (Photos by K. S. Hadley.)

these seedlings eventually develop tree canopies that increase the area of favorable microhabitat, improving subsequent regeneration success. This sequence of tree establishment and growth—microrelief feature–seedling regeneration–canopy development—is strongly suggestive of a positive feedback process that explains the spatial pattern of many local treelines (Alftine and Malanson 2004; Bekker 2005; Resler 2006).

Regardless of the role wind plays in shaping local treelines, it also cannot be considered the universal factor determining treeline positions, for two reasons. First, wind is relatively unimportant in tropical timberline areas. Second, any correlation between latitude, continentality, and wind velocity appears to be much weaker than that associated

with latitude and timberline (Fig. 7.6). Nonetheless, wind is clearly a major factor controlling local timberlines, as indicated by the marked depression of timberlines on the windward side of mountain ranges (Fig. 7.4).

SOLAR RADIATION

Several theories argue that the location of timberline is a consequence of too much or too little solar radiation (e.g., Bader et al. 2007). For example, high levels of incident UV impair the photosynthetic process and have other deleterious effects in plants (e.g., Billings 2000; Bader et al. 2007). High insolation levels can also lead to excessive heat build-up at the soil surface, limiting tree growth

(Aulitzky 1967). Nevertheless, major problems with the light intensity and heat build-up hypotheses include the presence of trees above 4,000 m (13,100 ft) in the dry subtropics of Tibet and South America and species-specific responses to both direct sunlight and clear-sky exposure that result in lower minimum nighttime temperatures (Germino and Smith 1999). Cloud cover has similarly been invoked to explain the upper limit of forests. Cloud cover does appear to correspond to low timberlines, since timberlines are invariably lower in marine than continental regions. However, in areas outside the tropics, it is difficult to separate the effects of clouds from those of wind, precipitation (especially snow), and temperature.

BIOTIC FACTORS

Species interaction among high-elevation trees is a relatively neglected but important factor in determining the composition and spatial pattern of upper treelines. Dullinger et al. (2005), for example, note that the presence of pine cover in the Austrian Alps has a negative effect on the *recruitment* (germination success and growth into larger size classes) of neighboring spruce and larch. This and similar competitive interactions influence both the local dominance and future composition of these treeline communities. Other research by Maher et al. (2005) in the Rocky Mountains found that tree and herbaceous plant cover increased seedling survival and improved photosynthesis rates relative to seedlings growing in open conditions, except during periods of low-water stress. These results suggest that tree seedlings benefit in a facilitative way similar to other types of plants growing under alpine and other extreme environmental conditions (e.g., Callaway et al. 2002; Callaway 2007) and are consistent with tree recruitment patterns occurring above treeline (e.g., Cuevas 2000) and within subalpine forests (Callaway 1998).

Seed, seedling, and tree predation by animals can also limit tree growth at high elevations, as noted at some locations in Switzerland after the successful reintroduction of the ibex (*Capra ibex*) to the Alps in the 1920s (Holtmeier 1973). Insects and diseases can also cause considerable destruction of trees at timberline. Bark beetle (*Dendrocotonus engelmannii*) and spruce budworm (*Choristoneura occidentalis*) are both serious pests at timberline in the Rocky Mountains (Johnson and Denton 1975; Schmid and Frye 1977), as is white pine blister rust caused by the exotic pathogen *Cronartium ribicola* (Tomback and Resler 2007; Resler and Tomback 2008; Smith et al. 2011). In the Alps and Scandinavian mountains, the *Oporinia autumnata* and *Zeiraphera grisena* moths periodically damage high-altitude forests (Holtmeier 1973). Insects also suppress growth rates of eucalyptus in the subalpine forests of the Snowy Mountains, Australia (Morrow and LaMarche 1978). The various blights, rusts, and other pathogens that live in snow also cause considerable damage to timberline species (Cooke 1955; Habeck 1969). In the Alps, needle rust (*Chrysomyxa rhododendri*) infects spruce, inhibiting photosynthesis and consequently reducing needle biomass (Bauer et al. 2000) and, presumably, tree growth.

Temperature

Low temperature has long been considered the primary cause of timberline. This conclusion, however, is largely based on the coincidence of treeless conditions in arctic and alpine climates, rather than on its demonstration through experimentation. Nonetheless, there seems to be a strong correlation between temperature and timberline (Körner and Paulsen 2004), especially in *mesic* (moderately moist) regions (see Lloyd 1997).

Low winter temperatures are less critical than summer temperatures, which control primary metabolic and growth processes. The exact mechanism whereby low temperatures stop tree growth is still unknown. Boysen-Jensen (1949) suggested that temperature plays an important role in the allocation of energy needed to maintain normal metabolic processes versus the production of new wood. This theory would explain why timberline trees minimize the development of nonproductive tissue and develop a small stature; the slow growth of trees at or near timberline; and their longevity (ca. 4,800 years for bristlecone pine) (Ferguson 1970; LaMarche and Mooney 1972). Slow growth rates may be a positive adaptation to a short growing season because they allow the shoot tissues to mature (Wardle 1974).

Wardle (1971, 1974) hypothesized that the inability of shoots to mature (ripen) is the principal cause of tree-growth cessation. This can occur when the shoots freeze or become desiccated during periods when transpiration requirements exceed moisture availability. Ripened shoots of timberline trees possess well-developed cuticles that can withstand low temperatures and desiccation, and their cells are not damaged by the growth of ice crystals (Wardle 1971). Upper timberline, according to this theory, represents the highest altitude at which woody shoots can grow and mature under the environmental conditions at the height of the tree canopy. The critical attribute of ripening is the protection it provides against winter desiccation. Temperature thus controls the time available for needle maturation, cuticle thickness (protective covering), and drought resistance (Tranquillini 1979). Following needle maturation, trees can withstand severe winter conditions, allowing them to reach higher elevations in extreme continental areas where winter conditions are more severe than in milder marine areas (e.g., Becwar et al. 1981). The capacity to withstand severe conditions is also influenced by the inherited tolerances of different species to various climatic extremes (Clausen 1963).

Körner (1998) hypothesized that tree growth is limited by a minimum soil temperature threshold that precludes root growth (cellular and tissue formation), thereby inhibiting shoot and canopy development. Recent evidence from root growth experiments by Alvarez-Uria and Körner

(2007) appears to support this hypothesis, finding a critical temperature of ~6°C for significant root growth in six tree species. This temperature threshold closely approximates the worldwide mean soil temperature (6.7°C) at climatic treelines (Körner and Paulsen 2004).

DISTURBANCE

Both natural and anthropogenic disturbances, including fire, grazing, logging, mining activities, avalanches, and mass wasting, play an important role determining the elevation and pattern of local and regional treelines (Jentsch and Beierkuhnlein 2003; Körner 2007; Young and Leon 2007). Treelines and ecotones are often lowered and lengthened following disturbances, creating more gradual treeline-to-alpine transitions (e.g., Daniels and Veblen 2003) and a patchy landscape characterized by individual trees and regeneration clusters (e.g., Cullen et al. 2001; Batllori et al. 2009; Coop and Schoettle 2009). Disturbance can also temporarily stabilize treeline where local factors, such as canopy cover, are a more important determinant of regeneration success than climate change (Cullen et al. 2001). Conversely, treeline sensitivity and change may be greatly increased following the cessation of human activities (Holtmeier and Broll 2005) and a decrease in disturbance frequency (e.g., Bolli et al. 2007).

Forest to Tundra Transition—Northern Hemisphere

Trees above forestline frequently grow in small clumps or "islands" with the tallest trees in the center, decreasing in height outward to the margins (Fig. 7.9). These "islands" typically begin as a single tree growing in a favorable microsite (see above). Once established, these trees subsequently modify the surrounding environment and enhance the probability of seedling establishment and the growth of the island. Environmental modification by early colonizing trees begins through their absorption of incoming solar radiation and the creation of small heat islands. These higher temperatures accelerate the rate of spring snowmelt and consequently extend the local growing season, which promotes seedling regeneration, growth, and needle maturation. The tree canopies also trap outgoing long-wave radiation during the summer and increase snow accumulation during the winter, helping to moderate seasonal temperature extremes. *Nurse trees* create wind shadows and contribute organic matter to the developing soils. Many timberline tree species also reproduce by layering—the establishment of new stems from lateral-growing branches that take root when they come into contact with the soil. This process is common in areas of high snow accumulation, where it aids the vegetative expansion and merger of tree clumps.

Krummholz forms initially develop a vertical habit, becoming more prostrate with increasing elevation. These small tree clumps typically have one or two primary stems surrounded by a dense skirt of limbs and shoots extending horizontally around the base. The height of this skirt reflects the depth of winter snowpack, as buds or shoots protruding above the snow are killed by wind abrasion and dehydration (Hadley and Smith 1983, 1986; Fig. 7.9). The aerodynamic structure of krummholz mats (Daly 1984), along with their high needle density, promote krummholz survival by increasing snow accumulation, minimizing snow abrasion, and increasing needle temperatures (Hadley and Smith 1987). While environmental factors are clearly important to the formation of krummholz forms, the relative contribution of genetics (ecotypes) remains poorly understood (e.g., Wardle 1974; Grant and Milton 1977).

Several other environmental and ecological factors have been implicated in the development and persistence of alpine krummholz. These include snow-related factors, such as frost damage, snow pressure, pathogens, and snow water content (Daly 1984), as well as variations in solar radiation, soil moisture, soil properties, and the autecology of the local timberline species (Weisberg and Baker 1995; Seastedt and Adams 2001). Hessl and Baker (1997) further found that the combination of warm temperatures and heavy snowpack occurring over several years are required for tree establishment in the Colorado Front Range. These factors suggest that the processes influencing tree establishment and the patterns of upper timberline are highly complex and dynamic. The dynamic nature of timberline is further evidenced by the downwind migration of krummholz (Ives 1973; Marr 1977; Benedict 1984).

Forest to Tundra Transitions—Tropics and Southern Hemisphere

Krummholz is generally not well developed in the tropics or some southern hemisphere mountains. In New Zealand, for example, the *Nothofagus* forestline occurs as an abrupt transition from erect trees to a zone of low-lying shrubs (Wardle 1998). These shrubs consist of non-forest species and are genetically predisposed to a low stature. This differs from northern hemisphere timberlines (Wardle 1965), where few species other than the dwarf mountain pine (*P. mugo*) found in the Alps, *Pinus pumilio* in eastern Asia and Japan, and green alder in several mountain systems of Eurasia have established ecotypes.

Abrupt forestlines in New Zealand and South America are related to the sensitivity of *Nothofagus* seedlings to sunlight beyond the closed forest (Wardle 1973b; Cuevas 2000). A similar pattern exists in the moist tropics of New Guinea, where treeline occurs at about 4,000 m (13,200 ft) (Wade and McVean 1969). These trees exhibit some stunting, but maintain their umbrella-shaped crowns and a 3–6 m (10–20 ft) high canopy. This forest boundary changes abruptly into grasses or shrubs with no marked tendency toward krummholz, reflecting the relative lack of strong prevailing winds. The abruptness of this forestline may also be the result of fire (Gillison 1969, 1970; Hope 1976, Smith 1975, 1977).

Summary

Although upper timberline occurs in every climate region, its exact cause remains elusive. Nonetheless, some generalizations are possible. The most obvious of these is the universal occurrence of short stature, which appears to be essential for high-elevation trees to survive. This physical acclimatization or genetic adaptation allows these plants to benefit from several near-surface conditions including warmer daytime temperatures, lower wind stress, and insulation provided by snow (Grace 1989). These and other factors far outweigh the negative consequences of higher summer leaf temperatures and increased evaporative stress experienced by shorter plants (Aulitzky 1967). Globally, treeline appears to be most strongly controlled by a common thermal threshold (Körner 1998; Körner and Paulsen 2004), but it varies at the regional scale, as higher treelines occur in continental than in marine climates (Caccianiga et al. 2008). At the landscape scale, timberline elevations exhibit a direct relationship with insolation; the influence of microtopography becomes increasingly important at the scales of communities and individual plants. All of these observations underscore the importance of temperature in establishing the elevation of the upper timberline. However, several decades of research regarding the energy relationships between trees and their net assimilation rates (e.g., Handa et al. 2005; Hoch and Körner 2005) have shown that the number of variables controlling local vegetation patterns increases with decreasing area (cf. Gosz 1993).

Alpine Tundra

Tundra is a Russian word meaning "treeless plain" and originally referred to arctic areas north of the timberline. More recently, "tundra" has also been used to refer to alpine vegetation because its appearance is similar to arctic plant communities. This similarity partly explains why several earlier studies compared various attributes of arctic tundra and alpine tundra vegetation (e.g., Bliss 1956, 1962, 1971, 1975; Billings and Mooney 1968; Billings 1973, 1974a; Table 7.1).

Arctic and alpine vegetation share many of the same species but become increasingly dissimilar with decreasing latitude. Alpine areas in the mid- and low latitudes often have less than one-half of the tundra species found in the Arctic (e.g., Billings 2000). Floristic differences increase in the tropics: The flora and life forms differ almost entirely, with few arctic representatives present (Van Steenis 1935; 1962; Hedberg 1951, 1961, 1965, 1995; Troll 1958; Cuatrecasas 1968). These differences mirror the global latitudinal diversity gradient, the diversity of mountain environments (Kikvidze et al. 2005), the isolation of alpine areas (Hadley 1987), and differences in their evolutionary histories (e.g., Weber 1965; Agakhanjanz and Breckle 1995). Floristic dissimilarity between the arctic and alpine floras is intensified in floristically rich mountain ranges, where the number of alpine plants can exceed the total number of plants found in the entire arctic tundra (Körner 1995; Virtanen 2003).

The total area of alpine tundra is comparatively small in the southern hemisphere, and it supports a dissimilar flora with the exception of a few bipolar plants. These communities remain, however, similar in function and structure to the arctic tundra (Ward and Dimitri 1966; Billings 1974a). Nevertheless, most writers have preferred the term *alpine vegetation,* although some prefer the term "*oreophytic*" (Körner 1995), to refer to plant communities growing above the climatic timberline in southern hemisphere and tropical mountains (Van Steenis 1935; Hedberg 1965; Costin 1967; Mark and Bliss 1970).

Arctic and alpine tundra are commonly divided into low, middle, and high tundra zones (e.g., Nagy and Grabherr 2009). The low alpine tundra is the zone immediately adjacent to the timberline, consisting of a nearly complete cover of low-lying shrubs, herbs, and grasses. The greatest number and diversity of species are found in this zone; it is generally the most productive area within the tundra. Most grazing above timberline takes place in the low alpine tundra (Bliss 1975). This zone is characterized by weakly developed soils stabilized by alpine turf under natural conditions, but sensitive to erosion under high grazing, turf extraction, and harvesting of alpine shrubs for fuelwood (Byers 2005; Nagy and Grabherr 2009). The lower alpine zone contrasts markedly with the highest alpine or nival zone, with bare and rocky ground with occasional mosses, lichens, and dwarfed vascular plants, similar to that found in high Arctic or polar deserts.

Similar to treeline, the highest elevations reached by vascular plants (5,800–6,400 m/19,100–20,100 ft) are located in the subtropics between 20° and 30° latitude in the Himalayas and Andes (Zimmermann 1953; Webster 1961; Swan 1967). In the equatorial tropics, vascular plants do not occur above about 5,200 m (17,100 ft) because of cloud cover and lower temperatures. The massive nature of the Himalayas and Andes also limits their exposure to moisture-laden air, resulting in higher snowlines and longer growing seasons (Swan 1967). Above the elevational limits of flowering plants is the eolian zone (Swan 1961, 1963a, 1963b, 1967; Mani 1962; Papp 1978; Spalding 1979). Life in this zone is dependent on organic materials—pollen, spores, seeds, dead insects, and plant fragments—carried to high altitudes by the wind (Edwards 1987). A similar life zone is thought to exist in the extreme desert and polar regions.

Floristics

The size and composition of alpine floras is a function of many factors including latitude, climate, size of mountain mass, continuity or isolation with respect to other mountains or the arctic tundra, age of the tundra surface, and the local and regional vegetative history (Nagy and Grabherr 2009). Nonetheless, alpine floras share the important characteristic of becoming less arctic-like with decreasing latitude and evolutionary history (e.g., Cooper 1989). Approximately 1,500 species make up the arctic flora; about 500 of these reach middle latitudes in the northern hemisphere (Murray 1995). A handful of the genera found in the Arctic also extend into the highest levels in the tropics, including *Draba, Arenaria, Arabis, Potentilla,* and some grass species such as *Poa* spp. and *Trisetum spicatum.* These species are bipolar in their distributions, extending from the Arctic to the sub-Antarctic. A slightly larger number of lichens and mosses are bipolar (Billings and Mooney 1968).

In some cases, the floras of midlatitude mountains consist mainly of arctic species, as on Mount Washington in New Hampshire, where most of the 75 alpine species also occur in the Arctic (Bliss 1963). In other mountains, such as the Sierra Nevada of California, the alpine flora contains over 600 species, but only about 20 percent have arctic affinities; most species are related to those found in the adjacent Mojave Desert, Great Basin, or Rocky and Cascades ranges (Went 1948; Chabot and Billings 1972). The Rocky Mountains occupy an intermediate position with respect to arctic and alpine species. The Beartooth Mountains on the Montana–Wyoming border contain 192 species, 50 percent of which also occur in the Arctic (Billings 1974b). Farther south in Colorado, the number of alpine species increases to about 300, while the proportion of arctic species decreases to 40 percent or less (Bliss 1962).

A similar pattern is present in Eurasia, where the Scandinavian and Siberian mountains have the highest number of arctic–alpine species, with approximately 233 of the 463 species (50 percent) of the alpine species in Scandinavia being circumpolar in their distribution (Virtanen 2003). This contrasts with the ca. 750–800 alpine species found in the Alps (Ozenda and Borel 2003); where about 35 percent of the flora has an arctic affinity. In the Altai Mountains of Central Asia, 40 percent of the 300 alpine species have an arctic affinity. Moving progressively north, this proportion increases to 50 percent in the Sayan Mountains, and 60 percent in the Stanovoi Ranges (Major and Bamberg 1967a).

The alpine flora of the central Asian mountains—Pamirs, Tien Shan, Altai, Himalaya, Karakoram, Hindu Kush, and Caucasus—is markedly larger and more distinctive than that of any other northern hemisphere mountain region (Agakhanjanz and Breckle 1995). This region was proportionately less glaciated than most other midlatitude mountains during the Pleistocene, leading to a greater accumulation of species through prolonged immigration and natural selection. The Himalayan region is believed to have served as a refugium for both plants and animals throughout the Pleistocene, and many species of the alpine tundra may have originated there (Hoffmann and Taber 1967). Although no exact figures are available, early estimates of the number of alpine plants in the high Central Asian ranges exceeded 1,000, with approximately 25–30 percent of these species also occurring in the Arctic (Major and Bamberg 1967a). More recently, Ozenda (2002) estimated that there are 3,000 Himalayan alpine plant species.

Tropical alpine floras are typically insular and disjunct. There are approximately 300 alpine species on the high mountains of East Africa; 80 percent of these are endemic. These Afroalpine plants have evolved largely through speciation from the surrounding lowland vegetation, and thus differ from most other midlatitude floras (Hedberg 1965, 1969, 1971, 1995). Other regional floras that have experienced a similar developmental process include those of the Andes and the mountains of Malaysia, Indonesia, and New Guinea (Van Steenis 1934, 1935, 1962, 1972; Troll 1958; Cuatrecasas 1968).

Nested within the broader context of regional climate and geology (e.g., Moser et al. 2005), local environmental factors largely determine alpine habitat and floristic diversity. These factors—including microclimatic conditions (temperature and moisture), soil characteristics (texture, pH, and nutrient status), snowpack (depth and duration), disturbance (grazing), and plant interactions (e.g., Callaway et al. 2002; Boyce et al. 2005; Volanthen et al. 2006; Pierce et al. 2007; Löffler and Pape 2008)—create mosaics of habitat types and local environmental gradients.

Phylogeography of Alpine Plants

Phylogeography is an emerging topic of biogeographic research that combines the spatial and environmental context of historical biogeography with the genetic composition of individual plant species. Based on many of the ideas developed by early biogeographers such as Marie Brockmann-Jerosch (Holderegger et al. 2011), recent advances in genetic research have deepened our ability to explore biogeographic patterns and origins of alpine plants (e.g., Comes and Kadereit 2003) and ecologically related organisms (e.g., Garnier et al. 2004). This merger of historical biogeography (dispersal and migration) combined with the discrete spatial nature of alpine environments (isolation), and a genetics-based alpine plant *genealogy* (the study of evolutionary lineages) provides researchers with an exciting new means to test a broad range of biogeographic assumptions and theories essential to our understanding of the past, present, and future distributions of alpine plants. Moreover, the recent availability of genetic information improves the accuracy of dating for significant evolutionary events related to the origin, migration, and diversification of alpine floras. As a result, biogeographers are now better able to relate episodes of plant migration and dispersal (e.g., Comes and Kadereit 2003), isolation (Schneeweiss and Schönswetter 2011), disjunctions (e.g., Schönswetter et al. 2003), and genetic diversification (e.g., Kadereit et al. 2004) to past climatic change (e.g., Paun et al. 2008) and orogenesis (Hughes and Eastwood 2006).

The advent of modern genetics has also provided new tools that allow researchers to test important hypotheses pertaining to biogeography of alpine plants. Phylogeography now allows us to better assess (1) the genetic contribution of arctic versus low-elevation

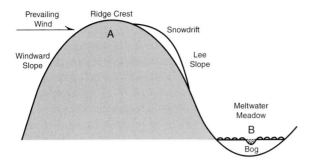

FIGURE 7.10 Environmental conditions distributed along a microto-pographic gradient. (Billings 1973.)

FIGURE 7.11 (A) Insect-pollinated globeflowers (*Trollius laxus*) and marsh marigolds (*Caltha leptosepala*) (photo by K. S. Hadley). (B) Alpine cushion plants alpine forget-me-nots (*Eritrichium aretioides*) and alpine phlox (*Phlox sibirica*) (photo by K. S. Hadley). (C) Glacier lily (*Erythronium grandiflorum*) growing through a late snowbank (photo by L. W. Price).

plants in specific alpine areas (e.g., Holderegger and Thiel-Egenter 2009); (2) the relative importance of immigration versus *in situ* diversification of alpine plant species (e.g., Paun et al. 2008); (3) the role of geology, soils, and other physiographic factors in promoting alpine plant biodiversity (e.g., Alvarez et al. 2009); and (4) the role of safe sites harboring plants during unfavorable conditions, such as habitat refugia and *nunataks* (isolated, nonglaciated areas that remained above Pleistocene glaciers and ice sheets) during oscillating periods of cool glacial and warm interglacial periods (Schneeweiss and Schönswetter 2011). Carefully designed phylogeography research can also address several of these types of questions within a single study. For example, recent DNA-based comparisons by Schönswetter et al. (2006) provide a compelling explanation of the divergent distributions between two ecologically similar, circumpolar arctic–alpine plant species resulting from postglacial diffusion, genetic drift, and long-distance dispersal.

Although most researched in Europe, alpine plant phylogeography has also been studied in Africa (e.g., Assefa et al. 2007; Kebede et al. 2007), Asia (e.g., Ikeda and Setoguchi 2007; Wang et al. 2008; Li et al. 2010), and South America (Hughes and Eastwood 2006), providing the potential to better understand how alpine plant distributions are historically linked at a hemispheric to global scale (e.g., Ehrich et al. 2007).

Environmental Gradients in Alpine Tundra

Close examination of alpine tundra reveals a variety of plant communities responding to different landscape features (e.g., Billings 1974b; Malanson et al. 2008). Many alpine plants are broadly distributed, yet vary considerably in their sensitivity to environmental conditions. Consequently, alpine plant communities often exhibit distinctive spatial patterns in response to steep environmental gradients over short distances.

The harshest environments in the alpine tundra result from the interaction of wind and snow (Billings and Mooney 1968). Exposed ridges, for example, are usually blown free of snow and offer little moisture or protection from wind (Fig. 7.10). Only lichens, mosses, and a

few prostrate cushion plants or dwarf shrubs are able to grow in such sites. Toe-slope positions and depressions can also present unfavorable growing conditions, if snow accumulation persists late into the season (Fig. 7.10). Midslope positions are often the most favorable and productive, affording the vegetation both winter protection and summer moisture. Thus, the pattern of snow accumulation and snowmelt is a primary factor in the establishment of alpine vegetative patterns (Billings and Bliss 1959; Billings 1973; Canaday and Fonda 1974; Webber et al. 1976; Boyce et al. 2005; Fig. 7.11a).

Under favorable microhabitat conditions, some vascular plants will grow above the lower boundary of the permanent snow zone. This generally occurs on rocky peaks where small crevices have collected sufficient fine materials to allow root growth (Fig. 7.11b). Miniature snow beds on these blocky surfaces protect these dwarfed plants during winter and provide a source of moisture the following summer. By taking advantage of such microhabitats, such snow-cranny plants reach exceptionally high elevations. Examples of these high-elevation plants include the miniature buttercups *Ranunculus glacialis,* in the Swiss Alps at 4,505 m (14,780 ft) (Körner 2011), and *R. grahami,* above the permanent snowline at 2,900 m (9,600 ft) in the New Zealand Alps (Billings and Mooney 1968). Ecological equivalents to these buttercups are the highest occurring vascular plants in the world, growing at approximately 6,400 m (20,100 ft) in the Himalaya.

Characteristics and Adaptations of Alpine Plants

The primary adaptations of alpine plants include their ability to survive low temperatures, short growing seasons, low nutrient availability, and surface instability. Consequently, alpine plant communities may exhibit higher levels of *facilitation* (the amelioration of extreme environmental conditions by neighboring plants) (but see Dullinger et al. 2007) and experience less resource competition than that experienced by plants in more moderate environments (Welden 1985; Choler et al. 2001; Callaway et al. 2002). Alpine plants exhibit many kinds of morphological, physiological, and ecological adaptations to extreme environmental conditions; only the most important are summarized here (Table 7.2).

LOW GROWTH FORM

One of the most striking characteristics of alpine plants is their low stature, which allows them to grow in more favorable microclimates. Wind is greatly diminished near the ground, resulting in less physical and transpirational stress. Temperatures are also higher near the surface because of the absorption and reradiation of solar radiation, and lower diffusion rates occur under the lower wind regime. An extreme example of differences between surface and air temperatures (30°C) occurs among the dense alpine carpets of azalea (*Loiseleuria procumbens*) in the Alps, where maximum canopy temperatures (47.1°C) combined with high canopy humidity (RH = 87 percent) create near-tropical microclimate conditions (Cernusca 1976). High temperature tolerance also varies by species, and is related to both growth form and habitat-related moisture availability (Buchner and Neuner 2003; Larcher et al. 2010). Similar to krummholz trees, the low growth form of alpine plants ensures that they remain snow-covered during the winter and the early growing season, when lethal frost can kill flowers (Inouye 2008) or entire plants during cold spells when temperatures drop below −5 to −8°C (Larcher et al. 2010).

A notable exception to the dominance of low growth forms in alpine environments is the giant rosette form characteristic of some plants in tropical mountains. These plants have a candle-like form that reaches heights of several meters. They include the giant *Senecios* and *Lobelias* of the East African volcanoes and convergent growth forms of *Espeletia* and *Lupinus* in the páramos of South America. The silversword (*Argyroxiphium sandwicense*) has a similar form and grows at high elevations in Hawaii (Troll 1958). Since tropical mountain environments allow metabolic processes to be carried out throughout the year, the principal adaptations of these large columnar plants are to protect the buds from nightly frost (Beck et al. 1982). They do this by developing large, woolly leaves that curve inward at night and cover the buds, by developing corky coverings, and by secreting liquid over the buds at critical times to prevent freezing (Beck 1994). The stem sheath, consisting of woolly and hairy leaves, also helps protect them from intense sunlight during the day (Hedberg 1965).

PERENNIALS

Most (98 percent) alpine plants in mid- to high latitudes are perennials, because of their greater energy efficiency during the cool, brief growing season (Billings and Mooney 1968). Whereas annual plants need to complete their entire life cycle (germinate, flower, and fruit) during the short growing season, perennial plants avoid germination and initial growth processes and begin to metabolize quickly once conditions permit. The lack of early life cycle stages among perennials presents the further advantage of being more hardy and resistant to the vagaries of the environment than young and immature annual plants. Another adaptation that allows perennials to complete their life cycle rapidly is the development of preformed shoots and flower buds (Billings 1974a; Moser et al. 1997). Many of these plants are also evergreen, thus eliminating the need to grow new leaves before they begin to photosynthesize during the growing season and creating additional biomass for energy reserves.

Tundra plants develop relatively large and extensive root or rhizomatous systems, with two to six times more biomass below the ground than above it (Daubenmire 1941; Billings 1974a; Webber and May 1977; Fig. 7.12). In some plants, such as the sedge *Carex curvula* in the Alps, root longevity can result in a below to above-ground biomass ratio as high as 18:1 (Grabherr 1997), and these plants can persist for hundreds to thousands of years (Grabherr et al. 1978; Steinger et al. 1996) as confirmed by DNA fingerprinting that detected clone systems (De Witte et al. 2012). These rhizome–root systems provide important food reserves in the form of carbohydrates and starches that the plants draw from during their rapid initial growth in the spring. Large root systems also enhance water uptake under dry alpine conditions. Moisture stress in many alpine areas is further substantiated by the presence of several other plant adaptations, including thick waxy leaves,

TABLE 7.2
Global Growth Form Classification of Alpine Plants

Growth Form	Structure	Key Traits	Distribution	Example
Shrubs (Nanophanerophytes)—Woody Species 0.5 to 6m Height				
Krummholz	Environmentally shaped (phenotypes) multistemmed, prostrate forms of subalpine trees	Wind-shaped trees requiring snow protection of canopy to avoid cuticle damage and desiccation	Midlatitude and boreal mountains	*Picea engelmannii* Rocky Mountains
Needle-leaved	Genetically fixed (genotypes) multistemmed, prostrate trees occurring as clonal populations with dense canopies	Requires snow protection to minimize cuticle damage and desiccation	Midlatitude and boreal mountains	*Pinus pumila* East Asia
Broad-leaved	Erect to prostrate deciduous shrubs; tolerant of heavy snow loads	Shed leaves in seasonally extreme climates to reduce frost damage	Tropical (evergreen), midlatitude and boreal mountains (deciduous)	*Alnus alnobetula* Eurasian Mountains
Giant Rosettes	Erect, soft-stemmed plants with wide pith; age can exceed 300 years	Night bud development and antifreeze compounds	Equatorial but absent in Southeast Asia	*Dendrosenecio adnivalis* East Africa
Scleromorphic	Shrubs with small, hard leaves or pubescent (hairy), soft leaves; other species are leafless with green branches or succulents (e.g., cacti) that occur up to lower alpine	Lack structural characteristics for frost protection; many species adapted for efficient cooling by transpiration	Subtropical and continental midlatitude mountains	*Spartocytisus supranubius* Teide Teberufe
Dwarf Shrubs (Chamaephytes)—Woody Species <0.5m Height				
Erect	Woody plants >0.1 m height, forming dense canopies	Require snow for frost protection and mycorrhizae to enhance nutrient uptake	Globally distributed in lower alpine zone, with Ericaceae a prominent family	*Rhododendron ferrugineum* Alps and Pyrenees
Prostrate to Semi-erect	Woody plants <0.2 m height, forming dense mats and carpets with creeping branches; present on exposed sites	Canopy structure decouples plant from temperature of free air; frost tolerant	Global distribution; rare in the tropics	*Dracophyllum muscoides* New Zealand Alps

TABLE 7.2 (continued)

Growth Form	Structure	Key Traits	Distribution	Example
Cushion	Woody plants <1 m tall, occurring as a dense canopy cushion	Growth form decouples plant from free-air temperatures; some species absolutely frost resistant; internal nutrient cycling	Common but most spectacular in the Andes and New Zealand	*Azorella compacta* Andes
Spiny Cushion	<0.5 m tall plants with a spiny form, more loosely packed than typical cushion shrubs	Heat-collecting but open canopy that decreases overheating	Common in Mediterranean mountains	*Echinacea anthylloides* Sierra Nevada, Spain
Woody-based (suffruticose)	Shrub-like or creeping canopies that die back in autumn	Woody-based plants protected by winter snow in seasonal climates	Widespread in seasonal climates	*Gypsophila repens* Alps
Succulent	<0.5 m tall plants occurring as small, single, or cluster forming plants with high moisture-storing capacity	Frost-sensitive plants protected by dense cover of hairs	Predominantly occur in subtropical mountains	*Austrocylindropuntia floccosa* Bolivian Andes
Forbs—Nonwoody Flowering Plants (Hemicryptophytes)				
Tall "Megaforbs"	>0.5 m (up to 4 m) tall forbs with leafy stem	Protected by snow; require nutrient surplus	Predominantly in warm temperate mountains	*Rheum nobile* Himalaya
Medium-Height Forbs	>0.25 m, <0.5 m erect plants with mostly leaved stems with or without rosettes	Protected by snow	Seasonal climates, rare in equatorial and subtropical mountains	*Phacelia sericea* Rocky Mountains
Small Forbs	<0.25 m erect stems mostly with leaves; with or without rosettes; caulescent (stemmed) or acaulescent (stemless) growth	Lack frost protection by structural means	Predominantly in midlatitude mountains	*Pedicularis oederi* Holarctic mountains
Dwarf Forbs	<0.08 m; mostly with rosettes, forming loose clusters; caulescent, some acaulescent; with or without leaves at stem	Exploit favorable conditions close to surface	Upper alpine/nival zone of midlatitude, arctic, and boreal mountains	*Draba aizoides* Alps

Graminoids (Hemicryptophytes)

Tall	0.5 to 2 m tall; occurs as tussocks with densely packed stems or dense stands (e.g., bamboo)	Snow-protected in seasonal climates	Subtropical to midlatitude mountains	*Chionochloa macra* New Zealand *Yushania niitakayamensis* Taiwan
Medium Stature	<0.5 m tall; dominant in alpine meadows; narrow to broad leaved	Meristem protected by leaves and leaf sheaths growing in clonal populations with high longevity (>1,000 years)	Midlatitude mountains	*Astelia alpine* Australian Alps
Short Stature	<0.10 m tall, forming from turf up to nival belt where they persist as "eternal" clonal populations; possess narrow- to broad-leaves and a creeping or densely packed	Meristem protected by leaves and leaf sheaths	Uppermost elevations; predominantly in midlatitude mountains	*Festuca halleri* Alps

Geophytes and Annual Plants

Bulb-Forming Plants	Forb-like above ground with onions or bulbs more or less deep in the soil	Buds protected in the soil; bulbs serve as an organ for water and nutrient storage	Predominantly found in subtropical and Mediterranean mountains	*Haberbaria dilatata* Appalachians

Annual (Ephemeral) Plants

Annual or Ephemeral Plants	Small plants with short life cycles	Avoid environmental stress as seeds	Rare but most common in midlatitude and subtropical mountains	*Gentiana nivalis* Alps

NOTE: Modified from Nagy and Grabher 2009.

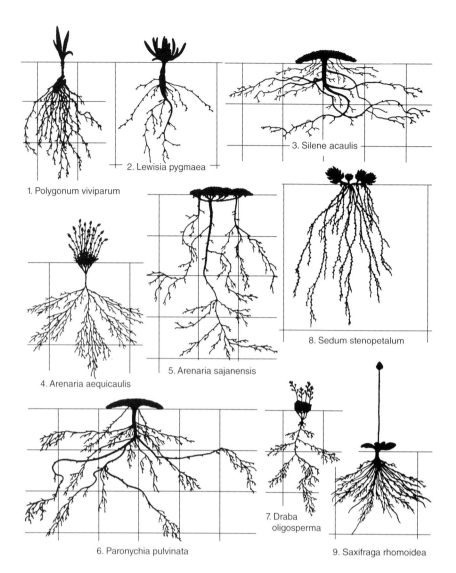

1. Polygonum viviparum

2. Lewisia pygmaea

3. Silene acaulis

4. Arenaria aequicaulis

5. Arenaria sajanensis

8. Sedum stenopetalum

6. Paronychia pulvinata

7. Draba oligosperma

9. Saxifraga rhomoidea

FIGURE 7.12 Silhouettes of alpine plants illustrating the relative size of their root systems relative to the above-ground shoot areas. (From Daubenmire 1941.)

corky bark, succulence, and high osmotic pressures, which enable the roots to take up any available ground moisture (Tranquillini 1964).

REPRODUCTION

The survival of any species ultimately depends on its ability to reproduce successfully. This becomes especially critical in harsh alpine environments, where growing conditions can be unfavorable for several years. Flowering and fruiting among alpine plants are often impaired, resulting in few or no viable seeds and poor conditions for germination. While this could result in highly variable vegetation patterns from year to year, this is rarely the case, because alpine plants are mostly perennials and often reproduce vegetatively or form extensive clonal populations.

Insect pollination is problematic for alpine plants, because low temperatures confine insect activity to brief periods during the summer (Billings 1974a). Some insect pollinators, such as bumblebees, are active only on sunny days

when the air temperature is 10°C (50°F) or higher (Macior 1970). As a result, at altitudes above 4,000 m (13,200 ft) in the Himalayas, bee-pollinated flowers disappear and are replaced by those pollinated by other insects (Mani 1962). Hummingbirds (or sunbirds in East Africa) are important in tropical mountains (Cruden 1972; Carpenter 1976) and, to a limited extent, midlatitude mountains where daytime temperatures exceed a minimum physiological threshold. The most important pollinators in mid- and high-latitude mountains are bees, flies, butterflies, and moths, but some ant species also act as pollinators (Puterbaugh 1998).

The comparatively large flower size and range of colors commonly associated with flowering alpine plants (relative to their small stems and short stature) may have several explanations, but it is at least partially a function of their ability to attract insects (Fig. 7.11c). An exception to this tendency occurs in New Zealand, where the alpine flowers are small, flat, and white and yellow in color. These traits appear to be related to pollination by flies and other small insects in the absence of long-tongued bees in New Zealand (Heine 1937; Mark and Dickinson 1997). Although most alpine plants are insect pollinated, wind and self-pollination become increasingly important under extreme environmental conditions. This alternative pollination strategy requires less energy expenditure to support elaborate adaptations among the plants and insects involved in insect pollination (Hocking 1968; Macior 1970; Heinrich and Raven 1972; Hickman 1974) and helps to compensate for the decreasing number of pollinators as environmental conditions become more extreme. A negative consequence of wind and self-pollination is less genetic diversity, which may be important for long-term fitness of the species (Bliss 1962), a tendency partially offset by a higher frequency of plants with multiple sets of chromosomes, or polyploids (Löve and Löve 1967).

Viable seed production following pollination depends largely on environmental conditions. Severe freezes or snowstorms during blooming can reduce seed set or inhibit germination until conditions become more favorable. Consequently, most alpine tundra plants have alternative methods of reproduction, the most common being through *rhizomes*, root-like stems that extend away from the center of the plant and are capable of growing new shoots. This process can result in single plants occupying several square meters, giving the appearance of many individual plants. This adaptation is advantageous where frost activity and mass wasting can lead to the partial destruction of the plant but leave the remainder to reproduce and spread.

Several other methods of vegetative reproduction, including layering, stolons, apomixis, and vivipary, are prevalent in the alpine tundra. These modes of reproduction become more prevalent as seed production and germination become less reliable with increasingly severe environmental conditions (Billings and Mooney 1968). Plants relying on vegetative reproduction may continue to produce seeds but rely on a variety of different reproductive strategies.

PLANT PRODUCTIVITY

Despite living under extreme environmental conditions and their reduced stature, alpine plants are highly productive during the growing season. Above-ground biomass production in midlatitude alpine tundra has been measured to average 0.20 to 0.60 g/m^2/day (0.0007 to 0.002 oz/ft^2/day) over the entire year, or 1 to 3 g/m^2/day (0.035 to 0.105 oz/ft^2/year) during the 30–70 day growing season (Bliss 1962). Since root biomass is commonly two to six times greater than shoot biomass, estimates of total productivity would be approximately three times greater (Billings and Mooney 1968; Webber 1974; Webber and May 1977). These annual productivity rates are far less than for temperate environments and similar to those found in desert environments. Daily growing season productivity rates, however, show that alpine tundra productivity exceeds that found in many temperate environments (Bliss 1962, 1966; Scott and Billings 1964; Bliss and Mark 1974; Webber 1974).

The key to the relatively high productivity rates of alpine plants is their adaptation to low temperatures and their ability to metabolize—at low rates—during subfreezing temperatures. This allows plant growth throughout the short growing season. During this period, photosynthates, the chemical products of photosynthesis, are initially used to produce plant biomass or later stored as energy reserves. This temporal allocation of resources is illustrated in the Alps by the high daily growth rates of *Carex curvula* during the first half of the growing season, followed by subsequent storage of starch and lipids in rhizomes (Grabherr et al. 1978).

Some alpine plants can also photosynthesize at temperatures as low as −6°C (21°F) (Billings 1974a), albeit at greatly diminished rates. Other alpine plants appear capable of withstanding frost conditions throughout their life cycle (Tsukaya and Tsuge 2001). While these characteristics improve the probability of survival and successful reproduction for some alpine plants, photosynthetic gains during frost periods are generally minimal, and frost damage during flowering often results in high stem, leaf, and flower mortality (e.g., Inouye 2008; Larcher et al. 2010).

Other adaptations of alpine plants include their rapid metabolic rates and their ability to use food reserves stored in old leaves and large root systems during initial spring growth. These reserves can then be replenished later in the season. Research on the dwarf shrub *Loiseleuria*, however, suggests that only a small proportion of photosynthates may be mobilized (Larcher 1977) and that unused reserves stored in branches and the evergreen leaves are shed after 3–6 years and lost to the system (Grabherr et al. 1978).

In general, alpine plant productivity and biomass accumulation are higher in areas of mild climates than in cold climates (Kikvidzek et al. 2005). Growth rates of alpine plants are also influenced by the length of the growing season, as determined both by seasonal changes in solar radiation and temperature and by the local persistence of snowpack. In areas of early spring snowmelt, plants can develop relatively slowly, taking the entire growing season for the various phenological processes. Conversely, in areas of late snowmelt, growth processes are compressed and the entire life cycle must be completed within a few weeks (Billings and Bliss 1959; Rochow 1969; Körner 2003). Spatially, plant biomass, productivity, and nitrogen use increase over alpine topographic sequences from dry to wet meadows, where snow contributes seasonal moisture but does not shorten the growing season (Fisk et al. 1998).

Summary

Depending on one's scale of interest, the plant geography and ecology of mountain forests and alpine tundra can be viewed as part of a broad environmental continuum, or as a complex mosaic of unique vegetation types reflecting highly localized environmental conditions. At the larger scales, mountain vegetation manifests itself as elevation-defined vegetation zones responding most closely to attitudinally controlled changes in temperature and precipitation. Locally, however, plant species and community structures vary dramatically in response to topo- and microscale environmental conditions and disturbance history.

The plant species in any mountain region are also the product of migration, adaptation, and evolution. In some cases, including most midlatitude mountains, subalpine and alpine plants are recent arrivals following the melting of extensive Pleistocene glaciers. These plants represent pre-adapted species moving among similar (cool) environments, often recolonizing formerly occupied areas. In more isolated ranges, mountain plants have evolved from ancient floras and have remained local or regional endemics. These plants have adapted slowly over the millennia as the mountains themselves were created and uplifted. In still other cases, the mountain vegetation is composed of surrounding lowland species that have been able to adapt to the conditions at altitude and have migrated slowly to higher elevations. These too may have become isolated and no longer interact with their distant relatives in the lowlands, especially in tropical regions.

Relationships between individual species and their environment (*autecology*) are highly specialized among most alpine tundra species, reflecting their adaptation to extreme environmental conditions. In some ways, they are similar to "weedy" plants common in midlatitude lowlands. Both types of plants exhibit extreme hardiness, the ability to live under extreme environmental conditions, and the ability to pioneer new sites. Alpine plants differ from weedy species by their perennial life cycle and other adaptations that allow them to survive repeated periods of extreme differences in temperature and moisture. Although some species are highly adapted to local conditions and limited in their distribution, others have broad ecological amplitudes and wide geographical ranges, living successfully from the lowlands to the highlands.

The Future of Mountain Vegetation

Humans have played a major role in altering the structure and spatial patterns of mountain vegetation for thousands of years. Since prehistoric times, people have used fire for hunting and to clear land for agriculture. In the tropics, where these activities have been practiced longest, it is likely that few, if any, tropical timberlines are truly "natural" (e.g., Ellenburg 1979; Horn 1998; Young and Keating 2001). This may be illustrated by the predominance of the heath family, Ericaceae, near tropical timberlines and its ability to regenerate after fire (Sleumer 1965; Janzen 1973; Wesche et al. 2000). Ericaceae is also well represented in the subalpine and alpine zones of northern hemisphere mountains, in part because Native Americans regularly set fires in the upper forest to maintain and increase their harvests of *Vaccinium* spp. berries (Boyd 1999).

Humans have also altered the character of forests and timberline through logging, agricultural clearing, firewood collection, and charcoal production (e.g., Zimmerer and

Young 1998; Bolton 2005). Intensive livestock grazing has further eliminated young trees and hindered seedling reestablishment through soil compaction (Byers 2005). These and other types of resource use have resulted in the lowering of natural timberlines by tens to hundreds of meters, especially in parts of the Andes, Alps, and Himalayas that have supported large human populations for centuries (Costin 1959; Molloy et al. 1963; Budowski 1968; Pears 1968; Holtmeier 1973; Eckholm 1975; Plesnik 1973, 1978).

Global climate change and atmospheric pollution represent two of the greatest concerns regarding the stability and function of mountain ecosystems. Upward shifts in treeline, the subalpine–alpine and alpine–nival ecotones, and alpine plant distributions are well documented (Grabherr et al. 1994; Gian-Reto et al. 2005; Grabherr 2009; Díaz-Varela et al. 2010; Harsch et al. 2009). Other drivers also need to be considered—for example, shifts in precipitation patterns, increased intensity of ultraviolet radiation, acidic conditions, and CO_2 concentrations (e.g., Kareiva et al. 1993; Lesica and McCune 2004; Handa et al. 2005; Gottfried et al. 2012; Pauli et al. 2012). Such changes will undoubtedly occur across a range of spatial and temporal scales, transforming relationships among species and plant communities. At the landscape scale, this could result in changing spatial patterns within the forest–alpine ecotone (Baker and Weisberg 1997), the reassemblage of plant communities (e.g., Lloyd and Graumlich 1997), and increasing (e.g., Körner 1995; Jentsch and Beierkuhnlein 2003) or decreasing (Thuiller et al. 2005) species diversity. Beyond certain critical climate thresholds, mountain areas could experience shifts in vegetation zones similar to those during the Hypsithermal (cf. Delting 1968; Markgraf and Scott 1981, Hadley 1984), causing the potential loss of local or regional alpine areas (e.g., Peters and Darling 1985).

Human land use is another factor that will continue to play a critical role in shaping mountain vegetation. Local and regional land-use decisions may mitigate local habitat losses (e.g., Dirnböck et al. 2003), but climate change will create new and complex relationships between fire and human disturbance (e.g., Veblen et al. 1999, 2000; Jentsch and Beierkuhnlein 2003; Westerling et al. 2006) among forests already experiencing ecological stress related to drought, insect outbreaks, competition, and growth decline caused by air pollution (e.g., Miller et al. 1963; Savage 1994; van Mantgem et al. 2009).

Although mountain forests and alpine tundra have always functioned as dynamic ecosystems, it now appears that their future will include ecological changes occurring at historically unprecedented rates. Some of these changes will reflect natural variations in climate and disturbance processes, but it is now apparent that humans have become the dominant agent of change in mountain regions. What remains less certain is how mountain vegetation will respond to the impending environmental changes across spatial scales (e.g., Gosz 1993; Malanson et al. 2011) in response to the most recent and ongoing global "experiment."

References

Agakhanjanz, O., and Breckle, S. W. 1995. Origin and evolution of the mountain flora in Middle Asia and neighbouring mountain regions. In F. S. Chapin and C. Körner, eds., *Arctic and Alpine Biodiversity: Patterns, Causes and Ecosystem Consequences* (pp. 63–80). Berlin: Springer-Verlag.

Agee, J. K. 1993. *Fire Ecology of Pacific Northwest Forests.* Washington, DC: Island Press.

Alftine, K. J., and Malanson, G. P. 2004. Directional positive feedback and pattern at alpine tree line. *Journal of Vegetation Science* 15: 3–12.

Alvarez, N., Thiel-Egenter, C., Tribsch, A., Holderegger, R., Manel, S., Schönswetter, P., Taberlet, P., Brodbeck, S., Gaudeul, M., Gielly, L., Küpfer, P., Mansion, G., Negrini, R., Paun, O., Pellecchia, M., Rioux, D., Schüpfer, F., Van Loo, M., Winkler, M., Gugerli, F., and IntraBioDiv Consortium. 2009. History or ecology? Substrate type as a major driver of spatial genetic structure in alpine plants. *Ecology Letters* 12: 632–640.

Alvarez-Uria, P., and Körner, C. 2007. Low temperature limits of root growth in deciduous and evergreen temperate tree species. *Functional Ecology* 21: 211–218.

Arno, S. F., and Hammerly, R. P. 1984. *Timberline: Mountain and Arctic Frontier Forests.* Seattle: The Mountaineers.

Assefa, A., Ehrich, D., Taberlet, P., Nemomissa, S., and Brochmann, C. 2007. Pleistocene colonization of afro-alpine "sky islands" by the arctic-alpine *Arabis alpina.* *Heredity* 99: 133–142.

Aulitzky, H. 1967. Significance of small climatic differences for the proper afforestation of highlands in Austria. In W. E. Sopper and H. W. Lull, eds., *International Symposium on Forest Hydrology* (pp. 639–653). Oxford: Pergamon.

Bader, M. Y., van Geloof, I., and Rietkerk, M. 2007. High solar radiation hinders tree regeneration above the alpine treeline in northern Ecuador. *Plant Ecology* 191: 33–45.

Baker, W. L., and Weisberg, P. J. 1997. Using GIS to model tree population parameters in the Rocky Mountain National Park forest-tundra ecotone. *Journal of Biogeography* 24: 513–526.

Batllori, E., Camarero, J. J., Ninot, J. M., and Gutiérrez, E. 2009. Seedling recruitment, survival and facilitation in alpine *Pinus uncinata* tree line ecotones: Implications and potential responses to climate warming. *Global Ecology and Biogeography* 18: 460–472.

Barry, R. G. 2008. *Mountain Weather and Climate.* 3rd ed. Cambridge, UK: Cambridge University Press.

Bauer, H., Plattner, K., and Volgger, W. 2000. Photosynthesis in Norway spruce seedlings infected by the needle rust *Chrysomyxa rhododendri.* *Tree Physiology* 20: 211–216.

Beals, E. 1969. Vegetational change along altitudinal gradients. *Science* 165: 981–985.

Beck, E. 1994. Cold tolerance in tropical alpine plants. In P. W. Rundel, A. P. Smith, and F. C. Meinzer, eds., *Tropical Alpine Environments* (pp. 77–110). Cambridge, UK: Cambridge University Press.

Beck, E., Senser, M., Scheibe, R., Steiger, H. M., and Pongratz, P. 1982. Frost avoidance and freezing tolerance in Afroalpine "giant rosette" plants. *Plant, Cell and Environment* 5: 216–222.

Becwar, M. R., Rajashekar, C., Hansen-Bristow, K. J., and Burke, M. J. 1981. Deep undercooling of tissue water and winter hardiness limitations in timberline flora. *Plant Physiology* 68: 111–114.

Bekker, M. F. 2005. Positive feedback between tree establishment and patterns of subalpine forest advance, Glacier National Park, Montana, U.S.A. *Arctic, Antarctic, and Alpine Research* 37: 97–107.

Belsky, A. J., and Blumenthal, D.M. 1997. Effects of livestock grazing on stand dynamics and soils in upland forests of the Interior West. *Conservation Biology* 11: 315–327.

Benedict, J. B. 1984. Rates of tree-island migration, Colorado Rocky Mountains, USA. *Ecology* 65: 820–823.

Billings, W. D. 1969. Vegetational pattern near alpine timberline as affected by fire–snowdrift interactions. *Vegetatio* 19: 192–207.

Billings, W. D. 1973. Arctic and alpine vegetations: Similarities, differences, and susceptibility to disturbance. *BioScience* 23: 697–704.

Billings, W. D. 1974a. Arctic and alpine vegetation: Plant adaptations to cold summer climates. In J. D. Ives and R. G. Barry, eds., *Arctic and Alpine Environments* (pp. 403–443). London: Methuen.

Billings, W. D. 1974b. Adaptations and origins of alpine plants. *Arctic and Alpine Research* 6: 129–142.

Billings, W. D. 1978. Alpine phytogeography across the Great Basin. *Great Basin Naturalist Memoirs* 2: 105–117.

Billings, W. D. 1979. High mountain ecosystems: Evolution, structure, operation and maintenance. In P. J. Webber, ed., *High Altitude Geoecology* (pp. 97–122). AAS Selected Symposia 12. Boulder: Westview Press.

Billings, W. D. 2000. Alpine vegetation. In M. G. Barbour and W. D. Billings, eds., *North American Terrestrial Vegetation*, 2nd ed. (pp. 537–572). Cambridge, UK: Cambridge University Press.

Billings, W. D., and Bliss, L. C. 1959. An alpine snowbank environment and its effect on vegetation, plant development and productivity. *Ecology* 40: 388–397.

Billings, W. D., and Mark, A. F. 1957. Factors involved in the persistence of montane treeless balds. *Ecology* 38: 140–142.

Billings, W. D., and Mooney, H. A. 1968. The ecology of arctic and alpine plants. *Biological Reviews* 43: 481–529.

Bliss, L. C. 1956. A comparison of plant development in microenvironments of arctic and alpine tundras. *Ecological Monographs* 26: 303–337.

Bliss, L. C. 1962. Adaptations of arctic and alpine plants to environmental conditions. *Arctic* 15: 117–144.

Bliss, L. C. 1963. Alpine plant communities of the Presidential Range, New Hampshire. *Ecology* 44: 678–697.

Bliss, L. C. 1966. Plant productivity in alpine microenvironments. *Ecological Monographs* 36: 125–155.

Bliss, L. C. 1971. Arctic and alpine plant life cycles. *Annual Review of Ecology and Systematics* 2: 405–438.

Bliss, L. C. 1975. Tundra grasslands, herblands, and shrublands and the role of herbivores. *Geoscience and Man* 10: 51–79.

Bliss, L. C., and Mark, A. F. 1974. High-alpine environments and primary production on the Rock and Pillar Range, Central Otago, New Zealand. *New Zealand Journal of Botany* 12: 445–483.

Bolli, J. C., Rigling, A., and Bugmann, H. 2007. The influence of changes in climate and land-use on regeneration dynamics of Norway spruce at the treeline in the Swiss Alps. *Silva Fennica* 41: 55–70.

Bolton, G. H. 2005. *Forest Structure under Human Influence Near an Upper Elevation Village in Nepal.* Ph.D. dissertation, University of Arizona, Tucson.

Bonde, E. K. 1969. Plant disseminules in wind blown debris from a glacier in Colorado. *Arctic and Alpine Research* 1: 135–140.

Boyce, R. L., Clark, R., and Dawson, C. 2005. Factors determining alpine species distribution on Goliath Peak, Front Range, Colorado, U.S.A. *Arctic, Antarctic, and Alpine Research* 37: 88–96.

Boyd, R., ed. 1999. *Indians, Fire, and the Land.* Corvallis: Oregon State University Press.

Boysen-Jensen, P. 1949. Causal plant-geography. *Det Kongelige Danske Videnskabernes Selskabs Biologiske Meddelelser* 21: 1–19.

Brockmann-Jerosch, H. 1919. Baumgrenze und Klimacharakter. *Bericht der Schweizer Botanischer Gesellschaft* 26; reviewed in *Journal of Ecology* 8: 63–65.

Brown, J. H. 1971. Mammals on mountaintops: Nonequilibrium insular biogeography. *American Naturalist* 105: 467–478.

Buchner, O., and Neuner, G. 2003. Variability of heat tolerance in alpine plant species measured at different altitudes. *Arctic, Antarctic, and Alpine Research* 35: 411–420.

Buchner, O., and Neuner, G. 2011.Winter frost resistance of *Pinus cembra* measured *in situ* at the alpine timberline as affected by temperature conditions. *Tree Physiology* 11: 1217–1227

Budowski, G. 1968. La influencia humana en la vegetación natural de montañas tropicals Americanas. In C. Troll, ed., *Geoecology of the Mountainous Regions of the Tropical Americas.* Proceedings of the UNESCO Mexico Symposium, August 1966 (pp. 157–162). Bonn: Ferd. Dümmlers Verlag.

Burke, A. 2005. Vegetation types of mountain tops in Damaraland, Namibia. *Biodiversity and Conservation* 14: 1487–1506.

Byers, A. 2005. Contemporary human impacts on alpine ecosystems in the Sagarmatha (Mt. Everest) National Park, Khumbu, Nepal. *Annals of the Association of American Geographers* 95: 112–140.

Caccianiga, M., Andreis, C., Armiraglio, S., Leonelli, G., Pelfini, M., and Sala, D. 2008. Climate continentality and treeline species distribution in the Alps. *Plant Biosystems* 142: 66–78.

Cairns, D. M. 2001. Patterns of winter desiccation in krummholz forms of *Abies lasiocarpa* at treeline sites in Glacier National Park, Montana, USA. *Geografiska Annaler* 83A: 157–168.

Caldwell, M. M. 1970. The wind regime at the surface of the vegetative layer above timberline in the central Alps. *Zentralblatt für die gestamte forstand Hofwirtschaft* 87: 193–201.

Callaway, R. M. 1998. Competition and facilitation on elevation gradients in subalpine forests of the northern Rocky Mountains, USA. *Oikos,* 82: 561–573.

Callaway, R. M. 2007. *Positive Interactions and Interdependence in Plant Communities.* Dordrecht: Springer.

Callaway, R. M., Brooker, R. W., Choler, P., Klkvidze, Z., Lortie, C. J., Michalet, R., Paolini, L., Pugnaire, F. I., Newingham, B., Aschehoug, E. T., Armas, C., Klkodze, D., and Cook, B. J. 2002. Positive interactions among alpine plants increase with stress. *Nature* 471: 845–848.

Canaday, B. B., and Fonda, R. W. 1974. The influence of subalpine snowbanks on vegetation pattern, production, and phenology. *Bulletin of the Torrey Botanical Club* 101: 340–350.

Carlquist, S. 1974. *Island Biology.* New York: Columbia University Press.

Carpenter, F. L. 1976. *Ecology and Evolution of an Andean Hummingbird (Oreotrochilus estella).* Berkeley: University of California Press.

Cernusca, A. 1976. Energie- und Wasserhaushalt eines alpinen Zwergstrauchbestandes während einer Föhnperiode. *Archiv für Meteorologie, Geophysik und Bioklimatologie,* B24: 219–241.

Chabot, B. F., and Billings, W. D. 1972. Origins and ecology of the Sierran alpine flora and vegetation. *Ecological Monographs* 42: 163–199.

Choler, P., Michalet, R., and Callaway, R. M. 2001. Facilitation and competition on gradients in alpine communities. *Ecology* 82: 3295–3308.

Clausen, J. 1963. Treelines and germ plasma: A study in evolutionary limitations. *Proceedings of the National Academy of Science* 50: 860–868.

Comes, H. P., and Kadereit, J. W. 2003. Spatial and temporal patterns in the evolution of the flora of the European Alpine System. *Taxon* 52: 451–462.

Cooke, W. B. 1955. Subalpine fungi and snowbanks. *Ecology* 36: 124–130.

Coop, J. D., and Schoettle, A. W. 2009. Regeneration of Rocky Mountain bristlecone pine (*Pinus aristata*) and limber pine (*Pinus flexilis*) three decades after stand-replacing fires. *Forest Ecology and Management* 257: 893–903.

Cooper, D. J. 1989. Geographical and ecological relationships of the arctic-alpine vascular flora and vegetation, Arrigetch Peaks region, central Brooks Range, Alaska. *Journal of Biogeography* 16: 279–295.

Costin, A. B. 1957. The high mountain vegetation of Australia. *Australian Journal of Botany* 5: 173–189.

Costin, A. B. 1959. Vegetation in the high mountains relative to landuse. In A. Keast, R. L. Crocker, and C. S. Christian, eds., 1959. *Biogeography and Ecology in Australia* (pp. 427–451). Monographiae Biologicae 8. The Hague: Junk.

Costin, A. B. 1967. Alpine ecosystems of the Australasian Region. In H. E. Wright and W. H. Osburn, eds., *Arctic and Alpine Environments* (pp. 55–58). Bloomington: Indiana University Press.

Covington, W. W., and Moore, M. M. 1994. Southwestern ponderosa forest structure: Changes since Euro-American settlement. *Journal of Forestry* 92: 39–47.

Cruden, R. W. 1972. Pollinators in high-elevations ecosystems: Relative effectiveness of birds and bees. *Science* 176: 1439–1440.

Cuatrecasas, J. 1968. Páramo vegetation and its life forms. In C. Troll, ed., *Geoecology of the Mountainous Regions of the Tropical Americas.* Proceedings of the UNESCO Mexico Symposium, August 1966 (pp. 163–186). Bonn: Ferd. Dummlers Verlag.

Cuevas, J. G. 2000. Tree recruitment at the *Nothofagus pumilio* alpine timberline in Tierra del Fuego, Chile. *Journal of Ecology* 88: 840–855.

Cullen, L. E., Stewart, G. H., Duncan, R. P., and Palmer, J. G. 2001. Disturbance and climate warming influences on New Zealand *Nothofagus* tree-line population dynamics. *Journal of Ecology* 89: 1061–1071.

Daly, C. 1984. Snow distribution patterns in the alpine krummholz zone. *Progress in Physical Geography* 8: 157–175.

Daly, C., and Shankman, D. 1985. Seedling establishment by conifers above tree limit on Niwot Ridge, Front Range, Colorado, U.S.A. *Arctic and Alpine Research* 17: 389–400.

Daniels, L. D., and Veblen, T. T. 2003. Regional and local effects of disturbance and climate on altitudinal treelines in northern Patagonia. *Journal of Vegetation Science* 14: 733–742.

Daniels, L. D., and Veblen, T. T. 2004. Spatiotemporal influences of climate on altitudinal treeline in northern Patagonia. *Ecology* 85: 1284–1296.

Daubenmire, R. F. 1941. Some ecological features of the subterranean organs of alpine plants. *Ecology* 22: 370–378.

Daubenmire, R. F. 1954. Alpine timberlines in the Americas and their interpretation. *Butler University Botanical Studies* 11: 119–136.

DeChaine, E. G., and Martin, A. P. 2005. Marked genetic divergence among sky island populations of *Sedum lanceolatum* (Crassulaceae) in the Rocky Mountains. *American Journal of Botany* 92: 477–486.

Delcourt, H. R., and Delcourt, P. A. 2000. Eastern deciduous forests. In M. G. Barbour and D. W. Billings, eds., *North American Terrestrial Vegetation*, 2nd ed. (pp. 357–395). Cambridge, UK: Cambridge University Press.

Delting, L. 1968. *Historical Background of the Flora of the Pacific Northwest*. University of Oregon Natural History Museum Bulletin 13. Eugene: University of Oregon.

De Witte, L. C., Armbruster, G. F. J., Gielly, L., Taberlet, P., and Stöcklin, J. 2012. AFLP markers reveal high clonal diversity, repeated recruitment and extreme longevity of four arctic-alpine key species. *Molecular Ecology* 21: 1081–1097.

Díaz-Varela, R. A., Colombo, R., Meronib, M., Calvo-Iglesiasc, M. S., Buffonid, A., and Tagliaferri, A. 2010. Spatio-temporal analysis of alpine ecotones: A spatial explicit model targeting altitudinal vegetation shifts. *Ecological Modelling* 221: 621–633.

Dirnböck, T., Dullinger, S., and Grabherr, G. 2003. A regional impact assessment of climate and land-use change on alpine vegetation. *Journal of Biogeography* 30: 401–417.

Dullinger, S., Dirnböck, T., Köck, R., Hochbichler, E., Englisch, T., Sauberer, N., Grabherr, G. 2005. Interactions among tree-line conifers: Differential effects of pine on spruce and larch. *Journal of Ecology* 93: 948–957.

Dullinger, S., Kleinbauer, I., Pauli, H., Gottfried, M., Brooker, R., Nagy, L., Theurillat, J. P., Holten, J., Abdaladze, O., Benito, F. L., Borel, J. L., Coldea, G., Ghosn, D., Kanka, R., Merzouki, A., Klettner, C., Moiseev, P., Molau, U., Reiter, K., Rossi, G., Stanisci, A., Tomaselli, M., Unterlugauer, P., Vittoz, P., and Grabherr, G. 2007. Weak and variable relationships between environmental severity and small-scale co-occurrence in alpine plant communities. *Journal of Ecology* 95: 1284–1295.

Eckholm, E. P. 1975. The deterioration of mountain environments. *Science* 184: 1035–1043.

Edwards, J. S. 1987. Arthropods of alpine aeolian ecosystems. *Annual Review of Entomology* 32: 163–179.

Ehrich, D., Gaudeul, M., Assefa, A., Koch, M. A., Mummenhoff, K., Nemomissa, S., Intrabiodiv Consortium, and Brochmann, C. 2007. Genetic consequences of Pleistocene range shifts: Contrast between the Arctic, the Alps and the East African mountains. *Molecular Ecology* 16: 2542–2559.

Ellenburg, H. 1979. Man's influence on tropical mountain ecosystems in South America. *Journal of Ecology* 67: 401–416.

Elliott, G. P., and Kipfmueller, K. F. 2010. Multi-scale influences of slope aspect and spatial pattern on ecotonal dynamics at upper treeline in the southern Rocky Mountains, U.S.A. *Arctic Antarctic and Alpine Research* 42: 45–56.

Eyre, S. R. 1968. *Vegetation and Soils*. 2nd ed. Chicago: Aldine.

Faegri, K. 1966. A botanical excursion to Steens Mountain, S.E. Oregon, U.S.A. *Blyttia* 24: 173–181.

Ferguson, C. W. 1970. Dendrochronology of bristlecone pine, *Pinus aristata*: Establishment of a 7484-year chronology in the White Mountains of east-central California, U.S.A. In I. U. Olsson, ed., *Radiocarbon Variations and Absolute Chronology* (pp. 237–259). New York: John Wiley and Sons.

Ferreyra, M., Cingolani, A., Ezcurra, C., and Bran, D. 1998. High-Andean vegetation and environmental gradients in northwestern Patagonia, Argentina. *Journal of Vegetation Science* 9: 307–316.

Fickert, T., and Richter, M. 2004. Nomenclature of altitudinal belts in tropical high mountains: It's time for a revision. In J. C. Axmacher and T. Gollan, T., *Biodiversity and Dynamics in Tropical Ecosystems*. Program and Abstracts, Society for Tropical Ecology 17th Annual Conference, Jan C. Bayreuther Forum Ökologie, University of Bayreuth. 18–20 February, Vol. 105 (p. 270).

Fischer, T., Lentschke, J., Küfmann, C., Haas, F., Baume, O., Becht, M., and Schröder, H. 2013. High-mountainous permafrost under continental-climatic conditions: Actual results of different mapping methods and an empirical-statistical modeling approach for the northern Tien Shan (SE Kazakhstan). *Geophysical Research Abstracts* 15, EGU2013-13074.

Fisk, M. C., Schmidt, S. K., and Seastedt, T. R. 1998. Topographic patterns of above- and belowground production and nitrogen cycling in alpine tundra. *Ecology* 79: 2253–2266.

Franklin, J. F., and Dyrness, C. T. 1988. *Natural Vegetation of Oregon and Washington*. Corvallis: Oregon State University Press.

Franklin, J. F., Muir, W. H., Douglas, G. W., and Wiberg, C. 1971. Invasions of subalpine meadows by trees in the Cascade Range, Washington and Oregon. *Arctic and Alpine Research* 13: 215–224.

Garnier, S., Alibert, P., Audiot, P., Prieur, B., and Rasplus, J.-Y. 2004. Isolation by distance and sharp discontinuities in gene frequencies: Implications for the phylogeography of an alpine insect species, *Carabus solieri*. *Molecular Ecology* 13: 1883–1897.

Germino, M. J., and Smith, W. K. 1999. Sky exposure, crown architecture, and low-temperature photoinhibition in conifer seedlings at alpine treeline. *Plant, Cell and Environment* 22: 407–415.

Gersmehl, P. 1971. Factors involved in the persistence of southern Appalachian treeless balds. *Proceedings of the Association of American Geographers* 3: 56–61.

Gersmehl, P. 1973. Pseudo-timberline: The southern Appalachian grassy balds (summary). *Arctic and Alpine Research* 5(Pt. 2): A137–138.

Gian-Reto, W., Beißner, S., and Burga, C. A. 2005. Trends in the upward shift of alpine plants. *Journal of Vegetation Science* 16: 541–548.

Gieger, T., and Leuschner, C. 2004. Altitudinal change in needle water relations of *Pinus canariensis* and possible evidence of a drought-induced alpine timberline on Mt. Teide, Tenerife. *Flora* 199: 100–109.

Gillison, A. N. 1969. Plant succession in an irregularly fired grassland area, Doma Peaks region, Papua. *Journal of Ecology* 57: 415–428.

Gillison, A. N. 1970. Structure and floristics of a montane grassland-forest transition, Doma Peaks Region, Papua. *Blumea* 18: 71–86.

Goldblum, D., and Rigg, L. S. 2002. Age structure and regeneration dynamics of sugar maple at the deciduous/boreal forest ecotone, Ontario, Canada. *Physical Geography* 23: 115–129.

Gosz, J. R. 1993. Ecotone hierarchies. *Ecological Applications* 3: 369–376.

Gottfried, M., Pauli, H., Futschik, A., Akhalkatsi, M., Barancok, P., Benito Alonso, J. L., Coldea, G., Dick, J., Erschbamer, B., Fernández Calzado, M. R., Kazakis, G., Krajci, J., Larsson, P., Mallaun, M., Michelsen, O., Moiseev, D., Moiseev, P., Molau, U., Merzouki, A., Nagy, L., Nakhutsrishvili, G., Pedersen, B., Pelino, G., Puscas, M., Rossi, G., Stanisci, A., Theurillat, J.-P., Thomaselli, M., Villar, L., Vittoz, P., Vogiatzakis, I., and Grabherr, G. 2012. Continent-wide response of mountain vegetation to climate change. *Nature Climate Change* 2: 111–115.

Grabherr, G. 1997. The high-mountain ecosystems of the Alps. In F. E. Wielgolaski, ed., *Ecosystems of the World,* Vol. 3: *Polar and Alpine Tundra* (pp. 97–121). Amsterdam: Elsevier.

Grabherr, G. 2000. Biodiversity of mountain forests. In M. F. Price and N. Butt, eds., *Forests in Sustainable Mountain Development: A State of Knowledge Report for 2000.* (pp. 28–38). Wallingford, UK: CABI Publishing

Grabherr, G. 2009. Biodiversity in the high ranges of the Alps: Ethnobotanical and climate change perspectives. *Global Environmental Change* 19: 167–172

Grabherr, G., Mähr, E., and Reisigl, H. 1978. Nettoprimärproduktion und Reproduktion in einem Krummseggenrasen (*Caricetum curvulae*) der Ötztaler Alpen, Tirol. *Oecologia Plantarum* 13(3): 227–251.

Grabherr, G., Gottfried, M., and Pauli, H. 1994. Climate effects on mountain plants. *Nature* 396: 448.

Grace, J. 1989. Tree lines. *Philosophical Transactions of the Royal Society of London, Series B: Biological Sciences* 234: 233–243.

Grace, J., Berninger, F., and Nagy, L. 2002. Impacts of climate change on the tree line. *Annals of Botany* 90: 537–544.

Grant, M. C., and Mitton, J. B. 1977. Genetic differentiation among growth forms of Engelmann spruce and subalpine fir at treeline. *Arctic and Alpine Research* 9: 259–263.

Griggs, R. F. 1938. Timberlines in the northern Rocky Mountains. *Ecology* 19: 548–564.

Grubb, P. J. 1971. Interpretation of the Massenerhebung effect in tropical mountains. *Nature* 229: 44–45.

Grubb, P. J. 1974. Factors controlling the distribution of forest types on tropical mountains: New facts and a new perspective. In R. Flenley Jr., ed., *Altitudinal Zonation in Malesia* (pp. 1–25). University of Hull Department of Geography Miscellaneous Series. Hull, UK: University of Hull.

Grubb, P. J. 1977. Control of forest growth and distribution on wet tropical mountains. *Annual Review of Ecology and Systematics* 8: 83–107.

Habeck, J. R. 1969. A gradient analysis of a timberline zone at Logan Pass, Glacier Park, Montana. *Northwest Science* 43: 65–73.

Hadley, J. L., and Smith, W. K. 1983. Influence of wind exposure on needle desiccation and mortality for timberline conifers in Wyoming, U.S.A. *Arctic and Alpine Research* 15: 127–135.

Hadley, J. L., and Smith, W. K. 1986. Wind effects on needles of timberline conifers: Seasonal influence on mortality. *Ecology* 67:12–29.

Hadley, J. L., and Smith, W. K. 1987. Influence of krummholz mat microclimate on needle physiology and survival. *Oecologia* 73: 82–90.

Hadley, K. S. 1984. *A Biogeographic Interpretation of Vascular Alpine Plant Distributions within the Rocky Mountain Cordillera: Preliminary Investigations.* M.A. thesis, University of Wyoming, Laramie.

Hadley, K. S. 1987. Vascular alpine plant distributions within the central and southern Rocky Mountains. *Arctic and Alpine Research* 19: 242–251.

Hadley, K. S. 1999. Forest history and meadow invasion at the Rigdon Meadows Archaeological Site, western Cascades, Oregon. *Physical Geography* 20: 116–133.

Hadley, K. S., and Savage, M. 1996. Wind disturbance and the development of a near-edge forest interior, Marys Peak, Oregon Coast Range. *Physical Geography* 17: 47–61.

Handa, I. T., Körner, C., and Hättenschwiler, S. 2005. A test of the treeline carbon limitation hypothesis by in situ CO_2 enrichment and defoliation. *Ecology* 86: 1288–1300.

Harsch, M. A., and Bader, M. Y. 2011. Treeline form: A potential key to understanding treeline dynamics: The causes of treeline form. *Global Ecology and Biogeography* 20: 582–596.

Harsch, M. A., Hulme, P. E., McGlone, M. S., and Duncan, R. P. 2009. Are treelines advancing? A global meta-analysis of treeline response to climate warming. *Ecology Letters* 12: 1040–1049.

Hättenschwiler, S., and Smith, W. K. 1999. Seedling occurrence in alpine treeline conifers: A case study from the central Rocky Mountains, USA. *Acta Oecologica* 20: 219–224.

Hedberg, O. 1951. Vegetation belts of the East African mountains. *Svensk Botanisk Tidskrift* 45: 140–202.

Hedberg, O. 1961. The phytogeographical position of the afroalpine flora. *Recent Advances in Botany* 1: 914–919.

Hedberg, O. 1965. Afroalpine flora elements. *Webbia* 19: 519–529.

Hedberg, O. 1969. Evolution and speciation in a tropical high mountain flora. *Biology Journal of the Linnean Society* 1: 35–148.

Hedberg, O. 1971. Evolution of the afroalpine flora. In W. L. Stem, ed., *Adaptive Aspects of Insular Evolution* (pp. 16–23). Pullman: Washington State University Press.

Hedberg, O. 1995. Features of afroalpine plant ecology. *Acta Phytogeographica Suecica* 49: 1–144.

Heine, E. M. 1937. Observations on the pollination of New Zealand flowering plants. *Royal Society of New Zealand, Transactions* 67: 133–148.

Heinrich, B., and Raven, P. H. 1972. Energetics and pollination ecology. *Science* 176: 597–602.

Hessl, A. E., and Baker, W. D. 1997. Spruce and fir regeneration and climate in the forest-tundra ecotone of Rocky Mountain National Park, Colorado, U.S.A. *Arctic and Alpine Research* 29: 173–183.

Hickman, J. C. 1974. Pollination by ants: A low energy system. *Science* 148: 1290–1292.

Hiemstra, C. A., Liston, G. E., and Reiners, W. A. 2002. Snow redistribution by wind and interactions with vegetation at upper treeline in the Medicine Bow Mountains, Wyoming, U.S.A. *Arctic Antarctic and Alpine Research* 34: 262–273.

Hill, R. S., and Brodribb, T. J. 1999. Southern conifers in time and space. *Australian Journal of Botany* 47: 639–696

Hoch, G., and Körner, C. 2005. Growth, demography, and carbon relations of Polylepis trees at the world's highest treeline. *Functional Ecology* 19: 941–951.

Hocking, B. 1968. Insect-flower associations in the high arctic with special reference to nectar. *Oikos* 19: 359–388.

Hoffman, R. S., and Taber, R. D. 1967. Origin and history of Holarctic tundra ecosystems, with special reference to their vertebrate fauna. In H. E. Wright and W. H. Osburn, eds., *Arctic and Alpine Environments* (pp. 143–170). Bloomington: Indiana University Press.

Holderegger, R., and Thiel-Egenter, C. 2009. A discussion of different types of glacial refugia used in mountain biogeography and phylogeography. *Journal of Biogeography* 36: 476–480.

Holderegger, R., Thiel-Egenter, C., and Parisod, C. 2011. Marie Brockmann-Jerosch and her influence on Alpine phylogeography. *Alp Botany*, 121: 5–10.

Höllermann, P. W. 1978. Geoecological aspects of the upper timberline in Tenerife, Canary Islands. *Arctic and Alpine Research* 10: 365–382.

Holtmeier, F. K. 1973. Geoecological aspects of timberlines in northern and central Europe. *Arctic and Alpine Research* 5(Pt. 2): 45–54.

Holtmeier, F. K. 2003. *Mountain Timberlines: Ecology, Patchiness, and Dynamics*. Dordrecht: Kluwer.

Holtmeier, F. K., and Broll, G. 2005. Sensitivity and response of northern hemisphere altitudinal and polar treelines to environmental change at landscape and local scales. *Global Ecology and Biogeography* 14: 395–410.

Hope, G. S. 1976. The vegetational history of Mt. Wilhelm, Papua, New Guinea. *Journal of Ecology* 64: 627–664.

Horn, S. P. 1998. Fire management and natural landscapes in the Chirripó National Park, Costa Rica. In K. S. Zimmerer and K. R. Young, eds. *Nature's Geography: New Lessons for Conservation in Developing Countries* (pp. 123–146). Madison: University of Wisconsin Press.

Huggett, R. 1995. *Geoecology: An Evolutionary Approach*. London: Routledge.

Hughes, C., and Eastwood, R. 2006. Island radiation on a continental scale: Exceptional rates of plant diversification after uplift of the Andes. *Proceedings of the National Academy of Sciences of the United States of America* 103: 10334–10339.

Ikeda, H., and Setoguchi, H. 2007. Phylogeography and refugia of the Japanese endemic alpine plant, *Phyllodoce nipponica* Makino (Ericaceae). *Journal of Biogeography* 34: 169–176.

Inouye, D. 2008. Effects of climate change on phenology, frost damage, and floral abundance of montane wildflowers. *Ecology* 89: 353–362.

Ives, J. D. 1973. Studies in high altitude geoecology of the Colorado Front Range: A review of the research program of the Institute of Arctic and Alpine Research, University of Colorado. *Arctic and Alpine Research* 5: A67–A75.

Iwasa, Y., Kazunori, S., and Nakashima, S. 1991. Dynamic modeling of wave regeneration (Shimagare) in subalpine *Abies* forests. *Journal of Theoretical Biology* 152: 143–158.

Jacoby, G. C., D'Arrigo, R. D., and Davaajamts, T. 1996. Mongolian tree rings and 20th-century warming. *Science* 273: 771–773.

Jakubos, B., and Romme, W. H. 1993. Invasion of subalpine meadows by lodgepole pine in Yellowstone National Park, Wyoming, U.S.A. *Arctic and Alpine Research* 25: 382–390.

Janzen, D. H. 1967. Why mountain passes are higher in the tropics. *American Naturalist* 101: 233–249.

Janzen, D. H. 1973. Rate of regeneration after a tropical high elevation fire. *Biotropica* 5: 117–122.

Jentsch, A., and Beierkuhnlein, C. 2003. Global climate change and local disturbance regimes as interacting drivers for shifting altitudinal vegetation patterns. *Erdkunde* 57: 216–231.

Johnson, P., and Denton, R. 1975. Outbreaks of the western spruce budworm in the American northern Rocky Mountain area from 1922 through 1971. General Technical Report INT-20. Ogden, UT: U.S. Forest Service.

Kadereit, J. W., Griebeler, E. M., and Comes, H. P. 2004. Quaternary diversification in European alpine plants: Pattern and process. *Philosophical Transactions of the Royal Society of London, Series B* 359: 265–274.

Kareiva, P. M., Kingsolver, J. G., and Huey, R. B., eds. 1993. *Biotic Interactions and Global Change*. Sunderland, MA: Sinauer Associates Inc.

Kauffman, J. B. 1990. Ecological relationships of vegetation and fire in Pacific Northwest forests. In J. D. Walstad, S. R. Radosevich, and D. V. Sandberg, eds., *Natural and Prescribed Fire in the Pacific Northwest* (pp. 39–52). Corvallis: Oregon State University Press.

Kebede, M., Ehrich, D., Taberlet, P., Nemomissa, S., and Brochmann, C. 2007. Phylogeography and conservation genetics of a giant lobelia (*Lobelia giberroa*) in Ethiopian and tropical East African mountains. *Molecular Ecology* 16: 1233–1243.

Kikvidze, Z., Pugnaire, F. I., Brooker, R. W., Choler, P., Lortie, C. J., Michalet, R., and Callaway, R. M. 2005. Linking patterns and processes in alpine plant communities: A global study. *Ecology* 86: 1395–1400.

Körner, C. 1995. Alpine plant diversity: A global survey and functional interpretations. In F. S. Chapin and C. Körner, eds. *Arctic and Alpine Biodiversity: Patterns, Causes and Ecosystem Consequences* (pp. 49–62). Berlin: Springer-Verlag.

Körner, C. 1998. A re-assessment of high elevation treeline positions and their explanation. *Oecologia* 115: 445–459.

Memorias del Museo de Historia Natural "Javier Prado" (Lima) 18.

Peet, R. K. 1981. Forest vegetation of the Colorado Front Range. *Vegetatio* 45: 3–75.

Peet, R. K. 2000. Forests and meadows in the Rocky Mountains. In M. G. Barbour and W. D. Billings, eds., *North American Terrestrial Vegetation* (pp. 75–121). Cambridge, UK: Cambridge University Press.

Perry, D. A. 1994. *Forest Ecosystems*. Baltimore, MD: John Hopkins University.

Perry, J. P., Jr., Graham, A., and Richardson, D. M. 1998. The history of pines in Mexico and Central America. In D. M. Richardson, ed., *Ecology and Biogeography of Pinus* (pp. 137–149). Cambridge: Cambridge University Press.

Peters, R. L., and Darling, J. D. S. 1985. The greenhouse effect and nature reserves. *BioScience* 35: 707–717.

Peterson, D. L., Schreiner, E. G., and Buckingham, N. M. 1997. Gradients, vegetation and climate: Spatial and temporal dynamics in the Olympics Mountains, U.S.A. *Global Ecology and Biogeography Letters* 6: 7–17.

Pierce, S., Luzzaro, A., Caccianga, M., Ceriani, R. M., and Cerabolini, B. 2007. Disturbance is the principle a-scale filter determining niche differentiation, coexistence, and biodiversity in an alpine community. *Journal of Ecology* 95: 698–706.

Plesnik, P. 1973. La limite supérieure de la forêt dans les hautes Tatras. *Arctic and Alpine Research* 5(Pt. 2): A37–44.

Plesnik, P. 1978. Man's influence on the timberline in the West Carpathian Mountains, Czechoslovakia. *Arctic and Alpine Research* 10: 491–504.

Price, L. W. 1978. Mountains of the Pacific Northwest: A study in contrast. *Arctic and Alpine Research* 10: 465–478.

Price, L. W. 1981. *Mountains and Man: A Study of Process and Environment*. Berkeley: University of California Press.

Puterbaugh, M. N. 1998. The roles of ants as flower visitors: Experimental analysis in three alpine plant species. *Oikos* 83: 36–46.

Qian, H., White, P. S., Klinka, K., and Chourouzis, C. 1999. Phytogeographical and community similarities of alpine tundras of Changbaishan Summit and Indian Peaks, USA. *Journal of Vegetation Science* 10: 869–882.

Raven, P. H. 1973. Evolution of subalpine and alpine plant groups in New Zealand. *New Zealand Journal of Botany* 11: 177–200.

Resler, L. M. 2006. Geomorphic controls of spatial pattern and process at alpine treeline. *Professional Geographer* 58: 124–138.

Resler, L. M., and Tomback, D. F. 2008. Blister rust prevalence in krummholz whitebark pine: Implications for treeline dynamics. *Arctic, Antarctic and Alpine Research* 40: 161–170.

Richards, P. W. 1966. *The Tropical Rain Forest: An Ecological Study*. Cambridge, UK: Cambridge University Press.

Riebesell, J. F. 1982. Arctic-alpine plants on mountaintops: Agreement with island biogeography theory. *American Naturalist* 119: 657–674.

Rochefort, R. M., and Peterson, D. L. 1996. Temporal and spatial distribution of trees in subalpine meadows of Mount Rainier National Park. *Arctic and Alpine Research* 28: 52–59.

Rochow, T. F. 1969. Growth, caloric content, and sugars in *Caltha leptosepala* in relation to alpine snow melt. *Bulletin Torrey Botanical Club* 96: 689–698.

Sarmiento, O., and Frolich, L. M. 2002. Andean cloud forest tree lines: Naturalness, agriculture and the human dimension. *Mountain Research and Development* 22: 278–287.

Savage, M. 1994. Anthropogenic and natural disturbance and patterns of mortality in a mixed conifer forest in California. *Canadian Journal of Forest Research* 24: 1149–1159.

Schaack, C. G. 1983. The vascular alpine flora of Arizona. *Madroño* 30: 79–88.

Schmid, J. M., and Frye, R. H. 1977. Spruce beetle in the Rockies. U.S. Forest Service General Technical Report RM-49.

Schneeweiss, G. M., and Schönswetter, P. 2011. A re-appraisal of nunatak survival in arctic-alpine phylogeography. *Molecular Ecology* 20: 190–192.

Schönswetter, P., Popp, M., and Brochmann, C. 2006. Rare arctic-alpine plants of the European Alps have different immigration histories: The snow bed species *Minuartia biflora* and *Ranunculus pygmaeus*. *Molecular Ecology* 15: 709–720.

Schönswetter, P., Tribsch, A., Schneeweiss, G. M., and Niklfeld, H. 2003. Disjunctions in relict alpine plants: Phylogeography of *Androsace brevis* and *A. wulfeniana* (Primulaceae). *Botanical Journal of the Linnean Society* 141: 437–446.

Scott, D., and Billings, W. D. 1964. Effects of environmental factors on standing crop and productivity of an alpine tundra. *Ecological Monographs* 34: 243–270.

Scuderi, L. A. 1987. Late-Holocene upper timberline variation in the southern Sierra Nevada. *Nature* 325: 242–244.

Seastedt, T. R., and Adams, G. A. 2001. Effects of mobile tree islands on alpine tundra soils. *Ecology* 82: 8–17.

Shankman, D. 1984. Tree regeneration following fire as evidence of timberline stability in the Colorado Front Range, U.S.A. *Arctic and Alpine Research* 16: 413–417.

Shaw, C. H. 1909. Causes of timber line on mountains: The role of snow. *Plant World* 12: 169–181.

Sleumer, H. 1965. The role of Ericaceae in the tropical montane and subalpine forest of Malaysia. In *UNESCO Symposium on Ecological Research in Humid Tropics* (pp. 179–184). Paris: Pegetalion.

Smith, E. K., Resler, L. M., Vance, E. A., Carstensen, L. W. Jr., and Kolivras, K. N. 2011. Blister rust incidence in treeline whitebark pine, Glacier National Park, U.S.A.: Environmental and topographic influences. *Arctic, Antarctic, and Alpine Research* 43:107–117.

Smith, J. M. B. 1975. Mountain grasslands of New Guinea. *Journal of Biogeography* 2: 27–44.

Smith, J. M. B. 1977. Vegetation and microclimate of east- and west-facing slopes in the grasslands of Mt. Wilhelm, Papua, New Guinea. *Journal of Ecology* 65: 39–53.

Smith, J. M. B., and Cleef, A. M. 1988. Composition and origins of the world's tropicalpine floras. *Journal of Biogeography* 15: 631–645.

Spalding, J. B. 1979. The aeolian ecology of White Mountain Peak, California: Windblown insect fauna. *Arctic and Alpine Research* 11: 83–94.

Steinger, T., Körner, C., and Schmid, B. 1996. Long-term persistence in a changing climate: DNA analysis suggests very old ages of clones of alpine *Carex curvula*. *Oecologia* 105: 94–99.

Stevens, G. C., and Fox, J. F. 1991. The causes of treeline. *Annual Review of Ecology and Systematics* 22: 177–191.

Swan, L. W. 1961. The ecology of the high Himalayas. *Scientific American* 205: 68–78.

Swan, L. W. 1963a. Aeolian Zone. *Science* 140: 77–78.

Swan, L. W. 1963b. Ecology of the heights. *Natural History* 72: 22–29.

Swan, L. W. 1967. Alpine and aeolian regions of the world. In H. E. Wright and W. H. Osburn, eds., *Arctic and Alpine Environments* (pp. 29–54). Bloomington: Indiana University Press.

Taylor, A. H. 1990. Tree invasion in meadow of the Lassen Volcanic National Park. *Professional Geographer* 42: 457–470.

Taylor, A. H. 1992. Tree regeneration after bamboo die-back in Chinese *Abies-Betula* forests. *Journal of Vegetation Science* 3: 253–260.

Taylor, D. W. 1977. Floristic relationships along the Cascade-Sierran axis. *American Midland Naturalist* 97: 333–349.

Templeton, A. R. 1980. The theory of speciation via the founder principle. *Genetics* 94: 1011–1038.

Thuiller, W., Lavorel, S., Araújo, M. B., Sykes, M. T., and Prentice, I. C. 2005. Climate change threats to plant diversity in Europe. *Proceedings of the National Academy of Science* 102: 8245–8250.

Tinner, W., and Theurillat, J.-P. 2003. Uppermost limit, extent, and fluctuations of the timberline and treeline ecocline in the Swiss central Alps during the past 11,500 years. *Arctic, Antarctic, and Alpine Research* 35: 158–169.

Tomback, D. F., and Resler, L. M. 2007. Invasive pathogens at alpine treeline: Consequences for treeline dynamics. *Physical Geography* 28: 397–418.

Tranquillini, W. 1964. The physiology of plants at high altitudes. *Annual Review of Plant Physiology* 15: 345–362.

Tranquillini, W. 1979. *Physiological Ecology of the Alpine Timberline*. New York: Springer-Verlag.

Troll, C. 1958. Tropical mountain vegetation. Proceedings of the 9th Pacific Science Congress, Vol. 20 (pp. 37–45).

Troll, C. 1960. The relationship between climates and plant geography of the southern cold temperate zone and of the tropical high mountains. *Proceedings of the Royal Society of Medicine, London, Series B* 152: 529–532.

Troll, C. 1968. The cordilleras of the tropical Americas: Aspects of climatic, phytogeographical and agrarian ecology. In C. Troll, ed., *Geoecology of the Mountainous Regions of the Tropical Americas*. Proceedings of the UNESCO Mexico Symposium August 1966 (pp. 15–56). Bonn: Ferd. Dümmiers Verlag.

Troll, C. 1973. High mountain belts between the polar caps and the equator: Their definition and lower limit. *Arctic and Alpine Research* 5(Pt. 2): 19–28.

Tsukaya, H., and Tsuge, T. 2001. Morphological adaptation of inflorescences in plants that develop at low temperatures in early spring: The convergent evolution of "downy plants." *Plant Biology* 3: 536–543.

Van der Hammen, T. 1968. Climatic and vegetational succession in the equatorial Andes of Colombia. In C. Troll, ed., *Geoecology of the Mountainous Regions of the Tropical Americas*. Proceedings of the UNESCO Mexico Symposium August 1966 (pp. 187–194). Bonn: Ferd. Dümmiers Verlag.

van der Maarel, E. 1990. Ecotones and ecoclines are different. *Journal of Vegetation Science* 1: 135–138.

Vander Wall, S. B., and Balda, R. P. 1977. Coadaptations of the Clark's nutcracker and the piñon pine for efficient seed harvest and dispersal. *Ecological Monographs* 47: 89–111.

van Mantgem, P. J., Stephenson, N. L., Byrne, J. C., Daniels, L. D., Franklin, J. F., Fule, P. Z., Harmon, M. E., Larson, A. J., Smith, J. M., Taylor, A. H., and Veblen, T. T. 2009. Widespread increase of tree mortality rates in the western United States. *Science* 323: 521–524.

Van Steenis, C. G. G. J. 1934. On the origin of the Malaysian mountain flora, Part 1: Facts and statements of the problem. *Bulletin du Jardin Botanique de Buitenzorg* 13: 135–262.

Van Steenis, C. G. G. J. 1935. On the origin of the Malaysian mountain flora, Part 2: Altitudinal zones, general considerations and renewed statement of the problem. *Bulletin du Jardin Botanique de Buitenzorg* 13: 289–417.

Van Steenis, C. G. G. J. 1961. An attempt towards an explanation of the effect of mountain mass elevation. *Proceedings Koninklijke Nederlandse Akademie van Wetenschappen, Series C* 64: 435–442.

Van Steenis, C. G. G. J. 1962. The mountain flora of the Malaysian tropics. *Endeavor* 21: 183–193.

Van Steenis, C. G. G. J. 1964. Plant geography of the mountain flora of Mt. Kinabalu. *Proceedings of the Royal Society of London, Series B* 161: 7–38.

Van Steenis, C. G. G. J. 1972. *The Mountain Flora of Lava*. Amsterdam: Brill.

Veblen, T. T. 1982. Growth patterns of *Chusquea* bamboos in the understory of Chilean *Nothofagus* forests and their influences in forest dynamics. *Bulletin of the Torrey Botanical Club* 109: 474–487.

Veblen, T. T., Hill, S. R., and Read, J., eds. 1996. *The Ecology and Biogeography of Nothofagus Forests*. New Haven, CT: Yale University Press

Veblen, T. T., Kitzberger, T., and Donnegan, 2000. Climatic and human influences on fire regimes in ponderosa pine forests in the Colorado Front Range. *Ecological Applications* 10: 1178–1195.

Veblen, T. T., Kitzberger, T., Villalba, R., and Donnegan, J. 1999. Fire history in northern Patagonia: The roles of humans and climatic variation. *Ecological Monographs* 69: 47–67.

Veblen, T. T., Young, K. R., and Orme, A. R., eds. 2007. *The Physical Geography of South America*. Oxford, UK: Oxford University Press.

Virtanen, R. 2003. The high mountain vegetation of the Scandes. In L. Nagy, G. Grabherr, C. Körner, and D. B. A. Thompson, eds., *Alpine Biodiversity of Europe* (pp. 31–38). Berlin: Springer-Verlag.

Virtanen, R., Dirnböck, T., Dullinger, S., Grabherr, G., Pauli, H., Staudinger, M. and Villar, L. 2003. Pattern in the plant species richness of European high mountain vegetation In L. Nagy, G. Grabherr, C. Körner, and D. B. A. Thompson, eds., *Alpine Biodiversity of Europe* (pp. 453–464). Berlin: Springer-Verlag.

Volanthen, C. M., Kammer, P. M., Eugster, W., Bühler, A., and Veit, H. 2006. Alpine vascular plant species richness: The importance of daily maximum temperature and pH. *Plant Ecology* 184: 13–25.

Wade, L. K., and McVean, D. N. 1969. Mt. Wilhelm Studies, 1: The alpine and subalpine vegetation. Research School of Pacific Studies Publication BG/1. Canberra: Australian National University.

Walker, M. D., Walker, D. A., Welker, J. M., Arft, A. M., Bardsley, T., Brooks, P. D., Fahnestock, J. T., Jones, M. H., Losleben, M., Parsons, A. N., Seastedt, T. R., and Turner, P. L. 1999. Long-term experimental manipulation of winter snow regime and summer temperature in arctic and alpine tundra. *Hydrological Processes* 13: 2315–2330.

Wang, F.-Y., Gong, X., Hu, C.-M., and Hao, G. 2008. Phylogeography of an alpine species *Primula secundiflora* inferred from the chloroplast DNA sequence variation. *Journal of Systematics and Evolution* 46: 13–22.

Ward, R. T., and Dimitri, M. J. 1966. Alpine tundra on Mt. Cathedral in the southern Andes. *New Zealand Journal of Botany* 4: 42–56.

Wardle, P. 1965. A comparison of alpine timberlines in New Zealand and North America. *New Zealand Journal of Botany* 3: 113–135.

Wardle, P. 1968. Engelmann spruce (*Picea engelmannii* engel.) at its upper limits on the Front Range, Colorado. *Ecology* 49: 483–495.

Wardle, P. 1971. An explanation for alpine timberline. *New Zealand Journal of Botany* 9: 371–402.

Wardle, P. 1973a. New Guinea: Our tropical counterpart. *Tualara* 20: 113–124.

Wardle, P. 1973b. New Zealand timberlines. *Arctic and Alpine Research* 5(Pt. 2): A127–136.

Wardle, P. 1974. Alpine timberlines. In J. D. Ives and R. G. Barry, eds., *Arctic and Alpine Environments* (pp. 371–402). London: Methuen.

Wardle, P. 1998. Comparison of alpine timberlines in New Zealand and the southern Andes. In R. Lynch, ed., *Ecosystems, Entomology, and Plants* (pp. 69–90). Journal of the Royal Society of New Zealand, Miscellaneous Series 48.

Waring, R. H., and Franklin, J. F. 1979. Evergreen coniferous forests of the Pacific Northwest. *Science* 204: 1380–1386.

Webber, P. J. 1974. Tundra primary productivity. In J. D. Ives and R. G. Barry, eds. *Arctic and Alpine Environments* (pp. 445–473). London: Methuen.

Webber, P. J., Emerick, J. C., May, D. C., and Komarkova, V. 1976. The impact of increased snowfall on alpine vegetation. In H. W. Steinhoff and J. D. Ives, eds., *Ecological Impacts of Snowpack Augmentation in the San Juan Mountains, Colorado* (pp. 201–264). Colorado State University, National Technical Information Service PB-255 012. Prepared for U.S. Bureau of Reclamation, Springfield, VA.

Webber, P. J., and May, E. E. 1977. The distribution and magnitude of belowground plant structures in the alpine tundra of Niwot Ridge, Colorado. *Arctic and Alpine Research* 9: 157–174.

Weber, W. A. 1965. Plant geography in the southern Rocky Mountains. In H. E. Wright and D. G. Frey, eds. *The Quaternary of the United States* (pp. 453–468). Princeton, NJ: Princeton University Press.

Webster, G. 1961. The altitudinal limits of vascular plants. *Ecology* 42: 587–590.

Weisberg, P. J., and Baker, W. L. 1995. Spatial variation in tree seedling and krummholz growth in the forest-tundra ecotones of Rocky Mountain National Park, Colorado, U.S.A. *Arctic and Alpine Research* 27: 116–129.

Welden, C. 1985. Structural pattern in alpine tundra vegetation. *American Journal of Botany* 72: 120–134.

Weltzin, J. F., and McPherson, G. R. 2000. Implications of precipitation redistribution for shifts in temperate savanna ecotones. *Ecology,* 81: 1902–1913.

Went, F. W. 1948. Some parallels between desert and alpine flora in California. *Madroño* 9: 241–249.

Wesche, K., Miehe, G., and Kaeppeli, M. 2000. The significance of fire for Afroalpine Ericaceous vegetation. *Mountain Research and Development* 20: 340–347.

Westerling, A. L., Hidalgo, H. G., Cayan, D. R., and Swetnam, T. W. 2006. Warming and earlier spring increase western US forest wildfire activity. *Science* 313: 940–943.

Whittaker, R. H. 1953. A consideration of climax theory: The climax as a population and pattern. *Ecological Monographs* 23: 41–78.

Wipf, S., Stoeckli, V., and Bebi, P. 2009. Winter climate change in alpine tundra: Plant responses to changes in snow depth and snowmelt timing. *Climatic Change* 94: 105–121.

Young, K. R., and Keating, P .L. 2001. Remnant forests of Volcán Cotacachi, northern Ecuador. *Arctic, Antarctic, and Alpine Environments* 33: 165–172.

Young, K. R., and León, B. 2007. Tree-line changes along the Andes: Implications of spatial patterns and dynamics. *Philosophical Transactions of the Royal Society, Series B* 362: 263–272.

Zimmerer, K. S., and Young, K. R., eds. 1998. *Nature's Geography: New Lessons for Conservation in Developing Countries.* Madison: University of Wisconsin Press.

Zimmerman, A. 1953. The highest plants in the world. In Swiss Foundation for Alpine Research, *Mountain World* (pp. 130–136). New York: Harper.

Zobel, D. B., and Singh, S. P. 1997. Himalayan forests and ecological generalizations. *BioScience* 47: 735–745.

CHAPTER EIGHT

Mountain Wildlife

LARRY W. PRICE and VALERIUS GEIST

The study of mountain animals is somewhat more difficult than the study of vegetation. Animals are mobile and may be hard to follow because of broken topography, dangerous weather, or sheer inaccessibility due to distance, terrain, and logistical problems—national bureaucracies included. The firmly rooted plants can be more easily studied and their distribution mapped. This is possible with animals only through year-round observations, often under trying circumstances. Animal distribution is complicated by daily or seasonal migrations. Moreover, their seasonal home ranges, breeding areas, and migration routes may be widely dispersed. Some are full-time residents; others are part-time residents, temporary visitors, or even unintentional migrants, such as the occasional swarms of lowland insects drawn up to high altitudes and deposited on snowfields by ascending air currents. Since the 1970s, the study of radio-tagged individuals has made the biology of many species much more accessible. Moreover, mountain terrain also offers advantages: Large animals on steep slopes, unobstructed by tall vegetation, can often be observed continuously or followed visually for extended periods of time. However, the difficulties of logistics remain, as do the dangers of working in all seasons in mountains. Some classic studies of animals in the mountains that illustrate these advantages and difficulties include Murie (1944) on Alaskan wolves and Dall's sheep; Welles and Welles (1961), Geist (1971, 2002), and Schaller (1975, 1998) on mountain ungulates; and Barash (1989) on marmots.

Altitude and latitude are often parallel: Most animals at high altitudes or latitudes are really summer residents that spend most of the year at lower altitudes or latitudes. These include many of the larger and more mobile species. Birds are the most mobile, of course, and some species, such as eagles and hawks, commute daily to the alpine tundra. They frequently nest in the subalpine or montane zone, riding the slope breezes and updrafts up the mountain to spend their summer days foraging for prey above the timberline. On the other hand, the species that live permanently, or at least reproduce, in high mountains are of greatest significance for our purposes, since they reveal most clearly the characteristics and adaptations necessary to survive in the mountain milieu. Seasonal latitudinal and altitudinal migrants take advantage of the summer's productivity at high elevations in order to reproduce, using lower elevations and latitudes mainly to overwinter.

In spite of the multifaceted nature of animal populations, they share many characteristics with mountain vegetation, as discussed in Chapter 7. One of these is the decrease in number of species with altitude, with the exception of some desert mountains that catch moisture at high elevations and may be more productive there. Thus, there are 96 species of butterflies in the coniferous forests of the Swiss Alps, but only 27 in the shrub and meadow zone, while eight species range into the high tundra (Hesse et al. 1951). There are 61 species of grasshoppers at the base of the Front Range in Colorado at 1,650 m (5,000 ft). At an altitude of 3,300 m (10,000 ft), there are 17 and, at 4,300 m (13,000 ft), only two (Alexander 1964: 79). In New Guinea, 320 species of birds live in the lowlands, but only 128 at 2,000 m (6,600 ft) and eight above 4,000 m (13,200 ft) (Kikkawa and Williams 1971). The entire fauna of the upper part of Kilimanjaro—birds, mammals, reptiles, amphibia, and insects—has been estimated as shown in Table 8.1.

The decrease in number of species with altitude is also accompanied by decreases in species diversity and may lessen interspecies competition. The primary adaptations are those that allow the organism to survive the rigors of the physical environment. As with vegetation, the lower number of species may be counterbalanced somewhat by an increase in the numbers of individuals within any

TABLE 8.1
Number of Species of Fauna in the Upper Part
of Kilimanjaro

Life Zone	Elevation (m)	Number of Species
Cloud forest	2,000–2,800	600
Lower moorlands	2,800–3,500	300
Upper moorlands	3,500–4,200	150
Alpine desert	4,200+	43

SOURCE: Salt 1954: 409.

given species, but the total biomass and productivity are, nevertheless, low. This is particularly true today, since the number of large mammals able to cope year round with the extremes of the alpine tundra is relatively low, largely due to postglacial extinctions of the once rich Pleistocene megafaunas (Martin and Klein 1984; Leaky and Lewin 1996, Barnosky et al. 2004). For the same reason, there is also a significant deficit in bird species, in all ecosystems, mountain ecosystems included (Steadman and Martin 1984, Tyrberg 1998). Smaller creatures have been less affected by extinctions, and are consequently better represented, including in the alpine.

High mountain environments contain relatively undeveloped, young ecosystems, continually set back to earlier successional stages by climatic factors such as severe frosts, avalanches and rock slides, glacial actions, altitudinal temperature inversions, flash floods, or rapid soil erosion. Consequently, the majority of inhabitants are pioneers/colonizers, opportunists, a rough-and-ready lot taking advantage of new and unpopulated land. Being pioneers, most species are able to cope with a broad range of environmental conditions, yet they are specialists, each highly skilled in some narrow spectrum of circumstances. Versatility and flexibility are the qualities with highest survival value. For this reason, the bulk of today's high-altitude faunas consist of mobile ungulates with broad tastes and food habits, versatile predators, many rodents, some scavengers, and unspecialized insects, just as the flora is made up primarily of adaptable, "weedy" species.

It may be instructive to look at the components of a typical recent alpine animal population. The Beartooth Mountains are located about 50 km (31 mi) east of Yellowstone National Park. They are roughly 160 km² (100 mi²) in area, with elevations of 3,300–3,800 m (10,900–12,500 ft). The birds and mammals of this alpine zone have been studied by Pattie and Verbeek (1966, 1967). Thirteen species of herbivorous mammals are more or less permanent residents of the alpine zone. These provide the base of the food chain and include the following species: one pocket gopher (*Thornomys talpoides*), one ground squirrel (*Sperophilus lateralis*), four voles (*Arvicola, Clethrionomys, Microtus,* and *Phenacomys* spp.), one pika

(*Ochotona princeps*), one jackrabbit (*Lepus townsendii*), one chipmunk (*Eutamias minimus*), one marmot (*Mannota flaviventris*), the deer mouse (*Peromyscus maniculatus*), the elk (*Cervus elaphus*), the bighorn sheep (*Ovis canadensis*), and the mountain goat (*Oreamnos americanus*) (the last is an old Pleistocene resident reintroduced by humans; see Fig. 8.1). Six species of herbivorous and insectivorous birds regularly breed above timberline in the area, including the horned lark (*Eremophila alpestris*), rosy-finch (*Leucosticte afrata*), and water pipit (*Anthus spinoletta*), but all abandon the land of their birth and migrate to warmer climates during winter. Carnivores include the weasel (*Mustela* spp.), pine marten (*Martes arnericana*), badger (*Taxidea taxus*), grizzly bear (*Ursus arctos*), red fox (*Vulpes vulpes*), coyote (*Canis latrans*), and bobcat (*Lynx rufus*). With the exception of the weasels and the martens, these are also primarily summer animals, because the smaller mammals on which they depend retreat into subterranean and under-snow habitats during winter. Gray wolves (*Canis lupus*) were here earlier in the twentieth century, were exterminated, but are making a comeback with and without our help. There are seven or eight species of predatory hawks and owls, including the golden eagle (*Aquila chrysaefos*), but most of these, too, abandon the area during winter (Hoffmann 1974).

In addition to those species which are truly alpine in nature, others occasionally wander up beyond timberline. These include about 10 mammals, including the snowshoe rabbit (*Lepus arnericanus*), porcupine (*Erethizon dorsaturn*), mule deer (*Odocoileus hermionus*), and moose (*Alces alces*). There are between 15 and 20 species of vagrant birds, including the robin (*Turdus migratorius*), Clark's nutcracker (*Nucifraga columbiana*), raven (*Corvus corax*), mountain bluebird (*Sialia currucoides*), and pine siskin (*Spinus pinus*). Domestic sheep should also be added to this list, since several hundred are still driven annually to the high country (Pattie and Verbeek 1967). Thus, about 30 species of mammals and 30 species of birds may be present at one time or another during the summer above timberline in these mountains. The relationships among these mammals and the trophic web they form are presented in Figure 8.1. This picture is altered considerably during winter when the total number of species present is reduced to 15–20.

The species composition, of course, reflects today's conditions. Had we looked at these mountains before humans came to North America, we would have seen huge, tall-legged Columbian mammoths (*Mammuthus columbi*); massive mastodons (*Mammut americanum*); squat ground-sloths (*Nothrotherium shastense*); ox-sized, large-horned shrub-oxen (*Euceratherium collinum*); long-horned bison (*Bison latifrons*); fleet-footed forest musk oxen (*Symbos cavifrons*); elegant Conclin's mountain pronghorn (*Tetrameryx conklingi*); large, stocky mountain deer (*Navahoceros fricki*); herds of fleet-footed guanacos (*Hemiauchenia macrocephala*) and large camels (*Camelops hesternus*); at least two species of native horses (*Equus occidentalis, E. conversidens,* and a large horse); as well as assorted large predators, such as the huge

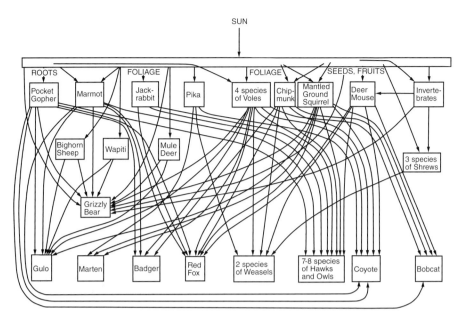

FIGURE 8.1 Schematic representation of the trophic web established among warm-blooded mammals living for at least part of the year (summer) above timberline in the Beartooth Mountains, Montana. (From Hoffman 1974: 515.)

short-faced bear (*Arctodus simus*); true lions (*Panthera leo atrox*) larger than the African variety; big American cheetahs (*Miracinonyx trumani*); and dire-wolves (*Canis diurus*) as tall as, but more massive than, today's gray wolves (*Canis lupus*). Missing from the assemblage would be elk (*Cervus canadensis*), moose (*Alces alces*), grizzly bear (*Ursus arctos*), and gray wolf, as these are post-Pleistocene Siberian immigrants that entered North America along with humans, following the massive collapse of the native megafauna (Ripple and Valkenburgh 2010) that so impoverished North America's species diversity. In addition to the later Pleistocene extinction of large mammals, there was also a major extinction of birds (Steadman and Martin 1984; Tyrberg 1998). These included not only several species of large vulture-like scavengers, but a broad spectrum of birds.

For millions of years, the megafauna had conditioned the composition and structure of the vegetation and molded the land surfaces to create unique habitats and shape the evolution of many mammalian and avian species. Large mammals diversify the landscape on a micro scale, with traditional dung deposits, deep cutting trails, bedding areas, sand-bathing, and feeding sites. These features can redirect and capture water, cause colonizing plants to sprout, open up closed forests into highly productive savannah, and reduce the fuel load on the land, thus restricting the size and intensity of lightning-set fires. In addition, the giants excavated drinking holes in dry creek and river beds, and thus also provided water for others. With the extinction of the large mammals and the widespread application of human-set fires, the landscape must have changed to the disadvantage of many species that had coevolved with the megafauna. Our current landscapes thus reflect not merely the forces of nature, but the severe distortions imposed deliberately and inadvertently by humans in prehistoric and historic times (Komarek 1981; Delcourt and Delcourt 1997).

Today, the exact number and composition of species vary for different mountain regions, depending on environmental conditions and the size, elevation, age, history, climatic patterns, and relationship to other mountain areas. Generally, the larger, taller, and thus more climatically diverse the mountains, the greater the number of species. Thus, the largest number of mountain species exists in the Himalaya, with its enormous canyons and its altitudinal and latitudinal striation of biotic communities from tropical to glacial and from rainforest to alpine deserts. There are no accurate surveys for the whole area, but species number in the hundreds (Hoffmann 1974). The opposite is also true: Small, isolated mountains display the least diversity. A good example is the Big Snowy Mountains, a small isolated range on the Great Plains of central Montana: Only seven species of mammals and four species of birds are known to breed in this alpine area (Hoffmann 1974). A number of other species visit in the summer, just as on the Beartooth Plateau, but the contrast in numbers is considerable.

Although alpine and arctic zones have much in common, the island-like, patchy distribution of mountains and the large, nearly continuous circumpolar belt of the Arctic constitute a fundamental difference. Arctic animals are able to maintain relatively large populations, and there is a good degree of continuity and interchange, with the same species occurring throughout. This is impossible in disjunct and widely scattered mountains because of habitat patchiness and the difficulty in dispersal and colonization. Both flora and fauna become increasingly diverse away from the polar regions, particularly mammals, to the point that "not a single tundra species is common to both the alpine [tundra] in temperate latitude mountains and the arctic tundra" (Hoffmann 1974). This is true, however, for species that live only in arctic or alpine tundra; other widely ranging species—that is, subarctic and subalpine—occasionally occupy both environments. Birds, insects, and vegetation are less exclusive; a number of species are shared by arctic and alpine tundra. Amphibians and reptiles are notably rare in both zones (Hock 1964a, 1964b).

Limiting Factors

In spite of differences in distribution, arctic and alpine animals have similar approaches to life. One early observation in biogeography was that of convergent tendencies in the environment, as expressed in the flora and fauna, with increase in latitude and altitude. We now know that there are a great many exceptions to this simple rule, but at the broadest level it is still useful.

Because it gets colder as altitude and latitude increase, it has also long been assumed that low temperature is one of the major limiting factors to life. Dunbar (1968) contends that, while low temperature may be the immediate or proximate cause for the decrease in numbers of species, it is not the ultimate reason. The ultimate limiting factors are, rather, large environmental oscillations, lack of nutrients, lack of habitat diversity, and youthfulness of the ecosystem (Dunbar 1968). Alpine environments have all of these problems, plus three other limiting characteristics of their own: the disjunct, island-like distribution of mountain areas; reduced oxygen with altitude; and the need to deal with gravity and broken landscapes, ascending or descending. Whether or not temperature is the ultimate limiting factor, each of the above-mentioned factors takes its toll and helps to establish the environmental framework in which mountain life exists.

Temperature

Although temperature has been singled out as being fundamental in controlling the distribution of organisms in mountains, all climatic factors are interrelated. The alteration of one factor alters all the others, and the impact of any one of these may, in specific cases, exceed the effect of temperature. This is implicit, but there is something compelling about narrowing the explanation for any given phenomenon to a single cause. A classic example is C. Hart Merriam's life-zone concept, in which he attempted to show that the flora and fauna of North America are distributed into distinct life zones based on temperature (Merriam 1890, 1894, 1898). This concept had a profound impact on North American ecology, but has fallen into disrepute specifically because of its singular dependence on temperature (Daubenmire 1968; Kendeigh 1932, 1954). Such an approach may be useful in a small area where the mesh of factors is kept fairly constant, but when the fundamental relationships change, so will the relative importance of each. Alexander von Humboldt's climatic zones *tierra caliente, tierra templada,* and *tierra fria* (see Chapter 11) work reasonably well in the humid tropics, but not elsewhere. Likewise, Merriam's life zones (e.g., arctic–alpine,

Hudsonian, Canadian) can be applied fairly usefully in the region of his original study, the San Francisco Mountains, Arizona (although the assumed convertibility of latitude and altitude is questionable), but not necessarily in other areas. These were pioneering attempts that, in modern times, blossomed into the use of satellite imagery, often incorporated in geographic information systems (GIS) (Aranoff 1989), as well as their application via ecosystems management (Boyce and Haney 1997).

Another major problem with temperature as a limiting factor is in ascertaining its effect on organisms. Humboldt's and Merriam's life zones, and those of many other studies, were based on correlations between temperature and the distribution of plants and animals. A certain species will reach a particular elevation but go no higher: The temperatures at that elevation are presumed to be responsible for this point of maximum distribution, but it is hard to identify the specific processes involved. For example, as discussed in Chapter 7, the upper timberline coincides closely with the 10°C (50°F) isotherm for the warmest month, though the reason for this relationship remains elusive. A number of animals reach their altitudinal limits at timberline, but it would be dangerous to say that these creatures are limited by temperature, when in fact they may simply be limited by the absence of trees. In this case, then, temperature may be more or less directly responsible for the location of timberline, but only indirectly responsible for the distribution of the associated animal species.

Temperature operates in several ways to limit life. The first and most obvious is where it exceeds the tolerance limits of organisms. All species have an upper and lower threshold of temperature that they can withstand and, within this range, an optimum or preferred level. The same can be said of any environmental factor, and animals with different potentials, tolerances, and abilities seek out, as near as possible, their own preferred range. This is one of the basic reasons for the distributions of species on Earth. All life exists within the fairly narrow temperature range of 0° to 50°C (32°–122°F). Temperatures on Earth far exceed these limits, but most organisms escape the extremes by one means or another. The absolute low-temperature limit is reached when body fluids begin to freeze, usually within a few degrees below the freezing point of pure water. The upper limit is reached when body fluids begin to undergo destructive chemical change. The two effects are not equal, however. There is much greater flexibility at the lower end of the spectrum than at the higher: High temperatures usually result in irreversible damage and death, but cells can freeze in some animals without irreversible damage, and activity may be resumed when temperatures rise (Hesse et al. 1951). The length of exposure and extremity of the temperature are important, of course, but a number of creatures, especially insects, do survive winters of subfreezing temperatures. During Sir John Ross's second voyage to the Canadian Arctic in 1829 to 1833, the larvae of butterflies were found frozen to the point of brittleness. Specimens were taken inside and

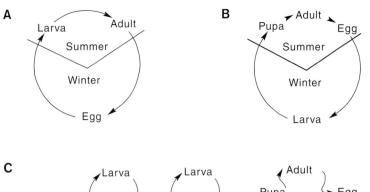

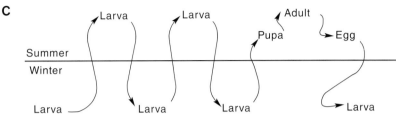

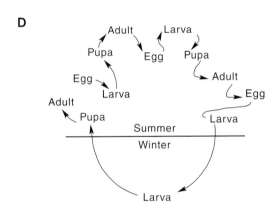

FIGURE 8.2 Diagrammatic representation of reproductive cycles displayed among mountain insects. (A) A normal one-year cycle with hibernation in the egg stage and adults emerging late in summer. (B) A one-year cycle with hibernation in the larval stage and adults emerging in midsummer. (C) A two-year cycle displayed among beetle at higher altitudes, with larval development taking place in summer, relapsing into dormancy in winter, and final development occurring in the third year. (D) An example of multiple-cycle reproduction in flies and mosquitoes where up to three or four generations develop within the brief summer. (After Mani 1962: 118.)

allowed to thaw. The larvae became active within two hours. They were then taken back outside into the −40°C (−40°F) weather and allowed to refreeze. The process was repeated three times and, although there was considerable mortality, some of the larvae still pupated in the spring (Downes 1964).

Temperature may limit life by slowing the rate of development, and by reducing metabolism and the rates of activity, fecundity, and reproduction (Andrewartha and Birch 1954). The deer mouse (*Peromyscus maniculatus*) in the Sierra Nevada of California produces three times as many young at 1,360 m (4,500 ft) as at 3,760 m (12,400 ft) (Dunmire 1960), primarily because low temperatures at high altitudes shorten the breeding season. The same is true with birds: Alpine species are generally limited to one nesting a year, whereas birds at lower elevations may nest two or three times (Lack 1954). Insects at high altitudes frequently take two or three years to complete their life cycle (Fig. 8.2). They may hibernate at different stages of development in succeeding years—egg, larvae, pupae, and adult—until complete metamorphosis is achieved (Mani 1962). This lower level of productivity in turn reduces opportunities for diversification and speciation.

Temperature also affects animal distributions indirectly through its influence on land form, vegetation, soil development, habitat diversity, and food supply. Temperature controls the distribution of permafrost, which limits root penetration or excavation by organisms. Warm temperatures ascend the slopes in spring, beginning on southern exposures and expanding to north-facing slopes. Consequently, a band of sprouting, highly digestible and nutritious vegetation moves uphill, and so do large herbivores. Therefore, the taller the mountains, the longer the availability of high-quality food for mountain herbivores. Similarly, vegetation sprouts at the edges of snowfields that are retreating as temperatures rise. Cliff faces exposed to the winter sun shed their snow earliest and become favorite places for mountain sheep and mountain goats in late winter. The body heat of resting sheep and goats accelerates the sprouting of plants rooted at the edge of such beds. Consequently, the first spots to green up in spring may be the habitual resting places of wild sheep. Since these animals defecate and urinate after rising, the beds tend to be well fertilized and humus-rich from millennia of feces deposition.

Daily temperature fluctuations can generate very hazardous snow conditions. In spring, snowfields may thaw in the afternoon but freeze solid overnight, forming a sun-crust, a steep ice surface on which no large animal can hold itself, sometimes leading to fatal accidents. On more level areas, sun-crusts allow sheep and goats to disperse before the snow melts at noon. Sun-crusts also limit the food supply for mountain ungulates to the afternoon and early evening, as their hooves cannot penetrate the ice crust when ambient temperatures are low. Consequently, wild sheep will rest until about noon before feeding, instead of feeding at dawn as they normally do. In cirques, the shadow effect may cause very cool summer temperatures, much sought out by caribou seeking relief from biting flies. Caribou also avidly seek out cool temperatures on snowfields and glaciers in summer to avoid insect pests, warble flies in particular. Slopes exposed to the sun generate upward-moving air currents that are used for lift by eagles, vultures, and condors, but may also discourage biting flies. Increasing daily temperatures in the spring lead to predictable avalanche activity in the afternoon, as well as accelerated rock fall. Both are mortality factors for the unwary. The warm *katabatic* winds (e.g., *Chinooks* or *foehn*) generated by mountains (see Chapter 3) clear exposed ridges of snow, thus exposing plants that mountain ungulates consume as food (Geist 1971, 2002). Very low temperatures along glacial fronts may maintain soft, powdery snow that animals can remove by pawing, allowing access to forage, even when elsewhere thawing and freezing have formed ice crusts on the snow.

Environmental Oscillation

Fluctuation of environmental conditions increases with latitude and altitude, greatest at the poles and least at the equator. Consequently, organisms at high latitudes and altitudes must adapt to wide ranges of climatic, ecological, and foraging conditions. Latitude controls the length of day and the seasons; combined with latitude, these effects are accentuated. For example, the growing season at any given latitude is generally shorter in mountains. Even in the humid tropics, where there is no marked seasonality, the range of daily environmental extremes increases with altitude—a species may have to cope with intense sunlight every afternoon and freezing temperatures every night. The number of species that can survive these conditions decreases with elevation, and each loss brings about concurrent changes in community structure and environment. The most striking change occurs at timberline, since the forest provides a buffer. Beyond timberline, species must adapt to the open habitat or be eliminated. The environment changes, and the soil becomes exposed to the full brunt of sun, rain, wind, and extremes of temperature. This may lead to other processes such as increased frost creep and solifluction (see Chapter 5), which in turn lead to greater erosion, stream infiltration, and habitat instability. Consequently, with the loss of trees, community structure becomes much less complex. There may be less food and cover, and greater environmental extremes.

Longer winters reduce the food supply, since the growing season is shorter and snow and ice cover the vegetation. While there may be ample food during the summer, little is available during the winter. Organisms must either leave the area in search of more favorable conditions (migrate), reduce their need for food physiologically (hibernate), or circumvent the need for food in some other way. For example, the pocket gopher (*Thomomys* spp.) stays active throughout the winter by burrowing and by harvesting plant roots in soil protected from freezing by deep snow. The strategies of animals for dealing with the lack of food also solve problems of dealing with great environmental extremes. With increasing environmental oscillation, there is increasing fluctuation of nutrients and of population numbers, and energy flow through the system is interrupted (Dunbar 1968). The opposite is true in humid tropical lowlands, where there is great environmental complexity of species and a maximum flow of energy through the system.

While environmental oscillations can limit life, climatic oscillations generating seasons also generate new and spectacular life forms. Because of the great discrepancies between productivity in summer and winter, animals at high latitudes in summer escape the limiting consequences of food competition and exploit the abundance in their seasonal food supply. The seasonal overabundance of food creates a vacation from want, allowing not only superior reproduction, but also superior body growth, the evolution of luxury organs, massive fat deposition, and the evolution of novelty. By contrast, tropical species may struggle with want year round, and their morphology and behavior reflect the severe scarcity of resources for body growth, reproduction, and innovation. This phenomenon has been explored in detail under the heading of *net primary production* and is significant here as net primary production latitudinally is mirrored altitudinally (Huston and Wolverton 2009, 2011; Wolverton et al. 2009).

Because the Ice Ages accentuated latitudinal habitat differentiation, they also generated sharply different adaptations with latitude and altitude. Moreover, because diversity of seasonal habitats generates new problems of adaptation with each season, the brain responds by growing larger. We therefore observe, within the same family, highly conservative, small-brained tropical species and highly evolved luxury species with large brains at high latitudes. The latter, because of large size, showy hair coats, large ornamental antlers and horns, fat deposits, and novel life strategies, may be called "Pleistocene grotesque giants." Compare the naked elephants and rhinos of the tropics with the woolly mammoth and rhinos of the north; the modestly antlered and furred deer of the tropics with the extravagant Irish elk, moose, reindeer, or wapiti; or the small tropical bears with the Kodiak or polar bears. Strong climatic oscillation greatly favors the evolution of spectacular novelty, while tropical conditions favor the long-term accumulation of

FIGURE 8.3 Mountain goats are found almost exclusively in rocky areas. (U.S. National Park Service.)

tried, successful adaptations. It thus is no accident that so spectacular a creature as the giant sheep or argali (*Ovis ammon*), with its enormous horns, occupies the highest mountains in Asia, while primitive ancestral forms, such as the serow (*Capricornis sumatrensis*), are found in the hot lowland jungles of Southeast Asia. This progression of adaptations is well illustrated in the deer family, and also in humans (Geist 1978, 1998).

Nutrient Availability

The availability of nutrients and amount of productivity generally decrease with elevation. This is largely a function of temperature and length of the growing season, coupled with the poor quality of most mountain soils and much exposed rock. The long dark winter of high latitudes has a counterpart in the higher reaches of midlatitude mountains where the surface is cut off from the sun by a heavy snowcover. While the daily productivity of high mountain plants during summer is fully equivalent to most lowland environments (Bliss 1962; Webber 1974), annual productivity is quite low because of the short growing season (see Chapter 7). Thus, there is abundant food during the growing season, and a large variety of animals spend their summers in high mountains, mainly to reproduce. The lack of nutrients is an almost insurmountable problem to year-round occupancy of alpine tundra and, in this respect, limits the utilization and efficiency of the ecosystem. However, in the vicinity of glaciers there can be a surfeit of nutrients, as glaciers grind rock into fertile silt and dust, which tend to accumulate downstream and downwind of the glacier. Thus silt flats and loess fields, formed respectively from rock dust transported by water and wind, are areas of high

nutrient availability, some of which crystallize out as salts and are used by animals as mineral licks (Geist 1978).

Lack of Habitat Diversity

The fundamental characteristic of high mountain landscapes is the openness of the habitat—the low-lying alpine vegetation and bare rocks. The tundra is essentially a two-dimensional habitat. The absence of trees eliminates the vertical component and reduces the number and diversity of ecological niches. Species that depend on the forest for food or shelter are eliminated unless they can adapt to the open country. Birds must build their nests on the ground and hover in the air to give their mating calls. Vegetation above timberline consists of grasses and herbaceous species with a simple community structure. Rocks usually abound, and these become increasingly important as habitats (Hoffmann 1974). Some creatures are found almost exclusively in rocky areas—for example, picas (*Ochotona* spp.), marmots (*Marmota* spp.), and mountain goats (*Oreamnos americanus*) (Fig. 8.3). Snowcover is also a prime consideration: A number of species depend on it for insulation and protection during winter. Patterns of population distribution for many creatures, large and small, are directly related to the distribution of snow (Pruitt 1960, 1970; Sleeper et al. 1976; Stoecker 1976; Geist 1971).

The most important source of habitat diversity in the alpine zone is the lay of the land itself. Topography largely controls the distribution of rocks and snow. Rocky crags, cliffs, valleys, and differences in slope gradient and exposure provide a variety of surfaces for the interplay of climatic, geomorphic, edaphic, and biologic factors, resulting in a mosaic of microhabitats. In general, areas of the least topographic diversity (e.g., smooth slopes) have the fewest ecological niches and least species diversity, whereas areas of maximum topographic diversity support the greatest number of ecological niches and maximum species diversity. Minimum complexity is found on small and simple mountains that slope directly to the summit (e.g., volcanic cones), whereas maximum complexity is found on large and diverse folded mountains, such as the Alps or Himalaya. Topographic diversity is a major reason why there are more species in alpine environments than in the arctic tundra (Hoffmann 1974). Nevertheless, habitat diversity is considerably less in alpine than in lowland environments.

Youthfulness of the Ecosystem

As discussed in Chapter 5, most mountains are young landscapes, both because they are relatively recent geological creations and because many have experienced severe glaciations. The Tertiary, which extended over 65 million years, was a period of major mountain building, while the Pleistocene—the last 2 million years—was a period of major climatic fluctuation. Although mountain building during the Tertiary was the initial event, and life had to adapt to the changing conditions of increasing altitude and relief with uplift, the Pleistocene has been by far the most important in terms of the present distribution of life in mountains. During the Ice Ages, the northern hemisphere was most affected because of its larger land masses and resulting continental climate; mountains served as centers of increased snow and ice accumulation, and huge lowland ice sheets extended from the north into the middle latitudes. Vast amounts of bedrock were removed, and the rock was ground into dust and redistributed by water (silt) and wind (loess).

While the centers of oceans and glaciers contain little life, the edges of oceans (intertidal zone) and of glaciers (periglacial environment) can be productive, pulse-stabilized ecosystems. Glaciers annually produce pulses of fertile rock dust and meltwater. These initially discharge into broad, braided streams that fill the ever-rising valley bottoms with glacial debris and silt. When flood waters recede, the fertile silt is dried and blown away to be massively deposited as fertilizing loess. The effects of fertility and water are enhanced plant productivity, which in turn attracts a diversity of wildlife, including thriving populations of large herbivores and carnivores. Moreover, there can be "land islands" protruding from oceans and from glaciers (the latter are called *nunataks*). These mountains may be vegetated and home to large herbivores and carnivores. Glaciers may have so much debris in and on the ice that even forests of spruce, poplars, willows, and birch may grow on the ice. On the Klutlan glacier of the western Yukon Territory, Canada, moose populations winter in such willow and dwarf birch flats, which are also visited in summer by caribou, mountain sheep and goats, grizzly bears, and wolves, while willow ptarmigan nest in bogs on the glacial ice.

The evidence of the most recent Ice Age (Wisconsin) is still relatively fresh, and the pattern of glacial development can be more readily reconstructed. Although the lowland ice retreated 10,000–15,000 years ago, mountain glaciers melted more slowly; the largest present-day glaciers may be remnants of the Pleistocene ice. Most small mountain glaciers, however, have melted entirely. Numerous small advances and retreats have occurred within the last 10,000 years, greatly affecting the fate of civilizations— studies have shown a slow rate of soil and vegetation development in some areas abandoned by these retreating glaciers. Some of the first animal inhabitants are primitive species of springtails and mites, followed by arthropods such as beetles and spiders. Most early insect occupants are carnivores (Brinck 1966, 1974) that live on algae, larvae, and other insects. Eventually, as the ecosystem develops, the various components and niches are filled with a variety of life forms, but on bare rock at high elevation, it can be a slow process.

Clearly, the age of the surface is a major factor in ecosystem development. This is demonstrated on a macro scale by the great floral and faunal richness of the Central Asian highlands (Sushkin 1925; Meinertzhagen 1928; Swan and Leviton 1962; Zimina and Panfilov 1978). This area escaped extensive glaciation during the Pleistocene owing to its aridity, so it served as a refugium for numerous species. In fact, it is believed that many present arctic and alpine species originated here and later spread to the Arctic and the mountains of Europe and North America (Hoffmann and Taber 1967; Hoffmann 1974).

The immaturity of mountain ecosystems can perhaps best be seen by comparing mountains with lowland tropical environments. The humid tropics are the oldest environments, containing the greatest number of species and organic complexity. The major problem for life in the tropics is scarcity of energy and nutrients, and thus severe competition within and between species. Consequently, organisms use their energies primarily for specialization, the corollary of which is ecological diversification. The result is a complex network of species occupying a maze of ecological niches for a near-complete utilization of available resources (Dobzhansky 1950). Just the opposite is true in mountains, where the bulk of energy is devoted to coping with difficult physical conditions and there is little diversification or specialization.

Island-like Distribution

As discussed in Chapter 7, mountains have been likened to islands, since they provide habitats unlike those of surrounding lowlands and support different floras and faunas. The analogy is only approximate, however. The surrounding lowlands present a less hostile environment to terrestrial alpine organisms than does the sea. An alpine bird tiring from flight between mountain areas can stop to rest in the lowlands, but one who tires over the sea is almost certainly doomed. Another difference is that the zone of transition between environments for oceanic islands takes place abruptly at the shoreline and always at the same elevation—that of sea level. The mountain equivalent of a shoreline occurs at various absolute elevations, depending on the climate, and the transition from the adjoining environment is more gradual.

Nevertheless, mountains and oceanic islands share a number of ecological characteristics (Diamond 1972, 1973, 1975, 1976; MacArthur 1972; Carlquist 1974; Mayr and Diamond 1976). In particular, the number of species and their diversity are closely related to the size of the island (mountain) and its distance from the mainland or other islands (mountains). The larger the island, the greater its resource base, carrying capacity, and variety and availability of habitats. The greater the distance from other islands, the fewer species are likely to be found. Both of these characteristics are demonstrated by the size and distance between enclaves of alpine vegetation in the northern Andes and the number of bird species they support.

Another characteristic of islands is their increasing taxonomic specificity with insularity. The more isolated and remote an island, the more unique its flora and fauna, since evolution has operated under local conditions to create endemic taxa. One has only to think of the exotic species on islands such as New Zealand or the Galápagos for confirmation. For high mountains, the greatest insularity exists in the tropics. As with plant species, high proportions of endemic animal species are found in the alpine areas of East Africa (Salt 1954; Coe 1967; Coe and Foster 1972).

The opposite situation exists in arctic and subarctic mountains: Because there is free interchange with lowland tundra species, virtually no endemic species occur (Brinck 1974). Midlatitude mountains are intermediate in this regard, their insularity being determined by their age and the nature of the surrounding habitats. For example, the alpine zone in the Rocky Mountains is surrounded by subalpine forests, with direct connections in the north to the eastern boreal forest and the western forests of the Pacific shores. The plants and animals thus have fairly easy access from arctic and subarctic to alpine and subalpine settings. Differences between the two tundras increase southward: In Colorado, more than half the plant species are non-arctic in origin, and the percentage for animals is even higher. But a few endemic species do exist. On the other hand, the Caucasus Mountains between the Black and Caspian Seas, separated from the Arctic by vast expanses of semiarid steppe, are much more isolated. Consequently, more than half the species found there are endemic (Zimina 1967, 1978). Another striking example is provided by the Basin and Range mountains of the American West (Fig. 8.4). These mountains exist as isolated alpine tracts surrounded by desert shrublands; as such, they are extremely insular. The alpine populations are totally isolated, and each peak has its own set of endemic species (Brown 1971; Johnson 1975).

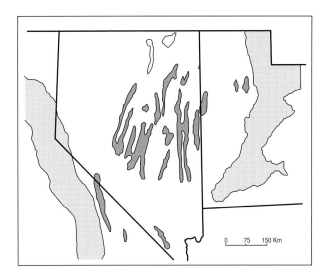

FIGURE 8.4 Mountain "islands" in the Great Basin between the Sierra Nevada (left) and Rockies (right). Each of these islands is at least 3,000 m high, surrounded by lower-lying country supporting desert shrubs. During the Pleistocene, there was relatively easy access from one mountain area to the other, but as the climate became warmer and drier, the species became isolated on individual mountain islands. The two top islands are unshaded because their faunas are poorly known and were not used in the original investigation. (After Brown 1971: 468.)

Theoretically, the number of species an island supports is set by the establishment of an equilibrium between rates of colonization of new species and extinction of other species. This principle also operates in mountains (Vuilleumier 1970; Brinck 1974). While true islands can only be colonized by immigration from the mainland or from other islands, mountains may be colonized by several other methods. One of these is by species from adjacent lowlands adapting to the habitats of higher elevations (Fig. 8.5a).

Speciation commonly occurs by stages: There may be a series of species up a mountain slope, each belonging to the same genus and fundamentally related, but isolated through morphology or behavior so that they may no longer interbreed. One good example from the mountains of East and South Africa is the hyrax, a small, rabbit-like ungulate, related to the coney of the Bible. Several species live on the mountain slopes. The lowland species is poorly insulated, has different food preferences, and lives entirely in trees, while the hyrax of the alpine zone has a heavy fur coat, burrows under the ground, and makes its home in rocky areas. The species living at different altitudes have become so differentiated that they no longer interbreed (Coe 1967).

In general, however, the alpine areas of the tropics have been colonized primarily through direct immigration from other alpine areas, not from the surrounding lowlands (Fig. 8.5b). This is because the lowland organisms evolved from very ancient stock in a specific and relatively stable environment. Very few apparently have the ability to pioneer new habitat, especially one that requires the ability to

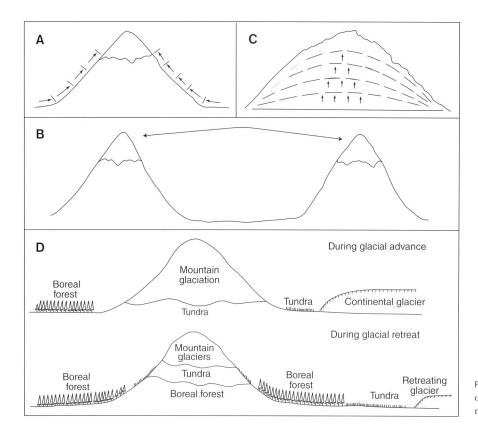

FIGURE 8.5 Schematic representation of different methods of colonizing mountains. (Figure by L. W. Price.)

withstand freezing, a phenomenon unknown in the tropical lowlands. Consequently, most high-altitude species in the tropics have been derived from species in the middle latitudes and are fundamentally different from anything found in tropical lowlands. Research in the Andes, however, indicates that climatic change in the tropics may have been considerable, possibly lowering the vegetation zones by as much as 1,500 m (5,000 ft) in some locations (F. Vuilleumier 1969, 1970; B. S. Vuilleumier 1971). Under these circumstances, the islands of alpine vegetation would have been greatly expanded, and in some areas they may have overlapped, allowing free interchange of species.

Another method responsible for the isolation of high-altitude species is the process of mountain uplift (Fig. 8.5c). Some plants and animals appear to have originally been lowland species but, as the land was uplifted, they were able to adapt to the slowly changing conditions. In this way, they became isolated from the lowland forms. One bird of the genus *Tinamous* in the páramos of the tropical Andes lives near and on the snow, breeding only above the timberline, while all other species of the genus live in the tropical lowlands. The bird resembles a northern hemisphere ptarmigan (*Lagopus* spp.) in appearance and behavior, but is far removed in terms of structure and taxonomy, an example of convergent evolution (Brown 1942).

A final method of colonization is through direct immigration during periods of climatic change (Fig. 8.5d). This has happened chiefly in the middle and high latitudes, where major migrations took place during the glacial and interglacial periods. Species moved in advance of the ice and established themselves in belts or zones around the ice in much the same way as happens today in arctic and alpine areas, that is, tundra first, then coniferous forest. As the climate warmed and the ice melted, these plants and animals began to reoccupy their former sites. Some moved northward, while others moved upward into mountains where suitable habitats existed. In some cases, as in the Rockies, the potential for exchange of species and colonization has continued, whereas in other smaller, isolated ranges, the alpine communities were cut off and became relict populations. Examples are the alpine species in the Basin and Range mountains. For this reason, they do not follow the equilibrium theory mentioned above; that is, there have been extinctions but no new colonization (Brown 1971).

The "creep" of habitats latitudinally and altitudinally is possible during glacials and interglacials if there are long stretches of unbroken land, as in Central Asia, with its huge mountains and long valleys. "Habitat creep" is not possible where the landscape is structured by many sharp, low mountain ranges with short distances between them, as in Europe, where interglacial habitats go extinct with glaciations, and reassemble again during interglacials. In Asia, this allows species to remain by merely following their habitats during the glacial cycles. This conserves primitiveness. Where habitats go extinct during glacial cycles, plastic generalists that can switch to different habitats are selected. Moreover, the huge amount of rock dust liberated by glacial grinding of bedrock forms itself into long, circumpolar stretches of periglacial loess-steppe (Geist 1978), which, due to its characteristic Ice Age fauna, has been labeled the "mammoth steppe" (Guthrie 1990). Rock dust deposited by water (silt) and wind (loess) formed deep, fertile soils close to glaciers along the shallow, braided streams, as well as shallow marches along the edges of the huge proglacial lakes. Only tiny fragments of these habitats exist during interglacials—that is, today. For the above reasons, Asia retained primitive or specialized mammalian species, while Europe generated assertive generalists. This extends even to humans. Thus, while our primitive parent species *Homo erectus* survived in Asia virtually from its inception some 1.6 million years ago, until displaced by "out of Africa" modern man about 50,000 years ago, in Europe there evolved an advanced new species, Neanderthal man, and later the modern Cro-Magnon of cave art fame. We see a similar pattern in bears and deer (Geist 1999). The size of mountains and valleys thus has a profound effect on speciation and survival during glacial cycles.

Lack of Oxygen

Free oxygen is essential for life. The composition of the atmosphere is relatively constant, containing about 21 percent free O_2 both at sea level and in the upper part of the atmosphere. Nevertheless, there is less oxygen in the air at higher altitudes, because air is compressible and has greater density and more molecules of oxygen per unit volume at lower elevations than at greater heights. The availability of oxygen is expressed as the partial pressure of oxygen (pO_2), which is derived by multiplying the total atmospheric pressure by 21 percent. Thus, normal atmospheric pressure at sea level is 760 mm, so the pO_2 is 159 mm. With increasing elevation and decreasing atmospheric pressure, the pO_2 decreases proportionately.

The reduced oxygen concentration at high altitudes does not appear to have a great effect on the biota. Oxygen deficiency has no noticeable effect on vegetation, insects, or reptiles and amphibians (Bliss 1962; Mani 1962; Hock 1964a). Little is known about its effects on birds. The South American condor (*Sarcorhamphus gryphus*) nests in the high Andes, but migrates daily to and from sea level, where it feeds on dead fish along the Pacific coast (Lettau 1967). The bar-headed goose (*Anser indicus*) winters in the lowlands of India, but flies over the summit of Everest on its way to nesting grounds in the high lakes of Tibet (Swan 1970). Many mammals have overcome the effects of oxygen deficiency so that they can occupy almost any environment where there is sufficient food. However, experience shows that for humans and their livestock, elevation does make a difference.

This is shown by the pronounced symptoms experienced when certain low-altitude mammals (including humans) go to high altitudes. With rapid ascent, acute mountain sickness or other maladies may develop, possibly resulting in death. If they go up slowly, acclimatization to the changing conditions takes place, but there are limits beyond which lowland mammals cannot go. For lowland cattle, this is approximately 3,000 m (10,000 ft) (Alexander and Jensen 1959).

Survival Strategies

Animals respond to environmental stresses first by behavioral adjustments, followed by physiological adjustments which, in consequence of natural selection, result in morphological adaptations, followed by possibly new community relations and interaction. Plants, being relatively immobile, respond primarily through morphological and physiological adaptations, while animals respond primarily through behavior (Kendeigh 1961). Animals often deal with environmental extremes by escaping them through migration, hibernation, burrowing, or the use of microhabitats. The nature and timing of growth and reproduction also comprise an important part of survival strategy in extreme environments. Only a few species expose themselves to the full brunt of the climate throughout the year; those that do, however, display the broad range of adaptations necessary for existence under such conditions. Escaping environmental extremes is primarily a function of behavior (although some metabolic adjustments are also involved), but for an animal to withstand extreme conditions, it must, like the plants, rely on morphological and physiological adaptations.

Migration

Migration allows some mobile species a way of utilizing the high mountain environment during favorable periods and lowland or other mountain environments during more stressful periods. Strictly speaking, migration is the "large scale shift of the population twice each year between a restricted breeding area and a restricted wintering area" (Lack 1954: 243). While some species practice a twice-yearly movement, others, like the mountain sheep, may shift between as many as six seasonal home ranges (Geist 1971). There may also be lesser movements in response to the sporadic occurrence of severe weather, as well as poor or exceptionally favorable food conditions.

In mountain sheep, knowledge of the home range is passed on from the older to the younger generation as a living tradition that eliminates the need for young sheep to explore for suitable living space on their own. The latter is the common mode for the young of many species to find a place to live. Because living space for sheep is reduced during interglacial times to widely spaced, tiny specks of grassland surrounded by timber, young sheep exploring for living space on their own would be faced with a daunting task with little payoff. It is more efficient to follow older individuals and thus learn quickly where to feed or to flee from predators. Unfortunately, harassment can make mountain sheep vacate home ranges and "forget" about them, leading to habitat loss at best and extinction at worst. A policy of aggressive reintroductions of sheep to abandoned habitats increased their number continent-wide by almost 50 percent (Toweill and Geist 1999). Similarly, the astute application of ecological and behavioral knowledge acquired by field observations led to the reclamation of abandoned coal strip mines in Alberta, Canada, into custom-built bighorn sheep habitat. It was searched out eagerly by the sheep, and led to quick population growth and the largest-bodied bighorns since Pleistocene times (MacCallum and Geist 1992, 1995).

Migration in middle and high latitudes generally coincides with the seasons, although some birds migrate daily during summer. In lower latitudes, migration is less important, but occurs to a certain extent (Moreau 1951, 1966). The island-like nature of mountains greatly facilitates part-time, seasonal occupancy, since animals can reach the area above timberline or escape it by relatively short vertical migrations. This is a fundamental difference between alpine and arctic environments: Daily occupancy is not possible in the Arctic, and migration is almost totally limited to birds since, with the exception of the caribou, mammals make no attempt to move the great distances required (Irving 1972). By contrast, mammals move relatively easily to and from the alpine tundra, although few migrant mammals breed there. A number of herbivores (e.g., rabbit, porcupine, deer, moose, caribou, and elk) and a variety of predators (e.g., weasel, badger, wolverine, fox, wolf, and bear) wander above timberline during summer. Some retreat to lower altitudes during winter, but late winter upward migrations are practiced also by some, such as mountain caribou (*Rangifer tarandus caribou*) and black-tailed deer (*Ococoileus hemionus columbianus*), which use the high snow levels at upper elevations to access tree lichens in the subalpine forests. In late winter, hard snow crusts and firm windblown snow may allow mountain goats and sheep to resume roaming and to disperse to higher elevations. Here they move about unimpeded, feeding on the sparse but nutrient-rich vegetation exposed by winds on very high mountain ridges. Wolves and wolverines (*Gulo luscus*) may also hunt at the highest elevations in late winter.

The number of birds in the alpine tundra rises exponentially as they arrive for the brief summer. Many are returning from overwintering in lower latitudes, while others are simply migrating upward from the surrounding lowlands. In the passing parade from early spring to autumn, the first birds in the high country are usually herbivores with catholic food habits which harvest dead insects on the snow. As conditions improve and insects become plentiful, insectivores become dominant. A fairly high percentage of these bird herbivores and insectivores breed in the alpine tundra. Later in the summer, raptors (birds of prey) move in to harvest the newborn crop of small rodents and birds. The raptors seldom breed above timberline, but are mobile enough to migrate daily to and from the alpine area. In the Beartooth Mountains of Wyoming, raptors usually do not

appear before the end of July, but are plentiful from then on (Pattie and Verbeek 1966).

Migration is primarily restricted to birds and mammals. Reptiles and amphibians are not mobile enough, and barriers such as waterfalls and beaver dams generally prevent mountain fishes from migrating. Although there are some remarkable movements of insects in mountains, true insect migration is rare and takes place primarily among lowland species. For example, insects exhibit a curious tendency to seek out the summits of mountains, known as summit swarming or hill-hopping (Hudson 1905; Van Dyke 1919; Chapman 1954; Edwards 1956, 1957; Shields 1967). This has been observed by many people in many parts of the world, but the exact reasons for such behavior remain elusive. Some consider the insects to be victims of the winds, blown there against their will; others maintain that the wind may aid them but they move primarily of their own volition. Both explanations appear to be true. A variety of lowland insects are occasionally drawn up to high altitudes by ascending air currents, and if sufficient heights are reached, they are frozen by low temperatures. The dead and stunned insects are then deposited in vast quantities on glaciers and snowfields, where they serve as an important source of food for resident birds and nival insects (Swan 1961; Mani 1962; Papp 1978; Spalding 1979). Such movement is unintentional, however, and is clearly a one-way trip.

The case of insects moving upward under their own volition is much closer to true migration. The principal insect groups displaying this behavior are the *Coleoptera* (beetles, weevils), *Diptera* (mosquitoes, flies), *Hymenoptera* (bees, ants), and *Lepidoptera* (butterflies, moths) (Shields 1967). Explanations include an innate urge to ascend the highest point, the search for food, attraction to the heat and light, and the use of the highest point as a meeting place for mating. There is no reason to choose among these, since they may all operate at one time or another, but the last explanation seems best borne out by the known facts. The tendency among many insects to seek out the highest available point in order to mate has a selective advantage among sparse and isolated populations, for it ensures the meeting of males and unmated females and helps to stabilize the gene pool (Shields 1967).

A specialized example of insect migration is that of ladybird beetles (*Coccinellidae*). Vast swarms of these colorful little creatures have been observed on mountain peaks around the world (Edwards 1956, 1957; Mani 1962). They apparently swarm up from the lowlands to certain peaks in summer and congregate in assemblages numbering up to thousands of individuals per square meter. The largest single mass assemblage ever documented was in the western Himalaya at an elevation of 4,200 m (13,850 ft). Discovered on a snowfield, where they covered a solid patch 10 m (33 ft) in diameter, they were estimated to number about 200,000 individuals per m^2 (11 ft^2). This was in May, and most of the beetles were alive but inactive. It was possible to scoop up handfuls of the creatures and they would stir only slightly (Mani 1962).

The exact purpose of ladybird-beetle migration to mountain peaks is unclear, since they apparently do not mate or feed there. Some scientists believe that they go there to hibernate (Mani 1962), while others do not believe that they spend the winter on the high peaks intentionally, but think that they simply get caught by cold weather and have no other choice (Edwards 1957). Whatever the case, it is not uncommon to find large numbers of dead beetles under rocks and in crevices in the spring, while others are alive and preparing to move into the lowlands (as demonstrated by the example in the Himalaya, cited above). Such mass assemblages serve as food for other high-altitude creatures—mainly birds and predatory insects, but also large mammals. The Himalayan bear (*Ursus thibetanus*) is known to overturn stones in search of ladybird beetles, and the grizzly bear (*Ursus arctos*) also feeds on them (Chapman et al. 1955; Mani 1962).

Other causes of migration among mountain animals are outbreaks of bad weather and poor food availability. Migrations for these reasons are well known for the Arctic—for example, the fluctuating populations of lemmings and voles and their predators, or the Arctic fox (*Alopex lagopus*) and snowy owl (*Nydea scandiaca*), but such cyclical fluctuations are not usual in alpine environments (Hoffmann 1974). Nevertheless, the downward migration of species is frequently observed when especially bad conditions exist (Verbeek 1970; Ehrlich et al. 1972; Geist 1971). A specialized illustration of this is provided by the Clark's nutcracker (*Nucifraga columbiana*), a large corvid that lives in the subalpine zone of many northern hemisphere mountains and feeds through the winter on pine seeds it has cached in the ground. During years of poor seed production, lack of food may force the nutcracker to leave the high country. In the Sierra Nevada of California, seven major invasions of these birds into the lowlands have taken place since 1898 (Davis and Williams 1957, 1964), apparently the result of poor seed production.

Hibernation

Another method of effectively escaping the winter cold and lack of food is hibernation. In this amazing adaptation, an organism becomes inactive and passes the stressful period in a state of dormancy. Hibernation is essentially a deep and prolonged sleep. The metabolic rate in warm-blooded mammals may be reduced by up to two-thirds, and the internal temperature may be lowered to within a few degrees above freezing. Many cold-blooded mammals and insects survive internal temperatures below 0°C (32°F) (Hesse et al. 1951; Mani 1962). This results in vast energy savings for the organism. Hibernation is an extremely efficient survival mechanism for species not mobile enough to migrate, or unable or disinclined to remain active all winter feeding themselves. This includes members of almost all divisions of animal life except for the birds, which, owing to their mobility, find it much more expedient to migrate (Carpenter 1974, 1976).

Among alpine mammals, hibernation is best developed in ground squirrels and marmots. Of these, the ground squirrel (*Citellus* spp.) is perhaps the most remarkable: Its hibernation period may last over eight months (Manville 1959). The timing of hibernation is greatly affected by environmental conditions, especially the length of the growing season. In the Montana Rockies, ground squirrels in mountain valleys emerge in late April or early May, while those above timberline emerge in mid-June. Similarly, valley populations disappear into their dens by mid-August, while those in alpine meadows may stay active until the end of September, when snow is beginning to fall (Manville 1959). The higher populations apparently need this extra time in the fall to provide for the sufficient maturation of the young, the accumulation of body fat to see them through the winter, and the storage of enough food for the spring. Similar behavior patterns have been observed among marmots (Pattie 1967; Hoffmann 1974; Barash 1989).

Among large mountain mammals, bears are hibernators, reflecting in their hibernating strategies the effects of their Pleistocene histories. Grizzly bears are newcomers to North America, spreading and multiplying only after the late-Pleistocene extinctions of the native megafauna, including the huge, predacious short-faced bears (*Arctodus simus*). In Europe, grizzly bears coevolved with the larger cave bear (*Ursus spelaeus*). As its name implies, this huge bear occupied caves for hibernation, thereby foreclosing caves as hibernating sites to the smaller brown bear, of which the grizzly is a mere subspecies. Consequently, grizzly bears excavate and occupy their own hibernating sites with great secrecy. They often choose gravelly moraines covered by willow bushes where large-scale excavation is possible and the willow roots keep the cave from collapsing. Here grizzlies retreat in late fall during a heavy snowfall that obliterates their tracks. They need to be secretive because of the threat of a larger bear finding the hibernating site and killing its occupant. Following a bear when it is heading for its den in a snowstorm can be lethal!

Black bears, which evolved in North America among many powerful Pleistocene predators, retained arboreal habits to escape their enemies. They are shy and exceedingly clever, hibernating in shallow nests built in hollow trees, under uprooted trees, in brush piles, or in dens carved in scree-slopes. In the mountains, small bears favor higher-elevation cliffs and late hibernating dates when snow is on the ground, apparently to minimize cannibalism. All bears awake from their winter sleep if disturbed, and may abandon their dens. Females give birth to and suckle cubs in their winter dens, living off large stores of body fat accumulated in fall. Bears have evolved a highly specialized physiology that allows both hibernating and gestating. Hibernation is triggered not by temperature, but by food shortages (Stirling 1993).

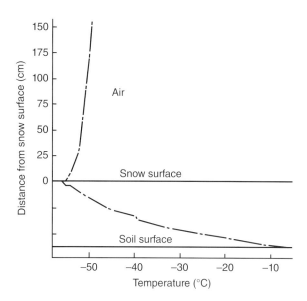

FIGURE 8.6 Generalized temperature gradient above and below a snow-covered surface in the subarctic. High mountain environments are similar. Note that the lowest temperatures occur at the snow–air interface, and temperatures increase rapidly within the snow down to the ground surface. This demonstrates the excellent insulative qualities of snow. (After Pruitt 1970: 86.)

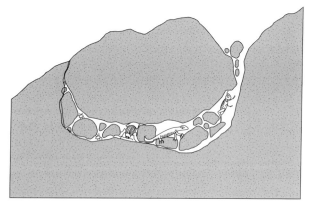

FIGURE 8.7 Various cold-blooded animals using the area underneath a rock as a microhabitat. The greatly tempered environmental conditions in such locations allow the survival of species in areas where surface conditions may be prohibitive. (After Mani 1962: 60.)

Cold-blooded animals—insects, reptiles, amphibians, and fish—all hibernate. Only warm-blooded animals are able to remain active during winter. Some invertebrates overwinter as eggs, while others hibernate as larvae, pupae, or adults. Species that normally complete their development in one season in the lowlands may take two or more seasons at higher elevations (Alexander and Hilliard 1969; Coulson et al. 1976). Butterflies in the Alps are known to overwinter as eggs in one generation and as pupae in the next. Carabid beetles in the western Himalaya were found hibernating as adults one year and as larvae the next (Mani 1962). A similar situation exists among the moths, flies, and spiders. On the other hand, high-altitude flies and

mosquitoes complete their development in one season (Smith 1966). The exact behavior depends on the insect type and environmental conditions, but since the growing season becomes progressively shorter at higher altitudes, overwintering in intermediate forms becomes increasingly characteristic (Fig. 8.2).

Insects hibernate by shallow burrowing or by crawling under rocks and into crevices. Snowfall must be adequate to cover the hibernation site during winter. Since the temperature under the snow remains fairly constantly at or near freezing (Fig. 8.6), hibernating insects need not endure extreme and fluctuating air temperatures. Insects usually do not emerge from hibernation in summer until the snow has melted. Thus, patterns of emergence depend on local conditions more than on altitude. Similarly, because hibernation is prolonged, development must be very rapid once conditions permit, in order for the insect to complete its life cycle or at least a segment of it. Since conditions are highly variable from year to year and from place to place, organisms must maintain considerable flexibility in the timing of their activities. Nature is not very precise or predictable in mountain environments; variability is the rule rather than the exception. Many animal forms show a kind of opportunism in their capacity to take advantage of favorable conditions when they appear and, equally, are able to withstand and persist through unfavorable periods (Downes 1964). For example, many high-altitude insects have the ability to begin hibernation at almost any stage of development and to remain dormant for prolonged periods. A case in point is seen when an avalanche traps insects under the snow for several years: They are usually not killed, but simply remain dormant until the snow melts and they are freed (Mani 1962).

Reptiles and amphibians do not generally inhabit the highest elevations, but some do range into the lower alpine zone (Hesse et al. 1951; Pearson 1954; Swan 1952, 1967; Karlstrom 1962; Campbell 1970; Bury 1973; Pearson and Bradford 1976). Being cold-blooded, their internal temperature is dependent upon the surrounding environment. They are largely limited to periods of activity when the sun is shining, and there is simply not enough heat or food during winter to maintain activity. Hibernation in cold-blooded animals is very different from that in warm-blooded animals. The latter experience a profound and controlled drop in internal temperature, but in cold-blooded animals it is difficult to detect any physiological difference between hibernating and non-hibernating individuals. Nevertheless, apparently, the heart rate and blood pressure do drop, and other subtle metabolic changes occur (Aleksiuk 1976).

In preparation for hibernation, snakes and lizards crawl under rocks or into animal burrows. They have a tendency to gather in groups, and it is not uncommon to find a variety of life forms, including insects, all congregated under one rock. This leads to very mixed bedfellows, with predator and prey gathering under the same roof, snuggled together for the duration (Fig. 8.7).

FIGURE 8.8 A pika (*Ochotona princeps*) in typical rocky habitat. (U.S. National Park Service.)

Snails, frogs, and salamanders bury themselves in mud at the bottom of ponds or in boggy areas, where they fall into a deathlike sleep that lasts until the snow and ice melt in the spring. As with other life forms, the development of amphibians may be delayed at higher altitudes. Thus, frogs in the Alps commonly pass the winter as tadpoles and require an extra year for full development (Hesse et al. 1951). Many high mountain fish hibernate once the ice forms over their heads and cuts off both their oxygen and food supply. They commonly gather in small schools at the bottom of the lake or stream and emit a protective ooze that enshrouds them all like a cloud. Fish living in shallow lakes or streams that freeze down to the bottom may actually be encased in ice without suffering damage.

Use of Microhabitats

Some species escape seasonal climatic extremes by burrowing or taking advantage of microenvironments under snow, rocks, and vegetation. The small mammals that remain active year round exhibit one of two basic types of behavior. In one group, the animal stores up food for use during winter; in the other, foraging continues much as usual. The first group includes pikas (*Ochotona* spp.), deer mice (*Peromyscus* spp.), wood rats (*Neotoma* spp.), various Eurasian hamsters (Cricetinae), and certain voles (*Microtus* spp.). These are all herbivores that gather bits of vegetation for their winter food. They construct nests in well-drained ground or amid rocks, and depend heavily on the presence of snow for insulation. Some species (e.g., lemmings) huddle together in subterranean nests and share body heat, while others (e.g., pikas) are strictly solitary. Chipmunks

and hamsters have the ability to become torpid for intermittent periods. This intermediate level of hibernation is a very useful adaptation, but they must occasionally awake and feed between periods of slumber. Other species such as the pocket gopher, deer mice, and pikas do not have this ability and must remain active throughout winter (Hoffmann 1974).

The pika is a particularly interesting little creature (Fig. 8.8). Pikas live in rocky habitats such as talus slopes and blockfields located near or above timberline. They are strongly territorial, with a definite spacing between their dens (Barash 1973; Smith 1974, 1978). Much of their summer activity consists of gathering twigs, which they deposit into a central "hay pile," for winter food. The exact winter behavior of pikas is unknown. They apparently do not build up stores of body fat or become torpid, so they depend heavily on their hay piles. Their dens are separated from the hay piles, so they must come up to the surface to feed, while other animals such as chipmunks use their winter supply of seeds as a bed and have only to turn their heads to feed. Another curious dietary characteristic of the pika is its tendency to reingest its own fecal matter. Like other members of the rabbit family, pikas excrete two kinds of feces. One is the familiar pellets, which are not reingested; the other is an elongated dark mass called "night" or "soft" feces, which is thought to come from excavation of a pouch in the large intestine. This material contains a high concentration of proteins and vitamins and its ingestion adds significantly to the pikas' nutrient intake (Johnson and Maxwell 1966).

The small mammals that continue foraging during the winter include the pocket gopher (*Thomomys* spp.), some voles (*Microtus* spp.), shrews (*Sorex* spp.), and weasels (*Mustela* spp.). Apart from the shrews and weasels, these all depend on seeds, roots, and other vegetable matter for food. The critical factor for their survival is the presence of snow, since they do not have adequate insulation to protect themselves from extremely low temperatures. Even when snow is present, temperatures remain near the freezing mark and conditions are suboptimal. The animals compensate by constructing warm nests, clustering together to share body heat, or, in the case of the weasel, by maintaining a high level of metabolism (Brown and Lasiewski 1972).

The pocket gopher is a good example of a small mammal that continues foraging throughout winter. These little creatures have well-developed front claws for digging and spend most of their time below ground or under snow. They are solitary and, during summer, each maintains its own underground burrow system, harvesting the plant roots that extend into the tunnels (Aldous 1951). During winter, however, the pocket gopher moves freely above the ground, under the protection of a deep snow cover, and harvests the surface vegetation. While most animals deposit their excavated material in mounds, the pocket gopher carries its refuse upward and stuffs it into snow tunnels. These are frequently quite extensive and, after the snow melts in spring, the material is deposited on the ground surface in a curious pattern of interwoven soil ropes.

There has been considerable concern over the question of whether the pocket gopher increases erosion through burrowing. While the close browsing of grasses and sedges under the snow may do temporary damage, it does not kill the plants. The annual deposition of fresh, loose soil on the surface may also result in some loss, but there is no evidence of major erosion (Ellison 1946; Stoecker 1976). In fact, it is generally believed that the fecal droppings and continual mixing of soil have a beneficial effect on soil and vegetation development in mountain meadows (Ingles 1952; Turner et al. 1973; Laycock and Richardson 1975).

These examples come from mid- and high-latitude mountains where strong seasonal contrasts exist and the chief problem for survival is posed by winter. No similar prolonged period of stress exists in tropical mountains, where the environmental extremes occur on a daily basis. In tropical mountains, there is very little migration or hibernation; instead, animals are heavily dependent on burrowing and use of microhabitats. Most animals are diurnal and are most active in early morning and late afternoon, between the periods of high sun intensity. At night, they retreat to sheltered sites amid rocks and vegetation or in shallow burrows. Most tropical mountain animals are not well adapted to cold and find alpine conditions only marginal; if small animals are live-trapped at night and left without bedding, they cannot survive freezing temperatures for more than an hour or two (Coe 1969). These animals capitalize on the fairly constant temperatures found in cavities under rocks or shallow burrows, however, and their own body heat raises the temperature of the confined space. The huddling

together of certain species, as in middle latitudes, also aids in heat conservation.

A good example of a high mountain mammal in the tropics is the Mount Kenya hyrax (*Procavia johnstoni mackinderi*) (Fig. 8.9), which lives above timberline on Mount Kenya, occupying rocky habitats below cliffs and in glacial moraines (Coe 1967; Roderick and Roderick 1973). Like many tropical mountain animals, it emerges early in the morning and basks in the sun by stretching out on its side to expose maximum body area to the sun; this compensates for the night chill. It then feeds for several hours, seeks cover during the hottest time of day, and appears again toward late afternoon to finish feeding (Coe 1969). Another animal on the same mountain, the groove-toothed rat (*Otomys orestes orestes*), finds shelter from night frosts and the intense sun by crawling underneath the drooping basal leaves of the giant *Senecios,* and frequently spends its resting hours in a cavity it has excavated in the trunk (Coe 1967).

Reptiles, being cold-blooded, are even more strongly controlled by the diurnal climatic regime in tropical mountains. The low night temperatures force them to be inactive and immobile. In early morning, they are barely able to crawl into the sunlight that will raise their internal temperatures and allow them to operate efficiently. When they emerge, they orient themselves to receive the maximum sunlight, and bask on rocks or dark surfaces so they are also heated from the underlying surface. Unlike warm-blooded mammals, they remain in the sun as much as possible. In fact, their periods of activity are almost entirely limited to sunshine; even a passing cloud will send them scurrying for cover. As a result, their feeding periods are generally limited to four hours or less per day. At night, they seek shelter under rocks, in shallow burrows, and in vegetation to escape the full extent of the cold (Pearson 1954; Pearson and Bradford 1976). Most reptiles select rocks at least 20 cm (8 in) thick—a size large enough to moderate the temperature extremes (Swan 1952, 1967; Coe 1969).

Invertebrates in tropical mountains also escape the daily extremes through timing of activities and clever use of microhabitats. In contrast to the highly active and visible insects of lowlands, those above timberline are sedentary and secretive. Consequently, there are very few insects visible at any given time. Insect collectors in tropical mountains have found most of their specimens beneath rocks and amid vegetation. This reclusiveness is viewed as a major adaptation to the rigors of the environment. Like mammals, insects largely restrict their activities to early morning and late afternoon. They are forced to this behavior because the low night temperatures make them lethargic and even comatose (Salt 1954).

An excellent example of vegetation serving as a microhabitat for insects is provided by the giant senecios and lobelias. These plants open their leaves each day and close them at night, and a number of insects take advantage of this habit to escape the nightly frosts. Temperature

FIGURE 8.9 A rock hyrax from South Africa. The body temperature of this species varies diurnally. (Photo by iStockphoto.com/ G Purdey.)

measurements on a giant lobelia revealed that, while the surrounding air cooled to −2.4°C (27.7 °F), the base of the flower remained at 3.3°C (37.9°F) and the center of the hollow stem was 4.0°C (39.2°F) (Coe 1969). Some insects live their entire lives without ever leaving the plant. This is true for a bibionid fly that lives within the flowers of the giant *Lobelia keniensis* on Mount Kenya. It feeds and reproduces in the flowers, where it may be found in all stages: adult, egg, larvae, and pupae. These flies form the primary food source for the scarlet-tufted malachite sunbird (*Nectarinia johnstoni johnstoni*), which hovers in front of the plant or moves up and down the trunk feeding on the insects (Coe 1969).

Another excellent plant shelter is provided by the large grass tussocks on Mount Kenya. Measurements within these tussocks reveal that temperatures fluctuate violently on the outer perimeter, less so halfway down the leaves, and remain fairly constant at the leaf base (Fig. 8.10). Temperatures on the outer leaf area during the measurement period showed a range of 13.3°C (23.9°F), while those at the leaf base revealed a range of only 2.1°C (3.8°F) (Coe 1969). A variety of insects live in these tussocks, regulating their temperature by moving up or down the blades of grass. One of the most interesting is a moth whose larva constructs silken tubes between the base and the outer leaves. The pupa has small spines on its side, which allow it to move up and down the tube to select the optimal temperature depending on the time of day. Similarly, several species of flies living in the tussocks display distinct diurnal movements (Coe 1969).

Mountain-dwelling large mammals are also masters at taking advantage of microclimates. During arctic cold spells, mountain sheep and goats may use caves at night. While these offer favorable microclimates, they can also act as death traps if predators clue in. Remains of sheep and goats have been found very deep in caves. Mountain sheep,

and also elk and moose, may be found in winter within strips of warm air at higher elevations, the *thermocline*. During deep snow, each mountain goat female with young becomes territorial and expels all other mature goats, including the largest males. These move off meekly ahead of the aggressive female—after all, they have just bred these very females, while the young may be their offspring from the preceding year; by meekly withdrawing they avoid competing with their own children. Only young males that have not bred may resist, but they are quickly overpowered by the aggressive females. Cliffs exposed to the sun not only shed snow, but also warm up, advancing seasonal plant growth (Geist 1971). In hot southern desert, it is shade that is vital during the day, and shady cliffs are much favored by desert sheep (Welles and Welles 1961). Mountain topography concentrates in tiny areas of fertile soil, mineral deposits, water seeps, snow pockets, and glaciers. All these may be important to the spatial and temporal distribution of the mountain biota.

Timing of Activities

A major survival strategy for organisms in high mountains is the timing of critical activities to coincide with the most favorable periods. This has already been discussed in connection with migration, hibernation, and escape into microhabitats; most small animals avoid the extremes of the seasons by employing one of these methods. Even during the favorable season, most activity is confined to the daylight hours. For this reason, predators such as the ermine, fox, and badger, which normally hunt at night in lowlands, hunt during the day in mountains (Zimina 1967). Restriction of activity to daylight hours is even more common in the tropics (Hingston 1925; Brown 1942; Pearson 1951; Salt 1954; Coe 1967, 1969). Unlike mountain sheep,

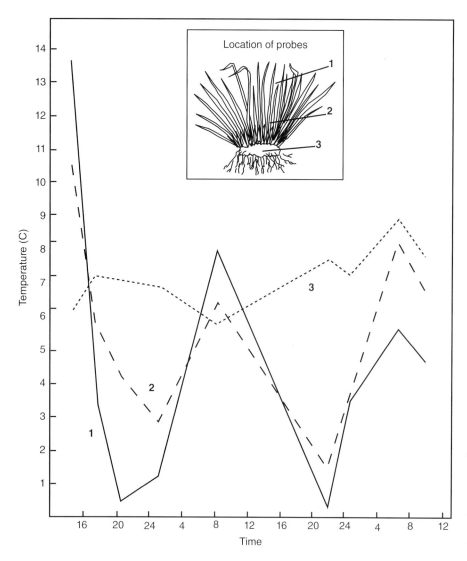

FIGURE 8.10 Temperature differences within a grassy tussock (*Festuca pilgeri*) in the alpine zone at 4,000 m (13,500 ft) on Mount Kenya, East Africa. Temperatures at the periphery of the tussock display much greater fluctuations than those within, indicating the more favorable microhabitat conditions within the tussock. (After Coe 1969: 115.)

mountain goats refuse to climb in cliffs at night. Mountain sheep have much larger eyes than mountain goats, and are probably well adapted to night vision.

The period of reproduction is of greatest interest and importance, since it is critical to the continuation of the species and is also a period of great vulnerability. The major problems for reproduction in high mountains are the same as those for other functions: brevity of the growing season and lack of food, contributing to severity of the environment. A number of adjustments are made in order to cope with these conditions. Migratory birds often arrive already paired, eliminating the need to spend time in courtship; others reduce the period of courtship (Hoffmann 1974). They arrive as early as possible in the spring (usually about the time the snow is beginning to melt), and occupy temporary roosts until appropriate nest building areas are bare (Pattie and Verbeek 1966). In the short alpine summer of mid- and high-latitude mountains, only one nesting attempt can be made. Because of the overwhelming urgency to complete the necessary functions within the

brief time allowed, the baby birds mature exceedingly quickly.

A wide variety of behaviors are displayed by different species of birds in relation to local environmental conditions. Often, either the male or the female does not stay and care for the young, but one or the other leaves as soon as the young are hatched. This is apparently an attempt to reduce competition for food, thereby increasing the chances of their offspring's survival (Hoffmann 1974). The timing of brood arrival in many species is closely synchronized with the seasonal peak in insect supply. This provides abundant and accessible food for the young at a time when they are poorly equipped to forage for themselves. From the insect's point of view, this approach also has survival value, since their avian predators are so swamped with insects that the survival of adequate numbers is assured (Maclean and Pitelka 1971; Hoffmann 1974).

Another carefully timed function is the molt (replacing feathers). It is usually coordinated with the breeding cycle and migration, so that these energy-expensive functions do not overlap. In some cases, molting may be compressed and take place before hatching occurs (Holmes 1966; Verbeek 1970), but more commonly it is delayed until after the breeding cycle has been completed (French 1959; Miller 1961). Long-distance migratory birds may leave the high country relatively early in the season and molt after arriving at their winter grounds, whereas short-distance migrants usually molt after the young arrive, and remain as long as possible before abandoning the alpine tundra (Hoffmann 1974).

Mammals show similar adjustments in their breeding cycles. The reproductive organs of many small animals enlarge and mature while the snow still covers the ground, so that breeding can take place during or immediately after snowmelt (Vaughan 1969). This is analogous to the tendency among alpine plants to begin growth under the snow in order to complete their life cycle in the short growing season. Another characteristic among mammals, as among birds, is that birth usually takes place when food is most readily available. Thus, animals living at higher altitudes breed later in the season than do those in the lowlands. This is partly because of the delayed snowmelt, but it is also because there would not be adequate food if they gave birth earlier (Pearson 1948, 1951; Geist 1971; Sweeney and Steinhoff 1976). The mule deer (Odocoileus hemionus) living at high altitudes in the Sierra Nevada of California fawn about the middle of July, while those at lower elevations fawn by the middle of May or earlier (Hoffmann 1974). Animals in tropical mountains do not have the problem of timing their breeding cycles, since food is available throughout the year. Consequently, reproduction can and does take place at any time (Coe 1967, 1969). It should be pointed out, however, that the tendency of mid- and high-latitude animals to reproduce in the favorable season of the year greatly reduces the environmental stress on them during this most vulnerable of periods. This is not true in tropical mountains, where climatic extremes occur on a daily basis; there, even newborn animals have to be capable of withstanding the entire range of environmental conditions present during the day and the night (Salt 1954).

Like birds, most mammals in mid- and high-latitude mountains are restricted to a single breeding attempt each year. Thus, while deer mice at lower elevations in the White Mountains, California, have two breeding seasons, those at high elevations have only one (Dunmire 1960). Similar behavior occurs among the reptiles and amphibians (Saint Girons and Duguy 1970; Goldberg 1974). In addition, the few reptiles that extend into the alpine zone (only three in the Alps) characteristically carry their young inside them and give birth to live young rather than laying eggs as lowland forms do. The reason for this is that there is simply not enough heat for cold-blooded animals to bring eggs to complete development and hatching in high mountains (Hesse et al. 1951). By carrying the young inside her and by keeping in the sun, the mother can put the heat she absorbs to maximum use. The ability to bear living young,

therefore, is fundamental to the entrance of reptiles into cold environments.

The reduction in the number of breeding attempts is offset somewhat by a tendency for litter size to increase with altitude and latitude (Lack 1948, 1954; Lord 1960; Spencer and Steinhoff 1968). The strategy behind this has been interpreted in different ways, but the most generally accepted explanation is that the shorter seasons at high altitudes limit the number of times an animal can reproduce in its lifetime compared to lowland environments, so it is advantageous to invest in a few large litters. This is despite of the facts that doing so reduces the life expectancy of the parents and that it is not as efficient as the production of several small litters (it is harder to care for a larger number of offspring). The production of a few large litters can be viewed as an all-or-nothing approach adopted because nothing is gained by being conservative (except in poor years, when they may not reproduce at all) (Spencer and Steinhoff 1968).

The timing of reproduction in relation to environmental severity is demonstrated by a comparative study of the woodchuck (Marmota monax) in southern Pennsylvania, the yellow-bellied marmot (M. flaviventris) in Yellowstone National Park, Wyoming, and the Olympic marmot (M. olympus) in Olympic National Park, Washington (Barash 1974, 1989). The lowland woodchuck, with the longest growing season, reproduces annually; the yellow-bellied marmot, with an intermediate growing season, also bears young annually but occasionally skips a year, while the alpine Olympic marmot has the shortest growing season and reproduces only in alternate years (Fig. 8.11). This is interpreted as an adaptation to the limited capacity of more extreme environments to support life.

The timing of reproduction is also tied to social behavior which, in some marmots, increases with environmental severity. Thus, the woodchuck is solitary, aggressive, and nonsocial, while the Olympic marmot lives in tightly knit colonies and is highly social. The yellow-bellied marmot in Wyoming and Montana is intermediate in sociability. This is a critical factor in determining when the young marmots disperse and form new colonies. If the young are forced to leave too early, their chances of survival decrease. Accordingly, woodchucks disperse the year they are born and become sexually mature as yearlings; yellow-bellied marmots remain with the parents for the first year and disperse the next, becoming sexually mature as two-year-olds. The Olympic marmot, however, remains with the parents for two full years and becomes sexually mature only in the third year. These characteristics are apparently the result of the time required in each environment for the animals to develop sufficient size and maturity to be able to disperse and reproduce successfully. For example, woodchucks achieve 80 percent of adult weight as yearlings, yellow-bellied marmots 60 percent, and Olympic marmots only 30 percent (Barash 1974). At the same time, increased sociability with increased environmental severity (which

Body size in large mammals varies in the same manner with altitude as it does with latitude. There is an initial increase, followed by a sharp reversal. That is, body size does not continue to increase with latitude as predicted by the invalid Bergmann's Rule. Since altitude and latitude compensate, the peak in body size is expected to be in mountains at lower latitudes, at an altitude equivalent in seasonality to 60–65°N (Geist 1999).

The fundamental problem for animals living in cold climates is that of maintaining their internal temperature. The body temperature of warm-blooded mountain and polar animals is essentially the same as that of tropical species; therefore, the contrast between internal temperature and environment is much greater in cold than in warm climates. At times, there may be a difference of as much as 100°C (180°F) between the interior of the body and the surrounding environment. Since internal temperature is relatively inflexible and must be maintained within narrow limits, the question is: How is this accomplished in animals that must live exposed to such conditions for weeks or months at a time? In answer, there are two main approaches: to reduce heat dissipation, or to increase heat production.

REDUCING HEAT DISSIPATION

Heat dissipation is reduced by increasing insulation and by reducing the temperature of extremities. Increased insulation is fundamental, and is a typical response of cold-climate animals. It is shown in the tropics by the Mount Kenya hyrax, which lives at 3,490 m (13,000 ft) and has thick luxurious fur, quite unlike the very thin-furred hyrax in the surrounding lowlands (Coe 1967). Perhaps the best example of insulation found anywhere is that of the arctic fox, which is so well protected that it can rest comfortably on the snow at temperatures down to −40°C (−40°F) before it has to raise its metabolism (Scholander et al. 1950). Similar abilities exist in larger animals such as mountain sheep and goats, wolves, bear, and caribou, but fur depth does not continue to increase on larger animals in proportion to their size; it appears to reach its maximum efficiency at about the size of the fox (Fig. 8.12).

On animals smaller than the fox, the fur becomes thinner; if it did not, these animals would not be able to move about. This is particularly true for the smallest forms such as shrews, lemmings, and weasels. The insulation value of the fur on these small animals is little more than that of tropical forms. Consequently, the only way they can survive is to escape into burrows and under snow. Bird species display no marked difference in plumage between warm and cold environments, apparently because of the restrictions imposed by flight requirements. However, the feathers of arctic and alpine birds are frequently structured to trap more air for insulation purposes than do the feathers of birds in warm climates (Irving 1972).

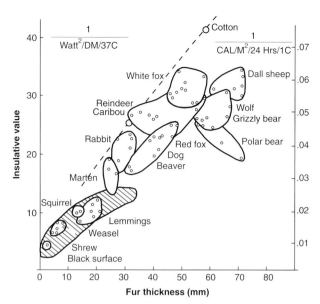

FIGURE 8.12 Insulation in relation to winter fur thickness in arctic and tropical animals. The insulative value of fur is roughly proportional to its thickness. Lemmings, squirrels, weasels, and shrews have a low degree of insulation, comparable to that of tropical animals, denoted by cross-hatched area. Consequently, they can only survive the cold by burrowing and sharing body heat. Heat transmission through the fur was measured in a room at 0°C (32°F), by stretching the fur over a ring and heating one side of it to 37°C (98.6°F) with a hot plate. Thermisters were used to ascertain the temperature on the other side of the skin. (After Scholander et al. 1950: 230.)

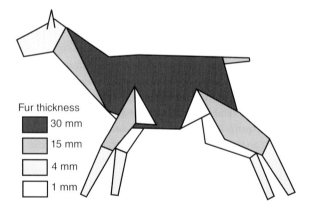

FIGURE 8.13 Diagrammatic representation of fur thickness on the guanaco (*Lama guanicoe*). The large variations in insulative quality from its back to its underbelly allow the animal to adjust to a wide range of temperatures (After Morrison 1966: 20.)

The amount of insulation on an animal is variable in space and time, owing to the need to both dissipate and conserve heat. Thus, certain areas of the body, especially the head, legs, and underbelly, may have thinner insulation than the rest. The guanaco (*Lama guanicoe*) of the high Andes has densely matted fur on parts of its body, while other areas are almost bare (Fig. 8.13). This animal lives in a dry environment, with intense sun and heat during

FIGURE 8.14 Ptarmigan display a continuously changing plumage from spring through autumn, and stay within either bare or snow areas depending on the condition of their plumage. (U.S. National Park Service.)

the day but rapid cooling and freezing at night. The variable insulation of the guanaco is designed to allow maximum flexibility (Morrison 1966). When the animal is hot, it can expose the more thinly insulated areas, but when cold, it can curl up and protect itself. This behavior is common in all animals. The mountain sheep rests with its legs extended during warm weather but, when it is cold, it tucks them under its body. Similarly, the fox or the wolf coils into a tight ball and wraps his long bushy tail across his face. Birds tuck their heads under their wings. The ability of fur or feather to insulate may also be controlled to a certain extent by flexing the fur or fluffing feathers to create more dead air spaces. Alternatively, the hair may be wetted or sleeked against the body, allowing greater heat loss. This is frequently associated with evaporative cooling through sweating and panting.

On a seasonal basis, animals commonly vary their insulation by molting or shedding fur or feathers. The mountain sheep's coat consists of a layer of dense woolly under-hair and an outer layer of long brittle guard hairs. Most of the guard hairs are rubbed off in spring during the major molt (Geist 1971). Similar behavior is observed in many cold-climate animals, and measurements made of the insulation ability of fur show marked differences between winter and summer (Hart 1956). The insulation value of the feathers of birds may be slightly less in summer, but the differences are not so marked as with animal fur (Porter et al. 2000). One of the most complex molts of any kind is that of the ptarmigan, which displays a continuously changing plumage from spring through autumn (Hoffmann 1974; Fig. 8.14). In this case, however, escape from detection by predators appears to be the selective agent.

The other major method of reducing heat dissipation is to lower the temperature of the extremities. This decreases the thermal gradient between the body and the surrounding environment so that less heat is lost from these less insulated areas, and it saves the energy that would be required to maintain the extremities at a higher temperature. Examples are the bare legs and feet of birds, and the legs, feet, and noses of animals such as the fox, wolf, caribou, or mountain sheep (Fig. 8.15).

The bare feet of birds and mammals standing in icy water or on cold snow are maintained just above the freezing temperature of the tissue, which may be as low as $-1°C$ ($30.2°F$) (Irving and Krog 1955; Irving 1964, 1966, 1972; Henshaw et al. 1972). This is accomplished by controlling the temperature and volume of the blood circulating to these areas. Many cold-climate animals have a circulatory system in which the blood coming from the extremities through the veins, and the blood leaving the heart through the arteries, passes through a series of vascular heat exchangers (Scholander 1957) (Figure 8.16). This lowers the temperature of the blood moving toward the extremities and raises the temperature of the blood returning to the heart, thus conserving heat within the body (Scholander 1955). If the animal suddenly has to increase its metabolism, for example, to escape predation, it has to dissipate heat. To accomplish this, it increases the flow of blood to the extremities, raising their temperature and therefore the air-to-body gradient.

Insulation, by contrast, is efficient and flexible. This is demonstrated in Figure 8.17, which shows the temperature sensitivity of selected tropical and cold-climate animals. Tropical animals increase their metabolic rate sharply when

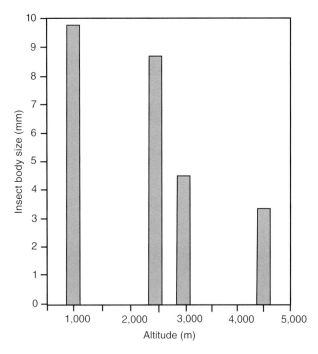

FIGURE 8.18 Decrease in beetle (*Carabidae*) body size with elevation in the Himalayas. (After Mani 1962: 97.)

everyone who has hiked in mountains in the spring has observed watermelon snow, the result of reddish-colored snow algae that accumulate in melt depressions in the snow (Fig. 8.20; Pollock 1970; Hardy and Curl 1972; Thomas 1972). The optimal growth temperature of most snow algae ranges from 0°C to 10°C (32°–50°F) (Hoham 1975).

Mention should also be made of snow worms, springtails, glacier fleas, and mites, all of which live in a constantly cold milieu (Marchand 1917; Scott 1962; Swan 1967). These are so cold-limited that they can be killed by the warmth of the human hand (Mani 1962). Of course, most organisms live at lower altitudes than snowline and can withstand much greater ranges of temperature. In arid mountains, moisture may be more important than temperature. Thus, frogs, snails, slugs, and many waterborne insects are limited to areas with adequate moisture and are not found in dry areas such as the Puna de Atacama, the inner ranges of the trans-Himalayas, or desert mountains.

insects can withstand being frozen solid, mainly in the intermediate stages (i.e., eggs, larvae, or pupae), but the major adaptation to subfreezing temperatures in insects is supercooling, achieved through the presence of glycerol, a natural antifreeze ingredient (Smith 1958; Salt 1961, 1969). Resistance to freezing is also related to moisture content and atmospheric pressure: The less body fluid an insect has and the lower the air pressure, the more supercooling the insect displays (Salt 1956; Crawford and Riddle 1974).

Adaptation to cold has become so complete in a few organisms that they are limited to environments with low temperatures. In the humid páramos of the Andes, frogs of the genus *Telmatobius* are found up to the snowline (4,550 m, 15,000 ft), but they seldom come below 1,500 m (6,000 ft) and then only along cold meltwater streams (Hesse et al. 1951). A cave beetle (*Silphidae*) lives in ice grottoes on glaciers at temperatures ranging from −1.7°C to 1.0°C (30.7°–33.8°F) (Hesse et al. 1951). Almost

Lack of Oxygen

The decrease in oxygen with altitude is more or less constant, so all mountain animals are subjected to oxygen-deficient conditions. Surprisingly little research has been done concerning the effects of lack of oxygen on naturally high-altitude animals; some of the earliest work was carried out in the mid-1930s (Hall et al. 1936; Kalabukov 1937).

The response of most lowland wild animals to altitude is basically similar to that observed in humans (Chiodi 1964; Timiras 1964; Hock 1964c, 1970). Curiously, however, some of the animals best adapted to high altitudes—for example, rodents, the Camelidae (llama, alpaca, vicuña, guanaco), and the yak, sheep, and goats—do not display these characteristics. The physiological characteristics of these high-altitude animals are often the opposite of those observed in sea-level animals exposed to high altitudes. The primary contrast is that they do not show an increase in red blood cells or hemoglobin (Hall et al. 1936; Morrison et al. 1963a, 1963b). They tend to have a larger blood-plasma volume, but there is less oxygen in their blood and they are able to function at a very low partial pressure of oxygen (Morrison and Eisner 1962; Bullard 1972). They do have larger hearts and lung volumes as

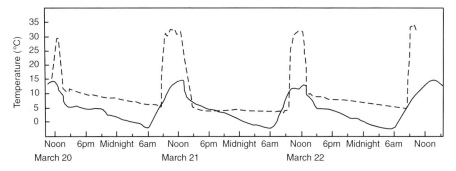

FIGURE 8.19 Body temperature of free-living lizard (*Liolaemus multiformis*) at an altitude of 4,300 m (14,200 ft) in the Peruvian Andes. This lizard is able to achieve and maintain a higher temperature (dashed line) than that of the surrounding air (solid line) by its behavior, for example, by basking or seeking shelter according to the weather conditions. (After Pearson and Bradford 1976: 157)

FIGURE 8.20 Snow algae in the high Cascades, Oregon. These algae accumulate in large numbers in the spring as the snow melts, turning pockets and depressions a deep red color. This is known as "watermelon snow" because of its distinctive color and odor. (Photo by L. W. Price.)

well as higher heart and breathing rates, just like animals transported to high altitudes, but there is no increase in the number of red blood cells or quantity of hemoglobin. Since this is one of the most conspicuous responses by lowland species taken to high elevations, its absence in native highland animals is striking. Although the increase in the production of red blood cells and hemoglobin appears to be the simplest means of improving oxygen transport, it must also present some disadvantage: perhaps that of the increased viscosity of cell-rich blood (Morrison 1964).

These differences in approach may be accounted for by two distinct processes: acclimatization and adaptation. Both meet the requirements of high altitudes, but the former is an adjustment or distortion of an existing pattern, whereas the latter is achieved through long-term selection and development. Acclimatization operates at the individual level on a short-term basis, whereas adaptation operates at the species or population level through time. A few wild mountain animals do display an increase in production of red blood cells and hemoglobin (Kalabukov 1937; Hock 1964c), but these occur in mountains where there is direct contact between valleys and peaks, facilitating intermixture. The Andes and trans-Himalaya, however, present broad areas at high altitudes which support a stable population separated from lowland habitats by barriers (Morrison 1964). The animal populations in smaller dissected mountains are apparently displaying acclimatizations to altitude rather than adaptations, while populations native to the higher and more massive uplands have developed their abilities through isolation and evolution, and their characteristics are viewed as adaptations (Morrison 1964). It has been suggested that even these mammals probably increased production of red cells and hemoglobin upon first inhabiting the mountain area but, through time,

this response was replaced by increased blood-plasma volume (Bullard 1972). It is unfortunate that the native highland animals most thoroughly studied are the rodents and camelids, since the rodents are essentially preadapted to lack of oxygen because of their ability to live underground and thus avoid costly flight from predators, while the camelids possess a number of unusual adaptations as a family even at sea level (Chiodi 1970–1971).

Thus, mammals appear to fall into two groups. Some species, such as mice, rats, and man, show an increase in the number of red blood cells and the quantity of hemoglobin at high altitudes. These are primarily low-altitude animals with the ability to acclimatize and live successfully at high altitudes. The other group consists of species that have been native high-altitude animals through geologic time, for example, certain rodents, the Camelidae, and other hoofed animals such as the yak, sheep, and goats; these do not display an increase in number of red cells or quantity of hemoglobin in their blood (Bullard 1972). This approach has apparently been favored by long-term selection and is viewed as adaptive. Although both approaches appear to achieve the same end, they are not equally efficient: The former, based on the need for greater oxygen in the system, is metabolically more expensive (Morrison 1964).

Clearly, animals deal with the stressful conditions and opportunities of high altitudes in numerous ways. Some make the high mountains a permanent home and display elaborate adaptations to escape predators, reduce the effects of low temperature and high snow accumulation, and deal with the brevity of the growing season and the seasonal scarcity of food. Most modify and escape these negative factors through behavior, followed by acclimatization and adaptations in their physiology and morphology. Similarly, as discussed in Chapters 10 and 11, human populations also extend into the high mountains, primarily through modification and amelioration of the environment, though there are biological adaptations to mountains in some human populations.

References

Aldous, C. M. 1951. The feeding habits of Pocket gophers (*Thomomys talpoides moorei*) in the high mountain ranges of Central Utah. *Journal of Mammalogy* 32(1): 84–87.

Aleksiuk, M. 1976. Reptilian hibernation: Evidence of adaptive strategies in *Thamnoplis sirtalis parietalis. Copeia* 1: 170–178.

Alexander, A. F., and Jensen, R. 1959. Gross cardiac changes in cattle with high mountain (brisket) disease and in experimental cattle maintained at high altitudes. *American Journal of Veterinary Research* 20: 680–689.

Alexander, G. 1964. Occurrence of grasshoppers as accidentals in the Rocky Mountains of northern Colorado. *Ecology* 45(1): 77–86.

Alexander, G., and Hilliard, J. R., Jr. 1969. Altitudinal and seasonal distribution of Orthoptera in the Rocky Mountains of northern Colorado. *Ecological Monographs* 39: 385–431.

Allen, J. A. 1877. The influence of physical conditions in the genesis of species. *Radical Review* 1: 108–140.

Andersen, D. C., Armitage, K. B., and Hoffman, R.S. 1976. Socioecology of marmots: Female reproductive strategies. *Ecology* 57(3): 552–560.

Andrewartha, H. G., and Birch, L. C. 1954. *The Distribution and Abundance of Animals.* Chicago: University of Chicago Press.

Aranoff, S. 1989. *Geographic Information Systems: A Management Perspective.* Ottawa, ON: WDL Publications.

Barash, D. P. 1973. Territorial and foraging behavior of pika (*Ochotona princeps*) in Montana. *American Midland Naturalist* 89: 202–207.

Barash, D. P. 1974. The evolution of marmot societies: A general theory. *Science* 185(4149): 415–420.

Barash, P. D. 1989. *Marmots: Social behavior and Ecology.* Stanford, CA: Stanford University Press.

Barnosky, A. D., Koch, P. L., Feranec, R. S., Wing, S. L. and Schabel, A. B. 2004. Assessing the causes of late Pleistocene extinctions on the continents. *Science* 306: 70–75.

Bergmann, C. 1847. Über die Verhaltnisse der Warmeökonomie der Thiere zu ihrer Grösse. *Göttinger Studies* 3: 595–708.

Bliss, L. C. 1962. Adaptations of arctic and alpine plants to environmental conditions. *Arctic and Alpine Research* 15: 117–144.

Boyce, M. S., and Haney, A., eds. 1977. *Ecosystem Management.* New Haven, CT: Yale University Press.

Brinck, P. 1966. Animal invasion of glacial and late glacial terrestrial environments in Scandinavia. *Oikos* 17: 250–266.

Brinck, P. 1974. Strategy and dynamics of high altitude faunas. *Arctic and Alpine Research* 6(2): 107–116.

Brown, F. M. 1942. *Animals above Timberline, Colorado and Ecuador.* Colorado Springs: Colorado College Studies 33.

Brown, J. H. 1971. Mammals on mountaintops: Nonequilibrium insular biogeography. *American Naturalist* 105(945): 467–478.

Brown, J. H., and Lasiewiski, R. C. 1972. Metabolism of weasels: The cost of being long and thin. *Ecology* 53(5): 939–943.

Brown, J. H., and Lee, A. K. 1969. Bergmann's Rule and climatic adaptation in wood rates (Neotoma). *Evolution* 23: 329–338.

Bullard, R. W. 1972. Vertebrates at altitude. In M. K. Yousef, S. M. Horvath, and R. W. Bullard, eds., *Physiological Adaptations: Desert and Mountain* (pp. 209–225). New York: Academic Press.

Bury, B. R. 1973. The Cascade frog, *Rana cascadae,* in the North Cascade Range of Washington. *Northwest Science* 47(4): 228–229.

Buskird, E. R., Kollias, J., Akers, R. F., Prokop, E. K., and Picon-Reategui, E. 1967. Maximal performance at altitude and on return from altitude in conditional runners. *Journal of Applied Physiology* 23: 259–266.

Campbell, J. B. 1970. New elevational records for the Boreal toad (*Bufo boreas boreas*). *Arctic and Alpine Research* 2(2): 157–159.

Carlquist, S. 1974. *Island Biology.* New York: Columbia University Press.

Carpenter, F. L. 1974. Torpor in an Andean hummingbird: Its ecological significance. *Science* 183: 545–547.

Carpenter, F. L. 1976. *Ecology and Evolution of an Andean Hummingbird* (Oreotrochilus estella). Berkeley: University of California Press.

Chadwick, D. H. 1977. The influence of mountain goat social relationships on population size and distribution. In W. Samuel and W. G. Macgregor, eds., *Proceedings of the International Mountain Goat Symposium* 1: 74–91.

Chapman, J. A. 1954. Studies on summit-frequenting insects in western Montana. *Ecology* 35: 41–49.

Chapman, J. A., Romer, J. I., and Stark. J. 1955. Ladybird beetles and army cutworm adults as food for grizzly bears in Montana. *Ecology* 36(1): 156–158.

Chiodi, H. 1964. Action of high altitude chronic hypoxia on newborn animals. In W. H. Weihe, ed., *The physiological Effects of High Altitude* (pp. 97–114). New York, Macmillan.

Chiodi, H. 1970–1971. Comparative study of the blood gas transport in high altitude and sea level Camelidae and goats. *Respiration Physiology* 11: 84–93.

Coe, M. J. 1967. *The Ecology of the Alpine Zone of Mount Kenya.* The Hague: Junk.

Coe, M. J. 1969. Microclimate and animal life in the equatorial mountains. *Zoologica Africana* 4(2): 101–128.

Coe, M. J., and Foster, J. B. 1972. The mammals of the northern slopes of Mt. Kenya. *Journal of the East Africa Natural History Society and National Museum* 131: 1–18.

Coulson, J. C., Horobin, J. C., Butterfield, J., and Smith, G. R. J. 1976. The maintenance of annual life-cycles in two species of *Tipulidae* (Diptera): A field study relating development, temperature, and altitude. *Journal of Animal Ecology* 45(1): 215–234.

Crawford, C. S., and Riddle, W. H. 1974. Cold hardiness in centipedes and scorpions in New Mexico. *Oikos* 25(1): 86–92.

Darlington, P. J. 1943. Carabidae of mountains and islands. *Ecology* 13: 37–61.

Daubenmire, R. 1968. *Plant Communities.* New York: Harper and Row.

Davis, J., and Williams, L. 1957. Irruptions of the Clark's nutcracker in California. *Condor* 59: 297–307.

Davis, J., and Williams, L. 1964. The 1961 irruption of the Clark's nutcracker in California. *Wilson Bulletin* 76: 10–18.

Delcourt, H., and Delcourt, P. A. 1997. Pre-Columbian Native American use of fire on southern Appalachian landscapes. *Conservation Biology* 11: 1010–1014.

Diamond, J. M. 1970. Ecological consequence of island colonization by southwest Pacific birds, I: Types of niche shifts. *Proceedings of the National Academy of Sciences of the USA* 67: 529–536.

Diamond, J. M. 1972. *The Avifauna of the Eastern Highlands of New Guinea*. Cambridge, MA: Nuttall Ornithology Club.

Diamond, J. M. 1973. Distributional ecology of New Guinea birds. *Science* 179: 759–769.

Diamond, J. M. 1975. The island dilemma: Lessons of modern biogeographic studies for the design of natural reserves. *Biological Conservation* 7: 129–146.

Diamond, J. M. 1976. Island biogeography and conservation: Strategy and limitations. *Science* 193: 1027–1029.

Dobzhansky, T. 1950. Evolution in the tropics. *American Scientist* 38: 209–221.

Downes, J. A. 1964. Arctic insects and their environment. *Canadian Entomologist* 96: 279–307.

Dunbar, M. J. 1968. *Ecological Development in Polar Regions*. Englewood Cliffs, NJ: Prentice-Hall.

Dunmire, W. W. 1960. An altitudinal survey of reproduction: *Peromyscus maniculatus*. *Ecology* 41: 174–182.

Edwards, J. G. 1956. Entomology above timberline. *Mazama Club Annual* 38(13): 13–17.

Edwards, J. G. 1957. Entomology above timberline, II: The attraction of ladybird beetles to mountain tops. *Coleopterists Bulletin* 11: 41–46.

Ehrlich, P. R., Breedlove, D. E., Brussard, P. F., and Sharp, M. A. 1972. Weather and the regulation of subalpine populations. *Ecology* 53: 243–247.

Ellison, L. 1946. The pocket gopher in relation to soil erosion on mountain range. *Ecology* 27: 101–114.

Festa-Bianchet, M., and Côté, D. 2008. *Mountain Goats: Ecology, Behavior and Conservation*. Washington, DC: Island Press.

French, N. R. 1959. Life history of the black rosy finch. *Auk* 76: 158–180.

Geist, V. 1971. *Mountain Sheep: A Study in Behavior and Evolution*. Chicago: University of Chicago Press.

Geist, V. 1978. *Life Strategies, Human Evolution, Environmental Design*. New York: Springer-Verlag.

Geist, V. 1987. Bergmann's Rule is invalid. *Canadian Journal of Zoology* 65:1035–1038.

Geist, V. 1990. Bergmann's Rule is invalid: A reply to J. D. Paterson. *Canadian Journal of Zoology*. 68: 1613–1615.

Geist, V. 1998. *Deer of the World: Their Evolution, Behavior, and Ecology*. Mechanicsburg, PA: Stackpole Books.

Geist, V. 1999. Adaptive strategies in mountain sheep. In R. Valdez and P. R. Krausman, eds., *Mountain Sheep of North America* (pp. 192–208). Tucson: University of Arizona Press.

Geist, V. 2002. *Mountain Sheep and Man in the Northern Wilds*. Caldwell, NJ: Blackburn Press.

Goldberg, S. R. 1974. Reproduction in mountain and lowland populations of the lizard *Sceloporus occidentalis*. *Copeia* 1: 176–182.

Green, C. V. 1936. Observations on the New York weasel, with remarks on its winter dichromatism. *Journal of Mammalogy* 17: 247–249.

Guthrie, R. D. 1990. *Frozen Fauna of the Mammoth Steppe*. Chicago: University of Chicago Press.

Hackman, W. 1964. On reduction and loss of wings in Diptera. *Notulae Entomologicae* 46: 73–93.

Häfeli, H. 1968. The alpine salamander. In M. Barnes, ed., *The Mountain World* (pp. 166–174). London: George Allen & Unwin.

Hall, F. G., Dill, D. B., and Guzman-Barron, E. S. 1936. Comparative physiology in high altitudes. *Journal of Cellular and Comparative Physiology* 8: 301–313.

Hammel, H. T. 1956. Infrared emissivities of some arctic fauna. *Journal of Mammalogy* 37(3): 375–381.

Hardy, J. T., and Curl, H., Jr. 1972. The candy-colored, snow-flaked alpine biome. *Natural History* 81: 74–78.

Hargens, A. R. 1972. Freezing resistance in polar fishes. *Science* 176(4031): 184–186.

Hart, J. S. 1956. Seasonal changes in insulation of the fur. *Canadian Journal of Zoology* 34: 53–57.

Heinrich, B. 1974. Thermoregulation in endothermic insects. *Science* 185(4153): 747–756.

Henshaw, R. E., Underwood, L. S., and Casey, T. M. 1972. Peripheral thermoregulation: Foot temperature in two arctic canines. *Science* 175(4025): 988–990.

Hesse, R., Allee, W. C., and Schmidt, K. P. 1951. *Ecological Animal Geography*. New York: John Wiley and Sons.

Hingston, R. W. G. 1925. Animal life at high altitudes. *Geographical Journal* 65(3): 185–198.

Hock, R. J. 1964a. Animals in high altitudes: Reptiles and amphibians. In D. B. Dill, ed., *Adaptation to the Environment* (pp. 841–842). Washington, DC: American Physiological Society.

Hock, R. J. 1964b. Physiological responses of deer mice to various native altitudes. In W. H. Weihe, ed., *The Physiological Effects of High Altitude* (pp. 59–72). New York: Macmillan.

Hock, R. J. 1964c. Terrestrial animals in cold: Reptiles. In D. B. Dill, ed., *Adaptation to the Environment* (pp. 357–360). Washington, DC: American Physiological Society.

Hock, R. J. 1965. An analysis of Gloger's Rule. *Hvalradet Skrifter* (Oslo) 48: 214–226.

Hock, R. J. 1970. The physiology of high altitude. *Scientific American* 222(2): 53–62.

Hoffmann, R. S. 1974. Terrestrial vertebrates. In J. D. Ives and R. G. Barry, eds., *Arctic and Alpine Environments* (pp. 475–568). London: Methuen.

Hoffmann, R. S., and Taber, R. D. 1967. Origin and history of holarctic tundra ecosystems, with special references to their vertebrate faunas. In J. W. H. Osborn and H. E. Wright, eds. *Arctic and Alpine Environments* (pp. 143–170). Bloomington: Indiana University Press.

Hoham, R. W. 1975. Optimum temperatures and temperature ranges for growth of snow algae. *Arctic and Alpine Research* 7(1): 13–24.

Holmes, R. T. 1966. Molt cycle of the red-backed sandpiper (*Calidris alpina*) in western North America. *Auk* 83: 517–533.

Houston, C. S. 1972. High-altitude pulmonary and cerebral edema. *American Alpine Journal* 18(1): 83–92.

Hudson, G. V. 1905. Notes on insect swarms on mountain tops in New Zealand. *Transactions of the Royal Society of New Zealand* 38: 334–336.

Huston, M. A., and Wolverton, S. 2009. The global distribution of net primary production: Resolving the paradox. *Ecological Monographs* 79: 343–377.

Huston, M. A., and Wolverton, S. 2011. Regulation of animal size by eNPP, Bergmann's Rule, and related phenomena. *Ecological Monographs* 81: 349–405.

Ingles, L. G. 1952. The ecology of the mountain pocket gopher, *Thomomys monticola*. *Ecology* 33(1): 87–95.

Irving, L. 1960. Human adaptation to cold. *Nature* 185: 572–574.

Irving, L. 1964. Terrestrial animals in cold: Birds and mammals. In D. B. Dill, ed., *Adaptation to the Environment* (pp. 361–378). Washington, DC: American Physiological Society.

Irving, L. 1966. Adaptations to cold. *Scientific American* 214: 94–101.

Irving, L. 1972. *Arctic Life of Birds and Mammals*. New York: Springer.

Irving L., and Krog J. 1955. Skin temperatures in the Arctic as a regulator of heat. *Journal of Applied Physiology* 7: 354–363.

Johnson, D. R., and Maxwell, M. H. 1966. Energy dynamics of Colorado pikas. *Ecology* 47: 1059–1061.

Johnson, N. K. 1975. Controls of number of bird species on montane islands in the Great Basin. *Evolution* 29: 545–567.

Kalabukov, N. J. 1937. Some physiological adaptations of the mountain and plain forms of the wood mouse (*Apodemus sylvaticus*) and of other species of mouse-like rodents. *Journal of Animal Ecology* 6: 254–274.

Karlstrom, E. L. 1962. *The Toad Genus* Bufo *in the Sierra Nevada of California*. Publications in Zoology 62. Berkeley: University of California.

Kendeigh, S. C. 1932. A study of Merriam's temperature laws. *Wilson Bulletin* 33(3): 129–143.

Kendeigh, S. C. 1954. History and evaluation of various concepts of plant and animal communities in North America. *Ecology* 35: 152–171.

Kendeigh, S. C. 1961. *Animal Ecology*. Englewood Cliffs, NJ: Prentice-Hall.

Kendeigh, S. C. 1969. Tolerance of cold and Bergmann's Rule. *Auk* 86: 13–25.

Kikkawa, J., and Williams, E. E. 1971. Altitude distribution of land birds in New Guinea. *Search* 2: 64–65.

Komarek, E. V. 1981. History of prescribed fire and controlled burning in wildlife management in the South. In G. W. Wood, ed., *Prescribed Fire and Wildlife in Southern Forests* (pp. 1–14). Georgetown, SC: Belle W. Baruch Forest Science Institute.

Lack, D. 1948. The significance of clutch size, III: Some interspecific comparisons. *Ibis* 90: 25–45.

Lack, D. 1954. *The Natural Regulation of Animal Numbers*. Oxford, UK: Oxford University Press.

Laycock, W., A., and Richardson, B. Z. 1975. Long-term effects of pocket gopher control on vegetation and soils of a subalpine grassland. *Journal of Range Management* 28(6): 458–462.

Leakley, R., and Lewin, R. 1996. *The Sixth Extinction: Biodiversity and its Survival*. London: Orion Publishing Group Ltd.

Lettau, H. H. 1967. Small to large-scale features of boundary layer structure over mountain slopes. In E. R. Reiter and J. L. Rasmussen, eds., *Proceedings of the Symposium on Mountain Meteorology* (pp. 1–74). Fort Collins: Colorado State University.

Lord, R. D., Jr. 1960. Litter size and latitude in North American mammals. *American Midland Naturalist* 64(2): 488–499.

MacArthur, R. H. 1972. *Geographical Ecology*. New York: Harper and Row.

MacCallum, B., and Geist, V. 1992. Mountain restoration: Soil and surface wildlife habitat. *Geographical Journal* 27(1): 23–46.

MacCallum, B., and Geist, V. 1995. Reclamation of a mountain coal mine: Designing habitat for bighorn sheep. In N. J. R. Allen, ed., *Mountains at Risk* (pp. 152–195). New Delhi: Manohar.

Maclean, S. F., Jr., and Pitelka, F. A. 1971. Seasonal patterns of abundance of tundra arthropods near Barrow. *Arctic* 24: 19–40.

Mani, M. S. 1962. *Introduction to High Altitude Entomology*. London, Methuen.

Mani, M. S. 1968. *Ecology and Biogeography of High Altitude Insects*. The Hague: Junk.

Manville, R. H. 1959. The Columbian ground squirrel in northwestern Montana. *Journal of Mammalogy* 40: 26–45.

Marchand, W. 1917. Notes on the habits of snowfly (Chiona). *Psyche* 24: 142–153.

Martin, P., and Klein, R., eds. 1984. *Quaternary Extinctions. A Prehistoric Revolution*. Tucson: University of Arizona Press.

Martof, B. S., and Humphries, R. L. 1959. Geographic variation in the wood frog, *Rana sylvatica*. *American Midland Naturalist* 6: 350–389.

Mayr, E. 1956. Geographical character gradients and climatic adaptation. *Evolution* 10: 105–108.

Mayr, E., and Diamond, J. M. 1976. Birds on islands in the sky: Origin of the montane avifauna of Northern Melanesia. *Proceedings of the National Academy of Science* 73(5): 1765–1769.

McNab, B. K. 1971. On the ecological significance of Bergmann's Rule. *Ecology* 52: 845–854.

Meinertzhagen, R. 1928. Some biological problems connected with the Himalaya. *Ibis* 4: 480–533.

Merriam, A. C. 1890. Telegraphing among the ancients. *Papers of the Archaeological Institute of America* 3: 1–32.

Merriam, C. H. 1894. The geographic distribution of animals and plants in North America. In *U.S. Department of Agriculture Yearbook* (pp. 203–214). Washington, DC: U.S. Government Printing Office.

Merriam, C. H. 1898. Life zones and crop zones of the United States. In *U.S. Department of Agriculture Biological Survey Bulletin 10* (pp. 9–79). Washington, DC: U.S. Government Printing Office.

Miller, A. H. 1961. Molt cycles in equatorial Andean sparrows. *Condor* 63: 143–161.

Moreau, R. E. 1951. The migration system in perspective. Proceedings of the 10th International Ornithological Congress, Uppsala (pp. 245–248).

Moreau, R. E. 1966. *The Bird Faunas of Africa*. New York: Academic Press.

Morrison, P. R. 1964. Wild animals at high altitudes. *Symposia of the Zoological Society of London* 13: 49–55.

Morrison, P. R. 1966. Insulative flexibility in the guanaco. *Journal of Mammalogy* 47: 18–23.

Morrison, P. R., and *Eisner, R. 1962*. Influence of altitude on breathing rates in some Peruvian rodents. *Applied Physiology* 17: 467–470.

Morrison, P. R., Kerst, K., Reynafarje, C., and Ramos, J. 1963a. Hematocrit and hemoglobin levels in some Peruvian rodents from high and low altitude. *International Journal of Biometeorology* 7: 51–58.

Morrison, P. R., Kerst, K., and Rosenmann, M. 1963b. Hematocrit and hemoglobin levels in some Chilean rodents from high and low altitude. *International Journal of Biometeorology* 7: 44–50.

Murie, A. 1944. *The Wolves of Mount McKinley.* Fauna of the United States, Fauna Series No. 5. Washington, DC: Government Printing Office.

Newman, M. T. 1956. Adaptation of man to cold climates. *Evolution* 10: 101–105.

Newman, M. T. 1958. Man and the heights: A study of the response to environmental extremes. *Natural History* 67: 9–19.

Papp, R. P. 1978. A nival aeolian ecosystem in California. *Arctic and Alpine Research* 10: 117–131.

Park, O. 1949. Application of the converse Bergmann principle to the carabid beetle, *Dicaelus purpuratus. Physiological Zoology* 22: 359–372.

Pattie, D. L. 1967. Observations on an alpine population of yellow-bellied marmots (*Marmota flaviventris*). *Northwest Science* 41: 96–102.

Pattie, D. L., and Verbeek, N. A. M. 1966. Alpine birds of the Beartooth Mountains. *Condor* 67: 167–176.

Pattie, D. L., and Verbeek, N. A. M. 1967. Alpine mammals of the Beartooth Mountains. *Northwest Science* 41: 110–117.

Pearson, O. P. 1948. Life history of mountain viscachas in Peru. *Journal of Mammalogy* 29: 345–374.

Pearson, O. P. 1951. Mammals in the highlands of southern Peru. *Bulletin of the Museum of Comparative Zoology* 106: 117–174.

Pearson, O. P. 1954. Habits of the lizard *Liolaemus multiformis multiformis* at high altitudes in southern Peru. *Copeia* 2: 111–116.

Pearson, O. P., and Bradford, D. F. 1976. Thermoregulation of lizards and toads at high altitudes in Peru. *Copeia* 1: 155–169.

Pollock, R. 1970. What colors the mountain snow? *Sierra Club Bulletin* 55: 18–20.

Porter, W. P., Budaraju, S, Stewart, W. E., and Ramankutty. N. 2000. Calculating climate effects on birds and mammals: Impacts on biodiversity, conservation, population parameters, and global community structure. *Integrative and Comparative Biology* 40(4): 597–630.

Pruitt, W. O. 1960. Animals in the snow. *Scientific American* 203(1): 61–68.

Pruitt, W. O. 1970. Some ecological aspects of snow. In *Ecology of the Subarctic Regions.* Proceedings of the Helsinki Symposium (pp. 83–89). Paris: UNESCO.

Rensch, B. 1959. *Evolution above the Species Level.* London: Methuen.

Ripple, W. J., and van Valkenburgh, B. 2010. Linking top-down forces to the Pleistocene megafaunal extinctions. *BioScience* 60(7): 516–526.

Roderick, J., and Roderick, D. 1973. Africa's puzzle animal. *Pacific Discovery* 26(4): 26–28.

Saint Girons, H., and Duguy, R. 1970. Le cycle sexuel de *Laceria muralis* L. en plaine et en montagne. *Bulletin du Museum National d'Histoire Naturelle* 42: 609–625.

Salt, G. 1954. A contribution to the ecology of upper Kilimanjaro. *Journal of Ecology* 42: 375–423.

Salt, R. W. 1956. Influence of moisture content and temperature on cold hardiness of hibernating insects. *Canadian Journal of Zoology* 34: 283–294.

Salt, R. W. 1961. Principles of insect cold-hardiness. *Annual Review of Entomology* 6: 55–74.

Salt, R. W. 1969. The survival of insects at low temperatures. *Symposia of the Society for Experimental Biology* 23: 331–350.

Schaller, G. 1975. *Mountain Monarchs. Wild Sheep and Goats of the Himalayas.* Chicago: University of Chicago Press.

Schaller, G. 1998. *Wildlife of the Tibetan Steppe.* Chicago: University of Chicago Press.

Schmidt, K. P. 1938. A geographic variation gradient in frogs. *Zoological Series, Field Museum of Natural History* 20: 377–382.

Schmoller, R. 1971. Nocturnal arthropods in the alpine tundra of Colorado. *Arctic and Alpine Research* 3(4): 345–352.

Scholander, P. F. 1955. Evolution of climatic adaptation in homeotherms. *Evolution* 9(1): 15–26.

Scholander, P. F. 1957. The wonderful net. *Scientific American* 196: 97–107.

Scholander, P. F., Walters, V., Hock, R., and Irving, L. 1950. Body insulation of some arctic and tropical mammals and birds. *Biological Bulletin* 99: 225–236.

Scott, J. D. 1962. What do snow worms eat? *Summit* 8: 8–9.

Shields, O. 1967. Hilltopping. *Journal of Research on the Lepidoptera* 6: 71–178.

Sleeper, R. A., Spencer, A. A., and Steinhoff, H. W. 1976. Effects of varying snowpack on small mammals. In H. W. Steinhoff and J. D. Ives, eds., *Ecological Impacts of Snowpack Augmentation in the San Juan Mountains, Colorado* (pp. 437–485). Fort Collins: Colorado State University.

Smith, A. T. 1974. The distribution and dispersal of pikas: Consequences of insular population structure. *Ecology* 55(5): 1112–1119.

Smith, A. T. 1978. Comparative demography of pikas (*Ochotona*): Effect of spatial and temporal age-specific mortality. *Ecology* 59(1): 133–139.

Smith, A. V. 1958. The resistance of animals to cooling and freezing. *Biological Reviews* 33: 197–253.

Smith, M. E. 1966. Mountain mosquitoes of the Gothic, Colorado, area. *American Midland Naturalist* 76: 125–150.

Spalding, J. B. 1979. The Aeolian ecology of White Mountain Peak, California: Windblown insect fauna. *Arctic and Alpine Research* 11(1): 83–94.

Spencer, A. W., and Steinhoff, H. W. 1968. An explanation of geographic variation in litter size. *Journal of Mammalogy* 49: 281–286.

Steadman, D., and P. Martin 1984. Extinction of birds in the Late Pleistocene of North America. In P. S. Martin and R. G. Klein, eds., *Quaternary Extinctions* (pp. 466–477). Tucson: University of Arizona Press.

Stirling, I., ed. 1993. *Bears.* Emmaus, PA: Rodale Press.

Stoecker, R. E. 1976. Pocket gopher distribution in relation to snow in the alpine tundra. In H. W. Steinhoff and J. D. Ives, eds., *Ecological Impacts of Snowpack Augmentation in the*

San Juan Mountains, Colorado (pp. 281–288). Fort Collins: Colorado State University.

Sushkin, P. 1925. Outlines of the history of the recent fauna of palearctic Asia. *Proceedings of the National Academy of Sciences of the USA* 11: 299–302.

Svihla, A. 1956. The relation of coloration in mammals to low temperature. *Journal of Mammalogy* 37: 378–381.

Swan, L. W. 1952. Some environmental conditions influencing life at high altitudes. *Ecology* 33: 109–111.

Swan, L. W. 1961. The ecology of the high Himalayas. *Scientific American* 205: 68–78.

Swan, L. W. 1967. Alpine and aeolian regions of the world. In J. W. H. Osborn and H. E. Wright, eds., *Arctic and Alpine Environments* (pp. 29–54). Bloomington: Indiana University Press.

Swan, L. W. 1970. Goose of the Himalayas. *Natural History* 79: 68–75.

Swan, L. W., and Leviton, A. E. 1962. The herpetology of Nepal: A history, checklist, and zoogeographical analysis of the herpetofauna. *Proceedings of the California Academy of Sciences* 32: 103–147.

Sweeney, J. M., and Steinhoff, H. W. 1976. Elk movements and calving as related to snow cover. In H. W. Steinhoff and J. D. Ives, eds., *Ecological Impacts of Snowpack Augmentation in the San Juan Mountains, Colorado* (pp. 415–436). Fort Collins: Colorado State University.

Thomas, W. H. 1972. Observations on snow algae in California. *Journal of Phycology* 8: 1–9.

Timiras, P. S. 1964. Comparison of growth and development of the rat at high altitude and at sea level. In W. H. Weihe, ed., *The Physiological Effects of High Altitude* (pp. 21–32). New York, Macmillan.

Toweill, D. E., and Geist, V. 1999. *Return of Royalty*. Missoula, MT: Boone and Crockett Club.

Turner, G. T., Hansen, R. M., Reid, V. H., Tietjen, H. P., and Ward, A. L. 1973. Pocket gophers and Colorado mountain rangeland. Fort Collins: Colorado State University Experiment Station Bulletin 5545.

Tyrberg, T. 1998. *Pleistocene Birds of the Palearctic: A Catalogue*. Cambridge, MA: Nuttall Ornithological Club.

Van Dyke, E. C. 1919. A few observations on the tendency of insects to collect on ridges and mountain snow fields. *Entomological News* 30(9): 241.

Vaughan, T. A. 1969. Reproduction and population densities in a montane small mammal fauna. *University of Kansas Miscellaneous Publication* 51: 51–74.

Verbeek, N. A. M. 1970. Breeding ecology of the water pipit. *Auk* 87: 425–451.

Vuilleumier, B. S. 1971. Pleistocene changes in the fauna and flora of South America. *Science* 173: 771–780.

Vuilleumier, F. 1969. Pleistocene speciation in birds living in the high Andes. *Nature* 223: 1179–1180.

Vuilleumier, F. 1970. Insular biogeography in continental regions, I: The northern Andes of South America. *American Naturalist* 104: 373–388.

Walshingham, L. 1885. On some probable causes of a tendency to melanic variation in Lepidoptera of high altitudes. *Entomologist* 18: 81–87.

Webber, P. J. 1974. Tundra primary productivity. In J. D. Ives and R. G. Barry, eds., *Arctic and Alpine Environments* (pp. 445–473). London, Methuen.

Welles, R. E., and Welles, F. B. 1961. *The Bighorn of Death Valley*. Fauna of the United States, Fauna Series No. 6. Washington, DC: Government Printing Office.

Wolverton, S., Huston, M. A., Kennedy, J. H., Cagle, K., and Cornelius, J. D. 2009. Conformation to Bergmann's Rule in white-tailed deer can be explained by food availability. *American Midland Naturalist*, 162: 403–417.

Zimina, R. P. 1967. Main features of the fauna and ecology of the alpine vertebrates of the U.S.S.R. In J. W. H. Osborn and H. E. Wright, eds., *Arctic and Alpine Environments* (pp. 137–142). Bloomington: Indiana University Press.

Zimina, R. P. 1978. The main features of the Causasian natural landscapes and their conservation, U.S.S.R. *Arctic and Alpine Research* 10(2): 479–488.

Zimina, R. P., and Panfilov, D. V. 1978. Geographical characteristics of the high mountain biota within nontropical Eurasia. *Arctic and Alpine Research* 10(2): 435–439.

CHAPTER NINE

Attitudes toward Mountains

EDWIN BERNBAUM and LARRY W. PRICE

Mountains today are almost universally viewed with admiration and affection. Positive attitudes toward mountains have not, however, always been universal. During the Middle Ages and much of the Renaissance, many people in Europe, the English in particular, shunned mountain ranges such as the Alps as demonic, abhorrent places to avoid whenever possible (Nicholson 1959; Mathieu 2006). However, earlier Europeans, such as the Celts and the Greeks, revered hills and mountains as divine palaces and abodes of deities whom they looked up to and worshipped (Bernbaum 1997). Positive attitudes have had a longer continuous history in other parts of the world, such as Asia (Mathieu 2011) and the Middle East, where they date back thousands of years. People have been attracted by mountains for millennia; as discussed in the following chapters, they have traveled through them, used different altitudinal environments on a seasonal basis for hunting and gathering and the grazing of their animals, built permanent homes in them, and tilled the soil. But throughout recorded history, humans have regarded mountains ambiguously, with both fear and fascination—a characteristic response to the experience of places regarded as sacred (Otto 1950). In order to understand our modern love of mountains, it is necessary to trace the development of these ideas through time and to place them in historical perspective.

The Prehistoric Era

Very little is known about early views of mountains. Much of the evidence is based on the study of indigenous societies who still follow prehistoric ways of life. Impressed by volcanic eruptions, storms, avalanches, and other physical manifestations of power, many of these societies view mountains as the homes of powerful deities and demons that have to be treated with great care and respect. Accordingly, various cultures in the prehistoric period probably established elaborate taboos, ceremonies, and sacrifices to appease the wrath of the gods and invoke their blessings.

Early societies probably identified mountains with the weather. Mountains are the homes of storms, lightning, strong winds, cold, and clouds. Mountains are also associated with snow, a phenomenon which may or may not occur in the lowlands; in any event, snow is much more persistent at high altitudes, transforming the mountain peak into an unearthly site (and sight), a natural abode of spirits and gods. The association with weather probably led to widespread reverence of mountains as sources of life-giving water in the form of rain and rivers—a reverence that continues in societies today, from the Andes to the Himalaya (Reinhard 2006)

Another observation probably made by early visitors to mountain peaks was of the reaction that took place within their bodies. Symptoms of high-altitude sickness such as stomachache, vomiting, dizziness, and shortness of breath may well have led early people to conclude that they were transgressing on hallowed ground and should go no higher. One notable exception was the ancient Incas, who regularly ascended to heights of 6,800 m (21,760 ft) and more for worship and human sacrifice (Reinhard and Constanza 2010; Reinhard 2006; Besom 2009).

While many of the features associated with mountains and weather evoked terror, mountains also had positive attributes. As noted above, they were considered sources of life and fertility, and as a major source of water through rainfall, clouds, streams, and rivers. They also provided sanctuaries and refuges from enemy attacks as well as cooler, more habitable climates than many of the plains and jungle areas below them.

Mountains were often considered the home of strange (sometimes mythical) and dangerous beasts. Some of these beasts were real, animals that lived in the dense mountain forests but occasionally wandered into the snow zone, such as the snow leopard and other large cats, bears, eagles, wolves, monkeys, and apes. Many were large predators that were elusive and seldom seen in the lowlands, made larger than life in legends and superstitions. Some of the legends, like that of the Yeti (Abominable Snowman) of the Himalaya and the Sasquatch (Bigfoot) in the mountains of western North America, have persisted to the present day.

We do not know just when human settlement of the mountains began. In the Alps and the mountains of the Middle East, archaeological sites indicate the presence of humans since at least the Stone Age (100,000 years ago). These include the alpine components of the so-called Mousterian and Paleolithic cultures (Charlesworth 1957; DeSonnerville-Bordes 1963; Young and Smith 1966; Schmid 1972, Champion et al. 2009). These mountain groups were composed primarily of transient hunting parties, but some also made permanent settlements. They lived in caves and manufactured stone hunting tools; later (about 40,000 years ago), they used tools of bone, ivory, and antlers. Eventually they began to paint on cave walls and to make a custom of burying their dead (often preserved by mummification in the dry alpine air). A steady cultural development continued in Eurasian mountains down through the Bronze and Iron Ages (Anati 1960; Reinhold 2003).

The Americas

Radiocarbon dating of bone, shells, and artifacts indicates human presence in the Rocky Mountains 10,000 to 11,000 years ago (Husted 1965, 1974; Benedict and Olson 1973, 1978; Kornfeld et al. 2001). The alpine tundra zone was used primarily by summer hunting parties, who apparently employed a technique of driving game resembling that used in the Arctic. The target of the game drives was probably mountain sheep. Since, unlike the arctic caribou, it is next to impossible to corral mountain sheep, it is thought that the drives were designed to direct the sheep to concealed hunters (Husted 1974). Archaeological findings show that late prehistoric peoples occupied village sites in subalpine zones near tree line (Adams 2010).

In North America, it seems fairly clear that migration to and from the mountains depended on favorable climatic conditions and the availability of food. There is evidence that some groups utilized the mountains and the plains on an annual cycle according to the resources available. In winter, they hunted antelope and bison on the plains; in summer, they went to the mountains to hunt and gather (Adams 2010).

Mountains played and continue to play an important role in the religion and culture of American Indians (Bernbaum 1997). Many mountains have Indian and Native Alaskan names; among the most famous are Tacoma or Tacobet (Mount Rainier) and Denali (Mount McKinley). The Hopi revere the San Francisco Peaks of Arizona as the abode of

the *Katsinas,* ancestral rain deities on whom they depend for their existence, while the Navajo or Diné include these remnants of an ancient volcano as one of four sacred mountains that enclose and protect the land where they dwell (Bernbaum 1997). Volcanic peaks in the Cascades, such as Mount Shasta and Mount Rainier, have inspired many myths and legends (Clark 1953).

Only a few North American alpine archaeological remains have religious significance. One possible candidate is located above timberline at an altitude of 2,940 m (9,640 ft) in the Big Horn Mountains of northern Wyoming. It consists of a crude circle of stones 25 m (80 ft) in diameter with a central cairn 4 m (13 ft) across from which 28 spokes radiate to the rim (Eddy 1974). Early observers thought that the structure was a medicine wheel constructed as a replica of the medicine lodge to allow the observance of the Sun Dance ceremony in the mountains (Grinnell 1922). However, later researchers have posited that it was an early astronomical observatory, but with some mystical and/or aesthetic connotations as well, since its astronomical purpose could just as easily have been served on the plains (Eddy 1974; Sliverman 1999).

Perhaps the most spectacular display the world has ever known of human settlement in mountains is found in the Andes. Here, thousands of years before the birth of Christ, at elevations up to 4,500 m (14,400 ft), there flourished civilizations that are still a wonder to the modern world. The culmination of these cultures is reflected in the ruins of Tiahuanaco and Machu Picchu (Fig. 9.1) and the Inca capital of Cuzco. Even today, it is difficult to imagine the techniques involved in the building of their famous stone structures.

Although Andean peoples were initially hunter gatherers, food was reasonably plentiful at the higher elevations, and the basis for the civilizations to come was agriculture. Several plant species, including potatoes, corn, squash, and beans, were first domesticated in the highlands of Central and South America (Sauer 1936; Linares et al. 1975; Iriarte and Vrydaghs 2009). The production of food, which released man from the constant burden of hunting, allowed greater numbers of people to settle in a small area and, eventually, to evolve the highly organized and complex cultures of Tiahuanaco and the Inca.

What attitude did these people have toward the mountains that were their home? We know that among their many deities were the sun, the moon, stars, and mountains. Like the ziggurats of Near Eastern cultures, the stepped pyramids of civilizations such as the ancient Maya and Aztec in Central America and Mexico were essentially man-made mountains with temples of deities on their summits (Quaritch-Wales 1953). It was usually on these artificial mountains that priests carried out sacrifices of foodstuffs, precious metals, animals, and humans. Pre-Columbian peoples in Central and South America also revered mountains as the abodes of deities who controlled the weather

FIGURE 9.1 Machu Picchu, ancient Inca settlement and religious site at 2,300 m (7,500 ft) amid precipitous terrain in the Peruvian Andes. Extensive terracing has increased the amount of usable land. Even the high peak on the right displays terracing near the summit. This was used as a lookout; soldiers stationed there grew their own food. (Photo by E. Bernbaum.)

and the water on which they depended for their survival (Reinhard 2006).

Mountains were frequently linked in legend with the origin of a tribe or people. The Panzaleo of highland Ecuador traced their descent from the volcano Tungurahua. Another tribe, the Puruha, believed that they were created by the union between two volcanoes, the feminine Tungurahua and the masculine Chimborazo (Trimborn 1969). Many cultures have viewed mountains as male and female, or have in other ways associated them with fertility and members of human families. An Algonquin legend from the northeastern United States provides a typical example: There was once an Indian girl gathering blueberries on Mount Ktaadn, and, being lonely, she said, "I would that I had a husband." Seeing the great mountain in all its glory rising on high, with red sunlight on the top, she added: "I wish Ktaadn were a man, and would marry me." Her wish came true, and she gave birth to a son who used his great supernatural powers to help his people (Bent 1913). Mountains could also be female: The Yakutat Tlingit regard Mount Fairweather as the estranged wife of Mount St. Elias, 240 km (150 mi) up the Alaskan Coast (Laguna 1972).

Many of the higher mountains of the Andes were considered to be the home of deities (Reinhard and Constanza 2010; Reinhard 2006). In the cosmology of a remote village located east of Cuzco, Peru, at 4,265 m (14,000 ft), a number of the surrounding peaks have special religious significance. The villagers still make offerings of coca leaves and foodstuffs every August to the gods of the mountains, as protection against disease and to ensure good crops. According to local tradition, the *Apu* or mountain lord of the highest peak, Ausangate, at 6,400 m (21,000 ft) resides in a palace inside the mountain; if he is not given enough food he becomes angry and wraps the mountain in clouds, sending down lightning and hail to destroy the fields (Mishkin 1940). He also watches over the wildlife and livestock of the region.

The Western Tradition

The Biblical Period

Mountains were objects of veneration and symbols of strength and peace to the people of the Old Testament. The three most important events in the *Torah,* the first five books of the Bible, the covenants with God, are all associated with mountains, beginning with the coming to rest of Noah's Ark on Mount Ararat after the flood (Genesis 8:4). Every Bible-school child knows the story of Moses receiving the Ten Commandments on Mount Sinai (Exodus 19, 20, 24) and how Abraham took his son Isaac to a mountain in the land of Moriah to sacrifice him to God (Genesis 22:2). In later books of the Bible, David established his capital on Mount Zion, the fortress hill where the Jebusite city that became Jerusalem was located (Psalms 78: 68–70). Tradition places the temple he and Solomon built, the sacred center of the Jewish people, on the site of Mount Moriah, the place of the primordial sacrifice performed by Abraham (Bernbaum 1997).

Other mountains, such as Mount Carmel, Calvary, and the Mount of Olives, were also considered sacred (Bernbaum 1997). It is important to realize that a number of these are no more than hills. Ancient Near Eastern religions referred to mountains as "the center of fertility, the primeval hillock of creation, the meeting place of the gods, the dwelling place of the high god, the meeting place of heaven and earth, the monument effectively upholding the order of creation, the place where god meets man, a place of theophany" (Clifford 1972: 5).

Many of the most important events in the New Testament also take place on mountains. Satan takes Jesus up on a mountain and tempts him with the power and wealth of the world. Like Moses on Mount Sinai, Jesus is transfigured with light on Mount Tabor and is there revealed, for Christians, as the Son of God. The most famous of his sermons is the Sermon on the Mount, delivered on a hill above Galilee. Jesus is crucified on the hill of Golgotha and ascends to heaven from the Mount of Olives (Bernbaum 1997).

Classical Heritage

GREEKS

For the Greeks, high peaks were primarily the abodes of gods and other deities. The twelve major gods and goddesses resided in a fortress paradise on top of Mount Olympus. Zeus, the king of the gods, was born and raised in remote mountain caves in Crete, and had numerous altars and shrines dedicated to him on the tops of mountains throughout Greece. The Muses, who inspired literature, art, music, drama, and science, lived originally on neighboring Mount Preiria and then moved in later mythology to Mounts Hellicon and Parnassus. Mountains were also the haunts of nymphs, wild beasts, and centaurs (Bernbaum 1997).

The wildness and isolation of mountains also impressed the ancient Greeks. Homer was very much aware of mountain weather and describes its force vividly:

> In spring, snow-water torrents risen and flowing down the mountainsides hurl at a confluence their mighty waters out of gorges, filled by tributaries, and far away upon the hills a shepherd hears the roar. As south wind and the southeast wind, contending in mountain groves, make all the forest thrash . . . swaying their pointed boughs toward one another in roaring wind, and snapping branches crack. (ILIAD, Book 16)

The mountain that figures most prominently in Greek mythology and literature is, of course, Mount Olympus in Thessaly. Olympus, a word that predated the Greeks, apparently meant "peak" or "mountain" in a generic sense, for a number of other Greek mountains are named Olympus. Several of these, like Olympus in Thessaly, were associated with weather cults. Olympus is often mentioned as the home of Zeus in his role as the god of storms and weather. Through his ability to strike with lightning and thunder, Zeus controls both gods and men from the mountaintop (Nilsson 1972).

Because of the association of mountain heights with deities, the ancient Greeks placed many of their shrines and temples on the slopes and summits of mountains, or oriented these structures with respect to sacred peaks. The early Minoan civilization of Crete tended to associate mountains with female deities and saw in them reflections of female shapes and body parts. They constructed a number of peak sanctuaries for making offerings to goddesses and other deities. Some of these peak sanctuaries appear to have been sites of human sacrifice (Sculley 1962).

The Greeks recognized the wild, rugged, and untamed nature of mountain scenery, but they preferred the more harmonious aspects of nature. They were engrossed with man and his works. Socrates, for example, was totally absorbed by the perplexities of the city. He is quoted as answering the reproach of his friend Phaedrus, who complained that he never left the city, by saying "he was fond of knowledge and could learn nothing from the trees and the country, but only from the people in the city" (Hyde 1915–1916: 71). The human form was considered the highest level of beauty, and even their gods appeared in human form. What was good in nature was that which provided comfort and harmony for man. Beauty was symmetry and order. Ruskin, in his interpretation of Greek art and literature, says, "Thus, as far as I recollect without a single exception, every Homeric landscape, intended to be beautiful, is composed of a fountain, a meadow, and a shady grove" (1856, Vol. 4, ch. 13, sec. 15).

The observations of mountain weather in *The Iliad* noted above prefigure the scientific curiosity of later Greeks about the origins of mountains and the causes of phenomena associated with them. Herodotus, for example, commented on the work of rivers and their ability to erode and deposit. To him is attributed the saying, "Egypt is the gift of the river." Having discovered fossil marine shells in the mountains, he speculated that the peaks had at one time been under water. He also thought it likely that earthquakes, rather than the wrath of the gods, were responsible for breaking apart the Earth and uplifting mountains. Aristotle observed the unequal distribution of mountains, the significance of springs flowing from mountainsides, and the changes in climate that occurred with altitude. He believed that earthquakes and volcanoes were closely related and that they were involved in the formation of mountains. One of Aristotle's students, Theophrastus, investigated mountain plants; another, Dicaearchus, attempted to calculate the heights of mountains. In his famous geography, Strabo described mountains of the ancient world, distinguishing them from plateaus (Sengor 2003).

ROMANS

Italy, like Greece, is a mountainous country: The Apennines run its entire length and the Alps form its northern border. Although some Romans—notably the philosopher Seneca and the encyclopedist Pliny—made important observations concerning mountains, on the whole the Romans did not share the Greeks' appreciation of mountains, except perhaps as distant vistas to be seen from the porches of their villas. Among the Latin poets, only Lucretius discerned a sublime beauty in the Alps (Geikie 1912; Nicolson 1959). These practical people viewed mountains primarily as wastelands and as obstacles to commerce and conquest. The Romans were regularly crossing the Alps by Caesar's time, but apparently never overcame their initial dread of them. To appease the primarily Celtic deities of the Alpine passes and to commemorate safe journeys, they made offerings of coins and small bronze tablets inscribed with the names of the deity and the traveler. The hospice museum at the Great St. Bernard Pass has gathered a large collection of these offerings from the surrounding area (Bernbaum 1997; personal observation).

The prevailing Roman attitude toward mountains was aptly expressed by Silius Italicus in his description of Hannibal's famous crossing of the Alps in 218 B.C.:

Here everything is wrapped in eternal frost, white with snow, and held in the grip of primeval ice. The mountain steeps are so stiff with cold that although they tower up into the sky, the warmth of the sunshine cannot soften their hardened rime. Deep as the Tartarean abyss of the underworld lies beneath the ground, even so far does the earth here mount into the air, shutting out with its shade the light of heaven. No Spring comes to this region, nor the charms of Summer. Misshapen Winter dwells alone on these dread crests, and guards them as her perpetual abode. Thither from all sides she gathers the sombre mists and the thunder-clouds mingled with hail. Here, too, in this Alpine home, have the winds and the tempests fixed their furious dominion. Men grow dizzy amidst the lofty crags, and the mountains disappear in the Clouds. (PUNICA 111: 479–495, in GEIKIE 1912)

Literally thousands of pages have been written concerning Hannibal's crossing, many of which debate the question of his exact route. DeBeer (1946: 405) mused:

I often wonder whether Polybius and Livy realized what a blessing they conferred on humanity by couching their accounts of Hannibal's passage of the Alps on a level of precision insufficient to make the tracing of his route obvious, but just enough to encourage their readers to think that there is sufficient internal evidence to give them a sporting chance of solving the puzzle of where he went.

The titles of works published by Freshfield are typical: "The Pass of Hannibal" (1883) and "Further Notes on the Pass of Hannibal" (1886) (both in the *Alpine Journal*), and his book *Hannibal Once More* (1914). DeBeer himself could not resist the temptation, producing *Alps and Elephants: Hannibal's March* (1955).

The Roman opinion of mountains remained almost consistently negative. They apparently never acquired a taste for mountain scenery, as the Greeks did. The implicit dualism in the attitudes of these two peoples toward mountains became part of the legacy for Western Europe (Nicolson 1959). Ultimately, as we know, the spirit of the Greeks, who worshipped their gods on Mount Olympus, and the Children of Israel, who lifted up their eyes to the everlasting hills, would triumph, but not before several centuries of antipathy toward mountains had passed.

From Medieval Fears to Romantic Enthusiasm

During the Middle Ages, mountains in Europe were primarily viewed as haunts of demonic beings such as dragons and witches. Medieval people, like their Roman predecessors, paid little attention to the grander aspects of nature, and there are few favorable references to mountains in either their literature or their graphic art (Mathieu 2006). What does exist is often distorted by allegory, abstraction, and moralization. Dante made mountains the guardians of hell, yet the central book of the *Divine Comedy*, the *Purgatorio*, describes the ascent of a mountain leading to the earthly paradise on its summit and Paradiso or Heaven beyond (Freshfield 1881; Noyce 1950; Schama 1995). Dante's ambivalent view of mountains reflects the fear and fascination characteristically evoked by sacred places charged with a power that can be experienced as both demonic and divine (Otto 1950). Records of Celtic beliefs preserved by monks in Ireland strongly suggest that, before Christianity took over Europe and demonized natural sacred sites central to the practice of pre-Christian religions, the Celts viewed the Alps and other European mountains in a positive, divine sense as the palaces and abodes of their gods (Bernbaum 1997).

Following conversion to Christianity, Europeans tended to regard mountains as dangerous places, sacred in a predominantly negative, demonic sense (Bernbaum 1997). As a consequence, medieval travelers disliked mountains, but nevertheless traversed them regularly. To ease the journey, Alpine villages provided inns and supplied guides; churches and hospices were constructed along the most popular routes. Pilgrims on their way to Rome from western and northern Europe favored the Great St. Bernard Pass, where a monastery has stood since A.D. 812, and a hospice since 859 (Coolidge 1889). Although August was considered the best month for mountain travel, the passes were attempted at all seasons (Tyler 1930). Master John de Bremble, a monk of Christ Church, Canterbury, England, who had been sent to Rome on business, sent a letter home describing his passage of the Great St. Bernard in February 1188:

Pardon me for not writing. I have been on the Mount of Jove [the Roman name for the Great St. Bernard Pass]; on the one hand looking up to the heavens of the mountains, on the other shuddering at the hell of the valleys, feeling myself so much nearer heaven that I was more sure that my prayer would be heard. "Lord," I said, "restore me to my brethren, that I may tell them, that they come not into this place of torment." Place of torment, indeed, where the marble pavement of the stony ground is ice alone, and you cannot set your foot safely; where, strange to say, although it is so slippery that you cannot stand, the death (into which there is every facility for a fall) is certain death. (COOLIDGE 1889: 8–9)

Tales of monsters and supernatural perils added to the fears of travelers and mountain dwellers. King Peter III of Aragon (b. 1236) set out to prove it was possible to climb Pic Canig (2,785 m, 9,135 ft), then believed to be the highest peak in the Pyrenees. Resting by a small lake near the summit, he absently threw a stone into the water. Suddenly, "a horrible dragon of enormous size came out of it, and began to fly about in the air, and to darken the air with its breath." The full account may be found in Gribble's *The Early Mountaineers* (1899).

Perhaps the most famous legend is that of Mount Pilatus (2,129 m, 6,985 ft) in the Swiss Alps. As the story goes, Caesar was angry with Pilate for crucifying Jesus, so he had Pilate brought to Rome to be put to death. His body was tied to a stone and dropped into the Tiber River, where it caused a great turmoil. The body was therefore retrieved, and was eventually placed in a small lake on Mount Pilatus, in the Swiss territory of Lucerne. From that time on, if anybody shouted or threw a stone into the lake, Pilate would avenge himself by stirring up a great tempest. He also rose from the water on each Good Friday and sat on a nearby rock; if anybody saw him, that person would surely die. So great was their fear of the tempests he might cause that the government of Lucerne forbade anybody to approach the lake; in 1387, six men who broke this regulation were imprisoned (Coolidge 1889).

An excellent collection of these beliefs is contained in Johann Jacob Scheuchzer's *Itinera per Helvetia Alpinas regionses*, published in 1723. Scheuchzer, a professor at the University of Zurich, was a highly respected botanist who was credited with being the first to attempt to formulate a theory of glacier formation and movement (Gribble 1899). He had a penchant for the extraordinary, however, and firmly believed that dragons lived in mountains. His book is a mixture of the real and unreal, containing many accounts of sightings of these creatures, with several illustrations of the various dragon forms.

Belief in dragons had almost died out before the time of Scheuchzer, however. In 1518, four scholars climbed Mount Pilatus and visited the lake with no ill effects and, in 1555, Conrad Gesner, a professor of medicine at the University of Zurich, climbed the mountain by special permission of the Lucerne Magistrates, to prove that there was nothing to fear. Only 30 years later, a group of villagers, led by the pastor of Lucerne, climbed to the lake, threw stones, and defiantly mocked the spirit of Pilate, chanting *"Pilat, wirf aus dein kath!"* ("Pilate, cast out your crud!") (Gribble 1899: 46–50).

It is probably fair to say that many medieval Europeans who had any acquaintance with mountains feared them or, at the very least, would have considered it a waste of time to climb to the top of one, but there were exceptions. In 1336, the poet Petrarch climbed Mont Ventoux in Provence simply "for the sake of seeing the remarkable altitude of the place" (Gribble 1899: 18–19). Petrarch's climb is often cited as the first evidence of Renaissance appreciation of natural beauty, but much of his account has so allegorical a cast that some scholars have suspected he never made the climb (Noyce 1950). Much more clear-cut evidence of a new interest in natural beauty and natural phenomena is Leonardo da Vinci's observations of mountains, both in his art and in his scientific notebooks, at the end of the fifteenth century (Schama 1995).

The person usually credited as being the first European to appreciate and love mountains for their own sake is the sixteenth-century Swiss naturalist Conrad Gesner. In a letter to a friend in 1541, Gesner wrote:

I am resolved henceforth, most learned Avienus, that as long as it may please God to grant me life, I will ascend several mountains, or at least one, every year, at the season when the flowers are in their glory, partly for the sake of examining them, and partly for the sake of good bodily exercise and of mental delight. For how great a pleasure, think you, is it, how great delight for a man touched as he ought to be, to wonder at the mass of the mountains as one gazes on their vastness, and to lift up one's head as it were amongst the clouds? The understanding is deeply moved, I know not wherefore, by their amazing height, and is driven to think of the Great Architect who made them. (COOLIDGE 1889: 12–13)

He not only carried out this resolve, but took other Renaissance naturalists along on his Alpine excursions, awakening their interest in mountain plants and opening their eyes to the glories of the mountains. His student and successor at the University of Zurich, Josias Simler, published a learned treatise in 1574 on snow and ice travel; in it, he discussed such things as crampons, alpine sticks, use of eye shades, and how to cross crevasses (Gribble 1899).

The hold of theology on science and philosophy was very strong throughout the Middle Ages, and the general antipathy felt toward mountains was reinforced by religious sanctions and ideas. When Christianity became the official religion of the Roman Empire in the first half of the first millennium, its leaders reduced the divinities of the natural landscape to demons antagonistic to the new religion. Christian missionaries deliberately cut down sacred groves where pagan rituals traditionally took place, as a means of putting such practices to an end. Inspired by the writings of early theologians such as St. Augustine, they tended to view the wilderness—and the mountains that formed a particularly wild and uncontrollable part of it—as the corrupt domain of the evil powers of nature that the Church had to suppress in order to establish the kingdom of heaven on earth (Bernbaum 1997).

An influential post-medieval spokesman for this idea was Thomas Burnet, who asserted in *The Sacred Theory of the Earth* (1684) that the Earth was originally a perfectly smooth sphere, the "Mundane Egg"; as punishment for man's sins, the surface was ruptured and the interior fluids boiled out as "vast and undigested heaps of stones and earth."

By the end of the seventeenth century, on the verge of the Enlightenment, publications began to appear supporting the idea of a purposefully designed Earth and the usefulness of mountains. Mountains were recognized as being valuable as wildlife preserves, as sources of minerals, and as a means of converting salt water to fresh (Rees 1975a). But mountains were still not generally appreciated for their beauty. The wild disarray of mountains and their utter lack of symmetry and proportion were difficult for the early modern mind to accept. Mountains represented confusion, and were quite unlike the order and uniformity these thinkers sought in the natural world. Their ideals were the classical ones of order, reason, and restraint. Yet, appalled as they were at these "warts, wens, blisters, and imposthumes" on the fair face of the Earth, they were also profoundly impressed by the vastness and enormity of mountains. This is well expressed in Burnet's *The Sacred Theory of the Earth:*

> The greatest objects of nature are, methinks, the most pleasing to behold; . . . there is nothing that I look upon with more pleasure than the wide sea and the mountains of earth. There is something august and stately in the air to these things, that inspires the mind with great thoughts and passions; we do naturally, upon such occasions, think of God and his greatness: And whatsoever hath but the shadow and appearance of the infinite, as all things have that are too big for our comprehension, they fill and overbear the mind with their excess and cast it into a pleasing kind of stupor and admiration. Still . . . although we justly admire its greatness, we cannot at all admire its beauty or elegancy for 'tis as deformed and irregular as it is great. (NICOLSON 1959: 214–215)

The feelings of "delightful horror" and "terrible joy" expressed by Burnet and his contemporaries are the first signs of the romantic enthusiasm that has typified European attitudes toward mountains since the eighteenth century (Mathieu 2011). In 1732, Albrecht Haller published *Die Alpen,* a book of poems in praise of the Alps and their inhabitants which became something of a best-seller in Europe. The journals and letters of another poet, Thomas Gray, describing his tour of the Alps in 1739, evoked a similar response in England. The most influential writer of all was Jean-Jacques Rousseau (Hyde 1917; Noyce 1950). In his *La Nouvelle Héloïse,* published in 1759, he enthuses:

> The nearer I came to Switzerland, the more were my feelings moved. The moment when from the heights of Jura I descried the Lake of Geneva was a moment of ecstasy and rapture. The sight of my country, that so-beloved country, where torrents of pleasure had overwhelmed my heart, the wholesome, pure air of the Alps; the soft air of home, sweeter than the perfumes of the East; this rich and fertile soil; this unrivaled landscape, the most beautiful that human eye has ever seen, this charming spot of which I had never beheld the like in my journey; the sight of a happy and free people; the softness of the season; the gentleness of the climate; a thousand delicious memories that recalled all the emotions I had felt,—all these things threw me into transports that I cannot describe. (PERRY 1879: 305)

Rousseau's writings had an almost revolutionary impact. Although the love of nature was not new, Rousseau's expression of it, particularly with respect to mountains, greatly increased popular appreciation of Switzerland as a place of beauty. Among those who came under Rousseau's spell were the famous German philosopher and poet Goethe,

and the English poet Wordsworth, who was perhaps the greatest interpreter of nature in all of literature. Rousseau also influenced Horace Benedict de Saussure, the Swiss doctor who offered a prize for the first ascent of Mont Blanc—accomplished by Jacques Balmat and Dr. Michel Paccard in 1786—and is considered the father of Alpinism. After four attempts between 1760 and 1787, he himself succeeded in reaching the summit in 1787. It is no coincidence that the birth of modern mountaineering coincided with the emergence of Romantic views of the Alps as symbols of the infinite and the sublime (Bernbaum 1997). In a sense, we can say that one sacred view of mountains—a positive, divine one—had replaced another, negative view of mountains as demonic places that had prevailed during the Middle Ages.

De Saussure initiated a great boom in scientific interest in mountains; in the following half-century, for scientific purposes, Swiss alpinists ascended many other mountains never before climbed (Noyce 1950). By the middle of the nineteenth century, however, the scientific focus gave way to the English sense of sport. When Hudson, Hadow, and Lord Douglas lost their lives on the Matterhorn in 1865, the disaster seemed to serve as a challenge rather than a deterrent, and in the years that followed, English climbers and tourists swarmed into the remote regions of the Alps (Peattie 1936; Schama 1995). The modern period of mountain adoration had begun.

The East

The development of attitudes toward mountains in the East contrasts greatly with that of the West. Attitudes in both civilizations changed from initial feelings of awe and aversion to admiration and love (Tuan 1974; Bernbaum 1997), but in the East, the appreciation of mountains began very early. According to the origin myth of the Korean people, they are descended from the union of a sky god and a bear woman on the sacred mountain of Paekdu (Henthorn 1971). In Japan, China, Tibet, and India, mountains have long been adored and worshipped. Mountains were considered sacred in China at least 2,000 years before the birth of Christ (Sullivan 1962; DeSilva 1967; Bernbaum 1997). Buddhism, Taoism, Confucianism, Shintoism, and Hinduism all incorporated mountain reverence into their beliefs.

The impact of mountains on early Chinese culture was profound. The mountain ranges that course through China were considered to be the body of a cosmic being (according to some, a dragon), the rocks his bones, the water his blood, the vegetation his hair, and the clouds and mists his breath (Sullivan 1962: 1). This belief probably sprang from the ancient cult of the Earth and, although largely replaced by other concepts, remains basic to Chinese philosophy. It also had to do with ancient views of mountains as divine sources of rain and water on which the agrarian society of China depended for its existence. Man is viewed as an integral part of nature. Inanimate objects have spirits and souls, just as do animate objects.

At first, there were four sacred mountains in China situated in the four quarters of the compass (later a fifth was added in the center): The eastern mountain, T'ai Shan, was the most holy and famous (Mathieu 2011). These mountains are usually associated with Taoist and Confucian thought, but were worshipped as far back as the Hsia Dynasty (2205–1176 B.C.) (Sowerby 1940). The ancient annals say that the legendary first emperors of China would go on ritual tours of inspection of the empire every five years and would climb these mountains and perform sacrifices on them to establish their sovereignty over the princes of the realm. Later historical emperors would climb T'ai Shan if they felt that they had brought their dynasties to the heights of glory and would perform sacrifices thanking Heaven and Earth for their successes (Chavannes 1910; Bernbaum 1997).

There are, in addition, four mountains of special significance to Buddhism, also situated in the four directions, of which Omei Shan in the west is probably the most famous (Shields 1913; Mullikin and Hotchkis 1973). Omei Shan was reputed to have over 50 pagodas and temples. Pilgrims still climb the mountain to see the Buddha's Glory from its summit: a Spectre of Brocken effect of a figure projected in mist surrounded by a rainbow halo. Another such effect was viewed, in a contrasting manner, as demonic in the Harz Mountains of Germany (Bernbaum 1997).

Many other mountains have local religious significance. One such peak is Dragon Mountain near the ancient city of Anking, celebrated in this poem:

FIGURE 9.2 Yamabushi descend Omine San, one of the most important sacred mountains for practitioners of the Japanese mountain climbing religion of Shugendo (Photo by E. Bernbaum.)

There is a dragon mountain in Hsu
With a spring which waters the fields
On the mountain is the spring
And upon the hill sides are tilled fields
From the earliest ages the men of Hsu have
    received this help,
In time of trouble all heads are turned toward the mountain.

High above the hills float the clouds,
And within them is a spirit who changes their shapes
    continually.

The people wondered who the spirit was
Until they found that it was the Lung Wang.

Hence they rebuilt the temple,
So that sacrifices might be made for a thousand years.

These sacrifices are still continued
And the people reap the reward.
(SHRYOCK 1931:118)

The mountain as a source of water is a religious motif found in many cultures: It was central to the Israelites of the Old Testament and to the Babylonians, as well as throughout Asia and Latin America (Van Buren 1943; Quaritch-Wales 1953; Reinhard and Constanza 2010). Clouds have a special fascination for the Chinese, and appear in their earliest art. Mountains are frequently shown rising out of clouds or enshrouded by them, not as a symbol of gloom or dreariness, but of beauty. Sometimes clouds are shown as dragons. To the Asian mind, dragons do not generally have an evil connotation as they do to the Westerner; they are benevolent creatures controlling the elements and guarding sources of wisdom (Sowerby 1940: 154).

Until the third century A.D., the Chinese regarded mountains as dangerous places of supernatural power that only those with proper spiritual preparation could enter safely to engage in religious practices. Around the fourth century, as result of a shift of the Chinese capital to more attractive mountains in the south and growing discontent with the confining strictures of imperial bureaucracy, literati from the court began going to the mountains for recreation, pursuing painting and poetry as they walked and sought inspiration in beautiful mountain landscapes. A similar transformation of attitudes took place in Europe more than a thousand years later (Mathieu 2011). In a very real sense, the modern practice of going to the mountains for sport and recreation actually had its birth in China. The following poem, composed in the fifth century, reflects this early shift in views of mountains:

> In the mountains all is pure, all is calm;
> All complication is cut off.
>
> Rare are they who know to listen;
> Happy they who possess wisdom.
>
> If the cold wind stings and bothers you,
> Sit in the sun: it is always warm there.
>
> Its hot rays burn like flames,
> While, opposite, in the shade, all is frost and snow.
>
> One pauses on ledges, one climbs to the foot of high clouds;
> One sits in the depths of a gorge, one passes windy grottos.
>
> Here is the realm of harmony and joy,
> Where the past and the present become eternal.
> (BERNBAUM 1997: 27)

Perhaps it is only natural, given East Asian sentiments toward mountains, that mountains should occupy a dominant position in their art. The very term for landscape in Chinese is *Shan Shui,* literally "mountains and water." Painting is considered a branch of calligraphy; the Chinese character for mountain 山 is a pictorial representation of a mountain (DeSilva 1967), and the characters for a hermit or Taoist immortal 仙 are those of a man (person) and a mountain. The mountain motif appears on the earliest known Chinese pottery and stone carvings, and in landscape paintings from the Han Dynasty (206 B.C.–A.D. 220) onward (DeSilva 1967).

The Japanese also view mountains as symbols of divine beauty and power. The use of stones to represent mountains is an ancient art form practiced in both Chinese and Japanese gardens. The culmination of this is what the Japanese call *Iskiyarna:* A natural stone about 15 cm (6 in.) high is placed vertically on a small wooden base. This simple piece of nature sculpture, a mountain landscape in miniature, is kept inside the house on a shelf or table, and often has great value and meaning to its owner. Japan has many sacred mountains, of which Fuji is perhaps the most famous. Up to 300,000 people climb the mountain each year during the July–August climbing season; it is still climbed annually by members of Fuji devotional sects as a sacred pilgrimage. When a commercial proposal was made for constructing a funicular railway to the summit, the Japanese angrily rejected the idea as a desecration of the holy mountain (Fickeler 1962). Until the mid-nineteenth century, perhaps the most widespread form of religion in Japanese village life was Shugendo, a blend of Buddhism and Shintoism based on the practice of climbing mountains as a metaphor for following the Buddhist path to enlightenment and acquiring spiritual powers. Practitioners of this religion are called *yamabushi,* meaning "those who lie down (or sleep) in mountains" (Fig. 9.2). Japanese also revere many of their sacred mountains as the abodes of ancestral spirits on whom they depend for water to grow their crops (Bernbaum 1997).

The people of Cambodia, Thailand, Bali, Java, and the Philippines also practice mountain worship (Quaritch-Wales 1953). Mount Popa in Burma has been considered sacred for over 2,000 years (Aung 1962). Rulers of many Southeast Asian kingdoms identified their capitals with Mount Meru—the mythical mountain at the center of the universe with the palace of the king of the gods on its summit—or with Mount Kailas, the abode of the Hindu deity Shiva, regarded by over a billion Asians as the most sacred mountain in the world (Bernbaum 1997).

Mountains have a particular significance in India (Saxena et al. 1998). The Himalayas, extending for 2,500 km (1,500 mi) along its northern border, have many religious and mythological associations. They are the source of major sacred rivers, such as the Indus, Brahmaputra, and Ganges, on whose waters hundreds of millions of people in the plains depend for their existence. The source of the Ganges is considered especially holy and is visited by many as a sacred pilgrimage. The Himalayas are the home of many Hindu deities, the most important of whom is Shiva, the archetypal yogi and one of the three forms of the supreme deity, who resides on Mount Kailas. Shiva's wife, Parvati, is the daughter of Himalaya (Bernbaum 1997). The range is also considered a favored place for meditation and the idyllic retreat of sages intent on attaining the ultimate goal of *moksha* or spiritual liberation.

In Tibet, as well as in the smaller Himalayan states of Kashmir, Nepal, Sikkim, and Bhutan, mountains have been natural shrines since very ancient times, even before the advent of Buddhism and Hinduism (Nebesky-Wojkowitz 1956). Cultural influences from both China and India are now evident, but the inhabitants have retained many of their indigenous beliefs. Thus, it is common for persons to make such pious gestures as tying strips of cloth on bushes, or placing stones or pieces of wood in sacred heaps at a pass they have reached after a steep climb along a mountain trail (Shaw 1872). Circumambulation of mountains is also widely practiced by Buddhists, particularly in Tibet. Among the most famous Tibetan mountains are Am-nye-rMachen and Kang Tise or Mount Kailas. Regarded as the most sacred mountain in the world by Hindus and Tibetan Buddhists, as well as by followers of two other religions, Jainism and the indigenous Tibetan tradition of Bon, Kailas is a dome-shaped peak of singular beauty and a favorite for circumambulation (Mathieu 2011; Fig. 9.3). The trip over the rocky trail around the mountain is nearly 50 km (30 mi) long and takes up to three days; many Tibetan pilgrims do the entire circuit, crossing a pass nearly 5,700 meters (18,240 ft) high, in one long day (Bernbaum 1997).

Many of the higher peaks are considered sacred by the people of the Himalaya. With the onslaught of modern mountain climbers, governments have had to restrict activities in certain areas. In Nepal, for example, the summits

FIGURE 9.3 Mount Kailas, 6,714 m (22,028 ft), the most sacred mountain in the world for more than a billion people in Asia. Followers of Hinduism, Buddhism, Jainism, and the indigenous Tibetan religion of Bon all revere the remote peak, which lies in northwest Tibet and is the focal point of one of the longest and most arduous pilgrimages in the world. (Photo by E. Bernbaum.)

FIGURE 9.4 Tibetan Buddhist Monastery of Tengboche near the foot of Mount Everest in the Nepal Himalaya at 3,873 m (12,715 ft). (Photo by E. Bernbaum.)

of Machapuchare (6,991 m, 22,371 ft) and Kangchenjunga (8,586 m, 28,169 ft) are both off-limits for religious reasons (Siiger 1955; Bernbaum 1997). Airplane flights are also prohibited over these peaks. This prohibition has been eased somewhat with familiarity and with the advent of high-altitude jets, but when aviation first began, a planned flight over Mount Everest in 1934 by two English airplanes raised quite a stir in India and Tibet (Fickeler 1962). Transcending the pure worship of ancient times is the zest for life in the mountains. This is exemplified by the Sherpa dance ceremony "Mani Rimdu," a three-day festival held during full moon in the spring. This ceremony has deep religious aspects, but it is also a vehicle for exhilaration and glorification of the way of life of a very proud people in the highest mountains in the world (Kohn 2001). Mount Everest itself is a sacred mountain, albeit a minor one. The Tibetan name of Mount Everest, Chomolungma or Jomolangma, is short for the name of the goddess of the peak, Jomo Miyolangsangma, one of the Five Sisters of Long Life, whom Tibetans and Sherpas invoke for the lesser mundane blessings of long life, food, and wealth (Bernbaum 1999; Fig. 9.4).

The Modern Period

The beginnings of the modern period of Western romantic adoration of mountains can be found in the writings of Albrecht Von Haller, Thomas Gray, Jean-Jacques Rousseau, and Horace Benedict de Saussure. By the nineteenth

FIGURE 9.5 An alpinist on the Aiguille du Midi, with Mont Blanc in the background. The first ascent of Mont Blanc in 1786 marked the birth of the modern sport of alpinism in the West. (Photo by E. Bernbaum.)

century, the beauty of mountains was a common theme for poets and philosophers; scientists began to take a serious interest in the origins of mountains and alpine phenomena, and popular accounts of scientific findings were published in newspapers and periodicals; and mountains became a favorite of landscape painters (Ruskin 1856; Lunn 1912; Rees 1975a, 1975b; Schama 1995). By the 1860s, railways provided relatively easy access to the Alps. Tourist resorts sprang up, and sanitariums were built to accommodate sufferers from consumption (tuberculosis), since the dry clean air of the mountains was found to have excellent therapeutic results (Barton 2008). The popular image of mountains was no longer that of a cold, inhospitable land of horrors, but that of an attractive, healthy environment (Mathieu 2011).

In 1857 the (British) Alpine Club was formed, and shortly thereafter other mountaineering clubs were established on the continent of Europe. It was the British, however, who inundated the Alps during the ensuing decades. An article on the history of the Alpine Club states: "It is impossible to dwell in detail on this wonderful period, and a mere enumeration of its first ascents and first passages would be intolerably tedious. At the end of it hardly any of the greater

summits of the Alps remained unconquered, and of the new ascents made, a very large share had fallen to the British climbers" (Mumm 1921). The Matterhorn tragedy of 1865 gave the Alpine Club, and the world, pause for thought but, after a few years, other younger members took up the challenge of unconquered peaks with renewed zeal (Fig. 9.5).

Alpine clubs gradually spread around the world (the first in the United States, the Appalachian Mountain Club, was formed in 1876), and many of these were quite exclusive: One had to be invited to join, and the requirements for admission were not easily met. Mountaineering became a consuming passion, almost a religion (Parker 2008). Peattie (1936: 6) called the development a "cult of mountains," a modern, more conscious phase of the ancient worship of mountains. An article entitled "Mountaineering as a Religion" was typical of this point of view:

[W]e must seek for some homogeneous and inward spiritual characteristics marking us off as a caste apart from other men. For myself, I find these characteristics in a certain mental predisposition, a distinct individual and moral bent, common to all mountaineers, but rarely found in those who are not addicted to mountain climbing. The true mountaineer is not a mere gymnast, but a man who worships the mountains. (STRUTFIELD 1918: 242)

The cult of mountains had many prophets and a large following. The journals published by the various alpine clubs are full of articles praising mountains (e.g., Freshfield 1904; Fay 1905; Lunn 1912; Godley 1925; Young 1943; Howard 1949; Vandeleur 1952; Thorington 1957). Perhaps the most far-reaching claim of any zealot is that advanced by Geoffrey W. Young, a respected and long-time member of the Alpine Club, at a lecture before the University of Glasgow in 1956, that mountains have been influential in the development of human intelligence:

It is a bold claim to make for mountains, that they contributed a third dimension, of height and depth, to man's intelligence; and, by means of it, adumbrated even a fourth dimension, that of spirit, not permeating it but placed above it. And yet, when mind first grew capable of comparison, when man's mastery began to move upon the earth, and he was released from labour only and from a surrounding darkness of fear, a mountain peak first sighted upon the skyline must indeed have seemed to belong to some sphere "visited all night by troops of stars," just as the first flash of sunrise upon a snow summit, for the first time realised, must have revealed the golden throne of a god. (YOUNG 1957: 14–15)

Further, in discussing the various components of landscape, he argued:

In all this visual balance, and in the influence it has exercised, the mountains play, and have played, the principal part. It is the heights which have given the measure. They are set like upright rulers, to mark the scale, against the perspectives of plain and sea and sky. In their constant contemplation, illuminated by a lighting definite and brilliant, upon colour and shadow positive and luminous, primitive mind had no alternative but to acquire, as part of its growth, laws of measure, of order, of proportion, in thought no less than in vision. From the acquired ability to compare, to discriminate, reasoning, speculation, with measure and proportion, dawning upon the human mind in the genius of the first Greek philosophers and in the sculpture and building of the first Greek artists, began or hastened the beginnings of civilisation and culture in every western race. (YOUNG 1957: 24–25)

Some of the alpine clubs became very powerful financially, socially, and politically, and it has been argued that expeditions in Africa, Asia, and South America were part of the era of neocolonialism in the early twentieth century (Ellis 2001). The (British) Alpine Club has supported a large number of projects, from polar exploration to the search for the Yeti or Abominable Snowman. Its members have traditionally been among the elite of British society. Hillary and Hunt, after climbing Mount Everest, were knighted by the Queen. Most of these clubs have diversified now, but they are still powerful. A good example in the United States is the Sierra Club, which now has a membership of about 1.4 million. It employs full-time lobbyists in Washington, D.C., and has considerable political clout. One of its most impressive achievements was to galvanize public opinion through a highly effective advertising campaign that forced the U.S. Congress to block the proposed construction of dams on the Colorado River that would have flooded parts of Grand Canyon National Park (Cohen 1988).

Mountains are no longer the private preserve of elitist clubs or of special-interest groups, but a "cult of mountains" continues. This has been beautifully expressed by René Dubos:

Man has now succeeded in humanizing most of the earth's surface but paradoxically, he is developing simultaneously a cult for wilderness. After having been for so long frightened by the primeval forest, he has come to realize that its eerie light evokes in him a mood of wonder that cannot be experienced in an orchard or a garden. Likewise, he recognizes in the vastness of the ocean and in the endless ebb and flow of its waves a mystic quality not found in humanized environments. His response to the thunderous silence of deep canyons, the solitude of high mountains, the luminosity of the deserts is the expression of an aspect of his fundamental being that is still in resonance with cosmic events. (DUBOS 1973: 772)

From a pastime pursued by a small group of aficionados, mountaineering has become a major sport that has captured the attention of numerous people and become a source of income for a growing number of guiding services who take clients to the summits of the world's highest peaks, including Mount Everest, sometimes with devastating results that get widespread publicity among the general public (Krakauer 1997). In addition to developing a commercial side, mountain climbing has also become a highly competitive endeavor, with races to see who can climb celebrated routes in the shortest time. Rock climbing has spun off as a sport in itself, often practiced exclusively indoors on artificial walls. People now climb mountains around the world for a plethora of reasons, ranging from the competitive to the contemplative, from the profane to the sacred (Bartlett 1993; Bernbaum 1997).

UNESCO's designation of "associative cultural landscapes" as a category of World Heritage Site in 1992 has raised to prominence the cultural and spiritual significance that people and traditions place on natural sites—in particular, sacred mountains such as Tongariro in New Zealand, Uluru or Ayers Rock in Australia, and Taishan in China. In the cases of Tongariro and Uluru, this designation has strengthened the role of indigenous peoples in managing places that have special value for them in their natural states (Bernbaum 2008). In addition, the growing impact of global cultural, political, economic, and environmental change, and particularly climate change, has focused attention on the cultural and spiritual importance of mountains and the responses by mountain communities to those changes (Bernbaum 2010).

As the world becomes increasingly populated and urbanized and the need to escape the pressures of the city grows, mountains become more and more a focus of attention (Macfarlane 2003). Mountains are now almost universally viewed as havens of retreat and symbols of freedom, and mountain tourism is one of the fastest growing industries in the world (see Chapter 12). Not content with brief visits, many people are taking advantage of technology and communications to move to and live in remote mountain communities like Aspen in the Colorado Rockies, participating in a modern phenomenon that scholars refer to as amenity migration, as discussed in Chapter 10. The city is no longer viewed as Socrates saw it, as the center of action where everything good happens, but more and more as the center of evil. This is particularly evident in the return-to-nature movement that took place among the young during the late 1960s and early 1970s, with its emphasis on casting off the artificiality of modern urbanized life. This trend, reminiscent of that advocated by Rousseau more than 200 years ago and by Thoreau a century later, is an important influence in our times.

Mountains are a favorite refuge for those seeking to commune with nature, whether they be motor tourists or backpackers (Parker 2008). This influx of tourists has created unprecedented pressures on mountain landscapes. Permits and other restrictions are being imposed in many areas, with waiting lists and reservations now required to trek and camp in some mountain areas. But mountains have never been in such demand or regarded with such favor in all the history of humankind. They comprise a major and praiseworthy theme in contemporary art, literature, and music.

Mountains are considered the embodiment of the good, the beautiful, and the sublime.

References

Adams, R. 2010. *Archaeology with Altitude: Late Prehistoric Settlement and Subsistence in the Northern Wind River Range, Wyoming.* Ph.D. dissertation, University of Wyoming, Laramie.

Anati, E. 1960. Prehistoric art in the Alps. *Scientific American* 202: 52–59.

Aung, M. H. 1962. *Folk Elements in Burmese Buddhism.* London: Oxford University Press.

Bartlett, P. 1993. *The Undiscovered Country: The Reason We Climb.* Leicester, UK: The Ernest Press.

Barton, S. 2008. *Healthy Living in the Alps: The Origins of Winter Tourism in Switzerland, 1860–1914.* Manchester, UK: Manchester University Press.

Benedict, J. B., and Olson, B. L. 1973. Origin of the McKean Complex: Evidence from timberline. *Plains Anthropologist* 18(62): 323–327.

Benedict, J. B., and Olson, B. L. 1978. *The Mount Albion Complex: A Study of Prehistoric Man and the Altithermal.* Research Report 1. Ward, CO: Center for Mountain Archaeology.

Bent, A. H. 1913. The Indians and the mountains. *Appalachia* 13(3): 257–271.

Bernbaum, E., 1997: *Sacred Mountains of the World.* Berkeley: University of California Press.

Bernbaum, E., 1999. A note on the Tibetan and Nepali names of Mount Everest. *American Alpine News* 8(227): 25–26.

Bernbaum, E., 2008. Mountains of spiritual world heritage. *World Heritage* 51: 36–45.

Bernbaum, E., 2010. Sacred mountains and global changes: Impacts and responses. In B. Verschuuren, R. Wild, J. McNeeley, and G. Oviedo, eds., *Sacred Natural Sites: Conserving Nature and Culture* (pp. 33–41). London: Earthscan.

Besom, T. 2009. *Of Summits and Sacrifice: An Ethnohistoric Study of Inka Religious Practices.* Austin: University of Texas Press.

Champion, C., Gamble, C., et al. 2009. *Prehistoric Europe.* Walnut Creek, CA: Left Coast Press.

Charlesworth, J. K. 1957. *The Quaternary Era.* London: Edward Arnold.

Chavannes, E. 1910. Le Tai Chan. Grimet: *Annales du Musée* 21.

Clark, E. E. 1953. *Indian Legends of the Pacific Northwest.* Berkeley: University of California Press.

Clifford, R. J. 1972. *The Cosmic Mountain in Canaan and the Old Testament.* Cambridge, MA: Harvard University Press.

Cohen, M. P. 1988. *The History of the Sierra Club, 1892–1970.* San Francisco: Sierra Club Books.

Coolidge, W. A. B. 1889. *Swiss Travel and Swiss Guide Books.* London: Longmans, Green.

DeBeer, G. R. 1930. *Early Travellers in the Alps.* London: Sidgwick and Jackson.

DeBeer, G. R. 1946. Puzzles. *Alpine* Journal 55(273): 405–413.

DeBeer, G. R. 1955. *Alps and Elephants; Hannibal's March.* London: Geoffrey Bles.

DeSilva, A. 1967. *The Art of Chinese Landscape Painting.* New York: Crown.

DeSonnerville-Borcles, D. 1963. Upper Paleolithic cultures in Western Europe. *Science* 142(3590): 347–355.

Dubos, R. J. 1973. Humanizing the Earth. *Science* 179(4075): 769–772.

Eddy, J. A. 1974. Astronomical alignment of the Big Horn medicine wheel. *Science* 184(4141): 1035–1043.

Ellis, R. 2001. *Vertical Margins: Mountaineering and the Landscapes of Neoimperialism.* Madison: University of Wisconsin Press.

Fagan, J. L. 1973. *Altithermal Occupation of Spring Sites in the Northern Great Basin.* Unpublished Ph.D. dissertation, University of Oregon, Eugene.

Fay, C. E. 1905. The mountain as an influence in modem life. *Appalachia* 11(1): 27–40.

Fickeler, P. 1962. Fundamental questions in the geography of religions. In P. L. Wagner and M. W. Mikesell, eds., *Readings in Cultural Geography* (pp. 94–117). Chicago: University of Chicago Press.

Freshfield, D. W. 1881. Notes on old tracts, IV: The mountains of Dante. *Alpine Journal* 10: 400–405.

Freshfield, D. W. 1883. The Pass of Hannibal. *Alpine Journal* 11(81): 267–300.

Freshfield, D. W. 1886. Further notes on the Pass of Hannibal. *Alpine Journal* 13(93): 29–38.

Freshfield, D. W. 1904. On mountains and mankind. *Alpine Journal* 22(166): 269–290.

Freshfield, D. W. 1914. *Hannibal Once More.* London: Edward Arnold.

Freshfield, D. W. 1916. The southern frontiers of Austria. *Alpine Journal* 30(211): 1–24.

Geikie, A. 1912. *Love of nature among the Romans.* London: J. Murray.

Godley, A. D. 1925. Mountains and the public. *Alpine Journal* 37: 107–117.

Gribble, F. 1899. *The Early Mountaineers.* London: Unwin, T. Fisher.

Grinnell, G. B. 1922. The medicine wheel. *American Anthroologist* 24: 299–310.

Henthorn, W. E. 1971. *A History of Korea.* New York: Free Press.

Howard, G. E. 1949. Alpine uplift. *Alpine Journal* 57 (278): 1–9.

Husted, W. M. 1965. Early occupation of the Colorado Front Range. *American. Antiquity* 30(4): 494–498.

Husted, W. M. 1974. Prehistoric occupation in the Rocky Mountains. In J. D. Ives and R. G. Barry, eds., *Arctic and Alpine Environments* (pp. 857–872). London: Methuen.

Hyde, W. W. 1915–1916. The ancient appreciation of mountain scenery. *Classical Journal* 11: 70–84.

Hyde, W. W. 1917. The development of the appreciation of mountain scenery in modern times. *Geographical Review* 3: 107–118.

Iriarte, J., and Vrydaghs, L. 2009. *Rethinking Agriculture: Archaeaological and Ethnoarchaeological Perspectives.* Walnut Creek, CA: Left Coast Press.

Kohn, R. J. 2001. *The Lord of the Dance: The Mani Rimbu Festival in Tibet and Nepal.* Albany: State University of New York Press.

Kornfeld, M., Larson, M. L., et al. 2001. 10,000 years in the Rocky Mountains: The Helen Lookingbill site. *Journal of Field Archeology* 28: 307–324.

Krakauer, J. 1997. *Into Thin Air: A Personal Account of the Mount Everest Disaster.* New York: Villard.

Laguna, F. de 1972. Under Mount Saint Elias: The history and culture of the Yakutat Tlingit. *Smithsonian Contributions to Anthropology* 7: 1–3.

Linares, O. F., Sheets, P. D., and Rosenthal, E. J. 1975. Prehistoric agriculture in tropical highlands. *Science* 187(4172): 137–145.

Lunn, A. H. M. 1912. An artist of mountains: C. J. Holmes. In A. H. M. Lunn, ed., *Oxford Mountaineering Essays* (pp. 3–34). London: Edward Arnold.

Macfarlane, R. 2003. *Mountains of the Mind: A History of Fascination*. London: Granta Publications.

Mathieu, J. 2006. The sacralization of mountains in Europe during the modern age. *Mountain Research and Development* 26(4): 343–349.

Mathieu, J. 2011. *The Third Dimension: A Comparative History of Mountains in the Modern Era*. Cambridge, UK: White Horse Press.

Mishkin, B. 1940. Cosmological ideas among the Indians of the southern Andes. *Journal of American Folklore* 53: 225–241.

Mullikin, M. A., and Hotchkis, A. M. 1973. *The Nine Sacred Mountains of China*. Hong Kong: Vetch and Lee.

Mumm, A. L. 1921. A history of the Alpine Club. *Alpine Journal* 34(223): 1–18.

Nebesky-Wojkowitz, R. M. de. 1956. *Where the Gods Are Mountains*. London: Weidenfeld and Nicolson.

Nicolson, M. H. 1959. *Mountain Gloom and Mountain Glory*. New York: W. W. Norton.

Nilsson, M. P. 1972. *Mycenaean Origins of Greek Mythology*. Berkeley: University of California Press.

Noyce, W. 1950. *Scholar Mountaineers, Pioneers of Parnassus*. London: Dennis Dobson.

Otto, R. 1950. *The Idea of the Holy*. Trans. J. W. Harvey. Oxford, UK: Oxford University Press.

Parker, P. 2008. *Mountains and Mountaineering: Their Spiritual Significance*. Michaelmas, UK: Farmington Fellowship.

Peattie, R. 1936. *Mountain Geography*. Cambridge, MA: Harvard University Press.

Perry, T. S. 1879. Mountains in literature. *Atlantic Monthly* 44: 302–311.

Quaritch-Wales, H. G. 1953. *The Mountain of God*. London: B. Quaritch.

Rees, R. 1975a. The taste for mountain scenery. *History Today* 25(5): 305–312.

Rees, R. 1975b. The scenery cult: Changing landscape tastes over three centuries. *Landscape* 19(3): 39–47.

Reinhard, J. 1985. Sacred mountains: An ethno-archaeological study of high Andean ruins. *Mountain Research and Development* 5(4): 299–317.

Reinhard, J. 2006. *The Ice Maiden: Inca Mummies, Mountain Gods, and Sacred Sites in the Andes*. Washington, DC: National Geographic.

Reinhard, J., and Constanza, M. 2010. *Inca Rituals and Sacred Mountains: A Study of the World's Highest Archaeological Sites*. Los Angeles: Cotsen Institute of Archaeology Press.

Reinhold, S. 2003. Traditions in transition: Some thoughts on late Bronze Age and early Iron Age burial costumes from the northern Caucasus. *European Journal of Archaeology* 6: 25–54.

Ruskin, J. 1856. *Modern Painters*. Chicago: Belford Clarke.

Sauer, C. 1936. American agricultural origins: A consideration of nature and culture. In A. L. Kroeber, ed., *Essays in Anthropology* (pp. 279–297). Berkeley: University of California Press.

Saxena., K. G., Rao, K. S., and Maikhuri, R. K. 1998. Religious and cultural perspective of biodiversity conservation in India: A review. In P. S. Ramakrishnan, K. G. Saxena, and U. M. Chandrashekara, eds. *Conserving the Sacred for Biodiversity Management* (pp. 153–161). New Delhi: Oxford and IBH Publishing.

Schama, S. 1995. *Landscape and Memory*. London: HarperCollins Publishers.

Schmid, E. 1972. A Mousterian silex mine and dwelling-place in the Swiss Jura. In F. Bordes, ed., *The Origin of* Homo sapiens (pp. 129–132). Paris: UNESCO.

Sculley V. 1962. *The Earth, the Temple, and the Gods: Greek Sacred Architecture*. New Haven, CT: Yale University Press.

Sengor, A. M. C. 2003. *The Large-wavelength Deformations of the Lithosphere: Materials for a History of the Evolution of Thought from the Earliest Times to Plate Tectonics*. Boulder, CO: Geological Society of America.

Shaw, R. B. 1872. Religious cairns of the Himalayan region. *British Association for the Advancement of Science Report* 42: 194–197.

Shields, E. T. 1913. Omei San: The sacred mountain of West China. *Journal of the North China Branch of the Royal Asiatic Society* 44: 100–109.

Shryock, J. 1931. *The Temples of Anking and Their Cults*. Paris: Librairie Orientaliste P. Guethner.

Siiger, H. 1955. A cult for the god of Mount Kanchenjunga among the Lepcha of northern Sikkim. In *Actes du IVe congrès international des sciences anthropoloqiques et ethnologiques*, Vol. 2 (pp. 185–189). Vienna: Verlag Adolf Holzhausens.

Sliverman, B. 1999. The history of astronomy: Medicine wheels. *Journal of the Royal Astronomical Society of Canada*. 93: 283–286.

Sowerby, A. 1940. *Nature in Chinese Art*. New York: John Day.

Strutfield, H. E. M. 1918. Mountaineering as a religion. *Alpine Journal*. 32: 241–247.

Sullivan, M. 1962. *The Birth of Landscape Painting in China*. Berkeley: University of California Press.

Thorington, J. M. 1957. As it was in the beginning. *Alpine Journal* 62(295): 4–15.

Trimborn, H. 1969. South Central America and the Andean civilizations. In W. Krickeberg, H. Trimborn, W. Müller, and O. Zerries, eds., *Pre-Columbian Religions* (pp. 83–146). New York: Holt, Rinehart and Winston.

Tuan, Y. 1974. *Topophilia: A Study of Environmental Perception, Attitudes, and Values*. Englewood Cliffs, NJ: Prentice-Hall.

Tyler, J. E. 1930. *The Alpine Passes, the Middle Ages (962–1250)*. Oxford, UK: Basil Blackwell.

Van Buren, E. D. 1943. Mountain-gods. *Orientalia* 12: 76–84.

Vandeleur, C. R. P. 1952. The love of mountains. *Alpine Journal* 58(284): 505–510.

Young, G. W. 1943. Mountain prophets. *Alpine Journal* 54(267): 97–116.

Young, G. W. 1957. The influence of mountains upon the development of human intelligence. W. P. Ker Memorial Lecture 17. Glasgow: Jackson, Son and Co.

Young, T. C., and Smith, P. E. L. 1966. Research in the prehistory of central western Iran. *Science* 153(3734): 386–391.

CHAPTER TEN

People in the Mountains

JAMES S. GARDNER, ROBERT E. RHOADES,
and CHRISTOPH STADEL

Many people have traveled to and through, settled in, moved from, and used mountain areas for a very long period of time. They may have been driven to the mountains seeking refuge from persecution elsewhere and, in the process, founded rich agriculturally based societies. They may have been pulled to the mountains in search of food and other resources, or they may have been attracted for spiritual purposes, as discussed in Chapter 9. They may have explored for ways through the mountains in search of more land and better livelihoods. They may have been driven to or from the mountains by shifts in weather and climate. The discovery of the "Iceman" (Fowler 2001) in a patch of melting ice high on the mountain border between Austria and Italy in 1990 raised speculation as to why he was there, and how he came to be buried by snow and ice approximately 5,000 years ago. Despite the long-standing connections between mountain areas and the wider world, we do know that groups with distinctive cultural and economic characteristics, and people with adaptations to altitude, cold, and steep slopes have occupied mountain areas for millennia. Current changes have pulled traditional mountain people into a globalized world (Cook and Butz 2011), where sharp distinctions between lowland and highland cultures and economies are increasingly blurred.

This chapter is the first of three that provide an overview of mountains and people, past and present, with emphasis on the status of mountain people and change in the early twenty-first century. This chapter focuses on people living in mountain areas. Chapter 11 considers land uses in mountain areas, particularly those relating to agriculture, which remains a primary source of livelihood for most of the world's mountain people. Chapter 12 addresses sustainable mountain development, which involves both mountain people and others living outside mountain areas; the chapter therefore stresses interactions between mountain and other regions.

Although mountains in the past have provided a refuge and a degree of isolation for their permanent inhabitants, the present era is a time when the tentacles of globalization reach to the most distant and marginal parts of the Earth, including mountain areas. New technologies, especially those related to mass transport and rapid communications, are increasingly facilitating the movement of people, goods, services, and information between the mountains and the lowlands, strengthening linkages and dependencies and driving changes in all aspects of life (Conover 2010; Stadel 1993; Fig. 10.1). Persistence and change among people in the mountains are a central focus of this chapter. With the past and the future in mind, human populations, their relationships with their environment, their livelihoods, and their life with hazards are described.

Mountain Populations

Having settled in and utilized the resources of mountain areas for many generations, people have adapted to the environmental conditions even while modifying those conditions to sustain their livelihoods. Most general references relating to mountain populations, and there are many, describe cultures, settlement patterns, economic activities, political issues, and development challenges (e.g., Price 1981; Stadel 1982a, 1982b; Stone 1992; Messerli and Ives 1997; Funnell and Parish 2001; Rhoades 2007; Löffler and Stadelbauer 2008). Some studies have focused on the impacts of global changes, rooted in natural and human causes, on mountain people and environment (e.g., Björnsen et al. 2005; Price 2006). Links between specific mountain populations and environmental factors, and their influence on livelihoods and development, have received extensive coverage (e.g., Kreutzmann 2006; Stadel 1985, 2003a, 2003b, 2006, 2008, 2010; Rhoades 2006; Sarmiento 2008; Borsdorf and Stadel 2013).

FIGURE 10.1 The world of mountain people has changed radically through the introduction of new transportation technologies such as the 'slope bridge' of the Hochtannberg Pass in Vorarlberg, Austria (Photo by C. Stadel)

About 10–12 percent of the global population lives in mountains, depending on the definition of "mountain" and recognizing that such data have limited validity. This would amount to a mountain population of about 700 million in 2011. This population is highly concentrated in developing and transitional countries such as Nepal, Peru, India, and China. Approximately half the mountain population is in Asia, followed by populations in the mountains of South and Central America. These regions have also witnessed the greatest population growth in mountain areas during the past 50 years. On a regional basis, the largest proportion of mountain dwellers is found in Central America (>50 percent). Although mountain populations in Asia account for only about 10 percent of the region's total population, they include nearly half of the world's mountain population (Huddleston et al. 2003). Globally, the distribution of the mountain population shows an association with altitude, referred to as "hypsographic demography" (Cohen and Small 1998). Most people (70 percent) live below an elevation of 1,500 m; less than 10 percent live above 2,500 m. Only in Latin American mountains is the trend somewhat different, with some 24 percent living at altitudes above 2,500 m. In the Himalaya and the Andes, a few people live permanently at altitudes above 4,500 m. The proportion of mountain people as a share of the national population varies greatly. Of developing and transitional countries, the following nations have the highest proportion of mountain people: Bhutan (89 percent), Rwanda (75 percent), Lesotho (73 percent), Armenia (70 percent), Guatemala (64 percent), Costa Rica (63 percent), and Yemen (61 percent) (Huddleston et al. 2003).

In developing regions, a significant number of mountain people are the rural poor who rely on scarce or dwindling resources and opportunities relative to demand, resources derived from agriculture, animal husbandry, forestry, mining, industries, and a variety of formal and informal service jobs. Many are unemployed or underemployed and migrate temporarily or permanently to seek work in lowland industrial agriculture, large cities, and abroad. The migration of mountain people may alleviate the population pressure on the scant resource base of rural areas and generate additional income in the form of cash remittances sent back to families. However, this may generate a sometimes precarious dependence on external resources, create social problems around divided families and communities, and place added pressure on the remaining women, children, and elders. In the economically developed mountain regions of Europe and North America, many people now enjoy a relatively high standard of living, though prior to the twentieth century, they generally experienced conditions of socioeconomic underdevelopment. Much of this has come about through the development of roads (Fig. 10.1), railroads, and air links, which have facilitated a variety of new livelihoods, a topic addressed below and in Chapter 12. Throughout the mountain world, in the twentieth century and continuing, there have been enormous shifts in population numbers and distributions, characterized by growth and urbanization in general, with some specific cases of rural depopulation.

The cultural fabric of mountain populations is diverse. Many cultural groups have been formally and informally identified in mountain regions. In most cases, they constitute a minority of national populations, but they may be a majority locally. Examples include: Quechua and Aymara in the Andes; Kurds in eastern Turkey and northern Iraq; Amhars in Ethiopia; Tibetans, Naxi, Miao, Yi, and Ughuri in China; numerous tribal groups in the Indian Himalaya;

FIGURE 10.2 A typical Himalayan village in which intensive terraced cultivation of grains (inset shows rice cultivation) has been a primary source of livelihood for generations in the Yamuna River Valley, Uttarkashi Himalaya, Uttarakhand, India. (Photo by J. S. Gardner.)

Sherpa and Bhoti in Nepal; and a multitude of First Nations and Aboriginal groups in the North American Cordillera. As national minorities, these distinct cultural groups have often suffered in their relationships with the majority through epidemic disease, discrimination and persecution, forced relocation and assimilation, genocide, and in-migration.

People living in and adjacent to the mountains amount to about 25 percent of the global population, or 1.7 billion people (Meybeck et al. 2001), who are wholly or partially dependent on mountain resources such as water, timber, minerals, and agricultural products. For example, the approximately 700 million people in the Indo-Gangetic Plain region of Pakistan, India, and Bangladesh rely on water from the Himalayan and trans-Himalayan ranges. Similarly high levels of reliance on mountain water are found in Europe, the Americas, East Africa, and China.

The population density of a place is a relatively sensitive measure of human population. Generally, the population density of mountain areas is relatively low. However, people are not distributed evenly, and specific regions are densely populated—for example, the East African highlands, Mesoamerican highlands, and parts of the central Andes. Grötzbach and Stadel (1997) draw a distinction between "physiological density," which is based on the area of agricultural land, and "arithmetic density," which is based on the total land area. In some areas, as in fertile and well-watered locations in the Himalaya and Andes, physiological density may be high while arithmetic density in the region is low. In the Alps, with an area of 190,568 km^2, the population is about 14 million people, giving an arithmetic density of 73 inhabitants/km^2. However, as only 17.3 percent of the Alpine area is suitable for permanent settlement, the physiological density is 414 inhabitants/km^2, comparable to other densely populated regions in Europe (Tappeiner et al. 2008). In a growing number of regions throughout the mountain world, both physiological and arithmetic densities are high as a result of urbanization and lifestyles that place demands on resources that far exceed the available resources of the region, creating large ecological footprints (Cole and Sinclair 2002).

The geography of mountain people may be described on the basis of their numbers and period of habitation. In mountain areas in the tropics and subtropics, such as the Andes, East African highlands, and South and Southeast Asian highlands, there are some relatively large, agriculturally based populations that have been in place for long periods of time (Fig. 10.2). In contrast, in the North American Cordillera, for example, large areas have been inhabited more sparsely by mobile people engaged in hunting, gathering, and fishing. These differences are important, but are insufficient for of a full understanding of mountain people today. As knowledge has advanced, we have come to know that, at one time, North American mountain areas supported relatively dense populations and that the region has been inhabited for over 30,000 years. In the Andes, large and dense populations with complex socioeconomic systems once inhabited areas which now support only small, mobile populations (Funnell and Parish 2001; Mann 2006). Populations have waxed and waned in response to global forces such as the introduction of new and highly productive crops, such as maize and the potato, and the arrival of devastating infectious diseases, like smallpox and measles. Even with spatial and temporal variability and diversity

FIGURE 10.3 Metropolization has crept into some high mountain environments, as shown by the city of Bogotá, Colombia, at an altitude of some 2,650 m, with a metropolitan population of close to 8 million (2005). (Photo by C. Stadel.)

among mountain populations and cultures, a close relationship with the environment and resulting livelihood opportunities and limitations have been important to mountain people (Rhoades 1979). By adapting to and altering environmental conditions, distinctive populations, cultures, languages, forms of governance, and economies have developed in the mountains. Today, many other global forces are impacting mountain people and, while distinctions and artifacts of the past are important and valuable, a more dynamic or process-oriented approach is needed for a full understanding of people in the mountains.

Urbanizing Mountain Populations

Urbanization and the rapid growth of cities were common features of the twentieth century and continue to the present, affecting mountain populations in many ways. Cities

within and at the edge of mountains are often integrated into the global economy, as Borsdorf and Paal (2000) demonstrated for the Alps. Close to a third of the world's mountain people live in urban areas. The largest urban areas are located on mountain margins and high plateaus, some at high altitudes. The degree to which they have been and are integrated with the mountain environment historically and socioeconomically varies, but all are influenced geoecologically by proximity and/or altitude. They differ geoecologically, historically, and socioeconomically somewhat from those cities located in the mountains, which tend to be smaller. Large mountain margin or plateau cities (population >1 million) include Mexico City, Carácas, Bogotá (Fig. 10.3), Quito, La Paz, Santiago, Denver, Seattle, Vancouver, Calgary, Geneva, Zurich, Addis Ababa, Nairobi, Tehran, Chandigarh, Dehra Dun, Siliguri, Kathmandu, Chengdu, and Kunming. Some, such as Vancouver and

FIGURE 10.4 Darjeeling, in the front ranges of the Sikkim Himalaya, India, is a ridgetop settlement founded in 1840 as an administrative, health, tourist, and tea cultivation center, attracting thousands of economic and amenity migrant/settlers from surrounding parts of India and Nepal. Kangchenjunga, third highest mountain in the world, is in the background. (Photo by J. S. Gardner.)

Chandigarh, are at a very low altitude, but nonetheless are bordered and influenced by mountains. Others, especially those in Latin America, are at very high elevations, including La Paz, Bolivia (3,500–3,800 m), and the new city of El Alto on the *Altiplano* (3,850–4,100 m), Quito, Ecuador (2,850 m), Bogotá, Colombia (2,650 m), and Mexico City (2,250 m) with a population of about 21 million.

Urban and suburban areas that are fully enclosed by and closely integrated with the mountain environment tend to be much smaller but nonetheless functionally diverse (Bätzing et al. 1996; Perlik and Messerli 2004; Stadel 1986). Most are located in valleys and a few on ridge tops (Fig. 10.4). Examples include Innsbruck, Austria; Chamonix-Mt. Blanc/ Sallanches, Grenoble, France; Trento and Bozen/Bolzano, Italy; the Martigny-Brig-Visp conurbation, Switzerland; Kullu-Manali, Shimla, and Darjeeling, India; Lijiang, China; Huaraz and surroundings, Peru; Aspen, Colorado; and Canmore-Banff, Canada. While some of these settlements have retained aspects of their original agricultural, mining, manufacturing, or transportation functions, many now have a multifunctional orientation and a polycentric spatial dimension (Dematteis 2009). Important driving forces in the urbanization process include: expanding tourism, recreation, and amenity migration; revitalization of resource extractive and processing industries; administrative functions; and the growth of service needs with population increase and socioeconomic change. Whereas traditional agriculturally based mountain settlements were relatively self-sufficient, the resource demands resulting from contemporary urban growth far exceed the resources of the immediate area, creating a large external dependence and ecological footprint.

Common to all urban areas, whether in the mountains or on their margins, are rapid population growth and densification, physical expansion, growth of infrastructure, and diversification of functions. The cities are powerful attractions for migrants: Many, though not all, come from the mountains in search of employment, security, education, and other services and opportunities. This has been especially the case in developing countries, leading to the growth of large, densely populated squatter or informal settlements on city margins and in deteriorating core areas. While the growth of the megacities is impressive, the proportional growth of secondary cities has often been greater, presenting a more attractive alternative to migrants and better options for sustainable urban living. Mountain and mountain-margin cities also serve as financial and labor feeders for resource exploitation and as sources of large numbers of recreationists, tourists, and semipermanent residents in the mountains.

Permanent Residents

Among the permanent residents are people who gain much of their family livelihood from activities in the mountain setting and have generational familial connections with a particular mountain area. As in any environment, some people are born, raised, live, and die in the same village or valley. Others have been forced to move for a variety reasons, but remain in a mountain setting. Examples include the hundreds or thousands of indigenous people in the Andes who were forced into resettlement schemes during the Inca and early part of the Spanish colonial periods (Rhoades 2006); large numbers of Kikuyu who were resettled on the western slopes of Mount Kenya during the British colonial period in East Africa (Elkins 2005); over 200,000 people forced from their homes in Garhwal Himalaya during the Gurkha invasion from Nepal in the early nineteenth century (Rangan 2000); and people displaced by large-scale hydroelectric schemes in mountain areas of northern India and southwestern China.

FIGURE 10.5 A nomadic Gaddi pastoralist and his flock returning from summer high pastures in the Pir Panjal Himalaya, Himachal Pradesh, India. (Photo by J. S. Gardner.)

In a different category are prospectors, miners, loggers, and, today, workers in secondary or tertiary forms of employment, who live their entire lives in the mountains but continually shift location within the mountains. The gold and silver rush era of the nineteenth century in the mountains of western North America provides examples. Prospectors examined a promising area, staked claims, maybe worked the site for a short period, and then moved on. Miners likewise tended to move from place to place, following the new finds and "rushes." More recently, the same pattern of shifting to sites of exploitation has existed within the community of loggers, tree planters, and construction workers. With the advent of more rapid and reliable transportation, such resource exploitation can take place from distant permanent homes, usually in established and diversified mountain communities.

It is not uncommon for some members of mountain families and settlements to have employment outside the mountains. Just as the Swiss and Scottish Highlanders were once the mercenaries of Europe, the Gurkha of the Indian and Nepal Himalaya have sought employment elsewhere as professional soldiers for several generations. Large numbers of young people in mountain areas of Afghanistan, Pakistan, India, and Nepal have gained employment as far away as the Gulf States, providing cash remittances to family members remaining in the mountain home. This pattern also is common in parts of the Andes, where males move annually to work on agricultural plantations on the coastal plains of Peru and Ecuador to harvest sugar cane, rice, bananas, and other products, or to work in the oilfields of the Amazon lowlands adjacent to the Cordilleras. The same pattern can be observed among residents of the Andes who work in the United States, Spain, and other European countries. They may periodically visit their home villages for *fiestas* or family events, and they also send remittances to family members left behind. Material testimonies to this practice are modern, but often empty, "urban-style" houses, built as showpiece investments and future retirement homes. The periodic absence of males from mountain communities obviously places new responsibilities, in terms of the family, the home, and the community, on the remaining members, many of whom are women, children, and older people. The process of out-migration with periodic return is described by the term *circular migration* or *circulation* (Flora 2006). Thus, it is important to recognize that permanence is a relative phenomenon and that, among the permanent people, there may be members of communities who are absent for long periods of time but who remain intimately connected to the mountain home through family connections and responsibilities.

In the mountains of the economically advanced countries, important daily commuter flows between the residences of workers and their places of employment can be observed (Tappeiner et al. 2008). Ever-expanding transportation infrastructure and services facilitate these daily movements.

Semipermanent Residents

The common types of semipermanent residents in mountain areas are people engaged in nomadic transhumance activities, economic migrants who take advantage of seasonal employment opportunities, and amenity migrants who maintain secondary residences in the mountains to take advantage of recreational opportunities and other amenities offered by the mountain environment. Nomads, such as Gaddis and Gujjars in the western Himalaya, are seasonal inhabitants who have traditionally taken advantage of the rich grazing resources offered by the forests, meadows, and pastures of the high mountain environments (Fig. 10.5). Similarly in the past, Aboriginal people in western North America utilized the high mountain areas

FIGURE 10.6 Construction projects, especially roads, have attracted large numbers of local and migratory workers such as these seasonal workers on NH 21, Pir Panjal Himalaya, Himachal Pradesh, India. (Photo by J. S. Gardner.)

on a seasonal basis for hunting and gathering. In both cases, livelihoods are or were tied directly to this periodic use of mountain resources.

Seasonal economic migrants also seek employment opportunities in tourism, agriculture, industry, and construction. In the European Alps and the North American Cordillera, many young people move to the summer tourist and winter ski resorts for seasonal employment. These might be considered amenity migrants seeking alternative lifestyles. They contrast to the hundreds of thousands of Nepalese, Biharis, and other poverty-stricken South Asians who migrate seasonally to construction projects and tourist sites at high altitude in the Himalaya. Such people are usually the primary source of manual labor in road building and maintenance and on construction projects (Fig. 10.6). A similar pattern has been more recently found in western China, where poor and usually landless peasants (from nonmountain areas) are employed in the construction of roads, railroads, and hydroelectric projects in the mountains. Seasonal or limited-term economic migrants are also found at construction sites in the Alps and the Cordilleras of the Americas. Other forms of seasonal economic migration into mountain areas include people engaged in reforestation or tree planting on a large scale, a very common occurrence in western North America, and those engaged to harvest cash crops. Many Sherpa people of the Everest region have become so wealthy through tourism that seasonal Rai laborers now perform the bulk of the agricultural, portering, lodge operation, and yak-herding work.

Seasonal economic migration that is persistent over time produces a form of semipermanent residency in the mountains. Although more common now than in the past, it was not unknown in the past. Trade across mountain ranges between people from different sides of the mountains often took place at well-established sites within the mountains that persisted for generations. Well-documented examples are found along the great trans-Himalayan trade routes between India and Tibet and Central Asia (Rizvi 1999). Wool, salt, hides, and borax from north of the Himalaya were traded for agricultural products, timber, and textiles from the southern Himalaya, or further south, at predetermined locations. Likewise, in western North America, prehistoric trade across the mountains took place at long-established locations within the mountains, creating a semipermanent footprint and some semipermanent residents there. In the Alps, since the Roman period, but especially since late medieval times, roads over mountain passes facilitated the trans-Alpine trade between the Mediterranean region and the dynamic commercial centers in Central and Western Europe. The Alpine passes served as routes for pilgrims and soldiers as well. Many of these routes remain in use to the present: Great St. Bernard Pass, Gotthard Pass, Simplon Pass, Brenner Pass, and Col d'Agnel are examples.

Affluence, particularly in Europe and North America, and increasingly in the emerging economies of India and China, has made possible widespread *amenity migration:* "migration to places that people perceive as having greater environmental quality and differentiated culture" (Moss 2006a: 3). This has produced forms of semipermanent residency in mountains around the world. The phenomenon has ancient roots, reaching back to early China and Greece; but it has expanded rapidly since the mid-twentieth century in the mountains not only of industrialized countries (Moss 2006b), but also of many developing countries including Argentina, Chile, China, Costa Rica, India, Indonesia, Laos, Malaysia, Mexico, Morocco, the Philippines, South Africa, and Thailand (Chaverri 2006; Glorioso 2006; Moss 2006a; Otero et al. 2006). Proponents of this trend argue that it brings affluence, enhanced infrastructure and services,

FIGURE 10.7 Mountains hold great spiritual value for residents and outsiders alike. Shown here is a sacrifice to Pachamama (Mother Earth) by native Uros on the Bolivian Altiplano near Lake Titicaca. (Photo by C. Stadel.)

and modernization to the mountains. Opponents warn of a speculative real estate market with exorbitantly rising housing prices, of potentially unsustainable economic growth, of cultural alienation, and of increased environmental stress (Sandford 2008). For example, in the Alps, there is considerable debate over the widespread proliferation of apartments and second homes and the ensuing seasonal influx of outsiders. Some communities are today seeking to curb the proliferation of second-home ownership by foreigners, and in 2012, following a national referendum in Switzerland, a law imposed a 20 percent ceiling on the number of second homes in any community.

In the Andes, the rural *hacienda* or *finca* of wealthy urbanites has a long tradition since colonial time and is seen as a symbol of social prestige and sophistication. Today, people build second residences in and around ski and golf resorts, and at or near natural amenities such as scenic river valleys and lakes. Verbier and Davos, Switzerland, Whistler, British Columbia, and Vail, Colorado, are examples of such settlements in Europe and North America. Indeed, the economic viability of some destination resorts is based on the sales of real estate to private individuals or groups. The owners visit their second homes on a periodic basis and form a distinctive social/cultural community that pays local taxes and feels it has a stake in local issues. Such groups may find themselves at odds with long-term traditional residents who rely more directly on local natural resources for their livelihoods. A new feature in this context is the emergence of "time share units," where clients purchase a certain period of time/year in a chalet or apartment for a period of 20 to 40 years. The semipermanent resident in the classical case may wish to preserve the amenities that attracted them, while the permanent residents may wish to preserve their traditional values and livelihoods, which may be dependent on the use, sustainable or not, of the

amenity resources. This type of conflict is common in western North America (Moss 2006b).

The so-called "hill stations" that developed in India during the British colonial period are early products of amenity migration to mountain and hill areas that led to semipermanent residency and permanent towns and cities of some size. The best known are located in the southern ranges and foothills of the Himalaya; they include Shimla, Murree (now in Pakistan), Mussoorie, Nainital, and Darjeeling (Kanwar 1990; Fig. 10.4). All were established and/or developed in the nineteenth century as retreats from the spring and summer heat in the Indo-Gangetic Plain region of British India. Colonial and state administrative services and personnel were moved annually to the hill stations in the April to September period. The annual migrations involved not only the British colonial officers and their families but office workers, servants, and entertainers, mostly of Indian origin, and their families. Significant numbers purchased property and settled in the adjacent areas upon retirement from the colonial service. Following Indian independence and the creation of Pakistan, many of the properties in the hill stations were purchased by wealthy and influential Indians and Pakistanis for use as second homes during holidays and the summer heat. Today, they also serve as retirement communities and tourist destinations, primarily for Indians and Pakistanis.

As discussed in Chapter 9, mountains hold significant spiritual value for mountain people (Fig. 10.7), as well as for many from outside the mountains. In addition to pilgrimage, this has given rise to a particular form of amenity migration and semipermanent residency. Most pilgrims travel only occasionally to sacred sites in the mountains in a manner akin to tourists, and thus are considered transients. Some make pilgrimages to mountain sacred sites on an annual basis and spend significant amounts of time in the mountains as *de facto* semipermanent residents. Many return annually to ashrams or teaching centers in places like Rishikesh on the Ganges River and Kullu on the Beas River, where they rent or build second homes.

Semipermanent residency in the mountains may lead to permanent residency. In fact, it is not uncommon for economic migrants working in tourism to stay on permanently because of a combination of economic opportunity and the amenities offered by the mountain environment. Increasingly, with the advent of home-based businesses and the use of the Internet, amenity migrants are choosing to establish permanency in their mountain homes. Others choose to retire to their mountain homes and establish a form of permanency in this way. This is particularly the case for scenic lake locations or for spas and health resorts, the latter offering a wide range of medicinal services.

Transient People

Transients are not considered people of the mountains in the strictest sense, but some spend large amounts of time in the mountains through frequent visits. However, they do not live there, nor do most gain their livelihoods directly from the mountain regions. Tourists, recreationists, businesspeople, and pilgrims are some of the important transient groups. They are discussed further in Chapters 9 and 12.

Environmental Relationships

The presence of people in, and their relationships with, the mountain environment are mediated by geoecological conditions that both support and constrain life (Stadel 1992). The geoecological conditions form the subject matter of most preceding chapters and therefore the focus here is on relationships with people. As both people and environment are spatially and temporally variable, so are these relationships. Three general characteristics of mountain geoecology are of particular importance in understanding people in the mountains: altitudinal variability; variability of microenvironments; and seasonality. To these may be added longer-term variation and changes in temperature and precipitation.

A constraint at one point may be useful at another point, the tipping point being determined by people and environment. For example, people travel to mountain areas for winter recreation, the attractions being scenery, terrain, snow, and social ambience. This was not always the case, as snow and steep terrain imposed many costs and inconveniences, so much so that some settlements were and are vacated during the winter in the Alps, North American Cordillera, and Himalaya (De Scally and Gardner 1994). Snow and cold make agriculture impossible, animals need to be protected and fed, travel may be difficult, heating of homes is expensive, and terrain, snow, and weather conditions interact to produce landslides, flash floods, or avalanche hazards. Yet today, millions of people are attracted, not repelled, by snow and cold for the recreational value, and their presence in the mountains is facilitated by many forms of modern transportation, housing, and hazard mitigation technologies. Other examples occur where forested land is cleared for the expansion of agriculture, roads, and ski slopes. The clearing may result in increased surface runoff of water, leading to floods, erosion, and damage to land, property, and people. This is a form of negative feedback in people–environment relationships, arising from actions that produce negative or costly outcomes.

Geoecological factors influence, but do not determine, the numbers and distribution of people and livelihoods and risks to which they are exposed. Important factors include geology, topography, climate and weather, and atmospheric pressure, all interacting to influence the distribution and characteristics of water, vegetation, and soils. The combination of factors produces a diversity of microenvironments, niches, and *ecological complementarity* in the relationships between people and environment (e.g., Moats and Campbell 2006).

Geological Conditions

Tectonics, geological structure, and rock types are conditions that have important implications for mountain people. Tectonic forces are relevant through processes such as earthquakes and volcanic eruptions. While contributing to mountain building and shaping the mountain stage with negative and positive effects, these processes also are part of everyday life for many mountain people (Hewitt 1997a), posing a hazard and, in the case of some volcanoes, augmenting the soil resource base, among other things. For example, the 2005 Kashmir and 2008 Wenchuan (Sichuan, China) earthquakes had devastating impacts on millions of mountain people but also set the stage for some positive redevelopment processes (Zimmermann and Issa 2009; Schutte and Kreutzmann 2011). The volcanoes of Java (Indonesia) are among the most densely populated mountain areas on Earth, in part because of their altitude-driven ecological complementarity and rich volcanic soils that contribute to an agricultural economy, even as they pose a hazard and a high risk of disaster. Geological structure primarily impacts human activities through its role in shaping mountain topography, a topic discussed below.

Rock types and their genesis have a strong bearing on the presence of valuable minerals. Mining for ores, gemstones, fossil fuels, and rock salt has been, and is, important in shaping human settlement in the Alps and other European ranges, the central and southern Andes, and the North American Cordillera. It has been associated with mountain areas since preindustrial times and continues today at an unprecedented scale (Fox 1997; Schweizer and Preiser 1997). The exploitation of rock, gravel and sand,

and related industries also continues to be of great economic importance in many mountain areas.

Topography

Mountain regions are defined by their altitude, local relief, slope gradient, slope aspect, and orientation and density of ridges and valleys, all of which influence other environmental factors. Two features of mountain topography, the magnification of gravitational effect and the barrier to movement, are of particular importance. Steep slopes magnify the influence of gravity: The steeper the slope, the greater is the gravitational effect on stationary and moving objects. Cultivation and construction of houses, roads, and all other structures must take into account this tendency for downslope movement. When cultivated or exposed during construction, soil moves downslope through tillage erosion and water erosion. Thus, steep slopes often require alteration of the slope morphology to produce stability. Examples include terracing for agriculture, contour plowing, and cut-and-fill landscaping for construction of roads and structures (Figs. 10.2, 10.20). These adaptations require large investments initially and for maintenance, but their widespread and longstanding use in mountain areas suggest that the benefits far outweigh the costs.

The mountain barrier effect influences the movement of air, water, plants, animals, and people in two main ways: physical obstruction to easy movement by high ridges, steep slopes, and deep valleys; and increasingly harsh weather and climate with increasing altitude. Historically, this has favored the isolation of mountain people and communities, while limiting their movement and constraining their use of higher altitudes. Isolation may have been protective and influenced the development of unique cultural, socioeconomic, and political arrangements. Examples include the diversity of indigenous cultural groups in the Canadian Cordillera and the Andes, and a diversity of languages, such as in the central Karakoram where Shina, Urdu, Waki, and Burashaski are spoken, and in the Swiss Alps, where many dialects of German, French, Italian, and Romansh can be identified. Relative isolation and local coherence have contributed to strong social systems and some political autonomy, the small independent state of Andorra in the Pyrenees, the states of the Caucasus, and semiautonomous areas such as South Tyrol in Italy being examples. Himachal Pradesh, Arunanchal Pradesh, and Uttarakhand, all in India, are examples of political entities that have been justified, in part, by their distinctive geographic and topographic settings. Isolation and restricted accessibility have, on the other hand, predestined such mountain areas as operational bases for guerilla groups, such as, historically, the *maquisards* in the Atlas Mountains during the Algerian uprising against the French rule of North Africa, the revolutionaries in the Sierra of eastern Cuba, or the Shining Path in the Cordillera of Peru.

Mountain topography presents challenges to movement and transportation that differ from those in other environments, requiring innovative and costly adaptations such as: graded and cobble-paved tracks, as in Inca and Roman times in the Andes and Alps; large cantilevered bridges on traditional trails in the Himalaya; cogwheel mountain railways, cable cars, and helicopter services; and long suspension bridges, pylon-supported over- and bypasses, and long tunnels (e.g., the 8-km Mt. Blanc Tunnel between France and Italy), all common on superhighways in the Alps, North American Cordillera, and, increasingly, in the Himalaya of India and southwestern China.

Many of the influences, constraints, and benefits produced by mountain topography are visible through their impact on other environmental elements. For example, with gain in altitude, atmospheric pressure and air temperature decrease; each has direct impacts on people. The orientation of mountain ridges and valleys controls exposure to sunlight, thereby affecting local energy balance. Topography influences water forms, sources, and movement. Likewise, it impacts airflow locally and atmospheric circulation regionally and globally, all of which influence the geography of precipitation. As discussed in Chapter 7, the gradations seen in the mountain environment with gain in altitude are very evident in the vegetation patterns so that distinct altitude-based vegetation or ecological zones are evident. Similar topographic zonation is evident in agricultural and other land uses and, thereby, the livelihoods of people.

Atmospheric Pressure

Atmospheric pressure declines with increased altitude, and therefore the oxygen available to humans and other aerobic life forms is also reduced. This produces short-term physiological changes and longer-term physiological adjustments and adaptations in people. Long-term high mountain residents have more hemoglobin in their blood, larger lung and heart volumes, shorter extremities, and more bone marrow. Fertility, growth and maturation, and morbidity and mortality in populations of mountain people are affected. Many essential human activities based on combustion—for example food preparation and operation of combustion-powered machinery—are also affected. Human oxygen deficit, or hypoxia, is not unique to altitudinal air pressure depletion, but is a condition found at all altitudes among people who are physiologically challenged in their oxygen uptake. In the mountains, it affects all people, whether permanent or semipermanent residents or transients. Our understanding of it comes from a very long scientific interest and, more recently, from intensive research on high-altitude mountaineers and permanent residents (Baker 1978; Houston 1998; West and Readhead 2004).

The most obvious and immediate effect on people as they gain altitude is shortness of breath. Other short-term effects may include dizziness, headaches, nausea, nosebleeds, and tiredness. People display altitude-based hypoxia symptoms differently. Some begin to feel these effects at altitudes under 1,500 m, while others do not experience ill effects even at 4,000 m. People who live permanently or for extended periods at high altitude display physiological adaptations through acclimatization that reduce hypoxia. Mountain climbers, for example, take extra time to ascend to higher altitudes so that their bodies adjust to reduced oxygen. Some climbers have briefly reached altitudes of over 8,000 m without supplemental oxygen. Acclimatization involves an increase in the number of red (oxygen-carrying) cells in the bloodstream, an increase in lung capacity to increase oxygen absorption, and an increase in the heart's ability to pump blood through the bloodstream. Understanding of altitude-driven hypoxia has increased over the past 200 years, as more people travel to the mountains on a short-term basis for work, recreation, and tourism (West 2004). Up to an altitude of about 5,000 m, with proper acclimatization, nourishment, and hydration most healthy people can maintain normal body functions, weight, and fitness over long periods. Above this altitude, however, continuous hypoxia-related deterioration occurs. This can be observed particularly at mining sites at extreme altitudes, such as the gold mine around La Rinconada in southern Peru at an elevation of 5,100 to 5,300 m.

The Himalayan border dispute between China and India, which started in 1962, provides a good example of hypoxia's impact on people not permanently resident at high elevation. In response to Chinese incursions in several border areas, Indian soldiers were moved quickly from low altitudes to 3,300 to 5,500 m. More than 2,000 soldiers were soon debilitated by acute mountain sickness and thus proved useless in the conflict. Chinese soldiers were acclimatized and less affected, having been stationed in Tibet for a considerable period. The dispute between India and Pakistan in Kashmir since the mid-1980s has taken place at altitudes of 3,500 to 7,000 m. At one continuing point of contact, the upper reaches of the Siachen Glacier, hypoxia and acute altitude sickness, along with cold temperatures and snow avalanches, have caused more injury, debilitation, and death than the military action.

Like the transitory effects of hypoxia on mountain visitors, the effects of low air/oxygen pressure are evident among permanent residents at high altitude (Baker 1978; West 2011). It has been known for some time that residents of Tibet, other peoples in the Hindu Kush–Karakoram–Himalaya, and the indigenous people of the high Andes have in some ways adapted to reduced oxygen pressure but remain affected in other respects, such as reduced fertility levels, reduced growth and maturation rates, and higher levels of morbidity and mortality. However, hypoxia is not alone in producing these effects; other factors common in mountain environments, including exposure to cold, poor housing, inadequate diet, genetics, and related socioeconomic determinants of poor health, such as poverty, play important roles.

The Spanish invasion and settlement of the central Andes during the sixteenth and seventeenth centuries provide some comparison between newcomers and permanent residents. In the seventeenth century, over 20,000 Spanish settled in Potosí (4,000 m), an important silver mining center. It was 53 years before the first Spanish child was born

FIGURE 10.8 The paradox of continuing domestic fuelwood use in a time of mobile phones, other electronic technologies, and modern construction, Yulongxue Shan, Hengduan Mountains, Yunnan, China. (Photo by J. S. Gardner.)

there, while the indigenous people gave birth at a normal rate. It was only after several generations of inbreeding with the indigenous population that Spanish birth rates increased. The reproductive capacity of imported livestock (horses, cattle, pigs, and chickens) was also impaired. In 1639, the capital of the Viceroyalty of Peru was moved from Jauja at 3,300 m to Lima on the coast, partly to address the problems imposed by altitude. The reasons for lower birth rates are unclear, but it is obvious that reduced oxygen pressure does influence the reproductive process through reduction of sperm quantity and quality, disturbance of menstruation and ovulation, hypoxic impacts on fetuses, and greater risk of miscarriage at higher altitudes.

Reduced atmospheric oxygen pressures affect other relevant aspects of human life at high altitudes, particularly the processes or devices that rely on combustion. Cooking food, heating homes, and operation of vehicles and other machinery are affected. As oxygen is reduced with the gain in altitude, combustion is less complete, producing less heat and more carbon monoxide. As less energy is imparted to cooking, heating, and powering machinery from a given amount of fuel, more fuel and time are required than at lower altitudes. This translates directly and indirectly into greater costs and greater environmental impacts (Fig. 10.8). Increasing air intake helps address this problem in vehicles and aircraft, but only to a point. Fixed-wing aircraft and helicopters with some adjustments can operate at altitudes above 4,500 m. Aircraft using jet or turbojet engines are not so affected by the problem of oxygen depletion. It is not uncommon now for military and civilian helicopters to operate at altitudes well above 6,000 m.

In the mountain villages, everyday tasks like cooking and heating are compromised by reduced air/oxygen pressure. In cooking, the problem is more complex than reduced combustion and energy transfer. Three complexities are involved: the increased expansion of leavening gases, the lowered boiling point of water, and the increased evaporation of water from food. Of these, the reduced boiling point is the most important, as it takes longer (4 to 11 percent) to cook food with each increase of 300 m. This translates to increased fuel expenditure and impact on local fuel resources (e.g., wood and dung), greater expense in purchasing other fuels, and increased emissions of harmful gases and particulates.

Weather and Climate

While the physical basis of mountain weather and climate is described in Chapter 3, key variables such as temperature, precipitation, sunlight, and wind that influence human occupancy are discussed briefly here.

TEMPERATURE

Altitude, latitude, proximity to warm or cold ocean currents, continentality, slope aspect, and cloudiness influence temperatures and thereby the ecological characteristics and processes of the mountain environment. Temperature generally declines with increasing altitude, but this environmental lapse rate varies considerably between different mountain regions and within a mountain complex, depending on location, topography, and aspect, and with changing weather situations. For example, as discussed in Chapter 3, mountain winds, such as *foehn* (Alps), *Chinook* (North America), and *zonda* (southern Andes), generated by regional weather systems, produce warmer than normal temperatures at low elevations, thus altering temperature lapse rates. In confined mountain basins and valleys, descending cold air may form low-lying pockets of cold air, thus temporarily reversing the normal altitudinal lapse rate. The cold air pockets may produce low-lying clouds and fog, which prolong the lapse rate reversal or inversion. Persistently recurring *foehn* and temperature inversions influence settlement, agriculture, and recreation possibilities, just as does the normal decrease in temperature with gain in altitude.

Superimposed on altitudinal effects are seasonal differences in temperature produced by latitude, being pronounced in the higher and midlatitudes and less so in the low latitudes. The seasonality of cold winters and warm summers, characteristic of the Alps, North American Cordillera, southern Andes, and much of the Himalaya, shapes the annual rhythms of life. Because of prolonged winter cold, high pastures and fields are productive only during the summer months, whereas, in the forest zone below, tree harvesting may be carried out preferentially during the winter months. Depending on latitude, valley locations may or may not permit agricultural activities. The altitudinal and latitudinal effects on temperature influence the locations of permanent settlements. Through time, social, economic, and technological changes have lessened this influence, as many higher settlements have expanded to form winter and summer resorts, cold and snow now being seen as a recreation resource. Medieval pastoral and agricultural settlements throughout Austria, Switzerland, northern Italy, and France are now famous resorts (e.g., Arosa, Davos, Zermatt, Lech, Zürs, Cortina, Chamonix, Les Contamines) (Fig. 10.9). Local and regional transhumance herding remain common adaptations to seasonal and altitudinal cold in the northwestern Himalaya, Karakoram, and Hindu Kush ranges. In some areas, such as Himachal Pradesh and Uttarakhand in India, further adaptations, such as seasonal cropping of potatoes, carrots, and other hardy vegetables and development of winter tourism, are occurring.

As discussed in Chapter 3, some altitudinal and seasonal temperature variations in the central Andes, East African, and other equatorial mountains are similar to those discussed above; others are distinctly different. Quito in Ecuador (2,850 m), for example, has an average annual temperature of 15°C and a January–July seasonal variation of only 0.6°C. In contrast, diurnal temperature fluctuations often exceed 10°C. Subfreezing temperatures are rare below 2,800 m, but significant diurnal fluctuations, including freezing and thawing, are common in the high Andes, where they are of great ecological and agricultural relevance. For example, permanent settlements are possible in the highest ecozones of the *Puna*, at altitudes exceeding 4,000 m, where agricultural activity consists of herding sheep, llamas, and alpacas, supplemented by some field cultivation of tuber crops and cold-resistant grains (e.g., quinoa). Regular night freezing makes agriculture and food storage difficult, but crop selection and preservation adaptations, such as the making of *chuño* (see Chapter 11), have sustained large populations and complex civilizations.

Mountain people and their activities have been greatly affected by temperature variations over time as part of shorter- or longer-term climate changes. While higher temperatures in medieval times allowed settlement, specific agricultural land uses and mining activities at higher altitudes, the subsequent Little Ice Age from the seventeenth to the late nineteenth century impaired human activities at higher elevations and restricted their altitudinal reach. Today, many mountain dwellers are concerned about the current warming trend, the melting of glaciers, and the related threat to a reliable water supply in some arid and semiarid regions (e.g., Orlove et al. 2008). In the Alps, winter resorts at altitudes below approximately 1,000 m fear for the future of a sufficiently long and reliable ski season and

FIGURE 10.9 Chamonix–Mt. Blanc is an example of a medieval agricultural settlement transformed into a modern year-round tourist and service center and expanded into areas of floods, snow avalanches, icefalls, and debris flow hazards. (Photo by J. S. Gardner.)

the greater economic and environmental costs of artificial snowmaking (Steiger and Mayer 2008).

PRECIPITATION

As discussed in Chapter 3, mountains increase precipitation amounts in windward locations (rainsheds) through the orographic (barrier) effects imposed by topography on moving air masses. In leeward locations, an opposite effect occurs (rainshadows). A pattern of windward wet/leeward dry is present at many scales across mountain systems. Also, precipitation tends to increase with altitude, though, in some higher ranges, such as parts of the Himalaya, this increase reaches a maximum below the summits and declines further upward. Such very general patterns may be altered at a smaller scale by more localized pockets of atmospheric instability that produce high-intensity rain or snow events in otherwise drier locations. A further complexity is that precipitation takes several forms in mountain environments—rain, snow, sleet, hail, rime, hoar frost, fog, and mist—and the relative importance of each varies geographically, examples being snow at high altitudes in the Cascade and Coast Ranges in the North American Cordillera, monsoon rain in the central and eastern Himalaya, and fog/mist at lower and mid-altitude in the south-central Andes (*garúa*). Precipitation also is variable through time, most obviously in the seasonality of wet, dry, snow, and so on. Seeming less regular are longer-term variations in precipitation from year to year or over many years. A good example of such variations is in the tropical Andes and the adjacent Pacific coast where the periodic occurrences of the warm *El Niño* and the cold *La Niña* currents result in rainfall irregularities.

The complexities of amount, type, and timing of precipitation are reflected in mountain ecosystems and in the lives of mountain people. Availability of precipitation-derived moisture, along with temperature, sunlight, and soil conditions, influences crop choice, timing of planting/harvesting, use of irrigation, and type of cultivation practices. Timing, location, and type of precipitation influence recreation and tourism, as in avoidance of the wet monsoon season in the Himalaya and avoidance of winter snow for some activities and its attraction for others. Precipitation in the form of snow is a very significant factor in people's relationships with the mountain environment, acting as both a resource and a hazard (Gardner 1986). However, its primary significance comes about through its derivatives, snowcover or snowpack and glacier ice, discussed below.

Longer-term variation and change in precipitation present mountain people with challenges of adaptation by adopting different crops and cropping systems, developing new forms of tourism and recreation businesses, seeking out new water sources, avoiding newly hazardous locations, moving away, and so on. Long-term adaptation to precipitation changes has been going on among mountain people for millennia. However, the present urgency related to climate change and variability is driven by the fact that economic and other dependencies on precipitation have grown with the populations of mountain people and of the many living outside the mountains.

SUNLIGHT

The amount of sunlight, the primary energy source, received at any location in the mountains depends on latitude,

FIGURE 10.10 An example of an alpine pastoral settlement, occupied seasonally, on sunny south-facing adrêt slopes, above the upper Rhine River Valley, Grisons, Switzerland. (Photo by C. Stadel.)

topography, and cloudiness. Thus, variation from place to place and over time can be considerable. As the primary source of energy, sunlight, or solar radiation, is fundamental in energy and moisture balances locally and regionally, and therefore exerts strong influences on vegetation and water, which are basic to human use and occupancy. Topography plays a very significant role locally, some slopes being more exposed to sunlight than others. On a slope facing into the sun, such as a south-facing slope in the northern hemisphere or a north-facing slope in the southern hemisphere, more direct solar radiation is received than on the slopes facing away from the sun. These slopes tend to be sunnier, warmer, and drier than shaded slopes. These differences are evident in the human use of opposing slopes. In the Alps, for example, in longitudinal, west–east extending valleys, the traditionally preferred settlement and agricultural sites are on the south-facing *adrêt* slopes (Peattie 1936; Fig. 10.10). The north-facing *ubac* slopes have been avoided as traditional settlement areas, but are favored locations for managed forests and ski runs. In equatorial mountains, where the sun is nearly overhead most of the year, the differences in sunlight received on south-facing and north-facing slopes are much less significant.

The quality of sunlight received in the mountains changes with altitude, with an increase in the proportion of short-wave radiation received, particularly ultraviolet (UV), because of the lower concentration of absorptive gases. This results in radiation burning of animal and plant tissue, producing sunburns, greater probability of skin lesions and cancers, and damage to crops at higher altitudes.

WIND

Wind is common at all scales in and around mountains, often with velocities and durations exceeding those found in other environments. Generally, average wind velocities

and gustiness increase with altitude, around exposed ridges, and in confined valleys. The effects, particularly those on people, are: accentuation of cold and dryness; transport of sensible heat or cold, gases, liquids, and particulates; and physical pressure on objects. Wind aids the evaporation of moisture from surfaces like skin and leaves, which, in turn, causes cooling and drying (i.e., wind chill and des-sication or "freeze drying"). On the positive side, this helps regulate heat in animals and plants, preventing damage from overheating. On the negative side, it may accentuate freezing and excessive drying, damaging exposed skin, crops, and trees. Wind, through warming and drying and transport effects, is a major factor in causing mountain forest fires and compromising control and mitigation efforts. Winds may have cooling or warming effects through the transport of colder or warmer air, as when cold air blows off a glacier surface to valley locations in a *katabatic wind*, which can limit agriculture and settlement downslope from glaciers. Earlier, we noted the *foehn, Chinook,* and *zonda,* downslope winds that produce warming and drying effects through compression of the moving air. "Valley winds," blowing upslope during the day, and "mountain winds," blowing downslope during the evening and night, are common in mountain areas. Wind acts as the transport agent for particulates, including snow, gases, allergens, and contaminants, creating a variety of hazardous conditions. In redistributing and altering snow, wind can contribute to the build-up of serious snow avalanche hazards. Very high-velocity *foehn,* katabatic, and storm-generated winds can generate very high-impact pressures that damage and destroy structures, transmission lines, forests, and crops, and interrupt essential services in mountain areas.

Mountain people have long experience in coping with and adapting to these conditions. Settlements, buildings, and other structures are built in wind-sheltered sites and are strengthened to withstand high-impact pressures. Mountain farmers may seek wind-protected niches for their crops and livestock. Along roads, snow fences and plantings are used to reduce wind-driven snow hazards. On the other hand, wind-driven snow is captured by snow fences and plantings to augment snow on ski slopes and water supply and, increasingly, the windiness of mountain areas and their immediate surroundings is being used for electricity generation.

Snow and Ice

Snowcover, glacier ice, and, to a lesser degree, ground ice are present in many high mountain areas and are embedded in the lives of people in and out of the mountains, serving as water supply sources, scenic enhancements, and recreation resources for tourism and recreation, as well as creating hazards in the form of snow avalanches, icefalls, rapid glacier advances (surges), and glacier lake outburst floods. As discussed in Chapter 4, some of the most obvious indicators of present global warming and other climatic variations have been reductions in mountain snow and ice cover over the past 150 years. The changes have been most pronounced in the shrinkage of glaciers, to the point where some small glaciers in tropical and temperate mountains, for example, on Mount Kenya in Africa and in Glacier National Park in the United States, representing equatorial and midlatitude locations, have disappeared (Watson et al. 2008). Continuation of this trend raises the specter of reduced water supply for people dependent on snow and ice sources. Depletion of the summit snow and ice cap of Kilimanjaro has become an icon for global climate change (Molg et al. 2010), while the shrinkage or disappearance of glaciers on nearby Mount Kenya threatens water supply to subsistence and commercial agriculture and horticulture on its lower slopes (Kaser et al. 2005). The equatorial Andes retain a greater but shrinking glacier ice cover, which has been important as a local and regional water source for generations. In the tropical Andes, the melting of snowcaps and glaciers is considered by the indigenous people to be a mythological and cultural loss. There, glacier shrinkage compromises future water supply for agriculture, industry, settlements, and power generation, as well as temporarily increasing the frequency of damaging snow, ice, and rock avalanches and glacier lake outburst floods (GLOFs) (Carey 2010). With some exceptions, the cover of mountain snow and ice increases in amount and extent and decreases in altitude from the equator to the poles. The very large glacier systems of the northwest Himalaya and Karakoram in subtropical latitudes, at very high altitudes, and driven by copious snowfall are among the exceptions in both their physical extent and their stability or growth in recent decades. Regardless of their size and extent, mountain snow and ice are and have been very important to people. However, in very large to subcontinental watersheds, the relative importance of glacier ice sources may have been overstated (Kaser et al. 2010). Nevertheless, at the local and regional scale in the Andes, parts of the Cordillera in

FIGURE 10.11 Hopar and Nagar villages are situated in close proximity to the Bualtar Glacier, on old moraines and ice-marginal lake beds. They are dependent entirely on snow and glacier meltwater for domestic, agricultural, and other uses in this semiarid region, central Karakoram, Pakistan. (Photo by J. S. Gardner.)

North America, and the dry trans-Himalayan ranges like the Karakoram (Fig. 10.11), snow and ice melt dominate stream hydrology, especially at the driest times of year, when they are critical water sources for human use. To date, there have been relatively few formal studies of the roles of glacier ice in the lives of people, communities, and cultures. Some that have been done (e.g., Staley 1982; Cruikshank 2005; Vivian 2005; Orlove et al. 2008; Carey 2010) describe deeply embedded relationships that have transcended generations. Staley describes efforts to create small glaciers in the Karakoram when snow and ice water sources have been depleted or demand has increased (Fig. 10.11), by combining the ice of "female" and "male" glaciers. Cruikshank describes the long-standing role of glaciers as avenues of transport and exchange among indigenous groups in northwestern British Columbia, the Yukon, and southeast Alaska. Both describe hazards such as glacier advances and floods of various types that impinged on land and settlements. This theme is pursued by Vivian in a detailed inventory of the glaciers of Mont Blanc in the French Alps, where, in response to cooling, many glaciers advanced into settlements and agricultural land in the late seventeenth century, and later produced outburst floods with, in some

FIGURE 10.12 The 50-km-long Tehri Reservoir and Dam (inset) on the Bhagirathi River, a major tributary of the Ganga River, Garhwal Himalaya, Uttarakhand. (Photos by J. S. Gardner.)

cases, even greater impact (Fig. 10.9). Carey's detailed case study in the Cordillera Blanca of Peru reveals a complex historical relationship between glaciers; their melting; the formation of lakes; water supply for agricultural, domestic, and power generation; flood, avalanche, and glacier lake outburst hazards; and the politics of it all.

Streams, Rivers, and Lakes

Mountain streams, rivers, and lakes have been and remain focal points and corridors for movement, settlement, and resource exploitation. Within the mountains, they have been primary water sources for people and settlements, avenues for transporting people and goods, and sources of mechanical and electrical energy. They are the conduits through which water is transported from the mountain "water towers" to high water-use areas beyond the mountains (Viviroli et al. 2007). All of the Earth's major river systems gain some or much of their water from mountain source areas. The importance of mountain streams and rivers thus cannot be overstated (Viviroli and Weingartner 2008). The hydrology of mountain surface water is described in Chapter 3, and its central role in sustainable mountain development is described in Chapter 12. Our objective here is to briefly summarize its complex relationships with mountain people and settlements. Streams, rivers, and lakes have influenced human use of the mountain environment, and they have been altered through their relationships with people.

The most obvious point of relationship lies in the fact that rivers and some lakes occupy valleys that, in many cases, are prime locations for settlements and movement of people. Floodplains and valley-side terraces and fans provide relatively flat ground for building, cultivation, and transportation, bringing people into easy contact with water sources suitable for domestic, agricultural, and industrial uses. Coincidently

advantageous is the fact that these lower-elevation sites generally have climate and moisture conditions that are relatively amenable to agricultural activity. Prior to development of the water turbine generator, rivers and streams provided mechanical power that was used for the milling of grains, among other things. Since the advent of the turbine generator and modern transmission and construction technologies, rivers have been favored sites for hydroelectric power generation to serve local, national, and international markets. Larger installations create new lakes, alter the levels of existing lakes, change streamflow patterns, destabilize slopes, and act as focal points for new temporary and permanent settlements. While serving a greater good, these projects often have negative local consequences in mountain valleys where long-established settlements, agricultural and other land uses, transportation, and other infrastructure are inundated or disrupted (Fig. 10.12). People are displaced from their homes and their livelihoods are compromised. Whether in the Himalaya, Andes, or the Cordillera of western Canada, large-scale projects have engendered widespread public protests and discontent, as in the cases of the Tehri Project on the Bhagirathi/Ganga River (Fig. 10.12) and the Lower Arrow Project on the Columbia River, for example. Compensation has rarely matched the measure of disruption in the mountain valleys. Micro-hydro installations that use run-of-the-river technologies are less controversial, but they do alter streamflow in specific reaches of rivers and streams. Rivers and river banks have served as means and locations for solid and wastewater disposal, and they continue to do so. Rivers, streams, and lakes are focal points for tourism, both recreational and pilgrimage, and thereby become embedded in the economy of mountain people and places.

Rivers, streams, and lakes may also act, in concert with topography, as barriers to movement in the mountains. They may have aided in protecting and defending settlements but,

FIGURE 10.13 Managed village forest areas and temple forests in the Beas River valley, Kullu-Manali, Pir Panjal Himalaya, India (photo by J. S. Gardner) and traditional fuel-wood gathering in the Chimborazo region, Ecuador (inset; photo by C. Stadel).

for the most part, every effort has been made to reduce the barrier effect through adaptive technologies ranging from ferries and suspension and cantilevered bridges to modern pylon-supported and elevated suspension bridges (Fig. 10.1). In the mountain world as a whole, all of these technologies, from the most rudimentary to the most complex, are still in use.

Mountain streams, rivers, and lakes generate floods of various types. This fact, plus the tendency for rivers to be focal points of settlement and transportation, produces a confluence of high hazard and elevated vulnerability of people and property which translates into high risk of disaster, and thus requires a high degree of cooperation at all scales from the local to the regional and international.

Vegetation and Wildlife

Plants, animals, fish, and birds are useful and central to the lives of mountain people. As discussed in Chapters 7 and 8, mountain areas are notable for their biological diversity, which reflects relationships with all other environmental factors, and with people. The diversity is driven by the altitude-based differentiation of ecosystems and the multitude of topographic niches. The relationships between people and vegetation have produced impacts through the harvesting of trees and other plants, alterations of plants and their habitat for agriculture and animal husbandry, introductions of new species, and habitat changes resulting from the growth of settlements, construction, and other forms of resource exploitation. For long-term mountain residents, the ecological complementarity and exchange relationships provided by vegetation and ecosystem diversity have been a basis for survival (Rhoades and Thompson 1975).

Under natural conditions, forests are or were extensive in many tropical and temperate mountain regions (Hamilton et al. 1997). They provide important products and services

locally, serve as hydrological regulators regionally, and continue to provide important wood. Today, very few mountain areas do not bear some evidence of the human use of plants, most obviously in the extensive deforestation of mountain regions. We tend to think of mountain deforestation and alteration as a recent phenomenon. This may be true for some mountain areas, such as in the Americas, but not elsewhere (Williams 2006). Mountain areas in the Mediterranean region and the British Isles were deforested by 500 C.E. Forests in the Alps have been cut and altered for 1,500 years, although many have been replaced and are sustainably managed today. The extensive mountain areas of China have been subject to cycles of deforestation beginning 3,000 years ago (Elvin 2004), during the Ming Dynasty (1300–1500 C.E.), and during the past 200 years with population and economic growth and political change. The 1998 floods on the Yangtze River resulted in, among other things, a prohibition on forest removal and agricultural use of steep lands, and massive reforestation. Similarly, the "Theory of Himalayan Environmental Degradation" (Eckholm 1975) attempted to explain damaging floods and erosion/sedimentation in the Gangetic–Brahmaputra plain as arising from deforestation by local people in the mountains. Again, we do know that Himalayan forests have been exploited since precolonial times (ca. 1750), particularly in what is now India, but we also know from more recent research that one cannot generalize about widespread impacts and their effects (e.g., Ives and Messerli 1989; Gardner 2002; Ives 2004; Hofer and Messerli 2006; Moseley 2006). Forests remain a central component of many mountain areas and, while they are continually stressed by human exploitation, diseases, and fire, increasing efforts are made to sustainably manage, replace, restore, and protect them for their economic, ecological, climatic, hydrological, and aesthetic benefits (Price and Butt 2000; Fig. 10.13).

Other mountain plants play important roles in the lives of mountain people. These include nontimber forest products and plants from grassland, meadow, and high-altitude tundra areas. A variety of mushrooms, fungi, and other plants found in the forests of the Himalaya, southwest China, the North American Cordillera, the Carpathians, the Alps, and many other ranges have had traditional food and medicinal uses locally and regionally for generations, and their harvest, export, and sale today generate substantial cash incomes. Forests, meadows, and tundra have yielded plants with medicinal qualities that are used in traditional and modern medicine or have been synthesized for commercial use. Their value as a source of income to mountain people has led to the domestication of some species and their rearing in gardens in some mountain areas. Opium poppies, cannabis, and coca, all of which thrive naturally in some mountain areas, have long been cultivated for their medicinal, spiritual, and recreational value locally and for export. The naturally occurring grasses, sedges, and other plants of the higher-altitude meadows and tundra have served as a seasonal food source for livestock. These uses have served to shape and support pastoral transhumance and nomadic lifestyles throughout the mountain world.

The diversity of mountain vegetation has made possible, through cultivation and selective breeding, a number of important food products. Maize (corn), potatoes, tomatoes, wheat, tea, coffee, and quinoa, among others, are derived from plants growing in mountain or mountain-margin areas (Laws 2010). Many are now grown commercially in places far removed from the mountains. Though domesticated in specific mountain areas (e.g., maize and potatoes in the Americas), the adoption of such food plants in other mountain areas has had profound and generally positive impacts on local populations. Maize, introduced to East and South Asia in the sixteenth century, has spread far and wide in the Hindu Kush–Karakoram–Himalaya, providing a nutritionally superior human and animal food product at higher altitudes than endemic food plants, such as millet. Maize accompanied the expansion of Han Chinese populations into the mountains of southwest China in the eighteenth century, extending cropland into previously forested areas, onto steeper slopes, and to higher altitudes, thus supporting further growth and expansion of local and immigrant populations. The introduction of maize and tea into the Indian and Nepal Himalaya in the nineteenth and twentieth centuries had similar effects (Subba 1989).

The advent of rapid and reliable transport in mountain areas has led to a widespread introduction of food plants that have thrived in the diversity of mountain microenvironments. Introduced as cash crops for distant markets, they have generated new income sources, entered local markets, and altered the diets of mountain people (Fig. 10.14). Carrots, cauliflower, cabbage, broccoli, beans, peas, garlic, grapes, apples, and other hard fruits are common cash crops throughout the mountain world, from India and China to East Africa to the Americas, Europe, and New Zealand.

FIGURE 10.14 Typical contemporary village market stall, Zilicun, Hengduan Mountains, Yunnan, China. (Photo by J. S. Gardner.)

In the Kullu Valley of Himachal Pradesh, on the slopes of Mount Kenya, and in the mountains of Yunnan there are thriving cut-flower and decorative nursery plant businesses. Thus, the natural ecological and environmental diversity of mountain environments that provided the ecological complementarity so important for subsistence of mountain people is today the basis for a commercial diversity that brings benefits to far wider populations.

Global plant hunting was an enterprise rooted in the colonial and commercial expansion of the eighteenth and nineteenth centuries. The diverse plants and vegetation of mountain areas, such as the eastern Himalaya, southwest China, parts of the Andes and the Coast Ranges of the North American Cordillera, were targeted (e.g., Christopher 2003). Driven largely by commercial drug and gardening interests, some of the most visible fruits of this enterprise are today seen in the great botanical gardens such as Kew in London, as well as in most garden plant shops, private gardens, balcony and window gardens, and indoor plant collections. Plant hunting in the mountains continues. The intense relationship between people and plants has changed both, at the global scale (Laws 2010).

Mountain wildlife, including fish and birds, also reflects the diversity of habitats. The presence of wildlife is embedded in the modern image of mountain wilderness to the extent that bears, snow leopards, wolves, giant pandas, mountain gorillas, and eagles, among others, are wilderness icons. Until about 300 years ago in Western society, the presence of large predatory animals, and some imagined ones, together with the gloom and darkness of mountain forests, sustained an aura of fear in and of the mountain environment (Nicholson 1963), especially among outsiders. Imagined icons like the Yeti or Abominable Snowman of the Himalaya and the Sasquatch or Bigfoot of the North American Cordillera remain part of the mountain wilderness image. Values and perceptions have changed to the point today that wildlife is seen by many people as an essential and valued component of the mountain environment. However, this view is not necessarily shared by mountain people in the Himalaya, East Africa, North American Cordillera, and New Zealand, where crops and flocks are preyed upon by endemic and introduced wildlife. The relationships between people and wildlife in the mountains have been and remain extremely complex.

Paramount among these relationships is the fact that wild animals, fish, birds, and insects have been, and remain, useful to people as both central and supplementary food sources throughout the mountain world. Bees, besides producing honey, are essential to the pollination of some food crops. Other animals, birds, and insects are, of course, damaging to crops, livestock, and homes. The result has been a favoring of some species over others and thereby a reshaping of mountain wildlife by people over time. Coupled with anthropogenic habitat changes, the effect has been to reduce the populations of most large animals, fish, and birds, with the result that dependence on wildlife as a food source in the mountains has declined through time. Niche activities such as sport hunting and fishing, bird watching, wildlife photography, and ecotourism now take place in some mountain areas, providing livelihoods for some. Nonconsumptive valuing of wildlife is today taking precedence in most mountain regions, with the result that protection, habitat restoration, population rehabilitation, and reintroductions are common. Ironically, the protection and restoration of mountain wildlife owes much to two "great white hunters" of the late nineteenth and early twentieth centuries, Theodore Roosevelt and Jim Corbett, both instrumental in the parks and protected areas movement discussed in Chapter 12. The relationships between mountain vegetation, wildlife, and people are so closely linked to livelihoods that they are discussed in more detail in the following section.

Livelihoods of Mountain People

Early livelihood systems were complex at the household level, as gathering, hunting, subsistence agriculture, and animal husbandry were practiced in combination, utilizing the diversity of ecosystems and microenvironments present. At the community level, the system was uniform in the sense that most households used similar combinations.

Subsequently, there has been a gradual transformation to greater simplicity at the household level and diversification/specialization at the community level. That is, more households engage in fewer livelihood activities and, within communities, a greater diversity of livelihoods is present through specialization. However, some mountain households have retained a complex livelihood system through tradition, limited opportunities, gender and age role differentiation within the household, and increased economic migration. It has been primarily over the past 400 years, with industrialization, colonialism, commercialism, and tourism that the livelihood diversity now present has developed, though this varies among mountain regions.

Livelihood diversification in the Alps, through the rise of industry, commercial agriculture, and tourism, has been taking place since the early nineteenth century, initially triggered by the application of steam engine technology. Many mountain people have sought seasonal or permanent employment outside the mountains or in new and expanding industries in the major mountain valleys, resulting in a "vertical migration" from the higher and often still remote mountain areas to the valleys, foothills, and plains at the margins of the Alps. This livelihood-based migration, in turn, started to depopulate some areas and heralded the process of urbanization in the emerging economic cores. The development of hydroelectric potential enhanced the process. The coincident development of transportation technology stimulated the further development of mass tourism, which opened a variety of new livelihood activities.

In the Hindu Kush–Karakoram–Himalaya, the last 200 years encompass the arrival of colonialism, which introduced strategic/military and commercial interests on a large scale and new ways of administering and regulating access to land and resources, both of which radically changed existing livelihood systems. In this period, the vast mountain areas of west and southwest China were subject to the in-migration of large numbers of Han (Chinese) people with their new and different technologies, commercial interests, administrative practices, and ways of life. In North American mountain regions, the past 250 years encompass the period of commercialization and industrialization based on extraction and trade of furs, forest products and minerals, new technologies, the influx of many outsiders of European heritage, and the demise of the indigenous populations through disease and genocide. The central Andes experienced similar changes over the much longer period of some 500 years. With the arrival of the Spanish "conquistadors," and the onset of the Spanish colonial hegemony that succeeded that of the Inca in the highlands, takeovers of land and the translocation of indigenous peoples, in part to work in the mines and colonial agriculture (the *mita* system), took place (Cole 1985).

The transformation of livelihoods in many mountain regions, while benefiting some, decimated indigenous populations and harmed their traditional cultures and livelihood systems. In general, there have been transformations in livelihoods and ways of life not seen before, and this process accelerated through the twentieth century to the present. This section describes four main sources of livelihood: hunting and gathering; mining; forestry; and trade and artisan livelihoods. Two other very important sources of livelihoods, agriculture and tourism, are addressed in Chapters 11 and 12, respectively.

Hunting and Gathering

Archaeological evidence indicates that Paleolithic hunters and gatherers were present in many mountain areas. The ecological diversity was ideal for hunters and plant gatherers. By locating along the lower mountain forest boundary, people found a variety of food types available within a relatively short distance. By following the migrating wildlife up and down the mountains as they moved from summer to winter pastures, hunters had an abundance of prey, which, along with the flora, provided an abundance of food throughout the year. Other advantages included plenty of firewood and shelter, fish, water available year round from mountain streams, and defensible locations. To early hunters and gatherers, the resources scattered over a variety of microecosystems with seasonal variability provided a rich environment. To take advantage of such an environment, hunting and gathering required a high degree of mobility, but, for the most part, it appears to have been highly systematic, territorialized, and practiced from permanent and seasonal settlements.

Although subsistence hunting and gathering as a widespread strategy has declined, pockets are still present in mountain areas. In many cases, it is supplemental to other livelihood activities. The collection of mushrooms, berries, and medicinal and decorative plants for domestic use or sale occurs throughout the mountain world. Hunting of wild game and birds and fishing are common in the North American Cordillera and parts of the Hindu Kush–Karakoram–Himalaya as a livelihood supplement. The mountain tribes of Kalimantan, Borneo, called the Dayak by outsiders, provide an example of a more traditional hunting and gathering society. Such groups, however, continue to feel pressure from the outside world, face considerable difficulties in protecting their way of life and environments, and, in some cases, have sought alliances with international conservation agencies and activists. Thus, in a few locations, hunting and gathering remains an old way of life, and continues widely in some mountain areas as part of the new way of life.

Mining

Mining has been and remains a widespread source of livelihood in many mountain areas (Fox 1997). Mountains offer ferrous and nonferrous ores, coal, stones, gravel and sand, gems, precious stones, and rock and evaporate salt.

All aspects of mining, from exploration and prospecting to extraction, processing, and transport, have occupied mountain residents and many more outsiders since Paleolithic times, when indigenous people and outsiders tapped mountain areas for tool and building materials, ornaments, pigments, and salt. In the Alps, important sources of rock salt were located in the provinces of Salzburg and Upper Austria, where it was mined on an industrial scale. Alkaline lakes north of the Himalaya and in the Atacama desert of the Andes have long been sources of evaporite salt. Industrial-scale silver and gold mining, supported by the forced labor or *mita* system, dates from at least Inca times (fifteenth century) in the Andes, continued through the Spanish colonial period, remains active in various places in Ecuador, Peru, and Bolivia (Cole 1985), and has expanded under multinational corporations for the extraction and processing of industrial minerals like copper, zinc, and tin (Fig. 10.15). In Canada and the United States, mining was the principal industrial activity that brought people and settlements to the western mountains from 1850 to 1930 (Harris 1997). Starting earlier, coal mining in Appalachia helped shape and support the unique mountain people there. The global economic depression of the 1930s brought mining everywhere to a temporary halt and left behind many abandoned settlements (ghost towns) and, in many cases, devastated and toxic environments.

Mining has retained its importance in the mountains as a major employer and source of environmental impacts, especially in the Americas, with many of the benefits accruing to large transnational mining corporations and their shareholders living far from the mountains. Growing global demand, high commodity prices, new technologies, and improved transportation have resulted in the expansion of open-cast and subsurface mining in the Andes and North American Cordillera at high altitudes and in arid and cold conditions. This has provided some employment for local people, but many workers are also from outside the immediate area. A result is the emergence of large temporary towns, like La Rinconada at 5,100 m in the Peruvian Cordillera, with a 30,000 mostly male population, repeating the boom-and-bust pattern of settlement and mining livelihoods that has characterized the modern history of mining in the mountains. Compared to the level and extent of mining in the Cordillera of the Americas and Appalachia, the vast reaches of the Hindu Kush–Karakoram–Himalaya have been relatively untouched, and mining has not constituted as significant a source of economic development and livelihood as yet. This is changing, with aggressive exploration in southwest China and increasing numbers of small-scale gemstone mining operations in the Karakoram and western Himalaya in Pakistan.

People in many mountain communities in mineral-rich areas believe they have been unjustly exploited and marginalized even within their own regions (Fox 1997). They have accused mining companies of exploiting common lands, especially at higher elevations, which also have deep religious significance for local communities. The contamination

FIGURE 10.15 La Oroya smelter, Río Mantaro Valley, Peru, one of the most polluted sites in the Andes. (Photo by C. Stadel.)

FIGURE 10.16 A former coal mining community, Canmore, Alberta, in the Canadian Rocky Mountains, has become an important tourist, recreation, amenity migration, and service center. (Photo by J. S. Gardner.)

of local land and water resources, and air pollution, which endanger the health and the bases of the livelihoods of local people, are also major concerns. In the United States and Canada, indigenous people have taken legal action to protect their interests and lands, forcing mining companies to accommodate themselves to indigenous cultures. In Bolivia, indigenous people, supported by the national government, have successfully sought to curtail the activities of foreign companies and to request a more equitable share of revenues.

An irony in the mountain world is that many old mine sites and communities, some long abandoned, have reemerged as important contributors to other forms of livelihood. Potosí still operates as a mine but is visited by as many tourists as miners. Former mountain mining towns

like Aspen, Telluride, Breckenridge, Copper Mountain, Canmore, and Fernie, to name several in the United States and Canada, are mainstays of tourism, recreation, and amenity migration (Fig. 10.16). Furthermore, the mining town of Potosí, Bolivia, the former mining towns of Røros, Norway, and Sewell, Chile, and the now abandoned Humberstone and Santa Laura mines, also in Chile, have been declared UNESCO World Heritage Sites.

Forestry

Most mountain regions are characterized by various types of forest, influenced by altitudinal ecological zonation and local topography, soil, climate, and moisture conditions. Mountain forests have been and remain one of the

landslide dam outburst floods which amplify the disaster impact spatially and temporally. The relationship between mountain areas and high earthquake frequency arises from the locational proximity between mountains and the edges of the Earth's crustal plates (e.g., the western Cordillera of the Americas and the Himalaya and Tibetan Plateau). Hewitt (1997a) demonstrated that some of the most disastrous earthquakes have affected not only the main mountain ranges but also mountain front ranges and margins. Thus, very high risk situations are found in the mountain regions of the Mediterranean, the Cordilleras of the Americas, the mountains of the Middle East, and the extended mountain systems of central, southern, and eastern Asia.

Two recent mountain earthquake disasters—Kashmir, Pakistan/India (8 October 2005) and Wenchuan/Sichuan, China (12 May 2008)—illustrate many of the elements of earthquake risk in mountain environments. Together, they resulted in more than 150,000 fatalities, many more injuries, the temporary and permanent displacement of millions of people from their homes, and billions of dollars of direct and indirect damage costs. Each occurred in seismically active regions on the Indo-Australian and Eurasian continental plate boundary where earthquake disasters have had a long history. They impacted cities and towns in valleys and mountain margin areas, remote mountain villages, communications and transportation infrastructure, homes, public buildings, and spaces such as schools, hospitals, and clinics, as well as a wide range of economic activities. Severe ground surface displacement and shaking in both cases produced an enormous cascade of landslides (over 4,000 in Kashmir and 3,800 in Sichuan) that destroyed and damaged roads and settlements, thus further delaying and complicating an emergency response already compromised by earthquake damage and adverse weather. Landslides in both cases also produced landslide-dammed lakes, which immediately posed a dam-burst flood hazard (Owen et al. 2008; Cui et al. 2011). Thirty-eight such lakes were identified in the case of Sichuan, where there has been a long and tragic history of dam-burst floods, requiring an immediate response of lake drainage and dam hardening. However, this case illustrates a further hazard cascade characteristic of mountain areas subject to intense monsoonal rainfall when, in mid-July/August 2010, exceptionally heavy rains generated large and extensive debris flows in the earthquake-exposed sediments, undoing much of the previous post-earthquake recovery and reconstruction efforts, and creating a delayed disaster (Tang et al. 2011; Fig. 10.18). The huge impacts, extensive and often isolated areas affected, and the frequent occurrence of some earthquake disasters in mountain areas call upon a combination of local, regional, and international resources.

Volcanic Hazards

Much, but not all, volcanic activity is expressed in the form of mountains, and some of the most densely populated

FIGURE 10.18 One of many hotels, restaurants, and shops constructed following the 2008 Wenchuan earthquake and subsequently destroyed by August 2010 debris flows and floods, Longmenshan, Hengduan Mountains, Sichuan, China. (Photo by J. S. Gardner.)

and intensively used mountain areas are on the slopes of volcanoes in Ecuador, Mexico, and Indonesia. Though the areas around volcanic mountains are most exposed to the hazards related to volcanic eruptions, the impact of eruptions can be global, in the form of ash clouds disrupting air travel and creating temporary decreases in atmospheric temperature. The spatial distribution of active and inactive volcanoes reflects that of crustal plate boundaries, particularly those where oceanic and continental plates converge (subduction zones) and where plates diverge (rifts). The "Pacific Ring of Fire," which includes the western Cordillera of the Americas, the Japanese Islands, the Philippines, and the Indonesian islands, is an example of the former; the Mid-Atlantic Ridge, with Iceland, the Canary Islands, and the Azores, is an example of the latter. The slopes of some volcanic mountains, particularly those in the subtropics and tropics, attract people in large numbers for two reasons: They may offer the best land available for settlement and agriculture as, under the prevailing weathering conditions, some volcanic materials (e.g., ash) rapidly transform to very productive soils. The agents of volcanic destruction include explosive blast effects, falling debris, volcanic ash falls, flows of lava, and rapid downslope flows of very hot debris, ash, and gases called pyroclastic flows. The volcanoes of the

FIGURE 10.19 Accelerated erosion and slope failure, a common consequence of road construction in the mountains, NH21, Pir Panjal Himalaya, Himachal Pradesh, India. (Photo by J. S. Gardner.)

Indonesian island of Java have very high population densities in an environment where periodic eruptions have produced death, severe damage, and disruptions. Yet people keep moving back, as the benefits far outweigh the costs of disaster in the long term. The primary means of mitigating risk in such areas today is by monitoring volcanic activity, eruption warnings, and evacuation of people and their mobile property. In places where there are relatively predictable paths of lava and pyroclastic flows, structural controls and land use prohibitions (zoning) have been successful mitigation methods. Naples, Italy, is not usually thought of as a mountain metropolis, but it rests on the lower slopes of Mt. Vesuvius, and together they represent one of the highest disaster risk areas on Earth, the mountain having a well-documented record of explosive eruptions and the metropolis having about 5 million people and structures sprawled over ground made from the remains of earlier pyroclastic and lava flows, ashfalls, and scoria.

Volcanic mountains that have a snow and/or glacier ice cover pose an additional hazard and a high level of risk because eruptions may produce lahars (mud and debris flows) that can extend some distance away from the mountains into densely populated areas. Attention was drawn to this type of hazard by the 1985 eruption of Nevado del Ruiz in the Colombian Andes (Sigurdsson and Carey 1986) when a summit area eruption melted snow and ice, generating a lahar that devastated a downstream area for a distance of 50 km. The greatest number of fatalities, approximately 20,000, was in the town of Armero, a community at a distance of less than 50 km from the mountains. Similar high-risk locations for this type of disaster are found through the western Cordillera of the Americas, a good example being the Cascadia volcanoes of southwestern British Columbia, Washington, and Oregon. The glacier- and snow-covered Mounts Hood, Rainier, Baker, and Meager,

among others, all with the imprint of prehistoric eruptions and lahars, loom over the densely populated lower slopes and lowlands, including the greater metropolitan areas of Portland, Seattle, and Vancouver (Clague and Turner 2003).

Slope Instability

Mountain topography, being dominated by very steep and long slopes, is prone to slope instabilities of several types, including large rock avalanches, smaller rockfalls, mass movements of rock, soil, and vegetative debris, debris and mudflows, and slow soil creep. Some occur very infrequently or very slowly, and therefore may not be perceived as important hazards. Others occur frequently in known locations, and are more easily planned for and mitigated. Many are damaging to people, land, and structures and are very disruptive to road and railroad traffic. Thus, on one hand, people avoid places where destructive processes are known to occur but, on the other hand, they use the resulting landforms and deposits for construction and agricultural purposes. While the primary cause of slope instability is the gravitational pull, inherent and prior conditions favoring movement and a trigger to set off movement are necessary. Such conditions include weathering of earth materials, a build-up of gravitational stress due to added material and water, and a reduction in slope stability and strength due to undercutting by erosion and construction and the addition of water to the material. Triggers include earthquakes, volcanic eruptions, sudden and intense rainfall, freezing and thawing temperatures, and construction activities. While slope instabilities of all types are endemic to mountain topography, it is clear that human alteration of slopes has increased their distribution, frequency, and magnitude. This is especially evident where deforestation and road building take place (Petley et al. 2007; Sidle et al. 2011; Fig. 10.19).

FIGURE 10.20 Snow avalanche prevention, control, and mitigation adaptations. (A) traditional summer pasture shelters built into an avalanche slope, Kaghan, Punjab Himalaya, Pakistan. (B) avalanche deflecting walls at La Lavanche. (C) snowpack stabilizers at La Tour, Haute-Savoie Alps, France. (Photos by J.S. Gardner)

The hazards of slope instability have been recognized through experience for generations. Modern research is adding to the body of knowledge. While some human activities and unfortunate locational decisions continue to exacerbate hazard and risk, a wide range of mitigative efforts are now in place in many mountain regions to reduce both. Avoidance of areas prone to or threatening slope instability is a traditional and most effective approach undertaken by free choice and/or regulation and restriction, the former common among longtime mountain dwellers and the latter now more common in highly developed and regulated jurisdictions in Europe and North America. Mitigation includes: strengthening of slopes using retaining walls, bolts, pylons, terracing, coatings, and vegetation; catching or diverting rockfalls and debris and mudflows using nets, walls, dikes, and sheds; draining water from slope using pipes, tiles, and diversions; strengthening buildings, bridges, and support structures against impacts; and forecasting and warnings to alert people to danger, close facilities, roads, and so on, and initiate evacuations if necessary.

As discussed in Chapter 6, soil erosion is a more subtle but, in the long term, very damaging process in mountain environments. It results primarily from water runoff on soils and is especially prevalent in areas of soil exposed for agriculture and construction.

Snow Avalanches

Snow avalanches, the sudden and rapid downslope movement of masses of snow, are the most widespread and common hazardous processes in high mountains. Millions occur globally every year, but relatively few pose a risk because they occur in remote locations. Nonetheless, they have been and remain a very significant hazard in populated and heavily used locations, as discussed in Chapter 4. The growth of snow-based tourism and recreation has dramatically elevated risk (McClung and Schaerer 2004). In addition to posing a hazard, snow avalanches transport sediment, influence vegetation, and redistribute water, soil, and wood in ways that are both harmful and beneficial to mountain people. For example, in the snow-covered parts of the northwest Himalaya in northern Pakistan, Kashmir, and Himachal Pradesh, avalanche-transported snow, sediment, organic material, and wood contribute to agriculturally productive fan deposits and to the local firewood supply (De Scally and Gardner 1994). On balance, however, avalanches are seen primarily as a hazard.

Advanced scientific understanding of snow avalanches has aided the effective management of the hazard, especially since the 1950s, when the combination of the rapid and widespread expansion of recreational skiing in Europe and North America, and very serious avalanche disasters in Europe during the 1950–51 winter coalesced to a crisis (Fraser 1966; McClung and Schaerer 2004). Previously, mountain people had reasonably effective traditional and local knowledge of avalanches that helped them to manage risk by winter evacuations, avoidance of dangerous locations for, and modifications of, permanent dwellings, maintaining protective forests, and forgoing travel, during high hazard episodes (Fig. 10.20). As discussed in Chapter 4, a better understanding of the interactions between topography, weather, and snow conditions, the three principal ingredients of avalanche conditions, are enabling more effective management of the hazard and the risk. The mitigating measures include: identification of avalanche slopes and event frequency and magnitude to produce effective hazard maps; using hazard maps to regulate the location of activities and structures; monitoring weather and snowpack conditions to forecast avalanche hazards; using forecasts to pre-release avalanches with explosives in selected locations; regulating highway, rail, and recreational traffic and activity during high hazard conditions; issuing public information bulletins indicating hazard levels and using public education to upgrade general knowledge of avalanche hazard; maintaining and developing protective forests to stabilize snow above settlements and infrastructure; using a variety of snow fences and buffers to stabilize snow in avalanche starting zones; building protective and deflecting walls and snowsheds where buildings and infrastructure are exposed; and strengthening structures to withstand the high impact pressures of avalanches (Fig. 10.20).

The vast effort made to understand avalanches, control them, mitigate their effects, and sensitize and manage people is a testament to the importance of avalanches in the life of people in the mountains. Snow avalanches remain a potent force in most high mountain areas, especially the heavily used and populated areas in the Alps and western Cordillera of Canada and United States.

Meteorological Hazards

Severe weather conditions in the mountains can produce various hazardous and even disastrous outcomes. In a general sense, these arise from extreme cold or heat, too much or not enough rain/snow, strong winds, and lightning. The impacts of these hazards on people, settlements, and land uses depend on diverse natural and human conditions. Drought, for example, may occur locally at small scales, but is often a large-scale phenomenon affecting not only mountains but surrounding regions as well. It usually is manifest as reduced and/or irregular precipitation that often lead to reduced stream and groundwater flow and reduced soil moisture, negatively impacting natural and human processes and conditions.

Drought conditions are particularly prevalent in interior basins and plateaus of tropical mountains (e.g., the Altiplano of southern Peru and Bolivia), in semiarid and arid mountains (e.g., the Karakoram; Fig. 10.11), those of a Mediterranean-type climate on a seasonal basis, and in rainshadow valleys (e.g., interior valleys and basins of the North American Cordillera). Lower than expected precipitation lessens agricultural and livestock production, causing food shortages, and creates power deficits and water supply shortages. Reduced snowfall in winter may have similar effects as well as limiting the snow-based recreation activities on which many mountain locations in Europe and North America are dependent, although the shortages of natural snow are today compensated to a certain degree by relying on artificial snowmaking. Besides impacting local people, drought in the mountains may create disastrous water and energy shortages for people, agriculture, and industry outside the mountains. Drought generally is caused by natural climatic variations but may be locally exacerbated by human activities (e.g., overutilization of soil and water resources, deforestation, water-demanding agricultural crops and techniques). The area around Mount Kenya in East Africa provides an example. Mount Kenya, with its small remaining glaciers, rainfall, and snowfall, is a water source for the semiarid surroundings. The traditional inhabitants of the southern and western slopes are the agrarian Kikuyu people who held *de facto* or customary land tenure rights prior to the British colonial period, which ended in 1963. Large areas of land remained in the hands of British settlers and corporations, and much of this land has been turned over to water-demanding commercial production of fresh vegetables and flowers for export. Additional factors influencing risk in the area relate to available land being designated within Mount Kenya National Park and demand for land rising dramatically as a result of Kikuyu population growth and resettlement in-migration of other groups. Not only does drought on Mount Kenya impact commercial production, it is disastrous for the small subsistence farms of the Kikuyu and for the supply of drinking and irrigation water in the lower parts of the Ewaso N'giro watershed. Traditional forms of drought mitigation include a greater reliance on animal husbandry as opposed to intensive field cultivation, use of drought-tolerant food crops (millet, sorghum) and livestock (goats), supplementary water-saving irrigation techniques, and appropriate forms of water storage (tanks and cisterns). Many introduced crops and livestock of less drought-tolerant species and varieties (e.g., vegetables, new forms of maize, dairy cattle), while more productive, have increased vulnerability to drought hazard. Also, open reservoirs increase evaporative water losses. Technological advances have supported increased productivity in drought-prone mountain areas, but vulnerability remains very high. This is even the case with tourism, where snow deficits are supplemented by snowmaking technologies, all of which are water and energy intensive.

Meteorological hazards in mountain areas are ingredients, triggers, and/or amplifiers of other hazardous processes. Heavy rain and snow contribute to floods, soil erosion, slope failures, and avalanches, and strong winds are also very effective in spreading forest fires, in addition to causing wind-impact damage to structures and infrastructure. Wind is a significant contributor to blizzards or whiteouts, avalanche hazard, extreme cold hazard, and drought conditions. Unexpected blizzards and heavy snowfalls severely disrupt travel of all forms and cause death and damage to people, crops, and livestock in the high subtropical, temperate, and subarctic mountains. Avoidance is about the only form of effective mitigation.

Floods

Mountain floods are particularly hazardous due to their rapid onset, high velocity, energy and impact pressures, erosive potential, and high sediment and debris load. Flash floods sometimes transform into very destructive debris flows. Not only do they inundate and deposit mud, rocks, debris, and seeds, they strike with force, eroding land, damaging and destroying buildings, bridges, roads, railroads, and agricultural crops. In high-magnitude flood events on major mountain rivers, such as the main stems and tributaries of the Yangtze River, China, the Rhine and Danube in Europe, and the Fraser in British Columbia, Canada, flood waters and sediments extend out of the mountains into surrounding piedmont and lowland areas. This causes widespread damage and disruption, but also provides new sediments to floodplain surfaces.

Well documented are the numerous and devastating rapid-onset floods in the Hindu Kush–Karakoram–Himalaya, triggered by steep slopes, heavy monsoon rains, and strong snow and ice melt. Flood-intensifying conditions include removal of vegetation, building of roads, nonterraced cultivation on steep slopes, and inadequate upkeep of water drainage systems. The high-magnitude floods caused by landslide-dam failures (Sichuan, China), glacier lake outbursts (Nepal, Peru), and dam overtopping in natural and constructed lakes caused by landslides, icefalls, and avalanches (Vaiont, Italy; Cordillera Blanca, Peru) have all been discussed earlier.

Today we find widespread settlement, dense populations, intensification of land use, high-value cash crops, roads, and other infrastructure in flood-prone areas, especially where regulation of land use is lacking or recent. Vulnerability is high and so is flood disaster risk. Mountain flood hazard mitigation includes a variety of traditional practices, such as: restricting construction to safe sites; building diversion channels, floodways, and retention basins; elevation of houses, roads, and trails; building and reinforcing protective dikes and dams; protecting forests; and reforestation. Mitigation relies on effective and enforced land-use regulation. Additional measures include flood forecasting based on expected precipitation, snowmelt, and other relevant variables, and emergency plans for protection or even evacuation of endangered people.

Biohazards

Biohazards are organisms and conditions or processes in the biological environment that cause damage to people, settlements, crops, livestock, and infrastructure. They range from microorganisms such as viruses and bacteria, to predatory and foraging animals such as bears and deer, to insects that attack crops and processes such as wildfires. All are present in mountain areas. Here we address two biohazard areas that have been and remain especially impactful in mountain areas: disease and wildfire.

Of all the hazards affecting mountain areas, infectious diseases have taken the greatest toll in human lives and on mountain societies, a condition shared by virtually all places on Earth. Infectious diseases, by their nature, are biological and social phenomena, and illustrate the importance of both understanding and managing the complexities of these biohazards. Usually their impact arises through social interactions, often in the context of inter-related factors like poverty, poor nutrition, physical and emotional stress, lack of access to safe water, and crowded and poorly ventilated housing, many of which may arise from lack of sustainable livelihoods, as well as economic, social and political marginalization. It is a complex web of causes and effects. HIV/AIDS has made its way into mountain communities through sexual activity and injected drug use, and spans the socioeconomic spectrum from high-end resorts to impoverished families and communities. Water-borne parasitic diseases such as dysentery, cholera, typhoid, and giardiasis take a heavy toll, particularly among children. Tuberculosis, including drug-resistant TB, other respiratory infections like bacterial pneumonia, and eye infections are common among mountain people living at higher altitudes in crowded quarters. Fortunately, some infectious diseases that have taken a gruesome toll in mountain areas in the past, namely smallpox, measles, polio, and some influenzas, have almost been eradicated through immunization, a frontline hazard mitigation. Other forms of mitigation, such as access to prevention and treatment services and access to safe water, often are lacking in remote, impoverished, and crowded mountain locations. Despite the relatively high prevalence of infectious diseases in mountain areas, the situation is much improved from some earlier times.

Historical estimates of mountain populations have been made difficult by the fact that the arrival of Europeans, usually as part of the colonial and commercial expansion from the sixteenth to the twentieth centuries, signaled the arrival and introduction of an array of exotic infectious diseases into mountain and other communities (Mann 2006). The lack of prior exposure, and therefore natural immunities, allowed a rapid and deadly spread of smallpox, measles, influenzas, and others, killing millions. An epidemic of sexually transmitted infections followed. This probably started with the Spanish incursions into the Andes in the sixteenth century. By the end of the nineteenth century, large areas of the North American Cordillera were devoid of indigenous peoples, or the remnant populations had retreated to remote locations or clustered around new settlements. The indigenous peoples of the Hindu Kush–Karakoram–Himalaya region were infected via the British commercial and colonial expansions through the late eighteenth to twentieth centuries. Some more isolated areas, such as northern Kashmir, Hunza, northern Garhwal, and parts of Nepal, did not bear the brunt of infectious disease until the mid-twentieth century. Other infectious diseases such as polio, some sexually transmitted diseases, and HIV/AIDS did not make their presence known until even later. Infectious diseases were effective in breaking through mountain barriers, and they did so in the company of a relatively small number of outsiders.

Wildfire hazard arises from the uncontrolled burning of forests, brush, and grassland. Most common in mountain areas are forest fires. Mountain topography amplifies wildfire potential by creating localized winds that drive the fire, and steep slopes that allow for rapid up-burning, making firefighting difficult. The fires have natural and anthropogenic triggers; favorable prior conditions are related to forest health, weather, and forest management practices. Fixed objects such as homes, buildings and their contents, crops, roads and railroads, bridges, and transmission lines are most vulnerable. People and livestock are less so, though deaths and injury among both do occur. While lightning is the most frequent natural trigger, anthropogenic triggers include campfires, waste fires, discarded cigarettes and matches, and various types of controlled burns. However, forest fires do not ignite and spread unless forest conditions are suitable. Extended periods of little precipitation, warm temperatures, low humidity, and wind favor ignition and spread. Forests that have been widely decimated by disease or insects, such as the mountain pine bark beetle in the interior mountains of British Columbia, provide large amounts of combustible material (Nikiforuk 2011). Forest management practices that suppress naturally occurring fires, reforest extensive areas with even-age, single species, and prohibit all forms of harvesting create forest monocultures that are susceptible to insect and disease attack and enhance the spread of wildfires. Very extensive areas in the North American Cordillera and the Himalayan front ranges, with seasonal "fire weather," have been so affected and therefore are today, and have been in the recent past, the mountain areas most impacted by and at risk from disastrous wildfire episodes. Recent trends in weather and land use suggest that wildfires may be increasing in all mountain regions where forests and bushland, however limited in extent, are present. In the high mountain Puna and Páramo areas of the Andes, grasslands and bush vegetation are frequently burned by rural people in a semi-controlled fashion, a controversial practice which local farmers regard as a measure to clear land for pastures and to renew soil fertility.

The phenomenon of interface wildfire has asserted itself in the rural–urban fringe of major cities and towns globally, as well-publicized examples in southeast Australia, southwest United States, Greece, and Russia have shown. Interface wildfire is an increasing problem in mountain areas as urban areas expand into surrounding forested land. In some cases, the expansion is driven by the positive aesthetics of forests, which attract investment in high-value real estate, as in the North American Cordillera, Southern Alps of New Zealand, parts of the Alps, the front ranges of the Himalaya, and the mountains of Yunnan and Sichuan. In other cases, urban fringe expansion involves low-income migrants from rural areas seeking a better life. The result is a potent mix of widespread high hazard and high vulnerability. In August 2003, over 2,000 wildfires were burning in mountainous British Columbia (Filmon 2004).

Several destroyed settlements, and one of the largest, the Okanagan Mountain Park fire, burned into the city of Kelowna, destroying and damaging homes and infrastructure. No people were injured or killed by the fire. It gained worldwide television attention, enhanced by spectacular images of towering flames and smoke clouds, as a possible indicator of things to come. Firefighters and experts converged from all over the globe. The wildfire bill for British Columbia in 2003 exceeded $1.5 billion. Whether the vulnerable properties are high-value real estate, or squatter settlements on the urban fringe, or dwellings scattered through the forest, the impact of wildfires on people's lives is substantial.

Wildfire hazard mitigation and disaster prevention in mountain areas is a complex enterprise involving the management of forests, fires, and people. Forest measures include the use of controlled burns to replicate small, naturally occurring fires that add diversity and open spaces to forest stands and prevent the spread of damaging insects and diseases, carefully controlled harvesting of trees to do the same, removal of dead material from forests, and cutting of firebreaks to protect vulnerable property. Fire hazard forecasting through the monitoring of weather and forest conditions is used to alert authorities and the public to the hazard and impose prohibitions on use of fire, travel, and other activities. Airborne and ground surveillance is used to spot fires and issue warnings and evacuations. Measures related to the wildfires themselves include efforts to control and extinguish using conventional techniques, including the use of backfires and airborne fire suppressants. In addition to prohibiting certain activities during high hazard periods, people are managed through public education, building restrictions, and building codes to reduce use of flammable materials. Thus, the mountain wildfire hazard is one of many hazards that impose costs on the lives and livelihoods of mountain people while creating new livelihoods for some.

The complexities of wildfire hazard and risk management in mountain areas are difficult to balance. Forests are inherently valuable in the lives of mountain people for their timber, fuelwood, nontimber products, aesthetics, recreational value, and the roles they play in microclimate, water regulation, and erosion prevention. Wildfires alter these benefits. Fire suppression is an answer, but it creates new problems. Maximizing the benefits and minimizing the costs presents one of the most perplexing management problems in those parts of the mountain world where forests are present.

For generations and in every mountain region, people have coped with and adapted to the challenges of complex hazards and risks. Coping involves short-term actions to reduce impact, but may not be sustainable or useful in the long term. Adaptations are longer-term strategies to reduce the risks, and may require substantial changes in how people use and manage their environment and lives. Adaptations may address the root causes of a hazard, intervene to reduce or eliminate the hazards, and/or compensate for the losses and restore damages. The ability to cope and adapt is referred to as resilience.

Resilience of Mountain People

Managing change requires resilience in individuals, families, communities, and large-scale social-ecological systems. Change comes in many forms, and we have reviewed many examples in this chapter. Some changes are gradual, others are sudden shocks, and all require that people and communities cope and adapt. The concept of resilience is often used to characterize the ability to successfully anticipate, cope with, and/or adapt to changes and shocks so as to enable livelihood sustainability (Berkes et al. 2003). Many mountain people and communities in different places and times have demonstrated resilience in the face of change in dealing with new environmental, cultural, socioeconomic, political, and institutional circumstances. We see numerous examples of this, as mountain people in many regions have successfully adjusted to the new processes, circumstances, and livelihoods that characterize the various and changing dimensions of mountain resource exploitation, economic and amenity migration, traditional and new market forces, and the growing importance of tourism. Other examples demonstrate less successful forms of resilience, the most devastating being the impact of introduced diseases on indigenous mountain people. Most diseases are examples of a rapid-onset hazard leading to disaster, and they often produce changes that are most difficult to adjust to in the short term. Earthquakes, volcanic eruptions, and other natural hazards, as well as global economic depression, persecution, conflict, and war are other examples of shocks that are difficult to cope with and adapt to. Building and enhancing resilience is an important step toward livelihood sustainability and development in mountain areas. Vibrant leadership, empowerment and participation of all segments of society, shared goals and values, established institutions and organizations, a stable and healthy population, a diversified economic base, constructive external partnerships, and local control of the availability of resources are all important ingredients for building resilience (Folke et al. 2003).

Using resilience as a concept in assessing the impacts of disasters and hazards and in planning for future shocks is increasingly common (Gallopin 2006; Gardner and Dekens 2007; Fuchs 2009). In traditional mountain communities, in an effort to minimize environmental risks, high levels of resilience were achieved through an optimal exploitation of spatial and temporal niches, the avoidance or attenuation of harmful developments and activities, a diversification or complementary utilization of agricultural crops and practices and of diverse microenvironments, and through community reciprocity. This worked well for certain local hazard events, but not so well for major regional hazard events such as earthquakes, prolonged periods of drought,

widespread flooding, epidemics, climate change, economic depression, and war. In these instances, local resilience may be insufficient, and other factors such as external partnerships and linkages with national and international organizations may become more important. Recovery, reconstruction, and redevelopment of livelihoods in the 2005 Kashmir and 2008 Wenchuan earthquake disasters illustrate the point.

Through time, the resilience of mountain people and communities has probably increased. However, resilience remains compromised and limited by complex terrain; variable and extreme weather; climate change; distance and isolation; social, political, and economic inequities; and poverty, marginality, and powerlessness.

Conclusion: Persistence and Change

The livelihoods of people in the mountains have changed much and will continue to do so. Physical isolation has decreased, accessibility has been improved, the spatial distribution and mobility patterns of people have been altered, and socioeconomic integration with the larger world has increased. However, other, less tangible forms of isolation persist. Political integration, in the sense of power sharing, is still lagging, and in some respects traditional mountain people may have lost power and autonomy in regard to many aspects of life. In most cases, centers of population and national economic and political power remain outside the mountains. Some mountain areas are inhabited by cultural minorities such as the Quechua and Aymara in the Andes, Tibetans, Yi, Naxi and other minorities in China, tribals and nonscheduled castes in India, and First Nations in North America, and these population groups remain largely marginalized. Many such groups seek to retain their cultural identity and claim traditional rights to land, water, and other resources, while aspiring for more political power and economic and social opportunities at local, regional, national, and international levels. There is a constant interplay between the forces and tendencies of integration into the larger world (change) and those of local tradition and desire for autonomy (persistence). This is nowhere more evident than in the day-to-day lives of individuals continuing traditional livelihood activities and practices while opening themselves up to new economic opportunities; making use of modern communications technologies such as smart phones, computers, and television; and being aware of and attracted to globalized cultures and lifestyles (Fig. 10.8).

Mountain populations have never been static and impervious to external economic and cultural influences. Over the past centuries, fundamental changes have occurred under the impact of colonialism and expanding commercialism, industrialization, and technological development. The importance of mountain areas as resource hinterlands for land, water, animals, plants, energy, and minerals, and as spiritual, recreational, and scenic spaces, has been a force for change. People living in mountain areas today are engaged in a variety of livelihoods. Their settlements offer a mosaic of characteristics: agricultural villages, marketplaces, forestry camps, mining settlements, refugee camps, tourist resorts, service centers, and major cities. Some settlements are new or ephemeral, whereas others have a long history, having experienced changes in size, structure, function, and socioeconomic characteristics. In some cases, the widespread expansion of the built-up areas and infrastructure has resulted in the destruction of habitats, the contamination of air, water, and soil, economic inflation, and enhanced risks and disasters.

The most obvious changes in mountain ecosystems are a product of anthropogenic and environmental factors. Forest cutting for timber, forest clearing for agriculture, river flow regulation for water supply and power generation, widespread extraction of minerals and fossil fuels, introduction of exotic plants and animals, expansion of settlements and infrastructure, and imposition of new forms of land tenure (e.g., parks and protected areas) are important anthropogenic factors. Increasing air temperatures and shrinkage of snow and glaciers over the past 150 years are examples and symptoms of change in environmental conditions, as were the lower temperatures and greater snow and ice cover in the previous 500 years. Less evident, as yet, are changes in the patterns of ocean currents and storm tracks, and in precipitation regimes which impact local and regional water balances. The changes in treeline, biodiversity, and plant and animal distributions reflect the multiple and complex interactions of environmental and anthropogenic factors. Also, the impacts on people in the mountains are complex and not always immediately evident. Resource extraction generally is destructive or damaging to local ecosystems and the services they provide. New forms of economic activities provide commodities and livelihoods for growing populations within and outside the mountains. Yet many mountain people remain impoverished, marginalized, and powerless. Finding a rational balance between the benefits and costs of new developments is difficult, and requires in-depth knowledge and understanding of the highly diversified, complex, and dynamic nature of mountain environments and mountain livelihoods.

What, then, are the key issues pertaining to mountains and people in the twenty-first century? This question has challenged scholars, planners, politicians, administrators, and activists as the ecological, economic, and social importance of mountain areas in themselves and beyond has become understood. Risk reduction, participatory governance, livelihood and ecological sustainability, and sociopolitical equity are at the core of twenty-first century issues. Accurate information and knowledge of the geo-ecological and socioeconomic conditions and processes operating in and affecting mountain areas are essential in addressing the issues.

References

Baker, P.T. ed., 1978. *The Biology of High Altitude Peoples.* Cambridge, UK: Cambridge University Press.

Bätzing, W., Perlik, M. and Dekleva, M. 1996. Urbanization and depopulation in the Alps. *Mountain Research and Development* 16: 335–350.

Berkes, F., Colding, J., and Folke, C. 2003. *Navigating Social-Ecological Systems: Building Resilience for Complexity and Change.* Cambridge, UK: Cambridge University Press.

Björnsen, G. A., Huber, U., Reasoner, M. A., Messerli, B., and Bugmann, H. 2005. Future research directions. In U. Huber, H. K. M. Bugmann, and M. A. Reasoner, eds., *Global Change and Mountain Regions: An Overview of Current Knowledge* (pp. 637–650). Dordrecht: Springer.

Borsdorf, A., and Paal, M. 2000. Die alpine Stadt zwischen lokaler Verankerung und globaler Vernetzung. *Beiträge zur Regionalen Forschung im Alpenraum.* ISR Forschungsberichte 20.

Borsdorf, A., and Stadel, C. (2013). *Geographie der Anden.* Heidelberg: Springer.

Braun, B. 2002. *The Intemperate Rainforest: Nature, Culture and Power on Canada's West Coast.* Minneapolis: University of Minnesota Press.

Brundl, M., Romang, H. E., Bischof, N., and Rheinberger, C. M. 2009. The risk concept and its application in natural hazards risk management in Switzerland. *Natural Hazards and Earth Systems Science* 9: 801–813.

Carey, M. 2010. *In the Shadow of Melting Glaciers: Climate Change and Andean Society.* New York: Oxford University Press.

Chaverri, P. 2006. Cultural and environmental amenities in peri-urban change: The case of San Antonio de Escazu, Costa Rica. In L. A. G. Moss, ed., *The Amenity Migrants: Seeking and Sustaining Mountains and Their Cultures* (pp. 187–199). New York: CABI International Publishing.

Christopher, T., ed. 2003. *In the Land of the Blue Poppies: The Collected Plant-Hunting Writings of Frank Kingdon Ward.* New York: Random House.

Clague, J., and Turner, B. 2003. *Vancouver, City on the Edge: Living with a Dynamic Geological Landscape.* Vancouver, BC: Tricouni Press.

Cohen, J. E., and Small, C. 1998. Hypsographic demography: The distribution of human population by altitude. *Proceedings of the National Academy of Sciences* 95: 14009–14014.

Cole, D. M. 1985. *The Potosi Mita, 1576–1700: Compulsory Indian Labor in the Andes.* Stanford, CA: Stanford University Press.

Cole, V., and Sinclair, A. J. 2002. Measuring the ecological footprint of a Himalayan tourist centre. *Mountain Research and Development* 22: 134–141.

Conover, T. 2010. *The Routes of Man: How Roads Are Changing the World and the Way We Live Today.* New York: Knopf.

Cook, N., and Butz, D. 2011. Narratives of accessibility and social change in Shimshal, northern Pakistan, *Mountain Research and Development* 31: 27–34.

Cruikshank, J. 2005. *Do Glaciers Listen? Local Knowledge, Colonial Encounters and Social Imagination.* Vancouver, BC: UBC Press.

Cui, P., Zhu, Y. Y, Su, F. H., Wei, F. Q., Han, Y. S., Liu, H. J., and Zhang, J. Q. 2011. The Wenchuan earthquake

(May 12, 2008), Sichuan Province, China and resulting geohazards. *Natural Hazards* 56: 19–36.

Dematteis, G. 2009. Polycentric urban regions in the Alpine space. *Urban Research and Practice* 2: 18–35.

De Scally, F., and Gardner, J. S. 1994. Characteristics and mitigation of the snow avalanche hazard in Kaghan Valley, Pakistan Himalaya. *Natural Hazards* 9: 197–213.

Eckholm, E. P. 1975. The deterioration of mountain environments. *Science* 189(4205): 764–770.

Elkins, C. 2005. *Imperial Reckoning: The Untold Story of Britain's Gulag in Kenya*. New York: Henry Holt.

Elvin, M. 2004. *The Retreat of the Elephants: An Environmental History of China*. New Haven, CT: Yale University Press.

Filmon, G. 2004. *Firestorm 2003: A Provincial Review*. Victoria: Government of British Columbia.

Flora, G. 2006. Circular migration and community identity: their relationship to the land. In R. E. Rhoades, ed., *Development with Identity: Community, Culture and Sustainability in the Andes* (pp. 271–286). Cambridge, MA: CABI International Publishing.

Folke, C., Colding, J., and Berkes, F. 2003. Synthesis: building resilience and adaptive capacity in social-ecological systems. In F. Berkes, J. Colding, and C. Folke, eds., *Navigating Social-Ecological Systems: Building Resilience for Complexity and Change* (pp. 352–387). Cambridge, UK: Cambridge University Press.

Fowler, B. 2001. *Iceman: Uncovering the Life and Times of a Prehistoric Man Found in an Alpine Glacier*. Chicago: University of Chicago Press.

Fox, D. J. 1997: Mining in mountains. In B. Messerli and J. D. Ives, eds. *Mountains of the World: A Global Priority* (pp. 171–198). New York: Parthenon.

Fraser, C. 1966. *The Avalanche Enigma*. London: John Murray.

Fuchs, J. 2008. *The Ancient Tea Horse Road: Travels with the Last of the Himalayan Muleteers*. Toronto, ON: Viking Canada.

Fuchs, S. 2009. Susceptibility versus resilience to mountain hazards in Austria: Paradigms of vulnerability revisited. *Natural Hazards and Earth Systems Science* 9: 337–352.

Funnell, D., and Parish, R. 2001. *Mountain Environments and Communities*. New York: Routledge.

Gallopin, G. C. 2006. Linkages between vulnerability, resilience and adaptive capacity. *Global Environmental Change* 16: 293–303.

Gardner, J. S. 1986. Snow as a resource and hazard in early-twentieth Century mining, Selkirk Mountains, British Columbia. *The Canadian Geographer* 30: 217–228.

Gardner, J. S. 2002. Natural hazards risk in the Kullu District, Himachal Pradesh, India. *The Geographical Review* 92: 282–306.

Gardner, J. S., and Dekens, J. 2007. Mountain hazards and the resilience of social-ecological systems: Lessons learned in India and Canada. *Natural Hazards* 41: 317–336.

Glorioso, R. S. 2006. A bioregion in jeopardy: The strategic challenge of amenity migration in Baguio, the Philippines. In L. A. G. Moss, ed., *The Amenity Migrants: Seeking and Sustaining Mountains and Their Cultures* (pp. 261–267). New York: CABI International Publishing.

Grötzbach, E., and Stadel, C. 1997. Mountain peoples and cultures. In B. Messerli and J. D. Ives, eds., *Mountains of the World: A Global Priority* (pp. 17–38). New York: Parthenon.

Hamilton, L. S., Gilmore, D. A., and Cassells, D. S. 1997. Montane forests and forestry. In B. Messerli and J. D. Ives, eds., *Mountains of the World: A Global Priority* (pp. 381–311). New York: Parthenon.

Harris, C. 1997. *The Resettlement of British Columbia*. Vancouver, BC: UBC Press.

Hewitt, K. 1997a. *Regions of Risk: A Geographical Introduction to Disasters*. Harlow, UK: Longman.

Hewitt, K. 1997b. Risk and disasters in mountain lands. In B. Messerli and J. D. Ives, eds., *Mountains of the World: A Global Priority* (pp. 371–408). New York: Parthenon.

Hofer, T., and Messerli, B. 2006. *Floods in Bangladesh: History, Dynamics and Rethinking the Role of the Himalaya*. Tokyo: UNU Press.

Houston, C. 1998. *Going Higher: Oxygen, Man and Mountains*. 4th ed. Seattle, WA: The Mountaineers.

Huddleston, B., Ataman, E., and Fe d'Ostiani, L. 2003. *Towards a GIS-based Analysis of Mountain Environments and Populations*. Environment and Natural Resources Working Paper 10. Rome: Food and Agricultural Organization of the United Nations.

Ives, J.D. 2004. *Himalayan Perceptions: Environmental Change and the Well-being of Mountain People*. New York: Routledge.

Ives, J. D., and Messerli, B. 1989. *The Himalayan Dilemma: Reconciling Development and Conservation*. London: Routledge.

Kanwar, P. 1990. *Imperial Shimla*. New Delhi: Oxford University Press.

Kaser, G., Georges, C., Juen, I., and Molg, T. 2005. Low latitude glaciers: Unique global climate indicators and essential contributors to regional fresh water supply, a conceptual approach. In U. Huber, H. K. M. Bugmann, and M. A. Reasoner, eds., *Global Change and Mountain Regions: an Overview of Current Knowledge* (pp. 185–196). Dordrecht: Springer.

Kaser, G., Grosshauser, M., and Marzeion, B. 2010. Contribution potential of glaciers to water availability in different climate regimes. *Proceedings of the National Academy of Sciences of the USA* 107: 20223–20227.

Kreutzmann, H. 2006: People and mountains: perspectives on the human dimension of mountain development. *Global Environmental Research* 10(1): 49–61.

Laws, W. 2010. *Fifty Plants That Changed the Course of History*. Toronto, ON: Firefly Books Ltd.

Löffler, J., and Stadelbauer, J. 2008. Diversity in mountain systems. *Colloquium Geographicum* 31. St. Augustin: Asgard Verlag.

Mann, C. C. 2006. *1491: New Revelations of the Americas before Columbus*. New York: Random House.

McClung, D., and Schaerer, P. 2004. *The Avalanche Handbook*. 3rd ed. Seattle, WA: The Mountaineers.

Messerli, B., and Ives, J. D., eds. 1997. *Mountains of the World: A Global Priority*. New York: Parthenon.

Meybeck, M., Green, P., and Vörösmarty, C. 2001. A new typology for mountain and other relief classes: An application to global continental water resources and population distribution. *Mountain Research and Development* 21: 34–45.

Moats, A. S., and Campbell, B. C. 2006. Incursion, fragmentation and tradition: Historical ecology of Andean Cotacachi. In R. E. Rhoades, ed., *Development with Identity:*

Community, Culture and Sustainability in the Andes (pp. 27–45). Cambridge, MA: CABI International Publishing.

Molg, T., Kaser, G., and Cullen, N. J. 2010. Glacier loss on Kilimanjaro is an exceptional case. *Proceedings of the National Academy of Sciences* 107(17): E68.

Moseley, R. K. 2006. Historical landscape change in northwest Yunnan, China. *Mountain Research and Development* 26: 227–236.

Moss, L. A. G. 2006a. The amenity migrants: ecological challenge to contemporary Shangri-La. In L. A. G. Moss, ed., *The Amenity Migrants: Seeking and Sustaining Mountains and Their Cultures* (pp. 3–25). New York: CABI International Publishing.

Moss, L. A. G. 2006b. *The Amenity Migrants: Seeking and Sustaining Mountains and Their Cultures*. New York: CABI International Publishing.

Nicholson, M. H. 1963. *Mountain Gloom and Mountain Glory*. New York: Norton.

Nikiforuk, A. 2011. *Empire of the Beetle: How Human Folly and a Tiny Bug Are Killing North America's Great Forests*. Vancouver, BC: Greystone Books.

Orlove, B., Wiegandt, E., and Luckman, B. H. eds., 2008. *Darkening Peaks: Glacier Retreat, Science and Society*. Berkeley: University of California Press.

Otero, A., Nakayama, L., Marioni, S., Gallego, E., Lonac, A., Dimitriu, A., Gonzalez, R., and Hosid, C. 2006. Amenity migration in the Patagonian mountain community of San Martin de los Andes, Neuquen, Argentina. In L. A. G. Moss, ed., *The Amenity Migrants: Seeking and Sustaining Mountains and Their Cultures* (pp. 200–211). New York: CABI International Publishing: 200-211.

Owen, L. A., Kamp, U., Khattak, G. A., Harp, E. L., Keefer, D., and Bauer, M. A. 2008. Landslides triggered by the 8 October 2005 Kashmir earthquake. *Geomorphology* 94: 1–9.

Peattie, R. 1936: *Mountain Geography. A Critique and Field Study*. New York: Greenwood Press.

Perlik, M., and Messerli, P. 2004: Urbanization in the Alps: Development processes and urban strategies. *Mountain Research and Development* 24: 215–219.

Petley, D. N., Hearn, G. J., Hart, A., Rosser, N. J., Denning, S. A., Oven, K., and Mitchell, W. A. 2007. Trends in landslide occurrence in Nepal. *Natural Hazards* 43(1): 23–44.

Price, L. W. 1981: *Mountains and Man*. Berkeley: University of California Press.

Price, M. F., ed. 2006. *Global Change in Mountain Regions*. Duncow, UK: Sapiens Publishing.

Price, M. F., and Butt, N., eds. 2000. *Forests in Sustainable Mountain Development: A State-of-Knowledge Report for 2000*. Wallingford, UK: CABI International Publishing.

Price, M. F., Gratzer, G., Duguma, L. A., Kohler, T., Maselli, D., and Romeo, R., eds. 2011. *Mountain Forests in a Changing World: Realizing Values, Addressing Challenges*. Rome: Food and Agriculture Organization of the United Nations.

Rangan, H. 2000. *Of Myths and Movements: Rewriting Chipko into Himalayan History*. London: Verso.

Rhoades, R. E. 1979. Cultural echoes across the mountains, *Natural History* 88(1): 46–57.

Rhoades, R. E., ed. 2006. *Development with Identity: Community, Culture and Sustainability in the Andes*. Cambridge, MA: CABI International Publishing.

Rhoades, R. E. 2007. *Listening to Mountains*. Dubuque, IA: Kendall/Hunt.

Rhoades, R. E., and Thompson, S. 1975. Adaptive strategies in alpine environments: Beyond ecological particularism. *American Ethnologist* 2: 535–551.

Rizvi, J. 1999. *Trans-Himalayan Caravans: Merchant Princes and Peasant Traders in Ladakh*. New Delhi: Oxford University Press.

Sandford, R. W. 2008. *The Weekender Effect: Hyper-development in Mountain Towns*. Vancouver, BC: Rocky Mountain Books.

Sarmiento, F. O. 2008. Andes Mountains and global change: An overview. *Pirineos* 163: 7–13.

Schutte, S., and Kreutzmann, H. 2011. Linking relief and development in Pakistan administered Kashmir. *Mountain Research and Development* 31: 5–15.

Schweizer, P., and Preiser, K. 1997. Energy resources for remote highland areas. In B. Messerli and J. D. Ives, eds. *Mountains of the World: A Global Priority* (pp. 157–170). New York: Parthenon.

Sidle, R. C., Furuichi, T., and Kono, Y. 2011. Unprecedented rates of landslide and surface erosion along a newly constructed road in Yunnan, China. *Natural Hazards* 57(2): 313–326.

Sigurdsson, H., and Carey, S. 1986. Volcanic disasters in Latin America and the 13 November eruption of Nevado del Ruiz volcano in Colombia. *Disasters* 10: 205–216.

Stadel, C. 1982a. Mountain regions: Their nature and problems. *Geographical Perspectives* 49: 26–33.

Stadel, C. 1982b. The Alps: Mountains in transformation. *Focus* 32(3): 1–16.

Stadel, C. 1985. Environmental stress and human activities in the tropical Andes (Ecuador). *Revista del Centro Panamericano de Estudios e Investigaciones Geográficos* 15: 33–50.

Stadel, C. 1986. Urbanization and urban transformation in a mountain environment: The case of the European Alps. In C. S. Yadav, ed., *Perspectives in Urban Geography,* Vol. 3: *Comparative Urban Research* (pp. 39–55). New Delhi: Concept.

Stadel, C. 1992. Altitudinal belts in the tropical Andes: Their ecology and human utilization. In T. Martinson, ed., *Benchmark 1990, Conference of Latin Americanist Geographers,* Vol. 14 (pp. 45–60). Auburn, AL: Conference of Latin Americanist Geographers.

Stadel, C. 1993. The Brenner Freeway (Austria/Italy): A mountain highway of controversy. *Mountain Research and Development* 13(1): 1–17.

Stadel, C. 2003a. L'agriculture andine: Traditions et mutations. In CERAMAC, ed. *Crises et Solutions des Agricultures de Montagne* (pp. 193–207). Clermont-Ferrand: Presses Universitaires Pascal.

Stadel, C. 2003b. Indigene Gemeinschaften im Andenraum. *HGG-Journal (Heidelberger Geographische Gesellschaft)* 18: 75–88.

Stadel, C. 2006. Entwicklungsperspektiven im landlichen Andenraum. *Geographische Rundschau* 58(10): 64–73.

Stadel, C. 2008. Agrarian diversity, resilience and adaptation: Rural development in the tropical Andes. *Pirineos* 163: 15–36.

Stadel, C. 2010. Vulnerabilidad resistividad en el campesinado rural de los Andes. In J. C. Tulet, ed., *Las nuevas figuras del mundo rural latinoamericano. Anuario Americanista Europeo* 6–7: 185–200.

FIGURE 11.1 Terraced fields above Dali, Cangshan Mountains, southern China. (Photo by S. F. Cunha.)

proximity to urban markets, further explains our emphasis on developing regions.

Going Vertical

The vertical distribution of different environments and, in extratropical areas, the different seasonal conditions at each level, demands a staggered schedule for exploitation. Each elevation is most ideal for growing specific crops or for certain animals to graze. This *verticality* concept occupies an important position in mountain studies, and was illustrated by Alexander von Humboldt's classic nineteenth-century *schemata* of altitudinal zonation, or upward progression of changing vegetation and landforms in the Andes (Helferich 2005). His early model (Fig. 11.2) identified Latin American subsistence adaptations to distinct life zones. Sugar cane, maize, poultry, and pigs flourished in the lowland *tierra caliente* below 900 m (3,000 ft). The higher and cooler *tierra templada* (900–1,800 m; 3,000–6,000 ft) is best suited for coffee, cut flowers, and short-horn cattle. Higher still, the thinning atmosphere produces a sharp cold season in the *tierra fria* (1,800–3,600 m; 6,000–12,000 ft) that requires hardier crops such as wheat, barley, apples, pears, and dairy cows. Above this, at the uppermost limit of agriculture near the treeline, the *tierra helada* (up to 4,600 m, or 15,000 feet) is best suited for crops that tolerate a short and frenetic growing season. These include grains and tubers, sheep, and the South American trio of camelids (llama, alpaca, and vicuña). Cold temperatures and a short growing season above the *tierra helada* preclude cultivation, although seasonal grazing of livestock occurs in many meadows during the brief summer. This change in land use with elevation can still be observed today, although some products, particularly maize and cattle, occur in more than one zone depending on the individual habitat, microclimate variation, and genetic manipulation.

Other scientists have since refined von Humboldt's early work in Latin America (Stadel 1992; Lauer 1993). In addition, analogous models from elsewhere include the *Staffelsysteme* (staggered exploitation) in the Karakoram Mountains (Kreutzmann 2000), *Almwirtschaft* in the Alps (Streifeneder et al. 2007), and variants in the Himalaya (Stevens 1993), Atlas Mountains (Miller 1984), and Papua New Guinea (Ohtsuka et al. 1995).

It is important to remember that, as with von Humboldt's scheme, the extraordinary diversity of mountain habitat results in significant digression from any theoretical model. As discussed in Chapters 7 and 8, mountains produce a remarkable array of ecological habitats resulting from the interplay of physical factors. These include geologic and edaphic conditions, temperature inversions, slope and aspect, valley flooding, elevation, latitude, mountain mass effect, relief, wind, and frost at any time of year. Zimmerer (1999), for example, believes that although altitude plays an obvious role in determining crop patterns in Latin America, agricultural landscapes are comprised of overlapping patchworks shaped by human and ecological processes over long periods. The advent of new roads, genetically altered crops and animals, mechanization, and the lure of external wage-earning jobs also sway the vertical distribution of crops and animal husbandry.

There are striking differences in settlement history between various mountain ranges, as discussed in Chapter 10. North America is unusual in that the highlands remain relatively unpopulated, except for rapidly expanding resort towns and some foothill regions adjacent to urban centers (e.g., Sacramento, California, and Denver, Colorado). In fact,

FIGURE 11.2 Von Humboldt's Andean vegetation zones.

much of this region, particularly the highest elevations, consists of protected parks and wilderness where most traditional economic activity is forbidden. Yet another trend here, and in other developed-world mountains such the Alps, is widespread migration to lowland cities that correlates with a deemphasis on mountain agriculture. In these regions, the explosive growth since 1950 in recreation and its associated service industries has almost totally replaced farming and herding as the economic mainstay. In contrast, Old World and South American mountains have long supported dense populations. The strategies used by various cultures to overcome the limitations of these often minimally productive landscapes are illuminating and provide insight into the nature of mountains themselves. Some may depend upon sedentary agriculture and livestock, while at the other extreme are nomads skilled in animal husbandry.

Generalists and Specialists

Mountain agriculturalists are either *generalists* or *specialists*. The generalists divide their effort into cultivation, tending livestock, and exploiting at least one other primary resource (usually forests). Specialists focus on either cultivation or animal husbandry, and are more reliant upon established trade links to procure the goods they do not produce. The size and scale of each region also influences the agricultural strategy. In smaller mountain ranges, like the Drakensberg (Southern Africa) or Pyrenees (Southern Europe), generalists may farm and graze every altitudinal zone during the course of a single year. On the other hand, in larger mountains such as the Himalaya–Karakoram and Andes, specialization is common, because the great distance between zones reduces the habitat variety available to any individual group.

Modernization

With few exceptions, mountain agriculture is undergoing rapid change. Academic, applied, NGO, and government specialists vigorously debate the external and internal forces driving this transformation. Distilling their theoretical and interpretive underpinnings would occupy a chapter in itself; readers should consult Jodha (1997), Kreutzmann (2000), Ives (2006), Hofer and Messerli (2006), and Aase et al. (2013) for more depth. Agrarian change seldom results from any single factor. As Ponte (2001) found in Tanzania's Uluguru Mountains, expanded cultivation on surrounding plains, migrant labor remittances, experimentation with alternative farming systems, and various schemes to increase nonfarm income combined to improve local people's livelihoods, enhanced by the fact that these highlanders took full advantage of well-established links to lowland markets.

Present-day modernization and improved production usually result from one or more of the following quantifiable influences that are either internal (from within the community and/or local environment) or external (from outside the community and/or local environment). The first is new road construction that brings previously isolated villages into the commercial fold. Reliable transportation induces farmers to abandon traditional subsistence crops for more lucrative cash monocrops that are trucked from small settlements to large lowland markets such as Bangkok (Thailand), Fez (Morocco), or Lima (Peru), or, where they exist, highland urban centers such as Quito (Ecuador), La Paz (Bolivia), or Maseru (Lesotho). This more intensive land use usually supports relatively larger populations. Farmers typically build permanent homes in compact agglomerations and go out from the village to work the fields. This arrangement increases protection from potential competitors, preserves

FIGURE 11.3 The Karakoram Highway near Gilgit, northwest Pakistan. (Photo by S. F. Cunha.)

the best land for cultivation, and facilitates communal activities. The Hunza region of northern Pakistan is an example of this transition. In 1978, the newly opened Karakoram Highway connected this secluded district to lowland cities (Kreutzmann 2004; Fig. 11.3). As the secondary and largely unpaved road network expands into the lateral valleys, the traditional subsistence collage of corn, wheat, barley, and vegetables gives way to cash monocrops such as potatoes, spices, opium, and coca. Better transport also ignited a tourist boom that generates service sector jobs (e.g., rooms, meals, crafts, transport, and guide services). The combined proceeds from cash crops and tourism allow farmers to purchase foodstuffs they no longer grow. This transition often increases household income to the point where, in regions such as Morocco's Atlas Mountains, tourism and crops compete for economic supremacy (Parish and Funnell 1996). In another example, new roads either introduced or improved better drinking water, electricity, telecommunications, and financial institutions in the Almora District of the Indian central Himalaya. In fact, a road network became "the most felt need for socioeconomic development in remote and inaccessible mountain areas that are cut off from mainstream development" (Rawat and Sharma 1997: 117). The implications of increased access are discussed further in Chapter 12.

The introduction of enhanced seed stock and chemical and/or organic fertilizers is a second input that increases agricultural yield and speeds modernization. This frequently involves external government or private funding and technical expertise. For example, the Swiss Agency for Development and Cooperation supports agrarian reform by providing seed stock and technical advice in the Himalaya, as does the private Aga Khan Rural Support Programme in eastern Tajikistan and northern Pakistan. This well-intentioned assistance must be administered

carefully. Introducing genetically modified (GMO) plants and animals to mountain regions is controversial because it alters the existing biodiversity of these longstanding agroecological systems. Though humans have modified plant cultigens and domestic stock for centuries, the recent advances in biotechnology and genetic engineering offer the prospect of more comprehensive and accelerated change. Although some introduced varieties increase food yields, they may not satisfy all the multiple uses that traditional crops provide. These include fodder for stock, raw material for weaving, and by-products used in indigenous medicines. Such local varieties emerge over time and thrive within the subtle variations of microclimate, soils, and storage requirements of each particular mountain environment. They also respond well to indigenous fertilizers (usually animal dung), irrigation, and harvesting strategies. Thus, introducing new varieties of seed and animal stock must be done carefully and involve local people, because increasing raw yields represents but one aspect of traditional overall productivity.

A third influence promoting modernization among traditional mountain cultures is additional income from migrant labor. This indirectly supports more people without increasing agricultural production on land that may already have reached or exceeded its capacity, given the existing environmental and cultural parameters. As discussed in Chapter 10, migration can be seasonal or permanent. For example, each year adult males from the Drakensberg Mountains in Lesotho contract to work for a specific time period in South African diamond mines. The host companies deduct room and board from the workers' wages, and remit part of the salary directly to the families in Lesotho. At least 70 percent of households have at least one migrant worker, and 35 percent count migratory earnings as their primary income (Cunha 2002). The shorter growing season of the Himalaya leads to an earlier harvest, freeing workers for supplemental work in lowland fields, factories, or the service industry. For some mountain villagers, the ability to work in the lower elevations provides insurance against crop failures in societies that otherwise lack any sort of welfare or disaster assistance. Papua New Guinea highlanders habitually seek employment with lowland tribes when their potato crops fail. In other cases, the demand for outside capital and the appeal of a more lucrative life elsewhere prompts a permanent departure. Each year the oil-rich Gulf States, Western Europe, and the United States host thousands of workers from the mountains of Southwest Asia, North Africa, and Latin America, respectively. The infusion of outside capital permits local investment in farms and livestock—essential during lean periods—and often puts education, health care, and consumer goods within reach of mountain residents for the first time. The material benefits, however, are not always evenly spread among the population. In the Karakoram of Pakistan, many male workers abandon their fields and pastures to seek wage-earning jobs in the plains or with mountaineering expeditions. The cash affords the men status and access to consumer goods, but also increases the subsistence workload on the women they leave behind (Azhar-Hewitt 1999).

A fourth and final influence promoting modernization is the astronomical rise in global tourism, as discussed further in Chapter 12. Although a tourist tradition arose in European and North American mountains in the late 1800s, this source of non-agrarian income was minor or nonexistent in the mountains of developing countries before the 1970s. In nearly all mountain regions, the income from migrant labor and tourism does not replace agriculture as the economic foundation, but instead provides investment for home and field, and a modest opportunity to engage in the global economy. There are exceptions, however, such as in the Everest region of Nepal, where, in the last two decades, tourism has enriched the local Sherpa to the extent that Rai migrants now perform most of the agricultural and grazing tasks, often under the supervision of the Sherpani, as well as running visitor lodges.

With this understanding of the verticality concept, the distinction between a generalist and a specialist lifeway, and an appreciation of how various internal and external influences promote modernization, we now turn to the heart of this chapter: the four generalized land-use strategies employed by mountain farmers.

Sedentary Agriculture

Sedentary agriculture represents almost complete and permanent reliance on farming for income, in more or less the same location. In the developing world, many farmers exploit small plots *intensively* by utilizing every sliver of available land (Figs. 11.4, 11.5). Sedentary specialists establish permanent settlements and enduring economic links with lowland populations. Consistently high population growth provides a dependable source of local labor. A growing number of this mountain populace now seeks work in lowland fields and factories for income that helps support their highland villages. Access to tertiary services such as stores, restaurants, and modern health care is minimal. Literacy rates, ranking among the lowest in the world, reflect dismal educational opportunities, especially for females. In most years, food surpluses are nonexistent, and per capita income amounts to less than two dollars per day.

Sedentary farming in mountain regions is most widely practiced within the tropics in two forms: shifting cultivation and terraced agriculture. In both strategies, it is an anomaly that the climate and soils at intermediate to high altitudes (900–1,800 m; 2,500–6,000 ft) can produce greater yields than in the lowlands. In addition, malaria and other infectious diseases are less common higher up. As a result, humid tropical mountains, such as those in Papua New Guinea and Central America, support denser populations than the lowlands. Weather conditions at any given altitude are much the same throughout the year, so there is little reason for the seasonal migration and movement within the system that occurs in midlatitude mountains. Consequently, these agricultural specialists focus tremendous effort on sedentary and intensive use of the land. This is different from the middle latitudes, where generalists make lowlands the focus of intensive sedentary agriculture, and utilize the uplands more intermittently to graze livestock. Moving animals to high pastures in the summer, followed by a return to the lowlands in the fall, is called *transhumance*.

Another important characteristic of tropical highland agriculture is that crops grow year round, except at very high altitudes. More or less continuous cropping is essential in areas where there are no provisions for the preservation or storage of food, an acute problem in warm and humid environments that lack refrigeration. Of course, the verticality concept is evident, as different crops have different requirements, growing better at one level than at another. In most areas, settlement concentrates in the middle highlands (*tierra templada* in Latin America), where the moderate temperature and precipitation regime permits a greater variety of crops—an important insurance buffer against disease, drought, or other disasters. Accordingly, most highlanders integrate many crops into their diet to complement one or two staples. Throughout much of Southeast Asia and Indonesia, the principal crop is rice. In the Ethiopian highlands, cereals and pulses dominate, while Andeans emphasize potatoes and maize.

Environmental conditions during a people's cultural and technological history influence agricultural choice. Isolated portions of highland New Guinea, Borneo, and Myanmar, for example, still rely on shifting cultivation, although surging populations and new roads will soon require more intensive farming. Elsewhere, aged agricultural landscapes reflect a relatively high level of

FIGURE 11.4 Intensive land use. Using a sickle to harvest wheat, Urubamba Province, Peru. (Photo by S. F. Cunha.)

FIGURE 11.5 Intensive Landuse. A Kalasha farmer plows in the Hindu Kush of North-West Frontier Province, Pakistan. (Photo by S. F. Cunha.)

technology, although with tremendous investment of hand-driven human labor. Among the best known are the dramatic terracing and irrigation of slopes by the Ifugao people in northern Luzon in the Philippines, the Hmong in Vietnam, the Balti in northern Pakistan, various Indo-Nepalese and Tibeto-Nepalese throughout Nepal, and historically, the Inca of the Peruvian Andes (Fig. 11.6). These cultures settle within circumscribed areas and support their populations through sedentary agriculture. Where animals exist, they are secondary to the cultivation of crops and usually browse or scavenge discarded plant parts, such as rice husks and corn rinds (except in highland East Africa, where cattle are more important). In most tropical highlands, crop and livestock production form a subsistence economy that supports only local settlements. As rural road networks expand to include these settlements, the transition from a diverse array of subsistence crops to one that favors cash monocrops occurs very quickly.

Other factors influence the settlement patterns of mountain farmers. When people rely on a restricted diet for cultural reasons, the ecological requirements of the crop can impose limits on human distribution. The highland tribes of New Guinea, for example, depend almost completely on the sweet potato or *kau kau* (*Ipomoea batatas*). Over 100 language groups occupy the high altitudes of this region, with population densities reaching up to 500 people per km^2 (1,300 per mi^2). This is a sharp contrast to the sparsely populated and isolated lowlands of the Sepik and Fly Rivers, where annual flooding, insects, malaria, and persistent heat and humidity complicate life. The highlanders practice a subsistence hoe culture, supporting themselves through intensive cultivation of active and fallow fields (Fig. 11.7). Fenced plots prevent incursions by pigs, which otherwise roam free. In the more accessible farms along the central highway, coffee and vegetable cash crops flank the sweet potato. The dependence on just one or two crops and the inability to store produce beyond a

FIGURE 11.6 Stepped rice terraces in northwest Vietnam, near the frontier with China. (Photo by S. F. Cunha.)

few days mean that the people consume food soon after harvest. Thus, growing conditions must be continuously favorable throughout the year, because interrupted production begets chronic food insecurity (Boyd et al. 2001).

Shifting Cultivation

Shifting cultivation (also known as *swidden, ladang,* and *milpa*) is common in remote portions of northern Thailand, Borneo, the eastern Andean foothills above the Amazon, Papua New Guinea, West Africa, Central America, the eastern Himalaya, and northern Myanmar, Vietnam, Cambodia, and Laos. It is particularly common in areas where limestone weathers into an infertile soil. Shifting cultivators the world over share common geographic traits: rugged terrain, isolation from markets and population centers, and a tropical rainforest biota and climate. They typically are politically marginalized minority populations with high fertility rates and little or no access to external health care, schools, and lending institutions.

Although many farmers live in permanent villages, they rotate fields every three to five years, followed by a decade of abandonment and rejuvenation. Although this fragmented ownership pattern requires laborious trekking between their soccer field–sized plots, it serves a positive adaptive function by dispersing the risks inherent to farming over several microenvironments. To prepare a field, farmers first clear the forest with a machete or hand axe. They ignite the downed slash when the temperature and humidity conditions are optimum, usually prior to an afternoon rain. The ensuing conflagration reduces the freshly cut trees and understory to a layer of nutrient-rich ash. They cultivate this enriched topsoil for the next three to five years. Little or no effort is expended on fertilizer,

weeding, or irrigation. Additional field maintenance such as fencing and furrowing is atypical (Fig. 11.8). This low upkeep reduces the labor input, allowing time for opportunistic hunting and transplanting of tree crops such as the plantain (banana). The high monthly rainfall quickly leaches the nutrients into the soil and eventually below the rooting zone. Without supplemental fertilizer or the copious infusion of organic matter characteristic of a natural forest, soil fertility drops to the point where, within four to six years, the farmer abandons the plot to begin the process anew at another location. Since shifting cultivators grow mostly starch and vegetables, they add protein by raising pigs, dogs, cats, and chickens. These animals require minimal care and feed because they scavenge for rice husks, roots, insects, and other foods that humans cannot digest.

Shifting cultivation is a land-use paradox. On one hand, it requires extensive terrain, since most of the land is in regenerative fallow condition. On the other hand, it requires less labor and technology. This strategy predates all others, having been employed in tropical mountains since the early Neolithic (or Stone) Age, 10,000 years B.P. Although it does alter the tropical rainforest, compared to other agricultural techniques there is lower net loss of biodiversity, soil erosion, and forest fragmentation. Shifting cultivators often maintain a healthy forest mosaic with high species diversity that is slowing gaining recognition as an excellent form of "human impacted" land offering significant conservation value (O'Brien 2002). However, it is increasingly blamed for the environmental decline of tropical mountain rainforests, though in many locations, such as mountainous northern Honduras (Jansen 1998), northern Thailand (Renaud et al. 1998), and southwestern China (Jianchu et al. 1999), the supposed environmental change is both poorly documented and inadequately understood.

FIGURE 11.9 Plowing rice terraces in the Hoang Lien Mountains, northwest Vietnam. (Photo by S. F. Cunha.)

FIGURE 11.10 Rice terraces and village, Dudh Kosi watershed, Nepal. (Photo by S. F. Cunha.)

to minimize erosion and efficiently decant water along slopes. They increase soil fertility with animal fertilizers and mulch. The problem of waterlogging is particularly acute in clay soils with limited permeability, such as those in the highlands of Ethiopia and Sudan. In these situations, farmers add gravel to improve soil texture and infiltration capacity, employ contour plowing, and block small channels to reduce the chance of slope failure. Permanent cropping and conservation techniques like these represent an advanced stage of land use, and may have evolved from shifting cultivation once the population became too great to support itself through this method.

Unlike shifting cultivation, terraced farming is increasing in the developing world in response to exponential population growth. Each year, farmers excavate ever-narrowing terraces higher up the mountainsides of Nepal, Pakistan, Sri Lanka, and elsewhere. The human excavation coupled with deforestation was thought to increase the catastrophic floods in the Granges–Brahmaputra Delta, and by extrapolation, in low-lying areas of the Mekong, Ayerwaddy, and Salween Rivers (Ives and Messerli 1989). This debate is discussed in more detail below under "Agroforestry."

Terracing is becoming more evident outside the tropics in midlatitude wine-producing regions, where escalating land rents justify the extraordinary labor costs. In contrast, most Andean and Southern European peoples abandoned terracing long ago, and the absence of human tending now results in landslides and gully erosion.

Cash Crop and Plantation Agriculture

An increasing number of sedentary tropical highland farmers supplement subsistence with one or more cash crops. These include various spices, nuts, and seeds that find their way into regional and international markets. For example, although most chili peppers are indigenous to South America, they are cultivated and used on every continent. The miniature Thai hot peppers (*Capsicum frutescens*) raised in Southeast Asian mountains appear in Latin American, African, and Asian cooking. Mountain farmers throughout India, Myanmar, Thailand, Indonesia, and Papua New Guinea supplement their food yields with nuts from the betel palm (*Areca catechu*). Chewing this walnut-sized kernel releases *arecoline*, a mild central nervous system stimulant that is also thought to counteract intestinal parasites. Although rare in the Americas and Europe, it may well be the world's most popular stimulant. Highlanders in Africa cultivate coffee, tea, tobacco, cacao, and cayenne pepper (*Capsicum frutescens*), while their Andean counterparts raise specialty tubers and vegetables.

Farmers usually consume a small portion of each cash crop locally. They transport the rest for sale in lucrative lowland markets. When roads permit, hired trucks transport the crop. Otherwise, animals and humans bear the loads. The considerable effort is worthwhile because cash crops offer a higher value to weight ratio. A kilogram (2.2 lb) of chili peppers or raw opium greatly outearns an equivalent weight of corn or potatoes. Most cash crops last longer and can withstand the heavy compression and pounding that accompany transport. They thrive in nutrient-deficient soils and are in great demand throughout the world.

Some cash crops carry a tainted reputation. The opium poppy (*Papaver somniferum*) is grown and consumed in many mountain regions of Asia (Fig. 11.11). However, most of the raw opium is sold to nonfarmers for processing into heroin bound for international markets. Afghanistan produces 85 percent of the world's opium (UNODC 2010). The rest originates in Southeast Asia's Golden Triangle, where the frontiers of Myanmar, Cambodia, and Thailand meet, although recently the poppy has also appeared in Colombia and California's Sierra Nevada. The South American coca plant (*Erythroxylum coca*) has been chewed since Incan times, but today finds its way into urban markets around the world as refined cocaine. This alkaloid was first isolated and purified in the mid-1800s, and is an important ingredient in medicines, drugs, poisons, and as a flavoring in cola drinks. Until 1906, cocaine alkaloids were not separated from the flavoring added to cola drinks in the United States. Since 1995, the global production of coca leaves has dropped due to government eradication efforts, especially in mature Andean fields. Though total planted coca acreage remains stable, the newer fields are pushed into more remote and less productive sites toward and into the Amazon. Peru is the major producer of illicit drug crops, followed by Bolivia, Ecuador, and Colombia.

FIGURE 11.11 A Karen farmer inspects an opium poppy (*Papaver somniferum*), Tanen Range, northwest Thailand. (Photo by S. F. Cunha.)

Similarly, marijuana (*Cannabis sativa*) from the highlands of Central and West Africa is smoked locally, but much of the crop makes its way into Europe as concentrated *hashish*. United States antidrug operations often target these isolated and mountainous source regions, especially Thailand and Colombia. Such efforts draw a mixed reaction from mountain farmers. Even though these crops are universally illegal in their refined state, governments rarely prosecute highlanders for traditional consumption. In this respect, they fill the social niche that legal alcohol and cigarettes do elsewhere. In 1992, Bolivian President Paz Zamora clarified the relationship between local use and largely urban-industrialized use of the coca leaf by stating that "coca is an Andean tradition while cocaine is a Western habit" (Tribune de Geneve 1992). Although narcotic crops also thrive in lowland environments, seclusion and skeletal law enforcement make mountains an attractive place to locate. Growing drug crops for export ultimately brings negative consequences for mountain farmers. These include severe deforestation, soil erosion, gangs, warlords, and government-sponsored eradication efforts with collateral damage to food crops (see Chapter 12).

In many tropical locations, traditional agriculture is evolving into a cash enterprise, including family farms that specialize in one or two crops and plantation farms that produce one main monocrop aided by inexpensive and often migratory labor. Such endeavors firmly trade subsistence strategy and social structure for the trappings of agribusiness: Chemical fertilizers replace manure, tractors supplant the animal-drawn plow, and seed is often purchased instead of culled from the prior years' effort. Farmers ship the harvest from seaports, road, or rail hubs to the global market. Many are processed into commodities highly desired in the lowlands, such as coffee from the highlands of Latin America and Vietnam, tea from the interior ranges of Malaysia and Myanmar, sugar cane from the Caribbean archipelago, and tobacco from the highlands of Papua New Guinea and East Africa.

Mountain Pastoralism

Pastoralists rely on animals for their principal means of support. Often, they do not sow any crops but, instead, gather wild plants or obtain what they need at periodic markets. Livestock grazing is most common in the middle and higher latitudes, where agriculture is more difficult because of shorter growing seasons and variable precipitation. The pastures are above the cultivated fields, where snow precludes farming and grazing for much of the year. Some foraging occurs year round in tropical mountains, except on the highest peaks where perennial snow and ice limit the terrain. Thus, pastoralism is a common strategy to utilize mountain areas that otherwise go uncultivated, if not unused altogether. Animal husbandry is vital to the success of highland societies. Livestock convert grass and sedge biomass that is not directly useful to humans into food products and cooking fuel. The animals are further exploited for labor, transport, textiles, hides, and the distribution of wealth.

The three general approaches to grazing stock in mountains are (1) nomadic pastoralism, (2) transhumance, and (3) mixed grazing with farming. All are found primarily in middle and high latitudes and represent cultural adaptation to the annual cycle of seasonal change in mountain and lowland grasslands. The first strategy, nomadic pastoralism, is a highly fluid lifeway in which small bands of people and their animals migrate between winter and summer pastures with no permanent settlement base (Fig. 11.12). This is common in Tibet and the high Mongolian steppe. A second strategy, transhumance, also requires migration between winter and summer pastures. However, most of the community remains in permanent settlements to raise crops and graze animals locally, while shepherds or a few families accompany other animals to distant high pastures. This occurs in midlatitude mountains the world over, and can be readily observed in ranges including the Alps, Andes, Carpathians, Himalaya, High Atlas, Pamir, and Rocky Mountains (Fig. 11.13). The third approach combines both grazing and farming, and is similar to transhumance but is more localized, as vertical migrations of livestock occur within the same valley or mountain slope. The subtle but important difference here is that animals are an adjunct to cultivation, as both herds and tenders return to the primary settlement each evening. This approach is covered in a separate section below.

Between these major types is found every transitional form, depending on environmental and cultural factors. As with mountain agriculture, though highland livestock strategies may appear ancient, they are very susceptible to social change. The internal and external modernizing influences described earlier affect highland herders in myriad ways. Various pastoral and cultivation systems often coexist even within a single valley, since one strategy seldom exploits all habitats. Each approach occupies its own ecological and social niche. In much of Europe and North America, transhumance is curtailed, if not eliminated, as competing uses preempt lowland winter pastures.

Coping with Risk

Pastoralists contend with significant risk and uncertainty. The quest for unexploited forage demands good geographic information about terrain, soils, vegetation, and seasonal weather. They must also be vigilant against potential competition from other humans (rival herders, government-sponsored initiatives, expansionist farmers, and immigrants). To assess these fundamental constraints, Bollig and Göbel (1997) distinguish between *individualistic* and *holistic* parameters. Individualistic constraints are internal, involving the economic and social choices made to survive in a landscape fraught with environmental uncertainty. While these dynamics influence pastoralists everywhere, the influence is greatest on more remote groups such as the Bakhtiari tribe in Iran's Zagros Mountains. More commonly, the holistic factors of population growth, market forces, armed conflict, government modernization programs, and changing social conditions provoke change. An example is found in Gaddi shepherds, who traditionally extracted a living from the arid and thus uncertain pastures of the Indian Himalaya by wide-ranging grazing over time and space (individualistic parameter). This strategy worked until the government built irrigation canals to expand farming onto the drier interfluves separating the river canyons (holistic parameter). The farmer-friendly policy reduced the Gaddi pastureland, disrupted their traditional foraging routes, and created the perception that Gaddis were "shiftless nomads who deforest and degrade the Himalayan environment" (Chakravarty-Kaul 1997: 133; Hoon 1996).

Another example of adapting to environmental and social uncertainty occurs in the High Sierra of the south-central Andes, where drought, declining soil fertility (from overgrazing and increased tilling), theft, and predation (by condors, pumas, and foxes) condition the behavior of Peruvian herders (Browman 1997). These pastoralists graze two distinct ecological zones: the drought-prone lower High Sierra (2,500–3,400 m; 8,200–11,150 feet), where

FIGURE 11.12 Pastoralists return goats and sheep to camp, Chinese Pamir, Xinjiang Uyghur Autonomous Region, western China. (Photo by S. F. Cunha.)

40 mm (1.5 in.) of seasonal rainfall produces sparse annual grasses; and the upper High Sierra, where 250 mm (9.8 in.) of seasonal rainfall produces a Matorral shrub forest and lush perennial pastures, but cold temperatures imperil stock during the winter. High mobility and occasional animal slaughter to reduce the herd size alleviate the climatic uncertainty. Dogs and constant alertness offset the negative impact of thieves and predators, although the combined losses from these two can still reach 10–20 percent per year (Kuznar 1991). The Peruvians graze the lower Sierra during the November to April wet season until the annual grasses are exhausted. In early spring, they ascend to exploit the perennial grasses of the upper Sierra. The constant mobility and family focus leave little sense of a larger community. Although families claim rights to each grazing zone, little documentation exists, and the resulting "first come, first served" strategy is a source of great tension among families.

Although sole reliance on animals is risky, pastoralists enjoy several advantages over sedentary farmers. Their highly mobile product (livestock) lets them transit from pasture to pasture to take advantage of fickle mountain precipitation. Walking (driving) their herds to market—instead of transporting them by truck or train—enables long-distance travel between markets. Unlike crops, animals represent an immediate food supply, available on demand throughout the year. Moreover, large beasts contribute valuable labor and transport functions. These *walking larders* or *fields on the hoof* comprise much of a family's wealth.

Unlike most lowland farm and livestock operations, mountain grazing is an open range affair where rivers, steep terrain, and snowline restrict movements. Managing access to the high pastures is a communal concern determined by environmental and cultural factors. Delineation of social groups and spatial boundaries depends on the ability to exclude outsiders and self-regulate shared pastureland, often without government jurisdiction. This establishes a paradox where, on one hand, individual herders want to maximize their income by increasing their number of livestock yet, on the other hand, they must employ a group conscience to conserve their collective resource (Fernández-Giménez 2002).

Swiss and Austrian pastoralists hold exclusive grazing rights to individual *Alps* (high meadows), with rights to sell products and grazing rights as they wish. In partial contrast, the Rereya and Ourika tribes graze the pastures of the Moroccan High Atlas according to a highly scripted *agdal* system of reserved rights. Each *agdal* contains various subsistence cultivation and grazing niches (riparian, slope, rainshadow, etc.). They tend sheep on open pastures, but cattle must be corralled and fed with locally harvested forage. The owners may partition unused allotments to others, but local feed must remain within the *agdal*. Unused corrals are dismantled, and the rights are assumed by others.

Foraging practices are dynamic and change over time. Although isolated and highly mobile, herders respond to changing market demands, improved transportation, technological improvements, and migration. They are similarly not ignorant or reluctant to pursue promising opportunities to improve their income, which alters the consumption of resources. For example, grazing *zomo* dairy cows in mid-elevation Himalayan *godes* (individual foraging plots) is currently giving way to less labor-intensive yak breeding at the higher elevations. The Nepali herders found that yaks earn more money with less effort. Although they still tend some *zomo* near their permanent foothill settlements, this new strategy of favoring yaks transfers grazing pressure from their former mid-elevation

FIGURE 11.13 A lone shepherd boy tends goats and sheep in the Pamir Mountains of western Tajikistan. (Photo by S. F. Cunha.)

godes to the lower and higher pastures. In the Tigray Highlands of northern Ethiopia, transhumant herders are adjusting their traditional routes to accommodate such disparate new demands as school enrollment for their young herd boys and reservoir development (Nyssen et al. 2009).

In North America and Europe, producing meat and fiber is the primary goal of commercial livestock ranching and grazing. In contrast, smallholder pastoralists of the developing world survive by maintaining gross livestock numbers. Although herders often cull animals for immediate consumption or sale, they protect and work most of the herd into old age because their multiple uses make them worth more alive than dead. Depending on the region, the milk of yaks, goats, llamas, and sheep is consumed daily, or manufactured into butter and cheese for the lean winter months. The dung from large stock such as camels, yaks, horses, and cattle is gathered for fuel—a crucial resource in semiarid steppe and high-elevation landscapes where wood is scarce or nonexistent. Shearing goats, sheep, llamas, and yaks provides raw material to weave into carpets, clothing, saddlebags, and other household necessities. These durable textiles are often an important, if not the primary, source of cash income. Large animals are also efficient and affordable beasts of burden for moving goods, people, and households from one location to another (Figs. 11.14, 11.15). Breeding the best specimens is yet another income source that accrues only from a live animal. In many nomadic societies, the number of animals owned, along with their specific attributes, influences social factors such as political lineage, marriage dowry, territorial rights, and overall wealth. When animals finally expire from old age, selective culling, or accidents, herders consume the meat and manufacture by-products into various domestic tools. Tibetan Kampa nomads, who fashion yak bladders into water bags and stitch hides into circular yurts with sinew (animal ligaments), are but one example.

In summary, pastoralism is found in mountain regions around the world. It is more important in middle and high latitudes, where shorter growing seasons and variable precipitation make sole reliance on crops more difficult than in the tropics. Obtaining grazing rights is vitally important to raising livestock and, despite the appearance of being age-old practices, the strategies evolve over time as herders adapt to internal and external changes. We now look more closely at two different forms of highland pastoralism.

Nomadic Pastoralism

Nomadic pastoralism is a highly organized type of social organization in which people move with their animals from place to place in search of pasture. Although many mountain nomads of the Central Asian steppe have no fixed abode, the majority in Africa, Europe, and South America spend part of each year near permanent settlements. Their movement either follows an elevation gradient—climbing higher in spring, and descending in fall—or is driven by latitudinal changes in temperature and rainfall. Nomads everywhere seek a precipitation regime that favors lush grassland, to the extent that land tenure, political boundaries, and terrain permit. Mixed herds of cattle/yaks, sheep, and goats are more common than dependence on a single beast. Camels, horses, and donkeys also transport tents (*yurt* or *ger* in Central Asia) and other household material.

This lifeway is largely a phenomenon of semiarid regions where agricultural pressure is less intense. Small bands of people and animals make use of extensive but nominally productive land that typically supports a meager population.

FIGURE 11.14 Kirghiz pastoralists milk goats in the Chinese Pamir. (Photo by S. F. Cunha.)

The diversity of mountain terrain, the variety of animals engaged, and the multiplicity of ethnic and social groups that comprise nomadic societies make it difficult, if not misleading, to categorize nomads. Each year, they migrate to high altitudes during spring and early summer, after the snow melts and relatively lush grasses develop in the mountains. Meanwhile, dry heat withers the lowlands. When winter approaches in the high country, autumn rains and clouds green the lowland grasses, spurring a return to the plains. This seasonal cascade of favorable conditions between high- and low-altitude pastures underlies the ecological basis for nomadic pastoralism. It should be stressed, however, that nomadism is a cultural decision, not one forced by environmental conditions. Other approaches may be equally successful, as demonstrated by the many sedentary agriculturists who share terrain with nomads. Examining history makes this fact even more astonishing. For example, before the twelfth century, Iran, Anatolia, and much of Afghanistan supported sedentary agriculturalists. It was only after the Turk–Mongol invasions of the Middle Ages ravaged farmland that nomadism became important here.

Today, nomadic pastoralism in mountain regions is found primarily on three continents. These include the Old World highlands of East Africa, the Atlas Mountains of North Africa, the Mediterranean Balkans, Scandinavia and Siberia, Zagros of Iran, the Afghan Hindu Kush, the western Himalaya, and Central Asia's Pamir, Tien Shan, and Altai Mountains. The common livestock includes sheep, goats, and cattle, although camels (North Africa), reindeer (Scandinavia and Siberia), horses (Mongolia), and yaks (Central Asia) are locally important. New World nomadic pastoralism is restricted to the Andes, where llamas and alpacas are raised. Pastoralism is not generally found in eastern China or Japan, owing to the traditional emphasis on intensive cultivation, or in North America, Western Europe (apart from the Balkans), Australia and New Zealand, where commercial cattle and sheep operations are the industrialized parallel to nomadism.

Nomadic peoples are usually ethnically and culturally distinct from those who practice sedentary agriculture, and usually comprise an ethnic minority population in their political states of residence. The Kirghiz in western China (Kreutzmann 2003) and Sami of Scandinavia are illustrative. The family is the principal social unit, and the collective group lives isolated from others since it is constantly on the move. Despite such self-reliant and solitary living, nomads are interdependent with sedentary farming and village peoples. They sell or barter milk, meat, wool, and skins in exchange for manufactured items (e.g., pots, knives, ammunition, and medicine), and foodstuffs (e.g., flour, tea, alcoholic beverages, and spices) they cannot produce themselves. *Nan* (circular bread) is an important staple for most nomads in Southwest Asia, obtained in trade with sedentary agrarians. In turn, farmers receive animal products and indirect benefits such as manure for fertilizer and extra labor at certain times of the year. Andean peoples depend on trade and reciprocity between herders and farmers. While milk and blood are both important foods throughout Asia and Africa, the Andeans use neither. They mitigate the absence of these nutritious foods by eating vegetable crops, especially tubers, which nomads gather or purchase in periodic markets.

The distance between the nomads' winter and summer pastures vary greatly from region to region. In the precipitous Himalaya and Andes, they often move less than 20 km (12 mi) up and down a major valley. In more arid terrain, such as Iran, the annual round trip may involve distances of over 1,000 km (600 mi). These long migrations own the livestock. Nomads socially bond into firm family or clan relationships. In contrast, many transhumant herders are hired from outside the family or social group to tend livestock they do not own.

Transhumance is more productive than sole reliance on either cultivation or pastoralism, and provides insurance when part of the operation fails due to weather, disease, or warfare. Consequently, it is more widely adopted in mountains than is pastoral nomadism. The term originates from the Latin *transhumer,* from the terms *trans* and *humus,* meaning land that sits beyond cultivated fields. Transhumance is common in the Alps, Pyrenees, Carpathians, Caucasus, Himalaya, Andes, Atlas Mountains of North Africa, New Zealand Alps, and Scandinavia. Within the tropics, transhumance is common in Colombia, the Andes (below the 27th parallel), Kenya, Ethiopia, and Rwanda, where migrations follow the onset of wet/dry cycles instead of temperature differences. Cattle and pigs predominate. Most of China (except Yunnan) and Japan are again notable exceptions. For specific examples around the globe, consult Uhlig (1995), Kreutzmann (2006), Robinson et al. (2010), and O'Flanagan et al. (2011).

Transhumance was first described as a regional pastoral practice in southern France. It evolved during the Middle Ages, only to wax and wane during periods of plague, warfare, and collapsing empires. Yet shepherds in the Pyrenees still lead stock to high *estives* (pastures) on both the French and Spanish sides of this southern European range. Until the 1960s, milk and cheese from cows were the primary outputs. Domestic cats accompanied the herds to protect the finished cheese from mice and rodents. Sheep foraging on high-angle slopes too steep for cows completed the ensemble. Families accompanied the animals to the summer pastures. They divided the labor, living in circular *orri* (mortarless stone huts with thatch roofing). This family-run strategy faded when the nineteenth-century Industrial Revolution sparked a decline in small-scale agricultural production. Livestock farms with hired *pâtre* (shepherds) now rule. In the last 40 years, the primary outlets for Pyrenean herders come from urban demand for meat and milk. Without human tending, secondary plant succession is rendering some high meadows unsuitable for cattle. The future here appears uncertain, as the revenue for mountain recreation (ski resorts, hiking trails, housing, etc.) may simply overwhelm the modest proceeds from transhumance, although the charisma of seasonal herding is an attractive cultural element popular with tourists. In addition, water projects in the Teno and Gallego Valleys of Spain have flooded field and winter grazing sites essential to seasonal transhumance (Garcia-Ruiz and Lasanta-Martinez 1993).

As with pastoral nomads, transhumant herders in China (since 1949) and the former Soviet Union (since the 1940s) are undergoing significant challenges. In the Qilian Mountains of northeastern Qinghai Province, China, the rights to land and livestock have been transferred to individuals since decollectivization began in 1984. While rural incomes have doubled in many areas, thanks to improved fenced winter grazing plots, the annual increase in sheep (3.9 percent) and yaks (2.4 percent) on the summer pastures above the Datong Valley degrades the rangelands (Cincotta et al. 1992). Thus herders must curb their individual appetite for growth and profit if they are to protect the collective "commons" upon which they all depend. Deregulation in the former Soviet Central Asian state of Kyrgyzstan had the opposite effect when transhumant herders quickly sold off their livestock for needed cash. This temporary population crash improved the rangeland, but it remains to be seen whether the herds can be reestablished at a sustainable carrying capacity that in turn protects the commons.

In the major longitudinal valleys of the western Himalaya–Karakoram, several distinct cultural groups have individual strategies to exploit the vertical environmental change. For example, in northern Pakistan, just east of the Khyber Pass, the Swat River rises among 5,500 m (18,000 ft) peaks in the Hindu Kush and flows southward through the Swat Valley, which gradually widens in its lower reaches until broadening into alluvial flats at 1,500 m (5,000 ft). Sedentary farmers in this low area depend on irrigated wheat, maize, and rice. To guard against crop failure, they settle and plant the valley only up to the elevation where they can raise two crops per year. Directly above this group is another cluster with a different strategy. Here the valley narrows, so irrigated terraces compensate for the lack of flat land. As the shorter growing season allows only one crop per year (usually maize, millet, wheat, or rice), transhumance with sheep, goats, and cattle enhances the single annual crop. The herds ascend to summer pastures at 4,200 m (14,000 ft), passing through seasonal camps along the way. Only a small contingent of people remains behind to tend the fields, or in some cases, they travel back and forth to maintain both endeavors. Thus, by employing two distinct strategies, they earn a satisfactory living from the higher and less productive environment. The uppermost occupants in the Swat Valley are nomadic pastoralists. They depend almost exclusively on livestock, and acquire some grain and necessities through trade with the lower groups. During winter, they graze livestock on surrounding hills that sedentary farmers in the lower watershed use only to gather fuelwood. The nomads pay grazing fees and gain tolerance because they provide milk, meat, and other animal products (including manure) necessary to the village economy. They also look after village animals and serve as laborers during the peak agricultural season.

This vertically integrated land-use system, practiced in Swat and throughout the Karakoram Mountains for centuries in one form or another, is currently reorganizing. Kreutzmann (2000) and Azhar-Hewitt (1999) link the decline in animal husbandry to new opportunities for off-farm resources earned from military and civil service, trade, tourism, and foreign migratory labor (i.e., everything from oil wells to domestic service). These more lucrative external opportunities, combined with a rise in school enrollment, leave the subsistence agricultural and pastoralism chores to women and the elderly who, on their own, cannot maintain the previous numbers of livestock.

In North America, transhumance began in the mid-1800s, about the time it started to decline in Europe, and thrived until the 1920s. It was common (and still occurs) where western mountains rise above semiarid lowlands. Immigrant shepherds, initially Basques from the Pyrenees, but later joined by Spanish, Peruvian, and Mexican men, drive cattle, sheep, goats, and horses to the largely unsettled high country during summer, and back to the lowlands in late fall. The San Juan Mountains of Colorado provide a good example. Winter grazing initially depended on the availability of extensive lowland areas. With the rapid population increase, crops and fenced pastures are used to feed and contain livestock in the valleys. The high mountain grazing occurs almost exclusively on government land, for which a small (and politically charged) fee is charged. By the early 1900s, it was apparent that sheep and cattle grazing were eroding alpine grasslands and threatening native species. In response, the U.S. Forest Service began restricting livestock permits and herd movements. The addition of more national parks between 1900 and 1970 and the passage of the 1960 Wilderness Act, which further limited commercial grazing in the mountainous west, all worked against transhumance in the United States. By the late 1980s, the economics of the operation began to swing in favor of year-round fenced pastures at lower elevations, some with supplemental feed. Today, it is the exception rather than the norm to see cattle and sheep moving to and from the high mountain pastures. Where this does occur—mostly in the Rocky Mountains and other western ranges in the Great Basin of Nevada, Utah, and eastern California—trucks often transport the stock. Eventually, these too seem destined to give way to recreation interests, summer homes, and *dude ranches* where mostly American, European, and Japanese city dwellers pay money to work a mountain ranch. This involves herding stock to pasture on horseback, complete with "chuck wagon" meals and tent camp. Aside from this profitable hybrid, transhumance in the United States is no longer economically significant.

In summary, pastoralism in all variants is a common land-use strategy found in mountains around the world. Though traditional in appearance, numerous external and internal influences are prompting social and agroeconomic change. Livestock are the pastoralists' economic mainstay, but they also depend on reciprocity with sedentary farmers for other goods and services they need. This is different from the next strategy, where cultivated crops are the primary source of income, but animals are an important adjunct.

Mixed Agriculture

The importance of mixed farming varies considerably from one mountain region to the next. It is most widely practiced in the middle latitudes, where it may be either a sporadic and localized phenomenon or the principal land use. The particulars and settlement vary widely across the mountain world. For example, llamas and potatoes dominate in the Andes, yaks and barley are important in much of Central Asia, and cattle and corn prevail in Africa. These differences aside, the ecological principles are similar, as are the midlatitude agricultural constraints of low temperature, a short growing season, sloping terrain, poor soil, unreliable moisture, and low plant productivity. Cultural strategies to ameliorate these conditions include irrigation, terracing, spreading animal manure, constructing buildings, introducing nonnative plant and animal cultigens, and the indirect cropping of marginal areas through animal husbandry. These tactics, along with constantly evolving social and behavioral adaptations, allow humans to support themselves at moderate to high elevations outside the tropics. Even in heavily industrialized regions of the Alps, Rocky Mountains, Australia and New Zealand, mountain farmers still abide by the basic principles of mixed agriculture, albeit with some government support.

Mixed farming and grazing is highly structured. As with the strategies discussed earlier in this chapter, farmers engage in vertical exploitation at different times of the year. They emphasize agriculture, since the production of supplemental feed becomes essential in areas without winter pasture. Accordingly, farmers drive livestock to high pastures during summer as a matter of expediency, and not because lowland grasses dry up, as is usually the case for nomadic pastoralism and transhumance. In contrast, the lush grasses typical of most midlatitude mountain valleys can easily support livestock during the summer. However, if the animals remain at low altitudes, it becomes difficult to cultivate fields or reap the hay that sustains livestock through the long winters. This harvesting of summer fodder—or *haying*—is a defining attribute of midlatitude mountains (Fig. 11.17).

Permanent settlement in this situation requires the preservation and storage of food through the nonproductive season. People living in Eurasian mountains rely heavily on milk and milk products because they can exploit well-fed and protected dairy animals throughout the winter. In the Andes, milk has never been an important food item because llamas and alpacas are relatively inefficient at lactation. For millennia, the Altiplano of Bolivia and Peru was used seasonally for cultivation, but permanent settlement became possible at these altitudes only after the Andeans learned to freeze dry tubers into *chuño*. This process takes advantage of the arid climate and nightly chill. Potatoes are spread on the ground in the open night air and allowed to freeze, then submerged in water the following morning. The process is repeated daily for several weeks until the tubers become hard and black, at which point they are stored intact or mashed into a white pulp for later use. In either case, they keep indefinitely. Since their introduction from the Andes in the sixteenth century, potatoes have become a major food source throughout mountainous Eurasia, from

FIGURE 11.17 Hay harvest near Zakopane, Tatra Mountains, Poland. (Photo by S. F. Cunha.)

FIGURE 11.18 Guambiano women inspect potatoes at a periodic market in Silvia, Colombia. (Photo by S. F. Cunha.)

the tropics to the higher latitudes. They are not freeze-dried in the more northern mountains, but instead are stored in underground cellars where cool, still air retards spoiling. Because it is one of the few dependable and productive crops that grow at demanding high altitudes, the nutritious and versatile potato is an important staple of traditional mountain societies the world over (Fig. 11.18).

Since mixed farming is particularly a midlatitude phenomenon, agrarians must contend with erratic and often chronically deficient rainfall. Population growth in many highland areas makes this problem more acute each year. The construction and upkeep of elaborate irrigation systems, and the fair adjudication of water rights, are cooperative endeavors involving the entire village. In many areas, they engage scores of villages within a single watershed. The methods to intercept, store, and transport water vary greatly, as discussed below and in Chapter 12. In northern Pakistan, irrigation canals siphon glacial meltwater into gravity-fed canals that descend to fields and villages below. A well-orchestrated system of sluice gates and settling ponds regulates the water. Villagers in the eastern Pamir simply shovel gravel to either block or direct irrigation water into small canals (Fig. 11.19). Farmers in the Al Hajar Mountains of Oman rely on a millennia-old system of gravity-fed *falaj* (canals). Before the arrival of modern clocks and wristwatches, they relied on the sun and stars to time water allocations. Although many villages still use sundials, reliance on stargazing and the star lore associated with it is rapidly disappearing (Nash and Agius 2011). One of the newest methods of harvesting water employs polypropylene mesh to intercept fog drip in arid mountains. Households channel the trapped water into storage tanks above or below ground for later irrigation of crops or other uses.

Mixed farming thrives in places such as the semiarid eastern Pamir of Tajikistan, where cultivating wheat, barley, apricots, and vegetables is the primary economic means of support (Cunha 2007). Yet these crops alone would leave the isolated Central Asians hungry without the proceeds and products from livestock. Each morning, herders leave their *kishlaks* (settlements) with 10 to 50 sheep and goats to forage local pastures. They return just before nightfall.

FIGURE 11.19 Controlling irrigation flow by shovel, eastern Pamir Mountains, Gorno-Badakshan Autonomous Oblast, Tajikistan. (Photo by S. F. Cunha.)

Others leave for a number of weeks to exploit distant pasture, while most of the villagers stay behind to tend fields. Mixed farmers here, as elsewhere, observe the following annual cycle:

- Spring: Manure fields, move livestock from village to lower fields, repair infrastructure (irrigation, farm implements, and structures), plow and sow fields;
- Summer: Progressively move animals to higher pasturage as snow recedes, weed and maintain fields, irrigate crops, prepare and, in some cases, begin the harvest;
- Fall: Complete the harvest, return livestock to lower pastures or the village, thresh harvest, dry fruit, grains, etc.;
- Winter: Maintain livestock, educate children, manufacturing (weave, spin wool, carvings, agricultural and household implements, etc.).

This composite strategy allows Tajiks to maximize production in their challenging environment. However, survival during the Soviet era required subsidies from the central government because trade with adjacent China and Afghanistan was forbidden, leaving them in a snowbound closed system for the five winter months. The support ended with the socialist collapse and Tajik independence in 1991. Hard times and chronic food shortages have plagued these Tajiks ever since. Yet even in this mountain outpost, an expanded road system that links their villages to the main highway and a new trans-frontier artery into China are prompting economic and social change.

In another example, Mexico's seminomadic Tarahumara Indians move both crops and animals up and down the steep *Baranca del Cobre* (Copper Canyon) in order to exploit various thermal and moisture regimes. The changing environments are known simply as *beterachi,* or "places to live." Depending on the season, the Tarahumara cultivate *milpa* (corn), squash, beans, potatoes, apples, and peaches on canyon bottoms or on mountain slopes, and supplement their diet by hunting small mammals.

Mixed farming in the European Alps thrived for centuries, but is now heavily supported by governments and the European Union, in an economy that is now dominated by the service sector (Price et al. 2011a). The juxtaposition of lofty peaks, green meadows and fields, and—particularly since the 1950s—tourist facilities (e.g., alpine huts, hotels, cable cars, and trains) creates a distinctive cultural landscape (Fig. 11.20). However, the current economy is very different from the one portrayed in travel literature, more accurately described as a rapidly evolving hybridized mixed farm-specialty (meat, dairy, organic) enterprise. Whatever the combination, the long-established alpine farm calendar begins with early spring snowmelt when livestock emerge from barns to graze low pastures. This is also maintenance time: removing avalanche debris from fields, tending vineyards, repairing barns, and servicing terrace and irrigation systems. As the snowline retreats, livestock are taken higher, but grazing is still supplemented with stall-feed, especially for milk cattle and draft animals kept near the village. Fields are tilled and planted with grains and potatoes by late April to early June. With most of the snowpack gone by late June, the livestock ascend to alpine pastures for the summer. The arduous hay harvest begins in July. The lowest south-facing slopes are cut first, then the higher meadows, followed by the same pattern on shaded slopes. Extra hay is stored in small barns near the fields or is offered for sale, and the rest saved for winter cattle feed. Any combination of family members or hired help tend crops and animals in summer. They reside in seasonal shelters at various elevations.

By late August or September, the rye and barley fields are ready for harvest and, shortly thereafter, the livestock descend the pastures to feed on stubble and grass around the village. Excess livestock are either butchered or bartered for other commodities at annual livestock fairs. Harvesting potatoes in late September represents the last major fall activity. By mid-October, all crops are securely stored and the household prepares for winter. With the return of snow and cold weather the long process of stall-feeding begins, continuing until the following spring, when the routine begins anew.

FIGURE 11.20 Farmhouse and fields above Grindelwald, Berner Oberland, Switzerland. (Photo by S. F. Cunha.)

Although traces of this idealized mixed farming scheme still exist throughout the Alps, it is most prevalent in the north. It survives as an economic entity only with substantial aid (e.g., tax incentives, price supports, and infrastructure projects) from governments and the European Union, partly because the picturesque rural agricultural landscape is a tourist gold mine, but also to maintain biodiversity as well as rural populations (Price et al. 2011a). Such backing is rare or nonexistent in the mountains of developing countries in Latin America, Asia, and Africa, where true subsistence mixed farming remains important. Throughout Europe, the term "Alpine" implies high-altitude landscapes with trees, villages, and pastures. This differs from standard biogeographical usage, where it denotes terrain above the treeline (see Chapter 7). The breakdown of traditional mixed agriculture in alpine Europe started in the mid-1800s, accelerating first after World War II and then again after the 1960s, when independent farmers could no longer match the highly mechanized productivity gains of lowland competitors who were also located closer to urban markets. A population exodus followed, led by young people seeking a more lucrative city life. With fewer hands to drive livestock, many abandoned farms are undergoing secondary plant succession, which promotes the return of wild ungulates (mostly deer). Browsing by these species is preventing the reestablishment of late-stage succession forests (Motta 1999, Soliva 2007). Alpine mixed farming today usually requires local people to combine government subsidies set by European Union Common Agricultural Policy (CAP) with industrial or service-sector jobs, often coupled to the exploding demand for mountain recreation (Tosi 2000; Garcia-Martinez et al. 2010). Most alpine farmers today are either older (>50 years) or combine part-time farming with another part- or full-time nonfarm occupation, of which renting rooms is but one example (Tappeiner et al. 2008). Since the mid-1990s, the emergence of specialty organic and GMO-free crops has brought a slight resurgence in mixed farming. The challenge is to incorporate modern technology with traditional crops and techniques, and to market these products on the basis of alpine "tradition" and "quality" (Schermer 2001). Even with this emerging market, abandonment of agricultural land along with the traditional farming practices described above remains a dominant land-use change in the mountain areas of Europe, as is generally the case in rural lowland areas as well (MacDonald et al. 2000).

In summary, mixed farming relies on both crops and livestock for economic and food security. This primarily midlatitude strategy is also found in arid subtropical highlands. The many forms are tailored to specific environmental and social conditions. In most locations, scant precipitation and cold winters limit plant and animal productivity. As modernization accelerates, many of the mountains of the developing world will likely follow the European alpine path, replacing traditional mixed farming with some combination of specialty crops, supplemental nonfarm and migratory employment, out-migration to urban centers, and tourism.

Agroforestry

Agroforestry is a composite of highly interdependent agriculture, grazing, and forestry. As discussed in Chapter 10, healthy forests play vital roles in productive mountain agroecosystems and the societies that depend on them. A significant change in forest biomass will influence an entire watershed. The harvesting of fuelwood, timber, nuts, fruit,

FIGURE 11.21 Yi women collect fodder (brush, needles, and leaves) in the Hengduan Mountains, southern China. This nontimber resource is gathered from the forest floor to use as roof thatch, animal feed, and fire starter. (Photo by S. F. Cunha.)

and other forest products helps sustain developing mountain economies; at the same time, good forest management benefits downstream residents by reducing soil erosion and flooding, and increasing overall water retention within the drainage. Although humans exploit mountain forests differently from place to place, the ensuing discussion presents broad categories of utilization, illustrated with recent examples of agroforestry management in highland subsistence economies. For more analyses, consult Hamilton and Bruijnzeel (1997), Price and Butt (2000), Stöhr (2009), and Price et al. (2011b).

Tree cover in *tropical* mountains is very diverse and includes cloud forests on rain-soaked tropical slopes, seasonally dry deciduous forests and woodlands, and pine forests at higher elevations. In midlatitude temperate zones, *montane* forests of mostly evergreen species (e.g., pine, fir, and cedar) cloak mountain slopes in predictable elevation zones. Within these two generalized classes—tropical and montane—the forest shows considerable variation as individual species adapt to the diverse microclimatic temperature and precipitation regimes that define mountain environments, as discussed in Chapter 7.

Mountain societies exploit forests in myriad ways. In addition to the obvious demand for fuelwood and construction timber, trees protect fields and dwellings from high winds and snow avalanches. Leaves, branches, and bark make excellent livestock fodder; some is used only for roofing, while other components boost the quality and quantity of milk, dung, and wool (Fig. 11.21). Mountain forests also harbor wildlife and feral animals that farmers pursue for meat, pelts, and body parts. Various tree species—especially nuts and fruits—are essential producers of food, medicine, and fiber. Forests store and purify water. The upper canopy filters

solar radiation, which augments soil moisture and groundwater storage. Vigorous tree cover decreases runoff and soil erosion. Mountain forests harbor important plant biodiversity, with the neotropical (New World) mountain forests alone accounting for half of the world's 90,000 species of higher plants. Most highland societies also now realize that standing forest cover attracts tourist revenue. Finally, mountains "covered in deep forest have generated an aura of hidden power, mysticism and spiritual authority on human life" (Mishra 2000: xxi). Table 11.2 summarizes the utilitarian uses of forest biomass.

Uses of Wood

Fuelwood for cooking and heating places the greatest demand on forest biomass in subsistence societies, accounting for two-thirds of global wood utilization. However, judicious collection of downed wood can promote overall forest health. In arid uplands, gathering reduces the fire hazard. Collection and pruning may impede the spread of disease. Moreover, selective cutting of branches and trees boosts sunlight to the upper canopy and the understory, which may harbor cultivated crops.

Gathering wood is relatively easy in tropical mountains, as vigorous growth sheds plenty of branches. The chore is more difficult around sedentary villages in montane forests, where amassing the weekly household supply often requires a full day of arduous labor (Fig. 11.22). Examples include the Nepalese Sherpa, who sometimes excavate juniper roots from deep within the soil; the Karen of Northern Thailand where young boys wield machetes with adult skill; and Xhosa women from the Drakensberg Mountains in Southern Africa who balance wood bundles

FIGURE 11.25 Xhosa women construct a *rondavel* with mud, dung, and sticks, Drakensberg Mountains, South Africa. (Photo by S. F. Cunha.)

capacities. They also provide more reliable income (Arnold and Ruiz Pérez 2000). The products are deeply woven into the socioeconomic interdependencies of mountain life, including diet, religious beliefs, building materials, household and textile goods, and livestock production. Peruvian anthropologist Lupe Camino identifies a central Andean "health axis" where the ongoing collection of medicinal plants dates from the first millennium B.C. (Bussmann et al. 2007). A 1994 study in Uganda's Mount Elgon National Park found that nontimber products, such as bamboo, fuelwood, medicinal plants, and rope, had seven times the value of a potential timber harvest (Scott 1998). There is mounting evidence that nonforest products generate more sustainable income than timber removal in the hilly terrain of India's Keonjihar District (Mahapatra and Tewari 2005). In southern China, the markets overflow with over 150 forest and forest-farm products (Hamilton et al. 1997). The overall social and economic value of nonforest products in the montane forests of the developed world is generally less, given the primary focus on the maximization of timber production and, particularly in densely populated areas, protection against natural hazards, as discussed in Chapter 10.

Domesticated fruit and nut orchards represent an important and growing sector of mountain agriculture

(Figs. 11.26, 11.27). These nonwood products are essential foods and medicines, having thrived since antiquity in mountains of Central Asia, the Middle East, and the Mediterranean. The High Atlas of Morocco is typical, with olives, almonds, walnuts, and tree fruit (oranges, lemons, cherries, and more recently apples) flourishing. The growing export of this fresh and dried produce throughout the Mediterranean and Europe heralds the transition from subsistence to a market economy (Parish and Funnell 1996). In the Pamir of Tajikistan, an important cultural hearth of arboculture, ubiquitous almond, pecan, and walnut orchards shade homes in the western foothills and on alluvial terraces east of the Pamir crest. In the tropics, papaya and pineapple are found in the highlands of New Guinea and Borneo, while betel and cashew nuts thrive in coastal mountains of India and Myanmar. The Central American cordillera and northern Andes are famous for bananas. Improved container shipping and genetic alterations tailored for specific markets bring formerly distant mountain farmers into the global economy.

Agroforestry Management in the Future

Agroforestry strives to increase timber and fuelwood production, while reducing declining soil fertility and erosion (Blatner et al. 2000). The uneven adoption of this management paradigm in subsistence mountain societies is generating much concern in international government and NGO circles. Some early programs designed to promote agroforestry fell short of their goals in part because well-intentioned outside support was compromised by inaccurate assumptions regarding the *traditional, and therefore inefficient*, mountain subsistence strategies (Donald 2004). Compared with cultivated ground crops and livestock grazing, the utilization of highland forests is more worrisome to lowland populations. In 1976, Eckholm (1976: 78) noted that "topsoil washing down into India and Bangladesh is now Nepal's most precious export, but one for which it receives no compensation." This opinion stoked the *theory of environmental Himalayan degradation*, according to which the agricultural, grazing, and forestry practices of allegedly ignorant Nepali hill farmers were to blame for periodic devastating floods in downstream India and Bangladesh. While a thorough analysis did hold some people in highland areas accountable for indiscriminate logging, overgrazing, and poor farming techniques, the claims of an imminent environmental *supercrisis* were both exaggerated and based on inaccurate and/or incomplete data. For example, evidence from the Likhu Khola drainage basin in the Middle Hills of Nepal proved that rates of soil erosion were less than commonly alleged. The investigators urged reconsidering "the common perceptions in an environment where farmers, for the most part, exercise great care and judgment in managing their limited land resources" (Gardner and Gerrard 2003: 43). Moreover,

FIGURE 11.26 Kalash women gather berries shaken loose by their grandfather in the treetop (top center), Rumboor Valley, Hindu Kush, North-West Frontier Province, Pakistan. (Photo by S. F. Cunha.)

involving local people in the planning, decision making, and implementation of development projects. This strategy reduced the rate of Himalayan forest loss that in the 1960s and 1970s followed a top-down agenda that failed miserably. Some of the community-based cooperative plans adopted in the 1980s met with more success and are partially responsible for the ongoing afforestation. This apparently intuitive idea is not new. The Swiss Alps were heavily deforested by the 1870s, at which point the first federal forest law was passed, stressing the importance of local management of the forests to ensure both productive and protective functions. Following similar legislation and action in other Alpine countries, the area and, more recently, the density of the forests have since increased significantly (Price et al. 2011a). In contrast, the weak communal organization in the northwestern Andean cordillera of Peru, combined with population growth and national economic and political upheaval, spurred overexploitation of forest and grassland resources (Cotler and Maass 1999). In 2004, Nepalese Sherpas instituted a community-based conservation program to mitigate and even reverse the degradation of three decades of "adventure tourism" in the alpine zone below the Everest massif. The impacts included overharvesting of alpine forest and shrub vegetation, accelerating overgrazing and soil erosion, and unregulated building of tourist lodges (Byers 2005).

The second initiative, as discussed earlier, recognizes that forests are more than just trees. That is to say, the other nonwood forest products, cultural and religious values, and overall watershed protection are equally important to mountain communities, so that sustaining them is vital to maintaining cultural identity and long-term economic viability. Moreover, since lowland populations benefit from mountain "environmental services" such as carbon sequestration, soil stability, and recreation potential, an emerging question is whether or not local people should receive payments for these services to incentivize environmental conservation (Franco-Maass et al. 2008).

Finally, agroforestry is more likely to be institutionalized when land tenure is clearly defined, and when local communities—instead of a central government or dominant landowner—share the political and management authority. Baker (2005), investigating India's rolling Chota Nagpur Plateau, found that integrating rural people with forest management can improve rural welfare and reverse environmental degradation. Such preconditions increase the likelihood that the entire socioeconomic and cultural fabric will contribute to sustainable agroecosystems. This is more common in mountain areas of East Africa and South Asia where forest and land resources are owned by families or communities. This is not the case in Colombia, Ecuador, and the Philippines, where the highland tropical forest diminishes 1 percent a year to accommodate commercial grazing, timber harvests, landless immigrants, native population growth, and illicit crops (Grau and Aide 2008).

government and international efforts to improve the lot of highland farmers and mitigate lower watershed catastrophe too often failed to seek local input. Worse still, their solutions did not necessarily serve local socioeconomic needs. This debate now frames the character of international aid and scholarship. For the history and analysis of the Himalayan degradation debate, consult Thompson et al. (1986), Ives and Messerli (1989), Forsyth (1996), Gerrard and Gardner (2002), Gardner and Gerrard (2002, 2003), Byers (2005), Ives (2006), and Hofer and Messerli (2006).

The deforestation issue of the last quarter century placed mountain agroecosystems at the epicenter of international controversy. In 1992, the United Nations Earth Summit officially recognized the crucial link between highland and lowland environments (Ives 1997; Sène and McGuire 1997; see Chapter 12). Several subsequent initiatives have addressed this inevitability. The first emphasizes community-based participation, or more simply,

FIGURE 11.27 Date palms below the ancient village of Misfat al Abreyeen, al Hajar Mountains, Oman. (Photo by S. F. Cunha.)

Conclusion

This chapter investigated four broad types of agricultural land use in mountains. The emphasis was on traditional (or indigenous) subsistence economies, where sedentary and mixed agriculture, pastoralism, and agroforestry account for the greater part of human endeavor. The rapid changes in temperature and precipitation resulting from changes in altitude, combined with an active geomorphic setting, require complex and staggered resource exploitation relative to the more homogenous lowlands. Mountain agriculturalists are either *generalists* who tend crops and livestock while exploiting at least one other resource (often forests), or *specialists* who focus on either cultivation or animal husbandry, and reciprocate with others to procure their remaining needs. However isolated and *traditional* in both technique and strategy, mountain agriculturists everywhere must cope with pressures exerted by modernization, population growth (except in developed economies), environmental sustainability, and surging economic globalization. Since one in eight humans lives in the mountains, and an additional 60 percent rely on mountain resources, the choices individual farmers make reverberate well beyond their highland homes. Yet, as the controversy surrounding the theory of Himalayan degradation illuminates, mountain agroecology revolves around a complex milieu of socioeconomic and cultural parameters, where the value system and objectives often differ—and may not be readily apparent—to those who live and work in a lowland economy.

Ellis-Jones (1999) identifies world bright spots where land improvement and poverty reduction are forging a sustainable future—even though sustainability is a *moving target* (see Chapter 12). Examples include agroforestry systems based on indigenous technologies in East African highlands; improved community forestry programs in the Middle Hills of Nepal, India, and Pakistan; and widespread use of cover crops/green manures, and farmer-centered community-based agricultural research and development programs in the Central American highlands. These mountain residents capitalize on rich natural resources, a well-developed social network of trust and shared expertise, and abundant labor with an excellent work ethic. People in areas endowed with these assets—including much of the developing world—are especially able to improve land care and per capita income if they have reasonable access to financial capital and basic infrastructure improvements.

For more information on this very interdisciplinary field, readers should consult the rich variety of academic journals in geography, anthropology, environmental science, soils and agronomy, political science, development, and forestry. In particular, *Mountain Research and Development, The Himalayan Research Bulletin,* and the *Journal of Sustainable Forestry* devote considerable focus to agroecological issues in mountain terrain.

References

Aase, T. H., Chapagain, P. S., and Tiwari, P. C. 2013. Innovation as an expression of adaptive capacity to change in Himalayan farming. *Mountain Research and Development* 33: 4–10.

Arnold, J. E. M., and Ruiz Pérez, M. 2000. Income from non-timber forest products. In M. F. Price and N. Butt, N., *Forests in Sustainable Mountain Development: A State of Knowledge Report for 2000* (pp. 300–306). Wallingford, UK: CABI Publishing.

Azhar-Hewitt, F. 1999. Women of the high pastures and the global economy: Reflections on Pakistan. *Mountain Research and Development* 19: 141–151.

Baker, J. M. 2005. *The Kuhls of Kangra: Community-Managed Irrigation in the Western Himalaya*. Seattle: University of Washington Press.

Banks, T. 1997. Pastoral land tenure reform and resource management in northern Xinjiang: A new institutional perspective. *Nomadic Peoples* 1: 55–76.

Barrow, C. J. 1999. *Alternative Irrigation: The Promise of Runoff Agriculture*. London: Earthscan.

Blatner, K. A., Bonongwe, C. S. L., and Carroll, M. S. 2000. Adopting agroforestry: Evidence from central and northern Malawi. *Journal of Sustainable Forestry* 11: 41–69.

Bollig, M., and Göbel, G. 1997. Risk, uncertainty and pastoralism: An introduction. *Nomadic Peoples* 1: 5–21.

Boyd, W., Prudham, W. S., and Schuman, R. A. 2001. Industrial dynamics and the problem of nature. *Society & Natural Resources* 14: 555–570.

Browman, D. L. 1997. Pastoral risk perception and risk definition for Altiplano herders. *Nomadic Peoples* 1: 22–36.

Brush, S. B. 1988. Traditional agricultural strategies in the hill lands of tropical America. In N. J. R. Allan, G. W. Knapp, and C. Stadel, eds., *Human Impact on Mountains* (pp. 116–126). Lanham, MD: Rowman and Littlefield.

Bussmann, R., Sharon, D., Vandebroek, I., Jones, A., and Revene, Z. 2007. Health for sale: The medicinal plant markets in Trujillo and Chiclayo, Northern Peru. *Journal of Ethnobiology and Ethnomedicine* 3: 37. <http://www.ethnobiomed.com/content/3/1/3>.

Byers, A. 1992. Soil loss and sediment transport during the storms and landslides of May 1988 in Ruhengeri prefecture, Rwanda. *Natural Hazards* 5: 279–292.

Byers, A. 2005. Contemporary human impacts on alpine ecosystems in the Sagarmatha (Mt. Everest) National Park, Khumbu, Nepal. *Annals of Association of American Geographers* 95: 112–140.

Chakravarty-Kaul, M. 1997. Transhumance: A pastoral response to risk and uncertainty in the Himalayas. *Nomadic Peoples* 1: 133–150.

Cincotta, R. P., Zhang, Y. Q., and Zhou, X. 1992. Transhumant alpine pastoralism in Northeastern Qinghai Province: An evaluation of livestock population response during China's agrarian economic reform. *Nomadic Peoples* 30: 3–25.

Clark, W. M. 1986. Irrigation practices: Peasant-farming settlement schemes and traditional cultures. In H. L. Penman, ed., *Scientific Aspects of Irrigation Schemes* (pp. 37–52). London: The Royal Society.

Cotler, H., and Maass, J. M. 1999. Tree management in the northwestern Andean cordillera of Peru. *Mountain Research and Development* 19: 153–160.

Cunha, S. 2002. Lesotho: A development challenge in South Africa's mountain rimland. In S. Pendergast and T. Pendergast, eds., *Worldmark Encyclopedia of National Economies*, Vol. 1: *Africa* (pp. 235–342). Farmington Hills, MI: Gale Group.

Cunha, S. 2004. Allah's mountains: Establishing a national park in the Central Asian Pamir. In B. Warf, D. G. Jannelle, and K. Hansen, eds., *WorldMinds: Geographical Perspectives on 100 Problems* (pp. 25–30). New York: Kluwer.

Cunha, S. 2007. The Badakshani of the Eastern Pamir. In B. Brower and B. R. Johnston, eds., *Disappearing Peoples? Indigenous Groups and Ethnic Minorities in South and Central Asia* (pp. 187–206). Oxford: Berg.

Dollfus, P. 1999. Mountain deities among the nomadic community of Kharnak (eastern Ladakh). In M. Van Beek, K. B. Bertelsen, and P. Pedersen, eds., *Ladakh: Culture, History and Development between Himalaya and Karakoram* (pp. 92–118). Aarhus: Aarhus University Press.

Donald, P. F. 2004. Biodiversity impacts of some agricultural commodity production systems. *Conservation Biology* 18: 17–38.

Eckholm, E. P. 1976. *Losing Ground*. New York: Worldwatch Institute/W. W. Norton & Co.

Ellis-Jones, J. 1999. Poverty, land care, and sustainable livelihoods in hillside and mountain regions. *Mountain Research and Development* 19: 179–190.

Fernández-Giménez, M. E. 2002. Spatial and social boundaries and the paradox of pastoral land tenure: A case study from postsocialist Mongolia. *Human Ecology* 30: 49–78.

Forsyth, T. 1996. Science, myth, and knowledge: Testing Himalayan environmental degradation in northern Thailand. *Geoforum* 27: 375–392.

Franco-Maass, S., Nava-Bernal, G., Endara-Agramont, A., and Gonzalez-Esquivel. 2008. Payments for environmental services: an alternative for sustainable rural development: The case of a national park in the central highlands of Mexico. *Mountain Research and Development* 28: 23–25.

Garcia-Martinez, A., Bernués, A., and Olaizola, A. M. 2010. Simulation of mountain cattle farming system changes under diverse agricultural policies and off-farm labour scenarios. *Livestock Science* 137: 73–86.

Garcia-Ruiz, J. M., and Lasanta-Martinez, T. 1993. Land-use conflicts as a result of land-use change in the central Spanish Pyrenees. *Mountain Research and Development* 13: 295–304.

Gardner, R. A. M., and Gerrard, A. J. 2002. Relationships between runoff and land degradation on non-cultivated land in the Middle Hills of Nepal. *International Journal of Sustainable Development and World Ecology* 9: 59–73.

Gardner, R. A. M., and Gerrard, A. J. 2003. Runoff and soil erosion on cultivated rainfed terraces in the Middle Hills of Nepal. *Applied Geography* 23: 23–45.

Gerrard, A. J., and Gardner, R. A. M. 2002. Relationships between landsliding and land use in the Likhu Khola drainage basin, Middle Hills, Nepal. *Mountain Research and Development* 22: 48–55.

Goldstein, M., and Beall, C. 1991. Change and continuity in nomadic pastoralism on the western Tibetan Plateau. *Nomadic Peoples* 28: 105–122.

Grau, H. R., and Aide, M. 2008. Globalization and land-use transitions in Latin America. *Ecology and Society* 13(2): 16. <http//www.ecologyandsociety.org/vol113/iss2/art16/>.

Hamilton, A. C., and Bruijnzeel, L. A.1997. Mountain watersheds: Integrating water, soils, gravity, vegetation, and people. In B. Messerli and J. D. Ives, eds. *Mountains of the World: A Global Priority* (pp. 337–370). Carnforth, UK: Parthenon.

Hamilton, A. C., Gilmour, D. A., and Cassells, D. S. 1997. Montane forests and forestry. In B. Messerli and J. D. Ives, eds., *Mountains of the World: A Global Priority* (pp. 281–311). Carnforth, UK: Parthenon.

Helferich, G. 2005. *Humboldt's Cosmos*. New York: Gotham.

Hofer, T., and Messerli, B. 2006. *Floods in Bangladesh: History, Dynamics and Rethinking the Role of the Himalayas*. Tokyo: United Nations University Press.

Hoon, V. 1996. *Living on the Move: Bhotiyas of the Kumaon Himalayas*. New Delhi: Sage Publications.

Ives, J. D. 1997. Comparative inequalities: Mountain communities and mountain families. In B. Messerli and J. D. Ives, eds., *Mountains of the World: A Global Priority* (pp. 61–84). Carnforth: Parthenon.

Ives, J. D. 2006. *Himalayan Perceptions: Environmental Change and the Well Being of Mountain Peoples*. 2nd ed. Kathmandu: HimASS.

Ives, J. D., and Messerli, B. 1989. *The Himalayan Dilemma: Reconciling Development and Conservation*. London: Routledge.

Jansen, K. 1998. *Political Ecology, Mountain Agriculture, and Knowledge in Honduras*. Amsterdam: Thela Publishers.

Janzen, D. 1998. Gardenification of wildland nature and the human footprint. *Science* 279: 1312–1313.

Jianchu, X., Fox, J., Xing, L., Podger, N., Leisz, S., and Xihui, A. 1999. Effects of swidden cultivation, state policies, and customary institutions on land cover in a Hani village, Yunnan, China. *Mountain Research and Development* 19: 123–132.

Jodha, N.1997. Mountain agriculture. In B. Messerli and J. D. Ives, eds., *Mountains of the World: A Global Priority* (pp. 313–335). Carnforth: Parthenon.

Kreutzmann, H. 2000. Livestock economy in Hunza: Societal transformation and pastoral practices. In E. Ehlers and H. Kreutzmann, eds. *High Mountain Pastoralism in Northern Pakistan* (pp. 89–120). Erdkundliches Wissen 132. Stuttgart: Franz Steiner Verlag.

Kreutzmann, H. 2003. Ethnic minorities and marginality in the Pamirian Knot: Survival of the Wakhi and Kirghiz in a harsh environment and global contexts. *Geographical Journal* 169: 215–235.

Kreutzmann, H. 2004. Accessibility for High Asia: Comparative perspectives on northern Pakistan's traffic infrastructure and linkages with its neighbours in the Hindukush-Karakoram-Himalaya. *Journal of Mountain Science* 1: 193–210.

Kreutzmann, H., ed. 2006. *Karakoram in Transition: Culture, Development and Ecology in the Hunza Valley*. Karachi: Oxford University Press.

Kuusipalo, J., Kangas, J., and Vesa, L. 1997. Sustainable forest management in tropical rain forests: A planning approach and case study from Indonesian Borneo. *Journal of Sustainable Forestry* 5: 93–118.

Kuznar, L. A. 1991. Transhumant goat pastoralism in the High Sierra of the South Central Andes: Human responses to environmental and social uncertainty. *Nomadic Peoples* 28: 93–104.

Lauer, W. 1993. Human development and environment in the Andes: A geoecological overview. *Mountain Research and Development* 13: 157–166.

MacDonald, D., Crabtree, J. R., Wiesinger, G., Dax, T., Stamou, N., Fleury, P., Gutierrez Lazpita, J., and Gibon, A. 2000. Agricultural abandonment in mountain areas of Europe: Environmental consequences and policy response. *Journal of Environmental Management* 59: 47–69.

Mahapatra, A. K., and Tewari, D. D. 2005. Importance of non-timber forest products in the economic valuation of dry deciduous forests of India. *Forest Policy and Economics* 7: 455–467.

Makino, Y. 2011. Lopping of oaks in central Himalaya, India: The link between the Garhwali people and their forests. *Mountain Research and Development* 31: 35–44.

Messerschmidt, D. A., and Hammett, A. L. 1998. Local knowledge of alternative forest resources: its relevance for resource management and economic development. *Journal of Sustainable Forestry* 7: 21–55.

Miller, J. A. 1984. *Imlil: A Moroccan Mountain Community in Change*. Boulder, CO: Westview.

Mishra, H. R. 2000. Foreword: Linking mountain forests conservation with sustainable human development. In M. F. Price and N. Butt, eds., *Forests in Sustainable Mountain Development: A State of Knowledge Report for 2000* (pp. xxi–xxvi). Wallingford: CABI Publishing.

Motta, R. 1999. Wild ungulate browsing, natural regeneration and silviculture in the Italian Alps. *Journal of Sustainable Forestry* 8: 35–53.

Nash, H., and Agius, A. 2011. The use of stars in agriculture in Oman. *Journal of Semitic Studies* 56: 167–182.

Nyssen, J., Descheemaeker, K., Zenebe, A., Poesen, J., Deckers, J., and Haile, M. 2009. Transhumance in the Tigray highlands (Ethiopia). *Mountain Research and Development* 29: 255–264.

O'Brien, W. E. 2002. The nature of shifting cultivation: Stories of harmony, degradation, and redemption. *Human Ecology* 30: 483–502.

O'Flanagan, P., Martinez, T. L., and Abad, M. P. E. 2011. Restoration of sheep transhumance in the Ebro Valley, Aragon, Spain. *Geographical Review* 101: 556–575.

Ohtsuka, R., Inaoka, T., Umezaki, M., Nakada, N., and Abe, T. 1995. Long-term subsistence adaptations to diversified Papua New Guinea environment: Human ecological assessment and prospects. *Global Environmental Change: Human and Policy Dimensions* 5: 347–353.

Parish, R. 2002. *Mountain Environments*. London: Prentice-Hall.

Parish, R., and Funnell, D. C. 1996. Land, water and development in the High Atlas and Anti Atlas Mountains of Morocco. *Geography* 81:142–154.

Ponte, S. 2001. Trapped in decline? Reassessing agrarian change and economic diversification on the Ulugurua Mountains, Tanzania. *Journal of Modern African Studies* 39: 81–100.

Price, M. F., Borowski, D., Macleod, C., Debarbieux, B., and Rudaz, G. 2011a. *From Rio 1992 to 2012 and Beyond: 20 Years of Sustainable Mountain Development: What Have We Learnt and Where Should We Go? The Alps*. Bern: Swiss Federal Office for Spatial Development.

Price, M. F., and Butt, N., eds. 2000. *Forests in Sustainable Mountain Development: A State of Knowledge Report for 2000*. Wallingford, UK: CABI Publishing.

Price, M. F., Gratzer, G., Duguma, L. A., Kohler, T., Maselli, D., and Romeo, R., eds. 2011b. *Mountain Forests in a Changing World: Realising Values, Addressing Challenges*. Rome: Food and Agriculture Organization of the United Nations.

Rawat, D. S., and Sharma, S. 1997. The development of a road network and its impact on the growth of infrastructure: a study of the Almora District in the Central Himalaya. *Mountain Research and Development* 17: 117–126.

Renaud, F., Bechstedt, H. D., and Udomchai, N. 1998. Farming systems and soil-conservation practices in a study area of northern Thailand. *Mountain Research and Development* 18: 345–356.

Robinson, S., Whitton, M., Biber-Klemm, S., and Muzofirshoev, N. 2010. The impact of land-reform legislation on pasture tenure in Gorno-Badakshan: From common resource to private property? *Mountain Research and Development* 30: 4–13.

Roder, W. 1997. Slash-and-burn rice systems in transition: Challenges for agricultural development in the hills of northern Laos. *Mountain Research and Development* 17: 1–10.

Schermer, M. 2001. GMO-free Alps: An alternative path in technology development? *Mountain Research and Development* 21: 140–147.

Scott, P. 1998. *From Conflict to Collaboration: People and Forests at Mount Elgon, Uganda*. Gland, Switzerland: IUCN.

Sène, E. H., and McGuire, D. 1997. Sustainable mountain development: Chapter 13 in action. In B. Messerli and J. D. Ives, eds., *Mountains of the World: A Global Priority* (pp. 447–453). Carnforth, UK: Parthenon.

Sillitoe, P. 1998. It's all in the mound: Fertility management under stationary shifting cultivation in the Papua New Guinea highlands. *Mountain Research and Development* 18: 123–134.

Soliva, R. 2007. Agricultural decline, landscape change, and outmigration: Debating the sustainability of three scenarios for a Swiss mountain region. *Mountain Research and Development* 27(2): 124–129.

Stadel, C. 1992. Altitudinal belts in the tropical Andes: their ecology and human utilization. In T. L. Martinson, ed., *Bench Mark 1990: Conference of Latin American Geographers* 17/18: 45–60.

Stevens, S. F. 1993. *Claiming the High Ground: Sherpas, Subsistence and Environmental Change in the Highest Himalaya*. Berkeley: University of California Press.

Stöhr, D. 2009. Is there a future for mountain forestry? In Global Change and Sustainable Development in Mountain Regions. *Alpine Space—Man and Environment* 7: 31–38.

Streifeneder T., Tappeiner, U, Ruffini, F. V., Tappeiner, T., and Hoffmann, C. 2007. Selected aspects of agro-structural change within the Alps: A comparison of harmonised agro-structural indicators on a municipal level in the Alpine Convention Area. *Journal of Alpine Research* 27(3): 41–52.

Tappeiner, U., Borsdorf, A., and Tasser, E., eds 2008. *Alpenatlas/Atlas des Alpes/Atlante delle Alpi/Atlas Alp/Mapping the Alps: Society—Economy—Environment*. Heidelberg: Spektrum.

Thompson, M., Warburton, M., and Hatley, T. 1986. *Uncertainty on a Himalayan Scale: An Institutional Theory of Environmental Perception and a Strategic Framework for the Sustainable Development of the Himalaya*. London: Ethnographica.

Tosi, V. 2000. Case study: Recreation and tourism in the Italian Dolomites. In M. F. Price and N. Butt, eds. *Forests in Sustainable Mountain Development: A State of Knowledge Report for 2000* (pp. 343–346). Wallingford, UK: CABI Publishing.

Treacy, J. M. 1989. Agricultural terraces in Peru's Colca Valley: Promises and problems of an ancient technology. In J. O. Browder, ed., *Fragile Lands of Latin America* (pp. 209–229). Boulder, CO: Westview Press.

Tribune de Geneve. 1992. Quoted in: Tupay Katan, *Coca: An Andean Tradition*. Paper presented to the UN Commission on Human Rights, Subcommission on Prevention of Discrimination and Protection of Minorities, Working Group on Indigenous Populations in July 1993.

Uhlig, H. 1995. Persistence and change in high mountain agricultural systems. *Mountain Research and Development* 15: 199–212.

UNODC. 2010. *World Drug Report 2010*. Vienna: United Nations Office on Drugs and Crime.

Wang, H., Pandey, S., Hu, F., Xu, P., Zhou, J., Deng, X., Feng, L., Wen, L., Li, J., Li, Y., Velasco, L., Ding, S., and Tao, D. 2010. Farmers' adoption of improved upland rice technologies for sustainable mountain development in Southern Yunnan. *Mountain Research and Development* 30: 373–380.

Zimmerer, K. S. 1999. Overlapping patchworks of mountain agriculture in Peru and Bolivia: Toward a regional-global landscape model. *Human Ecology* 27: 135–165.

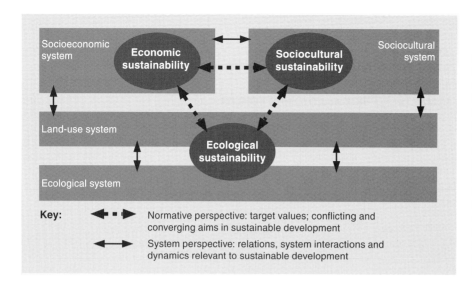

FIGURE 12.1 Sustainable development involves three basic dimensions: economy, society, and ecology, forming the "magic triangle" of sustainability. When applying the concept of sustainability, two perspectives have to be clearly distinguished: the systems perspective, which focuses on impacts (i.e., on processes within and among the three basic dimensions), and the normative perspective, which relates to the values assigned to these processes. The normative perspective always depends on a specific societal and temporal context. (From Wiesmann and Hurni 2011, adapted from Wiesmann 1998.)

a nongovernmental consultation took place in France in 1996. The 100 participants, from 24 countries across Europe, developed recommendations to governments and the European Union (ARPE/CIAPP 1996; Price 2003); the meeting also led to the establishment of the European Mountain Forum, one of five regional structures within the Mountain Forum during the 1990s and 2000s.

Citing the title of Chapter 13, most of these meetings, documents deriving from them, and many projects started in the 1990s identified *sustainable mountain development* (SMD) as an objective. However, no attempt was made to define SMD until the end of the decade. If it is to be more than a vague goal, agreement on its meaning, and then on priorities and means for its implementation, is essential. Accordingly, this chapter is divided into three parts: first, a brief overview of the concept of sustainable development, particularly in mountains, and discussion of the choice of appropriate indicators; second, key issues for SMD; third, a discussion on the implementation of SMD in terms of policies, institutions, and the outcomes of the IYM and subsequent processes and activities.

Sustainable Development: Definitions and Indicators

The concept of *sustainable development* was introduced in the *World Conservation Strategy* (IUCN 1980). It became fashionable in the 1980s, particularly through the report of the World Commission on Environment and Development (WCED), or "Brundtland Report," *Our Common Future,* which defined it as "development that meets the needs of the present without compromising the ability of future generations to meet their own needs" (WCED 1987: 43). This is probably the most cited of very many definitions: Over two decades ago, Pezzey (1989) had identified 190, although more recent texts only mention a smaller number (e.g., 57 in Rogers et al. 2008). Another commonly

used definition, agreed by three of the major international organizations working on the field, is "development which improves the quality of life, within the carrying capacity of the earth's life support system" (IUCN/UNEP/WWF 1991).

The establishment in the 1990s of many institutions and organizations whose name includes the words "sustainable development," and also journals such as *Sustainable Development* and *The International Journal of Sustainable Development and World Ecology,* show the persistence of the concept. Papers in these and other journals reveal continued debate about its meaning(s) and the contradictions that result from its use across many disciplines (e.g., van de Hamsvoort and Latacz-Lohmann 1998; Jabareen 2008). There is general agreement that sustainable development is a process that aims at ensuring that current needs are satisfied while maintaining long-term perspectives regarding the use and availability of natural (and often other) resources into the long-term future, and equity concerns about the well-being of future generations. However, sustainable development is "necessarily defined differently by each culture," and cultures may be defined not only in terms of location or ancestral lineage, but also of scientific discipline and worldview (Roe 1996; Fig. 12.1).

While sustainable development has been defined in many ways, with much discussion about its meaning and implementation, "sustainable mountain development" appeared first in the title of Chapter 13 of *Agenda 21.* Five years later, Sène and McGuire (1997: 447) noted that "the concept of sustainable mountain development has taken on new meaning" since UNCED and stated that "[a] multisectoral, more comprehensive approach to addressing mountain development issues is a relatively new concept, but one whose time has come." They contrasted this multisectoral approach with past approaches to the problems and needs of mountain areas which had largely been implemented within a sectoral context. They also noted many themes addressed at the regional intergovernmental consultations on SMD (Backmeroff et al. 1997; Banskota and Karki 1995; ILRI 1997; Mujica and Rueda 1996) and summarized by Price (1999). While all of these documents provide long lists of issues that are intended to contribute, or in some way are related, to SMD, they are not prioritized—which is appropriate given the very different characteristics of the world's diverse mountain regions, even on one continent.

Another key issue is the scale at which SMD should be implemented. For instance, one village or local community may be able to develop a strategy for its own future that appears to be viable in the long term, yet may have side effects that are unsustainable for neighboring or downstream communities. Along the many mountain ranges that form boundaries between countries and regions, there are particular needs for transboundary cooperation in SMD, given that ecological and societal processes and structures span these boundaries. The development of cooperative regional approaches is also important within the mountain massifs that are now divided between two or more nation states but have long-established cultural and economic identities distinct from adjacent lowlands in these states (e.g., Muhweezi et al. 2007). Thus, rather than proposing a precise definition of sustainable mountain development, it is probably better to recognize it as "a regionally-specific process of sustainable development that concerns both mountain regions and populations living downstream or otherwise dependent on these regions in various ways" (Price and Kim 1999: 205).

Defining Indicators

Sustainable mountain development is a complex process with uncertain objectives that are likely to shift over time. To assist in project development and wider planning, and to evaluate success, requires *indicators:* units of information measured over time that document changes in a specific condition. Sustainability indicators "should measure characteristics or processes of the human–environmental system that ensure its continuity and functionality far into the future . . . [and] be credible (scientifically valid), legitimate in the eyes of the users and stakeholders, and salient or relevant to decision makers" (Hak et al. 2007: 3–4). Van de Kerk and Manuel (2008: 229) have suggested five criteria for a good sustainability indicator: "(1) an indicator must be relevant for an issue according to the definition used; (2) an indicator must be measurable; (3) indicators have to be independent from each other and must have no mutual overlap; (4) data for the indicators must be available from public sources, scientific or institutional; (5) data must be available for all countries, at least for all but the smallest countries."

Various indicators for SMD have been proposed. At a global level, FAO (1996) proposed pressure and state indicators, to be used in a *pressure–state–response framework* (OECD 1993). Subsequently, Odermatt (2004) used the

driving force–pressure–state–impact–response (DPSIR) model (Jesingshaus 1999) to evaluate case studies of SMD. Such frameworks are based on the premise that human activities (driving forces) exert pressures (e.g., pollution emissions or land-use changes) on the environment, which can induce changes in the state of the environment (e.g., changes in ambient pollutant levels, habitat diversity, nutrient flows). These lead to impacts on populations, economies, and ecosystems, to which society then responds with environmental and economic policies and programs intended to prevent, reduce, or mitigate pressures and/or environmental change.

FAO (1996) suggested that the key pressure indicator for SMD is the population of mountain areas, to be measured in terms of population density, growth, and migration. Two key state indicators were proposed: (1) the welfare of mountain populations, to be measured in terms of nutritional anthropometry, that is, "measurements of the variations of the physical dimensions and the gross composition of the human body at different age levels and degrees of nutrition" (Jelliffe 1966); (2) qualitative assessment of the condition and sustainable use of natural resources in mountain areas, a composite of four sub-indices used to describe the state of the natural resource base of a watershed: extent of protection of soil; area of "hazard" zones; extent of degraded land; and an indication of productivity. Other proposals were made by Rieder and Wyder (1997), who, like many authors, suggested that sustainability should be measured in terms of three sets of indicators: ecological, economic, and social. Thus, they omitted a fourth dimension—ethics—identified by Ekins and Max-Neef (1992). Recognizing that indicators need to be tailored to specific circumstances, Rieder and Wyder (1997) discussed issues relating to economic, ecological, and social indicators for Bhutan, Encañada (Peru), Pays d'Enhaut (Switzerland), North Ossetia (Russia), and Puka (Albania). At the national level, Bulgaria, Hungary, Romania, Slovenia, and Switzerland suggested indicators of SMD in documents submitted to the second session of the European intergovernmental consultation.

Odermatt (2004) looked at 22 case studies of SMD from developing countries and 18 from industrialized countries, using key themes identified as being of specific importance in mountain regions around the world. These included freshwater (upstream–downstream cooperation), mountain forests and forestry, mountain agriculture and land management, poverty, local and indigenous knowledge, migration, mountain tourism, and legislation on mountains. He found that, while the DPSIR model was an adequate tool for deriving sustainability variables (i.e., Responses), it was better suited to the mountains of developing countries—but there was a serious lack of baseline data for these regions. He also noted that such frameworks can be challenging to use. For instance, depending on the particular geographical context, a variable such as out-migration may be seen as a Pressure, an Impact, or a Response.

Even at a regional or continental scale, agreement on priorities for SMD, and how they should be measured, will not be simple, as shown by a survey of key respondents working in government, NGOs, and scientific organizations in 30 European countries (Price and Kim 1999). Using a set of 36 possible indicators derived from meetings on SMD in Europe, they found that, for all respondents, ecological priorities ranked higher than sociopolitical or economic priorities. However, there were two highly ranked sociopolitical variables: the empowerment of mountain communities and the need for education and training in conservation and development. Respondents from Central/Eastern Europe placed greater emphasis on ecological indicators than those from Western Europe. The greatest similarities were with regard to sociopolitical variables, implying a common interest in the more equitable provision of benefits to people in mountain areas, to reduce marginality and ensure the long-term survival of populations in these areas. Finally, the most significant differences were found between government employees and those of NGOs and independent scientific organizations. Generally, the latter ranked ecological issues more highly than sociopolitical or economic issues. Two of the most significant differences related to compensation for sustainable management of mountain ecosystems by downstream populations, and the creation of new livelihood opportunities—both seen as more important by government employees, perhaps implying that they are more radical than suggested by the priorities of the organizations for which they work. Similarly, workshops of "specialists" and local stakeholders in the Cairngorms of Scotland found greater agreement between the two groups regarding indicators of "natural capital" than those relating to economic and social and political factors (Bayfield et al. 2000). Comparable research has not been done elsewhere. Nevertheless, it appears desirable that indicators for SMD should be appropriate to the region of concern and based on data that are measurable, available, easily understood, and meaningful (Rieder and Wyder 1997).

Key Issues in Sustainable Mountain Development

Given that SMD concerns both mountain people and those dependent, in various ways, on the environments they manage, many of the issues covered in earlier chapters are relevant in considering how to move toward this challenging goal. This includes concern for sustainable use of the various resources, as well as appropriate management of soils, geomorphic processes, vegetation, wildlife, and agricultural and forest resources. Equally, spiritual and cultural issues are embedded in the concept. A number of publications provide broad overviews of the various themes, and are pertinent to any consideration of SMD (e.g., Messerli and Ives 1997; Mountain Agenda 1997; Price and Butt 2000; Royal Swedish Academy of Sciences 2002). Another approach is that of Price and Messerli (2002), who focus

on five "pillars" of SMD: environment, culture and gender, risk, economics, and policies and legislation. The final section of this chapter considers the last of these. The remainder of this section considers the interactions of aspects of the first four in relation to a number of key issues that link mountain and other regions: access and communications; poverty, out-migration, and conflict; tourism and amenity migration; protected areas and conservation–development linkages; and water. The discussion builds particularly on Messerli and Ives (1997); reports prepared by Mountain Agenda (1998, 1999, 2001) for the annual meetings of the United Commission on Sustainable Development; background papers originally prepared for the Bishkek Global Mountain Summit, the final global event of the IYM and subsequently revised and published in Price et al. (2004); and more recent publications.

Access and Communications

From adjacent lowland areas, mountains have long been seen as obstacles to movement. Yet people have developed routes within, into, and across mountains since ancient times, initially for themselves and their pack animals and, since the mid-nineteenth century, by constructing roads and railroads for economic development and military, or "strategic," purposes, which are often linked. Ancient transit routes include those built by the Romans across the European Alps, the Silk Road linking China to Europe across the mountains of Central Asia, and others in the Andes established by the Incas. Long-established marketplaces show that mountain people have always been engaged in exchange, both within the mountains and outside them, and in areas where subsistence agriculture has dominated local economies. Exchange, and hence access, has always been integral to subsistence, even in seemingly isolated and secluded mountain areas whose people have relied on traders for salt and other essential goods.

ROADS AND RAILROADS

Increases in *accessibility* and the expansion of communications networks have been driven by industrialization, increasingly sophisticated technologies, and mass mobility at the global scale. In many mountain areas, access has been improved mainly by extensive road construction in recent decades, with high-capacity infrastructure linked to lower-capacity feeder networks. However, the density of access still differs greatly between mountain regions in industrialized and developing countries. Switzerland's road network, for example, is nearly 100 times denser per unit area, and 23 times denser per capita, than that of Ethiopia. One hundred percent accessibility means, in Switzerland, that every household in a given area reaches its home directly by car; in Ethiopia, that every household can walk to the next motorable road within a day (Schaffner and Schaffner 2001). In Ethiopia and other developing countries such

FIGURE 12.2 Flamsdalen Railway, Norway. Railways heralded the age of mass transport in mountain regions. They were instrumental in the development of mountain tourism. (Photo by M. F. Price.)

as Afghanistan, China, Nepal, and Peru, access to roads remains worse in mountain than in nonmountain areas (Huddleston et al. 2003). Railroads have also opened up parts of mountain areas to extract raw materials, for the development of tourism, and to link lowland centers of economy, for instance, through the Alps, Pyrenees, Rocky Mountains, Urals, and southern Andes (Fig. 12.2).

Increases in accessibility bring both benefits and negative effects (Mountain Agenda 2001). Mountain people can benefit from improved access to other parts of the mountains and to the lowlands through increased opportunities for employment and income, particularly for weekly or seasonal migration to surrounding lowland areas. Improved access facilitates the development of local markets, small and large industries, and services such as tourism—all creating local employment and increased economic diversification and provision of basic goods. Since the Karakoram Highway was constructed in 1959–1986 and the regional road network was improved, the mountains of northern Pakistan, notorious for periods of famine and starvation, no longer suffer from food deficits thanks to regular provision from the lowlands. The highway has also led to reduced prices for basic food and facilitated local production of seeds for potatoes and vegetables, which are exported to lowland areas (Kreutzmann 2004). Road connection can help stabilize population numbers and increase quality of life, as it makes possible the development of basic infrastructure such as clinics and schools, and increases access to consumer goods and exposure to the wider increases world, allowing the spread of new ideas. Access also increases opportunities for regional cooperation and economic exchange within mountain areas. In the Alps, for example, 72 percent of the total traffic volume (vehicle-kilometers) is local and regional inner-Alpine traffic, while tourism and recreational traffic accounts for 20 percent, and transit traffic for only

8 percent (Ackermann et al. 2006). Particularly in developing countries, adequate access is key to relief activities to mitigate the effects of natural disasters or famine.

Increased access by railroads and roads can also result in negative impacts, including brain drain and the overexploitation of resources, such as through logging and mining. For example, along the Karakoram Highway, forest stand density of highly accessible forests (closest motorable road within 2 km) decreased up to 85 percent, as opposed to 0–40 percent in less accessible forests (closest motorable road further than 8 km) (Schickhoff 2001). In South America, roads on the eastern slopes of the Peruvian Andes have created a rush for gold and timber in the eastern lowlands since the early 1990s. The completion of the Transoceanic Highway between Peru and Brazil across the Peruvian Andes is likely to increase levels of overexploitation and destruction of forests (Brandon et al. 2005). Other examples of overexploitation include tourism and forest use.

Road and railroad construction have led to fragmentation of mountain habitats. This can affect mammals or birds; for example, condors in the Peruvian Andes avoid areas cut by roads when searching for food (Speziale et al. 2008). Where road traffic is heavy, as in urban centers and along transit corridors, air pollution is often a problem, exacerbated by local topo-climatic conditions, with negative impacts on health, quality of life, and environment. Examples of such problems include the European Alps, Central America (e.g., Mexico City), the Andes (e.g., Santiago de Chile), and many larger tourist resorts and industrial centers in mountain areas around the world.

Roads and railroads can also increase erosion and sediment loads in rivers. The effects can be substantial; construction of a mountain road in Nepal, for example, increased the sediment load in the local stream by 300–500 percent (Merz et al. 2006). Though practices to

FIGURE 12.3 Construction of a stone sub-base layer in Tigray in the Ethiopian highlands for a rural access road. Labor-based road construction, such as that shown in this photo, is less expensive than machine-based construction, generates much higher income for local labor, and inflicts less damage on fragile mountain environment. (Photo by U. Schaffner.)

minimize impacts of construction, restore damaged vegetation, and secure slope stability exist, their application often increases construction and maintenance costs, which are already higher in mountain areas than in the lowlands, due to natural hazards and difficult topography, which are key determinants of the cost of transportation networks (Ramcharan 2009). While these costs are often prohibitive for poorer mountain countries, affordable approaches such as labor-intensive road construction often have great potential: Costs are lower than when heavy machinery is used, and much more of the investment is retained within mountains as wages for local labor. In Ethiopia, for example, this share was typically between 30 and 40 percent of total road construction costs. Labor-intensive methods also cause less environmental damage than construction by heavy machinery (Hartmann 2001; Schaffner and Schaffner 2001; Fig. 12.3).

OTHER MEANS OF ACCESS

Where railroads and roads are not economic, other technologies better adapted to dissected terrain and lower transport volume—typical traits of mountain transport in many places—such as ropeways, suspension bridges for footpaths (people and animal transport), or air transport can provide links. In Nepal, over 1,000 suspension and suspended bridges provide access to previously isolated settlements (Gaehwiler and Lamichaney 2001). Ropeways, of which over 10,000 are in operation, mostly in industrialized countries, and mainly in tourist areas, have great potential for hauling goods and people in mountains (Nikšić et al. 2010), but

are not adequately reflected in development policies. Typically, construction and maintenance costs are lower than for roads, especially for gravity ropeways. These installations are environmentally friendly and less susceptible to hazards, and local communities have greater economic control over trade and transport than with roads.

In many mountain areas, animals remain the most important means for moving goods and people; they represent a natural, renewable, widely available, and affordable energy source (Starkey 2001) (Fig. 12.4). However, government officials and development experts are largely unaware of the potential and of key issues, as the topic is omitted from their training. Animals are integrated into local subsistence production systems (Khoabane and Black 2009), are the key source of nonhuman power, consume local feed, reproduce, supply valuable manure, and minimize environmental damage. Animal transport is labor intensive, providing valuable employment. In the mountains of Ethiopia, 5 million donkeys carry water, fuelwood, and other goods. One-third of all donkeys are in the mountains of Asia, including China and Pakistan. Yaks and yak–cattle crossbreeds are important pack animals for trekking tourism in the Himalaya. Llamas carry small loads for tourists in the Andes, horses are used for local and tourist demands in many mountains, and camels are widely used for transportation in the mountains of the Middle East, Central Asia, and the Arab world (Mountain Agenda 2001).

INFORMATION AND COMMUNICATION TECHNOLOGIES (ICTS)

While animals, roads, ropeways, and railroads are still essential elements of SMD, modern communications technologies, such as the Internet, are growing in importance and, especially in developing countries, have reversed the "traditional" progression of access: road access, followed by electricity supply, and finally telephone. Increasingly, mobile or satellite telephones reach an area first, followed by electricity and, finally, roads, which are most costly (Montgomery 2002). Simple telephone access has the most impact, as users do not have to be literate and no specific language skills (e.g., in English) are required. However, linkage and exchange in the information technology sector are still affected by the "digital divide," as many people and institutions in the mountain areas of developing countries do not have access to these technologies. The reasons include lack of the requisite infrastructure, compounded by problems of connection due to the complex topography, and high initial costs for equipment such as computers. ICTs have proven their potential for economic development in mountain regions, in such diverse applications as telemedicine, distance education, tourism promotion, and marketing of local products. Farms selling salmon in the remote Western Highlands of Scotland, for example, reported a 30 percent increase in sales following the use of the Internet (Byers et al. 2001); in the southern highlands of Tanzania, mobile phones, e-mail, and

FIGURE 12.4 Animals are a widely used means of transport in mountain areas; especially in countries of the South and East, they dominate local movement of goods and people. From the Aconcagua region, Argentina. (Photo by H. Liniger.)

the Internet have enabled agricultural producers, processors, and traders to obtain better prices for their products (Lightfoot et al. 2008).

In both industrialized and developing countries, ICTs are often most effectively introduced to decentralized institutions, such as schools, health facilities, local governments and NGOs, or post offices. Such institutions generally find it easier than private households and small businesses to secure the means to obtain and maintain the requisite infrastructure for decentralized service centers—telecenters—which offer affordable telephone, e-mail, and Internet facilities to local people.

ICTs are relatively new, especially in developing countries. No overall conclusion can yet be made relating to their effect on development in general, and mountain development in particular. Recent studies show a mixed picture. In the Peruvian Andes, for example, telecenters only benefited specific user groups, such as teenagers and young farmers, who could maintain social networks and improve their farming practices where other information was not available (Heeks and Kanashiro 2009). Positive implications of rural telecenters on farming techniques and social welfare were also reported from Wu'an, China, but, as in the Andes, the changes observed cannot support the claims about an overall transformative role of such centers, and ICTs in general, especially for the rural poor (Soriano 2007).

Poverty, Out-migration, and Conflict

POVERTY

Poverty is a broad concept. While many criteria used to measure it are economic (e.g., per capita income in U.S. dollars), there is widespread recognition that it should refer more generally to quality of life and well-being, including basic needs (water, food, shelter, health, education) and capabilities (choice, empowerment, security) (Chambers 2005). Thus, the United Nations Development Programme (UNDP) developed the widely used *Human Development Index* (HDI), which also includes standard of education and quality of living conditions, and a *Human Poverty Index* (HPI), which recognizes that increases in cash income do not inevitably lead to decreased poverty (Kreutzmann 2001).

Most of the world's mountain people—63 percent in 2000 (Huddleston et al. 2003)—derive their livelihoods from subsistence agriculture, which may include the sale of crops, livestock, or other products. The physical geography of their environment can lead to severe limitations on agricultural productivity and options for economic development. Steep slopes and high altitudes, combined with low temperatures, result in soils that tend to be less developed, and with fewer nutrients than in lower and flatter areas. Complex topography leads to slope instability, increased frequencies of natural hazards, and lower accessibility. All of these physical factors have been widely cited as reasons for excessive poverty in mountain areas. Data from individual countries underscore the notion of mountains as pockets of poverty. To cite but two examples, in Kyrgyzstan poverty rates in mountains were found to be twice as high as the national average, and in Morocco up to three times higher (Padiukova and Padiukova 2007; Bürli et al. 2008).

Despite widespread consensus that mountain areas are characterized by high levels of poverty, solid evidence to back this general statement in a global perspective is lacking, due first to a lack of spatially disaggregated data, which would allow specific analysis for mountains; and second to the great variability of mountain environments. As noted by Ives (1997: 61), "to conclude that . . . mountains, relative to

FIGURE 12.6 Host to over 20,000 visitors on peak days, Grindelwald, in the Bernese Oberland, Swiss Alps, has largely maintained its rural character, while the local economy is totally dependent on tourism. (Photo by Grindelwald Tourism, Jungfrau Region.)

societies, the driving forces are usually not local, but external—typically at the national level. Consequently, a shift toward more sustainable development primarily requires critical decisions at this level, which need to focus on the linkages between mountain and lowland areas (and their centers of power) and on the particular potentials of mountain areas (Grau and Aide 2007). From their analysis of mountain regions mentioned above, Parvez and Rasmussen (2004) conclude that substantive national economic growth is a necessary prerequisite to developing the mountain regions of any particular country. A growing national economy means that the government is likely to provide greater subsidies or transfer payments (e.g., for ecosystem services) and other resources for mountain regions, and that the private sector is more likely to invest there. Steady economic growth also provides opportunities to absorb surplus mountain labor through out-migration, though this must be balanced with the increasing stresses on those who remain. Finally, a growing national economy creates demand and access to markets for mountain products, and promotes the diversification of mountain livelihoods. These issues are addressed further in the section on mountain policies, but a crucial conclusion is that the social and economic conditions of mountain areas are inextricably linked to their national contexts, and through these to an increasingly globalized world.

Tourism and Amenity Migration

TOURISM

Over the past century, and particularly since the Second World War, *tourism* has become a major force of change in mountain areas (Price et al. 1997; Godde et al. 2000; Dérioz and Bachimon 2009). Tourism has become one of the largest and fastest-growing economic sectors in the world, with particularly strong growth in emerging and developing countries. It has been estimated that 15–20 percent of this, or $128–170 billion a year, is associated with travel to mountain areas (Kruk et al. 2007; UNWTO 2011). With over 540 million overnight stays per year, the European Alps (Fig. 12.6) are the second largest tourist destination in the world after the Mediterranean region (UNWTO 2011). While tourism has roots in pilgrimage, a centuries-old phenomenon that is still important in many mountain areas, the rapid growth of tourism as an important element in the economy of many mountain areas derives from a number of linked factors. These relate to increased urbanization, discretionary time and income, and mobility; the need of an increasingly urban population to escape from cities for spiritual and recreational needs; and tremendous increases in accessibility to mountain areas. Many roads built for other reasons, such as the extraction of natural resources or military purposes, have also led to increases in tourism; other roads—to ski areas, for example—have been built mainly for this purpose. Railroads opened up the English Lake District and certain valleys in the Alps, the Rocky Mountains, and the Indian Himalaya from the mid-nineteenth century on; in the twentieth century, the "bullet train" system linked Japan's major cities to tourist destinations in the mountains. The real costs of international air travel have decreased and, within countries, small airlines and helicopters make it possible to reach almost any mountain area.

Tourism is regarded by many governments and communities in mountain areas around the world as vital for economic development and survival. Yet its distribution is very uneven within any given mountain region, and its benefits tend to be spread very unevenly at every scale, from the national to the local. At the scale of communities,

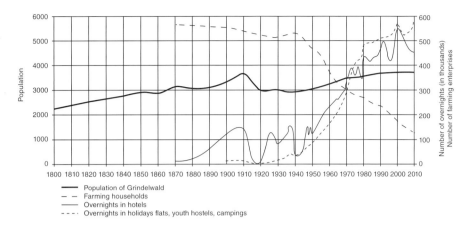

FIGURE 12.7 Boom and bust periods of tourism, Grindelwald, Swiss Alps: Tourism first peaked before World War I, when the village was still largely a farming community. As elsewhere in the Alps, tourism was heavily affected by the two World Wars and the Great Depression (1930s). After World War II, tourism grew rapidly to become the mainstay of the local economy, while the number of farms has been declining steadily. Tourism was instrumental in stabilizing local population figures—in many non-tourist Alpine regions, out-migration and abandonment of settlements is a serious problem. (From Wiesmann 1999, updated by Jungfrau Region 2012.)

the apparent benefits of tourism in terms of maintaining populations are shown by statistics from the Alps, which generate nearly 50 billion euros in annual turnover, about 8 percent of the annual global tourism turnover (Fig. 12.7). While tourism provides 10–12 percent of the jobs in the Alps, tourism-related activities are concentrated in only 10 percent of the communities. Generally, these communities have stable or growing populations, while other communities are losing population (Permanent Secretariat of the Alpine Convention 2011). Yet even in this global center of tourism, demand is not reliable; in the 1990s, there was a significant decrease in the number of overnight stays, although numbers recovered, particularly in the Austrian and Italian Alps (Macchiavelli 2009). Tourism is also highly sensitive to real and perceived risks to personal safety, as seen in the mountains of Afghanistan, Ethiopia, Kashmir, Nepal (Bhattarai et al. 2005), Pakistan, Rwanda, and Yemen in recent years.

Mountain tourism is a massive and complex interaction of people involved in a vast array of subsectors, from pilgrimage to mass tourism, cultural tourism, ecotourism, health tourism, and an incredible variety of types of sport tourism linked particularly to the development and marketing of new technologies, such as mountain bikes, carving skis, and paragliders in the 1980s and 1990s (Mountain Agenda 1999; Dérioz and Bachimon 2009). Each involves a different clientele in a highly competitive, unpredictable, and increasingly global market. Each type of visitor is attracted by different characteristics of mountain environments, landscapes, and/or cultures, and demands a specific range of services and facilities; those expected by one type may conflict with those of others, as shown by research on hiking in remote areas of Switzerland (Boller et al. 2010). Changes in fashion often mean that investments made to attract one type of tourist have to be supplemented

by new investments in new facilities to either encourage repeat visits or attract new types of tourists. Most types of mountain tourism are also highly seasonal. Skiing is only possible in cold temperatures, relying on natural and, increasingly, expensive artificial snow. Many types of ecotourism are related to the annual cycles of particular plants and animals (Nepal 2002). People coming for hiking holidays prefer the seasons with the least rain and few annoying insects. Operating in a competitive market, people in the tourism industry often find that services and facilities suitable for the tourists who come at one time of the year are not appreciated by those who come at other times or, for instance, that people who come to enjoy the summer landscape do not enjoy seeing ski lifts and other evidence of winter activities. Marketing is essential to attract tourists; but the image they leave with, and communicate to their friends, may be just as important in ensuring future visits.

Tourism is part of the process of *globalization* of mountain areas even though, in the majority of countries, most tourists are domestic rather than international; the few exceptions are countries such as Andorra, Austria, and Nepal, which are small and/or adjacent to countries with much larger populations. Tourism can provide important new sources of income for local people who provide goods and services. At the individual level, a few people get rich from tourism. They are often immigrants or others with access to outside capital, yet their success may depend on the majority who work for relatively low wages, and may not even be able to find affordable housing in the communities where they work. As tourism becomes dominant in the local economy, the costs of food, goods, and services tend to rise, sometimes making them unaffordable for local people. Yet participatory and well-planned tourism development can lead to benefits for a large proportion of

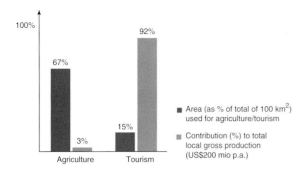

FIGURE 12.8 In Grindelwald, farming and tourism benefit from each other. Tourism is the mainstay of the local economy, generating over 90 percent of the local gross domestic product, much of which is reinvested in agriculture. On the other hand, farming largely dominates land use and is thus an asset for tourism. (From Wiesmann 1999.)

the local population, including women, whose status may increase as a result (Lama 2000; Valaoras 2000; Zhou and Liu 2008; Yang et al. 2009). Tourists also expect healthy living conditions; this can lead to public health benefits for local people if local authorities invest the profits from tourism wisely.

While the specific economic benefits and negative impacts of tourism are often hard to predict and assess (Fig 12.8), this is even more difficult for the inevitable cultural changes (Price 1992). Although the costumes and traditions of indigenous people are often among the reasons that tourists visit, these are often changed as "Western" clothing and footwear are adopted—often as status symbols, especially among young people—and cultural activities are adapted to the demands of tourists. Dances and singing traditionally performed at festivals can become mere performances, repeated often or taking place at an inappropriate time. To avoid such impacts, some communities, such as the Zuni of New Mexico (Mallari and Enote 1996), have decided that the disadvantages of tourists outweigh any potential benefits and have excluded them from festivals.

Tourists also expect to take home souvenirs. This can lead to the revitalization of local skills; yet souvenirs may be imported from far away for sale to tourists, and priceless cultural artifacts are stolen and sold. The physical changes resulting from tourism can also extend to the construction of buildings. Both the design and the ornamentation of traditional buildings are usually highly characteristic of a particular settlement, valley, or region; they are among the potential attractions for tourists. Tourism typically leads to homogenization and the construction of new buildings that are often not particularly well adapted to the rigors of the mountain climate. Yet while many mountain tourists, especially those most interested in sports activities, may not expect or particularly care about authentic food, buildings, or cultural activities, an increasing number do; within the growing ecotourism market, the value of diversity is explicitly recognized (Nepal 2002). Thus, while tourism can have many negative cultural impacts, it can also lead to the

rejuvenation of traditional skills (Creighton 1995; Pilgrim et al. 2010), providing employment, especially in the off-season, and also leading to the strengthening of cultural identity, as noted for the Sherpas of Nepal (Stevens 1993) and in the Rhodope Mountains of Greece (Valaoras et al. 2002).

Tourism can also bring many negative environmental impacts, both locally (e.g., Geneletti and Dawa 2009) and over a broader region, as shown for the Indian Himalayan resort of Manali (Cole and Sinclair 2002). Local environmental impacts mainly occur along the valley floors and on lower slopes, where most infrastructure is constructed. The impacts include the loss of agricultural and residential land, water pollution, and air pollution, particularly from traffic and in inversion conditions. However, as tourists visit higher and more dispersed locations, levels of fecal water pollution often increase (Monz 2000), and erosion occurs along trails and around campsites, requiring restoration (Good and Johnston 2004). In the mountains of developing countries, trees and shrubs are often used for cooking and heating, leading to changes in forest composition and structure (Chettri et al. 2002; Nepal et al. 2002; Byers 2005). In downhill ski resorts, there are impacts up to the summits of mountains, through the logging and bulldozing of ski runs, the construction of chairlifts and cableways, the installation and use of snow cannons, and the disturbance of vegetation and wildlife by skiers and machines. Yet, while all of these impacts have been recorded in mountain regions around the world, the awareness and potential resources that tourism brings can also provide opportunities to develop new technologies and strategies, for example, in energy production and utilization, water and waste treatment, traffic management, and trail design.

AMENITY MIGRATION

Associated with tourism is the phenomenon of *amenity migration*, as discussed in Chapter 10. Many people who first come to the mountains as tourists stay on, or return, as amenity migrants; many start tourism businesses. Like tourists, amenity migrants are often escaping urban environments, but at longer time scales. Yet their expectations are typically those of urban people, their lifestyles often change little, and, like tourists, they have diverse cultural, economic, and biophysical impacts (Moss 2006). Tensions often develop between them and longer-established residents, for various reasons. Their relative wealth allows them to purchase property and take it out of traditional uses such as agriculture, watershed protection, and residential use (Fig. 12.9). They tend to consume more of everything, both local and imported. Consequently, prices for property, goods, and services tend to rise, often beyond the capacity of local people. Amenity migrants also often have different worldviews, particularly with regard to the conservation rather than use of natural resources. Sometimes they foster indigenous cultures, though often their focus is on material objects rather than the culture as a whole.

FIGURE 12.9 Amenity migrants working in nearby towns gradually replace the local subsistence farming community, a process often accompanied by intensive construction and reconstruction activities, as in this small village near Kranj, northern Slovenia. (Photo by T. Kohler.)

Amenity migrants also provide new employment, particularly in the service sector, in which most of them start businesses and are employed. Yet many of the new jobs, particularly those available to indigenous people because of their lack of training, are not well paid.

DEVELOPING INTEGRATED AND LONG-TERM PERSPECTIVES

The long-term unpredictability and seasonality of almost every type of mountain tourism imply the need to develop it to be complementary to other economic sectors. Similarly, while amenity migrants may bring new resources to mountain communities, they will not necessarily stay for the long term; this is especially true of foreigners who move to attractive locations in developing countries, but leave when conditions do not turn out to be as they had expected, or change (Price et al. 1997; Moss 2006). The rapid development of both tourism and amenity migration typically creates great demands and stresses on local families, communities, and infrastructure (e.g., Gurung and DeCoursey 2000; Saffery 2000). Rather than shifting too far into providing goods and services for tourists, people working in tourism and for amenity migrants need to maintain their opportunities for employment and earnings in other ways, whether in agriculture, forestry, industry, handicrafts, tele-working, or commuting in the off-season.

A further reason to maintain diverse opportunities is the fact that many aspects of tourism may be influenced not only by changes in tourism markets, but also by climate change. For instance, the skiing industry is highly sensitive to climate change in at least two ways: first, if snow is not available, or cannot be made, particularly during major holiday periods (Agrawala 2007); second, if increases in natural hazards lead to the closure of access roads and the decommissioning of cable cars and ski lifts (Nöthiger and Elsasser 2004). More broadly, climate change may affect the attractiveness of mountain landscapes, a key reason for both tourists and amenity migrants to come to mountain areas. On the other hand, higher summer temperatures at lower altitudes may increase the attractiveness of mountain areas, with more people leaving cities to go to the cooler mountains (Serquet and Rebetez 2011).

To prosper in a competitive industry, those involved in tourism in every mountain community and region need to develop a unique image based on local environmental and cultural assets—one in which the income from tourism should be invested to ensure a sustainable future, even if numbers of tourists decline. As investments in constructing and maintaining physical infrastructure have to be made in any case—and may be needed more in order to adapt to climate change—these can be planned and implemented to address the new challenges and opportunities deriving from tourism and amenity migration. Local, regional, and national governments need to develop and implement policies and financial instruments to ensure that the economic benefits of tourism—not just locally, but to national economies—are reinvested in the resources that attract tourists—not just physical infrastructure, but also landscapes and cultures. Equally, governments and development agencies need to provide the means for training mountain people in skills necessary both in tourism and for other employment opportunities. Finally, tourists should be aware of the ways in which they can contribute to mountain environments and people. For instance, they can choose to travel with companies that employ local people, stay in locally owned accommodations, eat and drink

FIGURE 12.10 Mountain forest at Ruwenzori, Uganda, at about 3,300 m above sea level. Tropical mountain forests typically have more species per unit area than adjacent lowland rainforests. (Photo by R. Burtscher.)

local produce, insist on the use of alternative energy sources rather than firewood, and buy locally made souvenirs.

Protected Areas: Linking Conservation and Development

BIODIVERSITY HOT SPOTS

Many of the most biologically diverse parts of the Earth are in mountain areas, one reason that the Conference of Parties to the United Nations Convention on Biological Diversity approved a specific Programme of Work on Mountain Biodiversity in 2004 (Anonymous 2004). Globally, 25 of the 34 world centers of greatest *biodiversity,* or *biodiversity hot spots* recognized by Conservation International (Mittermeier et al. 2005) are wholly or partly mountainous. Most are in or include tropical mountains: the Atlantic forest of Brazil, northern Borneo, the eastern Andes, the eastern Himalaya–Yunnan region, and Papua New Guinea. Many secondary centers are in Mediterranean areas—which have the greatest number of tree species outside the tropics—as well as arid mountains, parts of the Rocky Mountains, and Central Asia (Barthlott et al. 2005).

As discussed in Chapters 7 and 8, many factors combine to create these high levels of biodiversity, including endemism: in particular, steep altitudinal gradients and opportunities for movement along corridors—or, conversely, isolation—and evolution over geological time (Fig. 12.10). A further set of factors leading to the high biodiversity of mountain ecosystems derives from human activities. While lowland areas have been cleared and cultivated for centuries, if not millennia, extensive parts of mountain areas were largely left alone because of the steep slopes and less attractive conditions for agriculture.

For instance, in Europe, mountains can immediately be recognized on a map of biodiversity hot spots (European Environment Agency 2010). The Alps, for example, host about 4,500 vascular plant species, more than a third of the entire European flora, of which about 15 percent are endemic (Ozenda and Borel 2003). However, human activities are not only a major reason that Europe's mountains have high levels of biodiversity relative to adjacent lowlands, but also a cause of high biodiversity within them. As in other mountain areas around the world, centuries—even millennia—of human activities including grazing, burning, irrigation, and the selective harvesting of species have had major influences on both flora and fauna, as, for instance, in New Guinea, where there is a strong relationship between biological and cultural diversity (Stepp et al. 2005). The continuation of such long-established practices is often essential not only to ensure the maintenance of these anthropogenic ecosystems and their constituent species, but also to preserve cultural heritage and food, soil, and water supplies for both mountain and downstream communities (Körner et al. 2005).

PROTECTED AREAS

Mountain areas have been a principal focus of the protected area movement since it began in the mid-nineteenth century. In 1864, the federal government of the United States gave the Yosemite Valley to the State of California as a park, and in 1872, declared Yellowstone, in the Rocky Mountains, as a "public park"; it is generally considered the first *national park* (Foresta 1982). Many of the other early national parks around the world, including Banff in Canada and Tongariro in New Zealand, are also in mountain areas. Since the 1980s, protected areas have been one of the fast-growing land-use types in mountains around the world (Kollmair et al. 2005); they now cover more than 5.6 million km² of the world's mountains—16.5 percent of the total mountain area outside Antarctica (Rodríguez-Rodríguez et al. 2011). Nevertheless, nearly 1 million km² of this total is in Greenland National Park, and protected areas have often been established where productivity and human use are low (Chape et al. 2008).

For about the first century of the protected area movement, "protection" often meant that local people were largely excluded from national parks. In some places, including Yellowstone, the Swiss National Park, the Gorge of Samaria on the island of Crete, and very often in developing countries, local people were encouraged to leave or were physically removed when parks were established, and their villages were left to decay or were destroyed. For some indigenous groups, this led to their near extinction (Higgins-Zogib et al. 2010). Since the 1980s, there has been increasing recognition that protected areas cannot be managed as "islands" separate from the surrounding landscape, and that the customary practices of local people can be complementary to, or even enhance, conservation goals (Stevens 1997; Stolton and Dudley 2010). One example is Thung Yai Naresuan Sanctuary, a World Heritage Site in Thailand, from which local Karen people were expelled, but later allowed to return (Thongmak and Hulse 1993). Twenty-first century conservation increasingly takes a wider, regional perspective, emphasizing both the conservation of biological diversity and the benefits of protection for local people (Stolton and Dudley 2010). This is becoming even more important in the context of climate change; one of many reasons why networks and corridors linking protected areas are being proposed within, and linking, many mountain regions (Worboys et al. 2010). This shift toward regional approaches means that the management of protected areas, key species, and their habitats should no longer be undertaken just by government agencies or organizations focusing on nature conservation (Borrini-Feyerabend et al. 2004).

PARTNERSHIPS FOR CONSERVATION AND DEVELOPMENT

To ensure the long-term survival of the species, ecosystems, and landscapes within a particular area, local people—and other *stakeholders* such as local governments, nongovernmental organizations, private companies, and, in some cases, the military—must be involved in their designation and management (Borrini-Feyerabend et al. 2004). Broad stakeholder involvement is a trend in mountain protected areas around the world, including the Amarakeri Communal Reserve in Peru (Alvarez et al. 2008), Simen Mountains National Park in Ethiopia (Hurni et al. 2008, Fig. 12.11), the Swiss Alps Jungfrau-Aletsch World Heritage Site (Wallner and Wiesmann 2009), and many mountain biosphere reserves around the world (Austrian MAB Committee

FIGURE 12.11 Trekking group in the Simen Mountains, a World Heritage Site in northern Ethiopia. Tourism can strengthen conservation aims if it is guided by clear rules and if it involves the local population as service providers or beneficiaries. (Photo by H. Hurni.)

2011). In addition, the emerging trend of privately managed protected areas should also be noted, as, for example, in the mountains of Central America (Chacon 2005), the Sierra Nevada of California (Balsiger 2010), and the mountains of Catalonia, Spain (Rafa 2004).

One characteristic of traditional mountain societies is their internal cooperation, built on *partnerships* involving different players in the common interest. Yet, in many mountain regions, such cooperative structures have decayed, or been marginalized, as regional and national governments have taken away many of the rights and responsibilities of mountain people over their resources, often giving them to government agencies staffed by people—often coming from far away—with scientific and technical training. The knowledge of local people, based on centuries of experience, has been ignored, and they have been excluded from using the resources of protected areas. This has sometimes led to the danger of losing some of the qualities for which an area was designated. For example, after a national park was declared in Poland's Tatra Mountains, local people were not allowed to graze their animals. Consequently, some rare plant species on formerly grazed meadows began to decline in number because they were shaded out by taller plants that had previously been grazed down. Once this was recognized, farmers were invited to bring their animals back into the park—and the populations of the rare plants recovered. Similarly, in the Vanoise National Park in the French Alps, local people are now paid to mow species-rich meadows, replacing grazing animals. In such ways, conservation is linked to the continuation of local practices and employment possibilities. These examples underline the key role of continued human intervention in maintaining the cultural landscapes that are essential elements of many mountain protected areas. This is also true in developing countries, where the indigenous knowledge, or traditional ecological knowledge, of local people has been recognized through their involvement in the management of resources, as in the Hunstein Range of Papua New Guinea, the Kilum and Ijim Mountains of Cameroon, the Kigezi Highlands of Uganda, and the Venezuelan Andes (Borrini-Feyerabend 1997; Boffa et al. 2005; Llambí et al. 2005).

Recognition of the expertise of local people also extends to environmental monitoring in mountain protected areas, which has involved local people in Western countries for some time (Peine 2004). In developing countries, indigenous knowledge is often complementary to, and sometimes more suitable and adaptable than, Western scientific methods. It may therefore be more appropriate to use indigenous expertise to monitor the success of nature conservation measures than expensive expatriate consultants or scientific equipment (Higgins-Zogib et al. 2010). Thus, cultural heritage is preserved and people gain employment as stewards of their own landscape. There can also be other incentives for local people to work in the long-term interests of conservation. For instance, tourism can contribute to poverty reduction by providing employment opportunities, which further contribute not only to biodiversity conservation, but also to the social responsibility commitments of national parks, as at Table Mountain National Park, South Africa (Ferreira 2011). Entrance fees can also be reinvested in local community facilities, as for instance in Bwindi National Park in Uganda, where the park authorities have also provided local people with bamboo rhizomes to plant on their farms (Borrini-Feyerabend 1997).

A further element in building partnerships is to ensure that all partners have access to the same information. For instance, in developing the master plan for Flathead County, Montana, residents used geographic information systems (GIS) in a privately funded process to select between various alternative futures (Culbertson et al. 1996). Similarly, the indigenous people of the Colombia's Sierra Nevada de Santa Marta were provided with GIS and trained to use them to map their landscape and evaluate the implications of different potential land uses (Johnson et al. 2001). Equally, when the dominant national language is different from the languages of mountain people, documents need to be made available in these languages, in forms that can easily be understood, and public meetings should be held in these languages and in appropriate cultural settings.

FIGURE 12.12 Provision of a sustained flow of freshwater is probably the most important contribution of mountain areas to global development. Huang He River in China. (Photo by H. Liniger.)

For government agencies and their staff, all of these examples require that they consider local goals and values, as well as those of the wider society, with a shift from "expert management" to management in partnership. This is usually not easy; it requires giving away power and authority, and recognizing that local people have complementary knowledge and the right to be involved in managing their region (Lockwood 2010). Overall, these examples also show that the conservation of biodiversity, just like tourism, is a complex use of resources with not just environmental, but also many economic, social, and cultural implications at many scales. The mere fact that people, wildlife, fires, diseases, and air pollution regularly cross administrative and ecological boundaries means that conservation has to be implemented at the regional scale. However, like tourism, the management of protected areas and conservation as a whole have to be considered within the broader context of SMD.

Water

As sustainable mountain development concerns both mountain regions and populations that depend on them, the water flowing from mountain areas is perhaps the most important element of SMD at the global scale (Fig 12.2). Mountains are often described as *water towers*—the sources of freshwater for billions of people (Viviroli et al. 2007). All of the world's major rivers originate in the mountains. Between a third and a half of all freshwater flows come from mountain areas; more than half of humanity relies on mountain water for drinking, domestic use, fisheries, hydroelectricity, industry, irrigation, recreation, or transportation (Bandyopadhyay et al. 1997; Viviroli and Weingartner 2008). Mountain and lowland people have long recognized the importance of mountains as sources of water by worshipping them as the home of deities and the source of clouds and rain that feed springs and rivers (Liniger and Weingartner 1998).

While mountain areas occupy only relatively small proportions of most river basins, a relatively large proportion of the precipitation falls in them due to the *orographic effect:* As air rises over the mountains, it cools, releasing the moisture it holds, as discussed in Chapter 3. The greater height of the mountains is important not only for triggering precipitation, but also because temperature decreases with altitude. Consequently, there is less evaporation once the precipitation has fallen, and it is also more likely to fall as snow than as water—one reason that many of the world's mountains have names that mean, or include, the local word for "white." For people living in the lowlands below, the storage of winter precipitation as snow or ice is crucial, because when temperatures rise in the spring and summer, the snow and ice melt. The water that is released enters the rivers, flowing downstream at exactly the time when it is most needed in the lowlands, sometimes thousands of kilometers away, for irrigation and other uses. A total of 65 countries use over 75 percent of available freshwater for food production, including China, Egypt, and India, all of which rely heavily on mountain water, much of it from distant mountain regions. The river basins of these 65 countries cover over 40 percent of the global land surface and are home to over 50 percent of the global population (Viviroli et al. 2003).

At a global scale, the mountain regions that are most critically important for providing water for human needs are in the Middle East, South Africa, the western and eastern Himalayas, the mountains of Central Asia, and parts of the Rocky Mountains and the Andes (Viviroli et al. 2007; Viviroli and Weingartner 2008). The greatest importance of mountain rivers is thus in arid and semiarid regions, where mountains are "wet islands"—often the only areas that receive enough precipitation to generate runoff and groundwater recharge (Fig. 12.13). Their importance increases in proportion to the durability and volume of their snowcover and the size of their glaciers (Viviroli et al. 2003). Mountains in semiarid and arid regions typically provide 70–95 percent of the flow to nearby lowlands (Fig. 12.13). For example, the watersheds of the Blue Nile and Atbarah rivers, which rise in

FIGURE 12.13 Valley in the western Pamir, Tajikistan. Agriculture and livelihoods in this semiarid environment depend entirely on irrigation from glacial rivers. (Photo by C. Hergarten.)

the Ethiopian highlands, occupy only 10 percent of the Nile River basin, but contribute 53 percent of the annual inflow to Lake Nasser—and 90 percent of the sediment input. The remainder of the inflow comes mainly from the White Nile, which flows from the mountains of East Africa (Hurni 1998). The semiarid Indus Basin in Pakistan, one of the largest irrigation areas in the world, which ensures the country's food supply and generates 23 percent if its GDP, derives 80 percent of its waters from the Hindu Kush–Himalaya (Immerzeel et al. 2010). In Central Asia, the Tien Shan and Pamir Mountains occupy only 38 and 69 percent, respectively, of the area of the basins of the Syr Darya and Amu Darya Rivers, which used to provide 95 percent of the water flowing into the Aral Sea. The sea has been rapidly drying since the 1960s, and split into 4 lakes in 2009, having lost 88 percent of its surface and 92 percent of its volume. The main cause of this decline was excessive irrigation development. In a bid to restore some of the water body, a dam was completed in 2005 and has been successful in reestablishing the smaller northern part of the Sea (Micklin 2010).

FIGURE 12.14 Owens Valley, California. The city of Los Angeles depends heavily on mountain water resources that have to be conducted over great distances. (Photo by P. Walther.)

It is not only in dry parts of the world that mountain water is essential to life and economies. Even in temperate humid areas, mountain water contributes 30–60 percent of the water flowing to the lowlands (Viviroli et al. 2003). The Alps cover only 23 percent of the area of the Rhine River basin, but they provide half the total flow over the year, varying from 30 percent in winter to 70 percent in summer. It is only in the humid tropics, where the lowlands receive at least 1,500 to 2,000 mm (600 to 800 in.) of annual precipitation, that the contribution of water from mountain areas is insignificant. Yet it should be noted that, in spite of the apparent accuracy of the figures cited above, our knowledge of the amounts and distribution of precipitation falling in, and the runoff deriving from, mountain areas around the world remains rather inadequate, especially in developing countries. At regional to global scales, this problem is exacerbated by the lack of availability of much of the data collected at national levels (Viviroli et al. 2007).

HARVESTING MOUNTAIN WATER

While most mountain water is used for lowland agriculture, it is also vital for mountain agriculture. *Irrigation systems* are found in mountain areas around the world, storing water and directing it to fields at the right time and place to allow crops to grow and optimize yields. The simplest systems involve blocking streams and allowing the water to flood over meadows, as for traditional hay meadows in the Alps (Netting 1981). More complex systems involve constructing channels to bring water from high springs and streams to the fields. These systems—such as the bisses of Valais in the Swiss Alps (Crook and Jones 1999) and the channels in the Pamir in Tajikistan and of the Pokot on the western rim of Kenya's Rift Valley—can extend for tens of kilometers, sometimes including channels that have been blasted from rock faces, or are made of planks suspended around cliffs. The most complex are the underground systems—*qanat* or *khettara*—a technology probably first developed about 2,500 years ago in Iran, then spread eastward to Afghanistan and westward through North Africa and across the Mediterranean to Cyprus (Cech 2009). They include water collection systems, storage reservoirs and cisterns, and underground pipes that carry the water to the fields. While these systems minimize evaporation, they also require high inputs of labor to build and maintain. Though many, some centuries old, are still in use in Iran, the Middle East, and North Africa, many others have fallen into disrepair because the required manpower is no longer available, or pumped groundwater is more easily acquired. Another traditional technology is spate irrigation, using floodwater from seasonal rainfall in mountains to irrigate fields in the piedmont areas, as practiced in Baluchistan in southwestern Pakistan, and in Yemen and Eritrea (Mehari et al. 2004).

A much more recent technology, fog water collection or fog harvesting, is used to *harvest water* in some of the world's driest mountain areas, such as Chile's Atacama Desert (Suau 2010). The water contained in the clouds rising over such mountains, especially in the afternoon and at night, does not always condense into rain, especially where there is little vegetation. However, by erecting high fences of polypropylene mesh, or "fog catchers"—or by planting stands of trees—this water can be harvested, stored, and piped to villages (Gischler and Fernandez 1984). Installation and maintenance costs are low. Similar projects have been installed in Cape Verde, Ecuador, Peru, Namibia, Oman, Yemen, and Eritrea. In

FIGURE 12.15 The reservoir of Lake Oberaar in the Swiss Alps. Here, as in mountain regions generally, hydropower is generated mainly to serve the needs of downstream areas. Mountain regions ought to be compensated adequately for such services. (Photo by M. F. Price.)

other dry areas, drip irrigation—another modern technology, though not mountain-specific—can increase the efficiency of water use and expand the cultivated area, so that smallholders can increase their food security and also grow crops for sale.

POWER FROM WATER

The steep gradients of mountain rivers and streams mean that they have great potential for generating energy. The simplest technology is the water mill, developed centuries ago and initially used mainly for grinding grain. The great advantage of traditional water mills is that they can be constructed from local materials, and are easily maintained. They are used around the world: In the Himalaya and Hindu Kush, for example, about 200,000 grind grain in mountain villages (Rijal 1999). A more recent innovation is upgrading these traditional mills to provide electricity. The simplest method is to fix a bicycle wheel to the grinding stone, so that as the stone rotates, the wheel can drive a belt to charge a battery using an alternator; again, all the components are locally available (Eagle and Olding 2001). To produce more electricity—and also to mill grain—turbines can be installed in streams and rivers. Simple and relatively inexpensive systems are now widely available (Agarwal 2006). The greatest challenge to making them economically viable is to ensure that the electricity, generated 24 hours a day, is used not just in the evenings for lighting and television, but also during the day to promote the local economy—for example, for small-scale industries such as mechanical workshops, processing agricultural products, handicrafts, and telecommunications. In areas with many tourists, the electricity can be used for cooking, decreasing demands on local forests for firewood.

Such small-scale initiatives are one end of the spectrum of *hydroelectricity* generation in the mountains. At the other end are large projects, often with sequences of large dams, such as the cascade of dams on the Columbia River system in Canada and the United States, one of the world's most altered river systems. *Hydropower* provides about 20 percent of the global electricity supply, in over 150 countries (Hailun and Zheng 2009). While the global potential for hydropower development is huge, levels of development vary greatly. For example, the countries of the Alps have developed 76 percent of their hydropower potential (Romerio 2008; Fig. 12.15). Nepal and Ethiopia have developed less than 1 percent of theirs (Mountain Agenda 2001). Currently, China, alongside India, Brazil, Pakistan, and Vietnam, has particularly high rates of dam construction, driven by its rapidly growing economy and based on its hydro resources, which are among the richest on Earth (Yonghui et al. 2006). Chinese hydropower development includes small, decentralized plants for rural electrification as well as large facilities, such as the Three Gorges scheme, providing power to urban centers and industries. Such large projects primarily benefit lowland people and economies, who often gain not only electricity, but also water for irrigation, flood control, and more reliable navigation. Mountain people often lose land—sometimes the most valuable for agriculture (Tefera and Sterk 2008)—as well as transport routes, requiring them to move when their settlements are flooded. While some people close to power stations may obtain the resulting electricity at reduced prices, many do not have access to this resource, especially in the

FIGURE 12.16 Painted by a monk from the local monastery, Salleri, Nepal, this picture symbolizes the integration of local culture and modern technology, in this case small hydropower. The plant is at the bottom centre of the painting. (Painting courtesy of ITECO Engineering, Switzerland.)

developing world. It is therefore often hard to argue that such large projects contribute to SMD. At a global scale, they have faced increasing obstacles (World Commission on Dams 2000), largely based on social motives in developing countries, such as inadequate compensation for resettled communities; and on environmental motives in industrialized countries, such as the fear of destroying the few remaining pristine mountain landscapes, or rare and endangered mountain ecosystems.

SHARING THE BENEFITS OF MOUNTAIN WATER

Sharing the benefits of mountain water has always been a challenge for all water users. For example, irrigation systems are usually constructed by an entire community, as these relatively large projects require major labor inputs and benefit everyone. Subsequently, *institutions* and sets of rules are necessary to ensure that the systems are maintained and the benefits fairly distributed (Shivakoti and Ostrom 2001; Kiteme et al. 2008). Generally, the village council or elders appoint someone to be responsible for maintaining the irrigation channels and opening the sluices according to a strict rotation to ensure the fair distribution of water. The rules governing distribution can be quite complicated. When several villages share a system, as in Pakistan's Hunza Valley, each village appoints a guardian for the purpose (Kreutzmann 1988). In traditional societies, payment for such services is often in crops and livestock, rather than money. Similarly, water mills are also often constructed communally, and each family has the right to a specified number of hours to grind their grain. User-managed systems consistently outperform agency-managed systems,

largely due to more effective social control and more stringent imposition of sanctions within the user communities (Ostrom 2009).

While small-scale hydroelectricity projects are mainly developed to benefit local people (Figs. 12.16, 12.17), they often impose negative impacts similar to large-scale projects, as mentioned above, although mutually beneficial arrangements can be developed between power-generating companies and irrigation institutions, as in the Swiss Alps (Crook 2001). An early example of the need to compensate mountain people for providing downstream benefits was the 1916 law in Switzerland, which entitles communities to substantial annual payments and quotas of free energy for granting the rights to hydropower development on their land (Mauch and Reynard 2002). Developing countries are also introducing similar compensation mechanisms (McNeely 2009). For instance, under Costa Rica's 1996 Forest Law, companies generating hydroelectricity from mountain water, or breweries depending on high water quality, pay mountain landowners for the appropriate management of watersheds (Pagiola 2008).

At even larger scales, it is notable that 214 river basins, home to 40 percent of the world's population and covering more than 50 percent of the global land area, are shared by two or more countries (Wolf 2002). Many have assumed that this would lead to increasing tension and conflict between nations, especially in water-scarce basins such as those of the Euphrates, Ganges, Jordan, and Nile, which all originate in mountains. Such prophecies have not proved accurate; if at all, water use conflicts occur within rather than between states. Since the early 1950s, only 37 acute international disputes have occurred, most between Israel

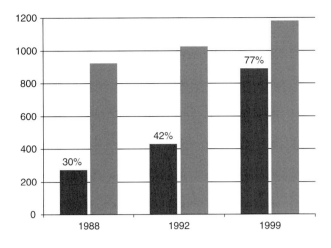

FIGURE 12.17 Initially designed to combat deforestation by replacing fuelwood with electric power, the small local hydropower plant in Salleri, Nepal, has been supplying energy to a steadily increasing number of households—thus presenting a case of successful local innovation. (T. Kohler, based on data from ITECO Engineering, Switzerland.)

and its neighbors, while over the same period more than 150 treaties were signed worldwide (Delli Priscoli and Wolf 2009; Weinthal and Vengosh 2011). However, the threat of conflicts over water both between and within states cannot be ruled out as populations grow, demands for all uses of water increase, and climate change leads to changes in both the timing and amount of precipitation and generally to increased evaporation. In this century, appropriate technologies and new institutions will be needed at all scales, from villages to international regions, to ensure that the benefits of mountain water are shared as fairly as possible among both the guardians of the water towers and people living downstream.

Implementing Sustainable Mountain Development

Over the decade following UNCED, interest in SMD grew in the very different contexts of the world's mountain regions, which include—in economic terms—some of the richest and some of the poorest communities in the world. One type of evidence for this growing awareness is the number of laws, policies, and institutions with a mountain focus that have emerged (Lynch and Maggio 2000; Castelein et al. 2006). While there is no global convention for mountain areas, they are specifically mentioned in three of the global conventions arising from UNCED: the Convention on Biological Diversity, the Convention to Combat Desertification, and the Framework Convention on Climate Change. In 1995, the Global Environment Facility (GEF) identified mountain ecosystems as the subject of one of its ten operational programs. By 2002, GEF had committed over $620 million and leveraged about $1.4 billion of additional funding for at least 107 mountain-related projects in 64 countries (Walsh 2002).

In 1996, the Government of Kyrgyzstan suggested that there should be an International Year focusing on SMD. This was proposed to the United Nations Economic and Social Council (ECOSOC) in 1997, leading to a resolution

to the UN General Assembly, sponsored by 105 member countries, that the year 2002 should be the *International Year of Mountains* (IYM). The General Assembly proclaimed the IYM in 1998, with support from 130 countries, the most ever to support such a resolution (Price and Messerli 2002). The objective of the IYM was to "promote the conservation and sustainable development of mountain regions, thereby ensuring the well-being of mountain and lowland communities" (FAO 2000b), explicitly recognizing the upland–lowland linkages embodied in the concept of SMD. The year 2002 was also the year of "Rio + 10"—the World Summit on Sustainable Development (WSSD), held in Johannesburg, South Africa. As at UNCED 10 years before, the meeting's final document specifically referred to mountains—this time, in paragraph 42 of the Plan of Implementation.

Regional and National Initiatives

In addition to the global initiatives that contribute to SMD in various ways, others have emerged at regional and smaller scales. At the regional scale, in addition to the Alpine Convention, signed in 1991, a Carpathian Convention was signed in 2003 (Quillacq and Onida 2011), and range-wide agreements have been discussed for the Altai, the mountains of southeastern Europe, and the Caucasus. Similarly, new regional institutions and initiatives complemented those existing before UNCED, such as the Andean Community of Nations (Communidad Andina), established in 1969 as the Andean Pact, and currently with four members (Bolivia, Columbia, Ecuador, Peru); various "working communities" in the Alps and the Pyrenees; the International Centre for Integrated Mountain Development (ICIMOD), established in 1983 for the Himalaya–Hindu Kush; and the Consortium for the Sustainable Development of the Andean Ecoregion (CONDESAN), established shortly before UNCED in 1992. These new initiatives also included the African Highlands Initiative (German et al. 2012), established for eastern Africa in 1995 as part of the

Global Mountain Program of the Consultative Group on International Agricultural Research (CGIAR), which also included programs in the Andes and Himalaya. However, these initiatives did not continue into the present decade.

At the national scale, a number of countries have passed laws and/or established policies or institutions with direct relevance for SMD (Castelein et al. 2006). A key issue with all laws and policies is the definition of the area defined as "mountain," a topic that continues to receive significant interest in Europe (Price et al. 2004; European Environment Agency 2010). While France, Greece, Italy, and Switzerland had mountain laws before UNCED, Cuba, Georgia, and Ukraine developed and passed laws subsequent to it; and Bulgaria, Kyrgyzstan, Morocco, Poland, Romania, and the Russian Federation have also begun this process. There is also subnational legislation for mountain areas, for example, for North Ossetia–Alania in the Russian Federation, most of Italy's mountain regions, and the mountains of Catalonia in Spain. The protection and sustainable development of mountain areas are the primary aims of most of these laws, as well as of the Alpine and Carpathian Conventions. Some also underline the importance of maintaining the cultural identity of mountain people; the law for North Ossetia–Alania gives them priority in rights to use natural resources. Other countries have implemented laws or policies that address specific economic sectors—particularly agriculture, tourism, and biological or landscape conservation—in mountain areas (European Environment Agency 2010).

To implement these laws and specific actions in mountain areas, various new institutions have been created, including a National Board for Mountain Regions in Bulgaria; an interministerial commission to promote socioeconomic development in mountain areas in Cuba; and a National Agency for the Mountain Area in Romania. Other countries have created institutions that are mainly concerned with mountain areas, such as the Agency for the Development of Economically Underdeveloped and Frontier Areas in Macedonia, and the Agency for Rural Development and Assuring Sustained Presence of Rural Populations in Slovenia (Price 1999; Castelein et al. 2006); and special funds that contribute to SMD in mountain areas, both generally (e.g., Bulgaria, France, Italy) and with respect to individual sectors. These include agriculture (e.g., France, Georgia, Italy, Switzerland, and also the European Union), tourism (e.g., France, Georgia, Ukraine, Vietnam), infrastructure (e.g., Bulgaria, France, Georgia, Italy, Switzerland, Ukraine, Vietnam), and environmental protection, especially soil erosion (e.g., Burundi, Italy, Uganda) (Castelein et al. 2006).

Principles for Sustainable Mountain Development

Most of the laws, policies, and institutions with a mountain focus are in Europe; relatively few are in developing countries. However, many GEF-funded and World Bank projects at the subnational, and in some cases national, level have aimed at creating effective institutions in these countries for the conservation of biodiversity, or managing ecosystems in transboundary watersheds, within the broader context of SMD (MacKinnon 2004; Walsh 2002). From 2000 to 2010, the World Bank made investments relevant to SMD of approximately US$66 billion globally (UN General Assembly 2011). Yet, whether in European or developing countries, such national or subnational approaches are often piecemeal, without an overall strategic approach at any scale, and many continue to serve "downstream" interests.

To address these issues, Ives et al. (1997a: 456) proposed an "agenda for sustainable mountain development," with "seven prerequisites for a twenty-first century mountain agenda":

1. Mountain Perspective: The characteristics of mountain people and their environments must not be stereotyped, especially by outside/lowland experts or institutions; development processes must integrate mountain specificities, and mountain people must be respected as equal partners in development, possessing key cultural and physical resources;

2. Mountain Reciprocity: Lowland people have vital needs for the resources of the mountains and should compensate and work with mountain people to ensure the long-term provision of these resources, with benefits to both lowland and mountain communities;

3. Mountain Devastation: International organizations and states must work together to bring an end to warfare, the greatest obstacle to SMD—a goal that the authors recognized as possibly utopian;

4. Mountain Hazards: Mountain people are vulnerable to natural and human-induced hazards, as well as those deriving from warfare and associated risks to welfare, and these vulnerabilities vary greatly with respect to "age, gender, ethnicity, affluence, and political influence" (p. 457). Socioeconomic factors are at least as important root causes of these risks as geophysical factors, and therefore inter- and transdisciplinary approaches are needed;

5. Mountain Awareness: "United Nations and national governmental agencies, private foundations, universities, non-governmental organizations worldwide, are proclaiming the importance of mountains as vital to world security" (p. 457); but will the benefits outweigh the costs, and how will the mountain perspective be included in institutional agendas?

6. Mountain Knowledge and Research: Basic, applied, and participatory research—and particularly the dissemination of results, which often remain in the "gray" literature—are all needed; inter- and

transdisciplinary approaches are essential and perhaps a "policy-sensitive science of montology" (p. 460) is needed;

7. Mountain Policy: Workable policies informed by the best possible science are needed, sensitive to the trade-offs of sustainable development and open to innovations; both "grassroots-focused" policy directions, complementing indigenous resources, and "macro-focused" policy directions, based on "equitably-oriented ecological economics" (p. 461), may be appropriate.

Reflecting an increasing emphasis on policies for SMD, in 2002, Mountain Agenda produced a report as a contribution to both IYM and WSSD, outlining seven key principles for policies to foster SMD (Mountain Agenda 2002):

1. Recognition of mountains as important and specific areas of development: In contrast to the typical political marginalization of mountain areas, in which policies are largely externally driven, they need to be recognized as equal partners in development, a strategy that requires decentralization, the building of local institutions, recognition of local rights to natural resources, and platforms/networks to give a "voice" to mountain populations;

2. Restitution for goods and services rendered to surrounding areas: Starting by defining the rights and obligations of mountain communities to natural resources, political discussion is needed to establish their rights to be compensated for the goods and services they contribute downstream—literally or figuratively—and compensatory mechanisms (e.g., investment and compensation funds, subsidies, royalties, taxes) need to be developed and implemented;

3. Diversification and the benefit of complementarities: Multifunctionality has long been a key principle of mountain economies, providing stability in marginal environments. Modernization, leading to monofunctionality, is often not realistic, so that policies should build on the strength of diversified systems with characteristics including the use of land at different altitudes over the year; the production, processing, and sale of quality products; decentralized infrastructure; complementary income from different sectors (e.g., agriculture, small industry, tourism); and seasonal migration;

4. Taking full advantage of the local potential for innovation: If employment and income opportunities are unavailable, the younger and more innovative people leave. To foster innovation, both basic education and targeted education that is both based on the needs of mountain areas and takes into account developments in the wider world are needed; curricula should be developed jointly by both local and external individuals and institutions, both public and private. To retain trained people, both basic and financial and business-related services must be available;

5. Cultural change without loss of site-specific knowledge: Mountain cultures have always been changing; the challenge is how to allow exposure to new ideas while avoiding tension, conflict, and the loss of long-held values and practices relevant to SMD. Governments can officially recognize the value of cultural diversity and strengthen mountain societies through sectoral policies, opportunities for self-determination, and awareness raising;

6. Conservation of mountain ecosystems and early warning function: The steep altitudinal gradients of mountain areas are reflected in their high biodiversity; but this also makes them vulnerable to global change, including both climate change and changes resulting from local human activities. Conservation and development should be undertaken jointly. Mountain ecosystems are also unique because they occur in all of the Earth's climatic zones, and thus provide a global early warning system for global change;

7. Institutionalizing sustainable development of mountain areas: National-level institutions need to be established in order to develop and implement sectoral and/or integrated policies that support SMD through means including broad stakeholder involvement, exchange of information, and awareness raising (Fig. 12.18).

These two sets of principles overlap significantly, forming part of a larger debate on SMD that includes some of the background papers prepared for the Bishkek Global Mountain Summit, the culminating global event of the IYM (Parvez and Rasmussen 2004; Pratt 2004; Starr 2004) and more recent proposals for strategic action (e.g., Messerli 2012). One key theme running through these documents is the need for mountain people to have a greater involvement in defining their futures, recognizing especially the many benefits they provide to wider societies and focusing on opportunities rather than constraints. Thus, Parvez and Rasmussen (2004) propose a "sustainable livelihoods" approach (Carney et al. 1999) that is people centered, based on an understanding of how people and communities "make use of their existing human, social, natural, physical and financial capital to adopt livelihood strategies to overcome vulnerability associated with exogenous and endogenous shocks." They note that one of the key attributes of many mountain societies is high levels of social capital (Coleman 1988) based on "cohesive communities with well-established traditions of cooperation and collective work." This is no longer true of all mountain communities, but the roots often remain—and one reason that some people become amenity migrants is to live in such a cohesive

focusing on integrated ecosystem management, partnerships, and innovative funding mechanisms such as payment for environmental services, debt-for-nature swaps, and environmental trust funds (MacKinnon 2004; Walsh 2002). The World Bank, UNEP, UNDP, FAO, and the Asian Development Bank are both GEF Agencies, which manage GEF projects, and members of the Mountain Partnership, as is PlaNet Finance, which supports the development of sustainable microfinance institutions worldwide.

From Rio 1992 to Rio 2012 and Beyond

Twenty years after UNCED, the United Nations Conference on Sustainable Development took place, again in Rio de Janeiro. In preparation for this "Rio + 20" conference, members of the Mountain Partnership prepared assessments of progress with regard to SMD, both for regions where there has been considerable attention to the concept and its implementation—such as the Alps, Andes, and Hindu Kush–Himalaya—and others where there has been less progress in this regard to date, such as Central America, Central Asia, Africa, and the Middle East (Maselli 2012). A synthesis of these reports concludes that water supply, food security, and clean energy are likely to become the major challenges for humanity in the near future, and that SMD is key to addressing these global challenges (Maselli et al. 2011). A second global synthesis report (Kohler et al. 2012) focuses on the two themes that dominated the Rio + 20 agenda: green economy (and green energy), and institutions for sustainable development.

In preparation for Rio + 20, the members of the Mountain Partnership focused particularly on the conference's outcome document. In the zero draft, released for comment in January 2012, mountains were mentioned in one paragraph; by June, when the conference took place, the outcome document, *The Future We Want* (UN General Assembly 2012) included three paragraphs (210–212) specifically on mountains. The fact that mountains are mentioned at all resulted from considerable lobbying at all levels and from all major mountain areas worldwide, including a preparatory global conference in Lucerne, Switzerland, organized by the Mountain Partnership and the Swiss Agency for Development and Cooperation (SDC) in October 2011, attended by members of the Secretariat of the Rio + 20 conference (Maselli 2012).

In summary, the last two decades, and particularly the IYM, have been a period of increasing momentum, enthusiasm, and attention for SMD, during which diverse organizations and governments have launched and realized many joint initiatives in favor of mountain regions, their inhabitants, and their resources (Rudaz 2011; Messerli 2012). Perhaps this is appropriate, because cooperation is a distinguishing characteristic of mountain societies; in such uncertain environments, it has long been recognized that sharing resources and working together is essential for long-term survival. The integration of mountain areas into regional and global economies has often decreased the effectiveness of such cooperative structures, as the interests of lowland areas—with their much larger weight in terms of political and economic influence and population numbers, or private interests for profit making—come to dominate. In many ways, mountain regions act to magnify the uncertainties of the modern world, of which the most profound worldwide manifestations are the globalization of economies, a looming multiple resource crisis, gloomy outlooks for energy and food supplies, and climate change. A key indicator of the long-term success of the process toward SMD that began at Rio in 1992 will be the number of effective partnerships developed to avoid conflicts and to increase cooperation—both among mountain people themselves and between them and the great diversity of other stakeholders who are concerned with the long-term security of mountain environments and the billions who depend on them.

References

Ackermann, N., Hiess, H., Simon, C., Schreyer, C., Weninger, A., and Zambrini, M. 2006. Future in the Alps: Report. Project Question 4: Leisure, Tourism and Commuter Mobility. Vienna: CIPRA.

Agarwal, S. K. 2006. Re-energizing watermills for multipurpose use and improved rural livelihoods. *Mountain Research and Development* 26: 104–108.

Agence Régional pour l'Environnement/Conseil International Associatif pour la Protection des Pyrénées (ARPE/CIAPP). 1996. *Recommendations of NGOs and Mountain Populations to Governments and to the European Union.* Toulouse: ARPE/CIAPP.

Agrawala, S., ed. 2007. *Climate Change in the European Alps: Adapting Winter Tourism and Natural Hazards Management.* Paris: OECD.

Akramov, K. T., Yu, B., Fan, S. 2010. Mountains, global food prices, and food security in the developing world. IFPRI Discussion Paper 00989. Washington, DC: International Food Policy Research Institute.

Allan, N.J. R. 1986. Accessibility and zonation models of mountains. *Mountain Research and Development* 6: 185–194.

Álvarez, A., Alca, J., Galvin, M., and García, A. 2008. The difficult invention of participation in the Amarakaeri Communal Reserve, Peru. In M. Galvin and T. Haller, eds., *People, Protected Areas and Global Change: Participatory Conservation in Latin America, Africa, Asia and Europe.* Perspectives of the Swiss National Centre of Competence in Research (NCCR) North-South, University of Bern, Vol. 3 (pp. 111–144). Bern: Geographica Bernensia.

Anonymous. 2004. *Programme of Work on Mountain Biodiversity.* Montreal: Secretariat of the Convention on Biological Diversity.

Austrian MAB Committee. 2011. *Biosphere Reserves in the Mountains of the World: Excellence in the Clouds?* Vienna: Austrian Academy of Sciences Press.

Backmeroff, C., Chemini, C., and La Spada, P., eds. 1997. *European Intergovernmental Consultation on Sustainable Mountain Development.* Proceedings of the Final Trento Session. Provincia Autonoma di Trento, Trento.

Balsiger, J. 2010. *Uphill Struggles: The Politics of Sustainable Mountain Development in Switzerland and California.* Köln: Lambert.

Bandyopadhyay, J., Rodda, J. C., Kattelmann, R., Kundzewicz, Z. W., and Kraemer, D. 1997. Highland waters: A resource of global significance. In B. Messerli and J. D. Ives, eds. *Mountains of the World: A Global Priority* (pp. 131–155). Parthenon, UK: Carnforth.

Banskota, M., and Karki, A. S., eds. 1995. *Sustainable Development of Fragile Mountain Areas of Asia.* Kathmandu: International Centre for Integrated Mountain Development.

Barthlott, W., Mutke, J., Rafiqpoor, M. D., Kier, G., and Kreft, H. 2005. Global centres of vascular plant diversity. *Nova Acta Leopoldina* 92: 61–83.

Bärtschi, S. 1998. The Aral Sea basin: Overuse of mountain water resources. In *Mountain Agenda. Mountains of the World: Water Towers for the 21st Century.* Bern: Mountain Agenda.

Baumgartner, M. F., Spreafico, M., and Weiss, H. W. 2001. Hydropower and conflicts over water in Central Asia. In *Mountain Agenda. Mountains of the World: Mountains, Energy and Transport.* Bern: Mountain Agenda.

Bayfield, N. G., McGowan, G. M., and Fillat, F. 2000. Using specialists or stakeholders to select indicators of environmental change for mountain areas in Scotland and Spain. *Oecologia Montana* 9: 29–35.

Becker, A., and Bugmann. H. 2001. *Global Change and Mountain Regions: The Mountain Research Initiative.* Stockholm: IGBP Secretariat.

Bhattarai, K., Conway, D., and Shrestha N. 2005. Tourism, terrorism and turmoil in Nepal. *Annals of Tourism Research* 32: 669–688.

Björnsen Gurung, A., Wymann von Dach, S., Price, M. F., Aspinall, R., Balsiger, J., Baron, J. S., Sharma, E., Greenwood, G. and Kohler, T. 2012. Global change and the world's mountains: Research needs and emerging themes for sustainable development. *Mountain Research and Development* 32(S1): 47–54.

Boffa, J. M., Turyomurugyendo, L., Barnekow-Lillesø, J. P., and Kindt, R. 2005. Enhancing farm tree diversity as a means of conserving landscape-based biodiversity: Insights form the Kigezi Highlands, southwestern Uganda. *Mountain Research and Development* 25: 212–217.

Boller, F., Hunziker, M., Conedera, M., Elsasser, H., and Krebs, P. 2010. Fascinating remoteness: The dilemma of hiking tourism development in peripheral mountain areas: Results of a case study in southern Switzerland. *Mountain Research and Development* 30: 320–331.

Borrini- Feyerabend, G., ed. 1997. *Beyond Fences: Seeking Social Sustainability in Conservation.* Gland: IUCN.

Borrini-Feyerabend, G., Kothari, A., and Oviedo, G., eds. 2004. *Indigenous and Local Communities and Protected Areas Guidelines: Guidance on Policy and Practice for Co-managed Protected Areas and Community Conserved Areas.* Gland: IUCN.

Borsdorf, A., and Braun, V. 2008. The European and global dimension of mountain research: An overview. *Revue de géographie alpine/Journal of Alpine Research* 96(4): 117–129.

Brandon, K., Da Fonseca, G. A. B., Ryland, A. B., and Cardoso de Sila, J. M. 2005. Introduction: Special section: Brazilian conservation: Challenges and opportunities. *Conservation Biology* 19: 595–600.

Bürli, M., Aw-Hassan, A., and Lalaoui Rachidi, Y. 2008. The importance of institutions in mountainous regions for accessing markets. *Mountain Research and Development* 28: 233–239.

Byers, A. 2005. Contemporary human impacts on alpine ecosystems in the Sagarmatha (Mt. Everest) National Park, Khumbu, Nepal. *Annals of the Association of American Geographers*, 95: 112–140.

Byers, E. 1998. The Mountain Forum: Networking for mountain conservation and development. *Unasylva* 49(195): 13–19.

Byers, E., Camino, A., Price, M. F., and Nelson, S. 2001. Using communications technologies to promote access for mountain people. In *Mountain Agenda. Mountains of the world: Mountains, Energy and Transport.* Bern: Mountain Agenda.

Carney, D., Drinkwater, M., Rusinow, T., Neefjes, K., Wanmali, S., and Singh, N. 1999. *Livelihood Approaches Compared: A Brief Comparison of the Livelihoods Approaches of the UK Department for International Development (DFID), CARE, Oxfam and the UNDP.* London: Department for International Development.

Castelein, A., Dinh, T. T. V., Mekouar, M. A., and Villeneuve, A. 2006. *Mountains and the Law: Emerging Trends.* Rome: Food and Agriculture Organization of the United Nations.

Cech, T. V. 2009. *Principles of Water Resources.* New York: John Wiley & Sons.

Chacon, C. M. 2005. Fostering conservation of key priority sites and rural development in Central America: The role of private protected areas. *Parks* 15(2): 39–47.

Chambers, R. 2005. Participation, pluralism, and perceptions of poverty. Conference paper for the International Conference: The Many Dimensions of Poverty, Brasilia, 29–31 August 2005.

Chape, S., Spalding, M., and Jenkins, M. 2008. *The World's Protected Areas: Status, Values and Prospects in the 21st Century.* Cambridge, UK: UNEP World Conservation Monitoring Centre and University of California Press.

Cheryll, R. R., and Soriano, C. R. R. 2007. Exploring the ICT and rural poverty reduction link: Community telecenters and rural livelihoods in Wu'an, China. *Electronic Journal of Information Systems in Developing Countries* 32(1): 1–15.

Chhettri, N., Sharma, E., Deb, D. C., and Sundriyal, R. C. 2002. Impact of firewood extraction on tree structure, regeneration and woody biomass productivity in a trekking corridor of the Sikkim Himalaya. *Mountain Research and Development* 22: 150–158.

Clark, T., and Lichtman, P. 1994. Rethinking the "vision" exercise in the Greater Yellowstone Ecosystem. *Society and Natural Resources* 1: 459–478.

Cole, V., and Sinclair, A. J. 2002. Measuring the ecological footprint of a Himalayan tourist center. *Mountain Research and Development* 22: 132–141.

Coleman, J. S. 1988. Social capital in the creation of human capital. *American Journal of Sociology* 94(Suppl.): S95–S120.

Creighton, M. R. 1995. Japanese craft tourism: Liberating the crane wife. *Annals of Tourism Research* 20: 463–478.

Crook, D. S. 2001. The historical impacts of hydroelectric power development on traditional mountain irrigation in the Valais, Switzerland. *Mountain Research and Development* 21: 46–53.

Crook, D. S., and Jones, A. M. 1999. Design principles from traditional mountain irrigation systems (bisses) in the Valais, Switzerland. *Mountain Research and Development* 19: 79–99.

Culbertson, C., Snyder, D., Mullen, S., Kane, B., Zeller, M., and Richman, S. 1996. Finding common ground in the last best place: The Flathead County, Montana, master plan. In M. F. Price, ed., *People and Tourism in Fragile Environments* (pp. 139–158). Chichester, UK: Wiley.

Debarbieux, B., and Price, M. F. 2008. Representing mountains: From local and national to global common good. *Geopolitics* 13: 148-168.

Debarbieux, B., and Price, M. F. 2012. Mountain regions: A global common good? *Mountain Research and Development* 32(S1): 7–11.

de Haas, H. 2007. The impact of international migration on social and economic development in Moroccan sending regions: A review of the empirical literature. Working Paper 3. Oxford, UK: International Migration Institute.

Delli Priscoli, J., and Wolf, A. T. 2009. *Managing and Transforming Water Conflicts*. Cambridge, UK: Cambridge University Press.

Denniston, D. 1995. *High Priorities: Conserving Mountain Ecosystems and Cultures*. Worldwatch Paper 123. Washington, DC: Worldwatch Institute.

Dérioz, P., and Bachimon, P. 2009. Preface: Mountain tourism and sustainability. *Journal of Alpine Research* 97(3): 1–14.

Eagle, N., and Olding, B. 2001. The watermill battery charger. *Boiling Point* 46: 35–36.

Ekins, P., and Max-Neef, M. 1992. *Real-life Economics: Understanding Wealth Creation*. London: Routledge.

Epprecht, M., and Heinimann, A., eds. 2004. Socioeconomic atlas of VietNam: A depiction of the 1999 population and housing census, Swiss National Centre of Competence in Research (NCCR) North-South. Bern: University of Bern.

European Commission. 2004. *Mountain Areas in Europe: Analysis of Mountain Areas in EU Member States, Acceding and Other European Countries*. Brussels: Directorate-General for Regional Policy, European Commission.

European Environment Agency. 2005. *Vulnerability and Adaptation to Climate Change in Europe*. Technical Report 7/2005. Copenhagen: European Environment Agency.

European Environment Agency. 2010. *Europe's Ecological Backbone: Recognising the True Value of Our Mountains*. Copenhagen: European Environment Agency.

Ferreira, S. L. 2011. Balancing people and park: Towards a symbiotic relationship between Cape Town and Table Mountain National Park. *Current Issues in Tourism* 14(3): 275–293.

Food and Agriculture Organization of the United Nations (FAO). 1996. *Criteria and Indicators for Sustainable Mountain Development*. Internal Report. Rome: FAO.

Food and Agriculture Organization of the United Nations (FAO). 2000a. Integrated planning and management of land resources: Report of the Secretary-General, Addendum: Sustainable mountain development. Document E/CN.17/2000/6/Add.3. New York: Economic and Social Council, Commission on Sustainable Development, United Nations.

Food and Agriculture Organization of the United Nations (FAO). 2000b. *International Year of Mountains Concept Paper*. Rome: FAO.

Foresta, R. A. 1982. *America's National Parks and Their Keepers*. Washington, DC: Resources for the Future.

Franco-Maass, S., Nava-Bernal, G., Endara-Agramont, A., and Gonzalez-Esquivel, C. 2008. Payments for environmental services: An alternative for sustainable rural development? *Mountain Research and Development*, 28: 23–25.

Gaehwiler, F., and Lamichaney, M. N. 2001. Bridges for rural access in the Himalaya. In Mountain Agenda, *Mountains of the World: Mountains, Energy and Transport*. Bern: Mountain Agenda.

Geneletti, D., and Dawa, D. 2009. Environmental impact assessment of mountain tourism in developing regions: A study in Ladakh, Indian Himalaya. *Environmental Impact Assessment Review* 29: 229–242.

Geoffre, M. 1997. Consultation of NGOs and involved populations. In C. Backmeroff, C. Chemini and P. La Spada, eds., *European Intergovernmental Consultation on Sustainable Mountain Development*. Proceedings of the Final Trento Session, Provincia Autonoma di Trento (pp. 55–57).

German, L. A., Mowo, J., Amede, T., and Masuki, K., eds. 2012. *Integrated Natural Resource Management in the Highlands of Eastern Africa*. London: Routledge.

Gischler, C., and Fernandez, C. 1984. Low-cost techniques for water conservation and management in Latin America. *Nature and Resources* 20(3): 11–18.

Godde, P. M., Price, M. F., and Zimmermann, F. M., eds. 2000. *Tourism and Development in Mountain Regions*. New York: CABI.

Good, R., and Johnston, S. 2004. Ecological restoration of degraded alpine and subalpine ecosystems in the Alps National Parks, New South Wales, Australia. In D. Harmon and G. L. Worboys, eds., *Managing Mountain Protected Areas: Challenges and Responses for the 21st Century* (pp. 306–312). Colledara: Andromeda.

Grau, H. R., and Aide, T. M. 2007. Are rural–urban migration and sustainable development compatible in mountain systems? *Mountain Research and Development* 27: 119–123.

Grêt-Regamey, A., Brunner, S. H., and Kienast, F. 2012. Mountain ecosystem services: Who cares? *Mountain Research and Development* 32(S1): 23–34.

Grötzbach, E. 1988. High mountains as human habitat. In N. J. R. Allan, G. W. Knapp, and C. Stadel, eds., *Human Impact on Mountains* (pp. 24–35). Totowa, NJ: Rowman and Littlefield.

Gurung, C., and DeCoursey, M. A. 2000. Too much too fast: Lessons from Nepal's lost Kingdom of Mustang. In P. M. Godde, M. F. Price, and F. M. Zimmermann, eds., *Tourism and Development in Mountain Regions* (pp. 239–254). New York: CABI.

Haile, A. M., Depeweg, H., and Stillhardt, B. 2003. Smallholder drip irrigation technology: Potentials and constraints in the Highlands of Eritrea. *Mountain Research and Development* 23: 27–31.

Hailun Huang and Zheng Yan. 2009. Present situation and future prospect of hydropower in China. *Renewable and Sustainable Energy Reviews* 13: 1652–1656.

Hak, T., Moldan, B., and Dahl, A. L., eds. 2007. *Sustainability Indicators: A Scientific Assessment*. Washington, DC: Island Press.

Hartmann, P. 2001. Human power instead of machines: Rural access roads in West Flores, Indonesia. In *Mountain Agenda,*

Mountains of the World: Mountains, Energy and Transport. Bern: Mountain Agenda.

Heeks, R., and Kanashiro, L. L. 2009. Telecentres in mountain regions: A Peruvian case study of the impact of information and communication technologies on remoteness and exclusion. *Journal of Mountain Science* 6: 320–330.

Higgins-Zogib, L., Dudley, N., and Kothari, A. 2010. Living traditions: Protected areas and cultural diversity. In S. Stolton and N. Dudley, eds., *Arguments for Protected Areas* (pp. 165–187). London: Earthscan.

Hoermann, B., and Kollmair, M. 2009. *Labour Migration and Remittances in the Hindu Kush–Himalayan Region.* Kathmandu: International Centre for Integrated Mountain Development.

Huddleston, B., Ataman, E., de Salvo, P., Zanetti, M., Bloise, M., Bel, J., Francheschini, L., and Fe d'Ostiani, L. 2003. *Towards a GIS-based Analysis of Mountain Environments and Populations.* Rome: Food and Agriculture Organization of the United Nations.

Hurni, H. 1998. The Nile basin: An international challenge to mountain water management. In *Mountain Agenda. Mountains of the World: Water Towers for the 21st Century.* Bern: Mountain Agenda.

Hurni, H., Abunie, L., Ludi, E., and Woubshet, M. 2008. The evolution of institutional approaches in the Simen Mountains National Park, Ethiopia. In M. Galvin and T. Haller, eds., *People, Protected Areas and Global Change. Participatory Conservation in Latin America, Africa, Asia and Europe.* Perspectives of the Swiss National Centre of Competence in Research (NCCR) North-South, University of Bern, Vol. 3 (pp. 287–324). Bern: Geographica Bernensia.

Immerzeel, W. W., van Beek, L. P. H, and Bierkens, M. F. P 2010. Climate change will affect the Asian water towers. *Science* 328(5984): 1382–1385.

International Livestock Research Institute (ILRI). 1997. *Sustainable Development in Mountain Ecosystems of Africa.* Proceedings of the African Intergovernmental Consultation on Sustainable Mountain Development. Addis Ababa: International Livestock Research Institute.

International Union for the Conservation of Nature (IUCN). 1980. *World Conservation Strategy.* Gland: IUCN.

International Union for the Conservation of Nature/United Nations Environment Programme/World Wildlife Fund (IUCN/UNEP/WWF). 1991. *Caring for the Earth: A Strategy for Sustainable Development.* Gland: IUCN.

Ives, J. D. 1997. Comparative inequalities: Mountain communities and mountain families. In B. Messerli and J. D. Ives, eds., *Mountains of the World: A Global Priority* (pp. 61–84). Carnforth, UK: Parthenon.

Ives, J. D., and Messerli, B. 2004. Mountain geoecology: The evolution of intellectually-based scholarship into a political force for sustainable mountain development. In M. Sala, honorary theme editor, Geography, in *Encyclopedia of Life Support Systems* (EOLSS), developed under the auspices of the UNESCO. Oxford, UK: EOLSS Publishers. <http://www.eolss.net>.

Ives, J. D., Messerli, B., and Rhoades, R. E. 1997a. Agenda for sustainable mountain development. In B. Messerli and J. D. Ives, eds., *Mountains of the World: A Global Priority* (pp. 1–16). Carnforth, UK: Parthenon.

Ives, J. D., Messerli, B., and Spiess, E. 1997b. Mountains of the world: A global priority. In B. Messerli and J. D. Ives, eds. *Mountains of the World: A Global Priority* (pp. 455–466). Carnforth, UK: Parthenon.

Jabareen, Y. 2008. A new conceptual framework for sustainable development. *Environment, Development and Sustainability* 10(2): 179–192.

Jelliffe, D. B. 1966. The assessment of the nutritional status of the community. WHO Monograph 53. Geneva: World Health Organisation.

Jesinghaus, J. 1999. *Indicators for Decision-Making.* Brussels: European Commission.

Johnson, N., Miller K., and Miranda, M. 2001. Bioregional approaches to conservation: Local strategies to deal with uncertainty. In J. W. Handmer, T. W. Norton, and S. R. Dovers, eds., *Ecology, Uncertainty and Policy* (pp. 43–65). Harlow, UK: Pearson Education.

Khoabane, S., and Black, P. A. 2009. The effect of livestock theft on household poverty in developing countries: The case of Lesotho. Stellenbosch Economic Working Papers 02/09. Working Paper of the Department of Economics and the Bureau for Economic Research at the University of Stellenbosch.

Kiteme, B. P., Liniger, H. P., Notter, B., Wiesmann, U., and Kohler, T. 2008. Dimensions of global change in African mountains: The example of Mount Kenya. *IHDP Update* 2: 18–22.

Kohler, T., Hurni, H., Wiesmann U., and Kläy, A. 2004. Mountain infrastructure: Access, communications, and energy. In M. F. Price, L. Jansky, and A. A. Iatsenia, eds., *Key Issues for Mountain Areas* (pp. 38–62). Tokyo: United Nations University Press.

Kohler, T., Pratt, J., Debarbieux, B., Balsiger, J., Rudaz, G., and Maselli, D., eds. 2012. *Sustainable Mountain Development, Green Economy and Institutions: From Rio 1992 to Rio 2012 and Beyond.* Final Draft for Rio 2012. Bern: Swiss Agency for Development and Cooperation (SDC) and Centre for Development and Environment (CDE), University of Bern.

Kollmair, M., Gurung, G. S., Huri, K., and Maselli, D. 2005. Mountains: Special places to be protected? An analysis of worldwide nature conservation efforts in mountains. *International Journal of Biodiversity Science and Management* 1: 181–189.

Körner, C. 2004. Mountain biodiversity, its causes and function. *Ambio,* Special Report 13: 11–17.

Körner, C., Ohsawa, M., Spehn, E., Berge, E., Bugmann, H., Groombridge, B., Hamilton, L., Hofer, T., Ives, J., Jodha, N., Messerli, B., Pratt, J., Price, M., Reasoner, M., Rodgers, A., Thonell, J., and Yoshino, M. 2005. Mountain systems. In R. Hassan, R. Scholes, and N. Ash, eds., *Ecosystems and Human Well-being: Current State and Trends,* Vol. 1: *Millennium Ecosystem Assessment* (pp. 681–716). Washington, DC: Island Press.

Kreutzmann, H. 1988. Oases of the Karakorum: Evolution of irrigation and social organization in Hunza, north Pakistan. In N. J. R. Allan, G. W. Knapp, and C. Stadel, eds., *Human Impact on Mountains* (pp. 243–254). Totowa, NJ: Rowman and Littlefield.

Kreutzmann, H. 2000. Improving accessibility for mountain development: Role of transport networks and urban settlements In M. Banskota, T. S. Papola, and J. Richter, eds., *Poverty Alleviation and Sustainable Resource Management in the Mountain Areas of South Asia* (pp. 485–513). Feldafing: Deutsche Stiftung für internationale Entwicklung.

Kreutzmann, H. 2001. Development indicators for mountain regions. *Mountain Research and Development* 21: 132–139.

Kreutzmann, H. 2004. Accessibility for High Asia: Comparative perspectives on northern Pakistan's traffic infrastructure and linkages with its neighbours in the Hindukush-Karakoram-Himalaya. *Journal of Mountain Science* 1: 193–210.

Kruk, K., Hummel J., and Banskota, K. 2007. *Facilitating Sustainable Mountain Tourism*. Kathmandu: International Centre for Integrated Mountain Development.

Lama, W. B. 2000. Community-based tourism for conservation and women's development. In P. M. Godde, M. F. Price, and F. M. Zimmermann, eds., *Tourism and Development in Mountain Regions* (pp. 221–238). New York: CABI.

Libiszewski, S., and Bächler, G. 1997. Conflicts in mountain areas: A predicament for sustainable development. In B. Messerli and J. D. Ives, eds., *Mountains of the World: A Global Priority* (pp. 103–130). Carnforth, UK: Parthenon.

Lightfoot, C., Gillman, H., Scheuermeier, U., and Nyimbo, V. 2008. The First Mile project in Tanzania: Linking smallholder farmers using modern communication technology. *Mountain Research and Development* 28: 13–17.

Lindner, C. 1997. Agenda 21. In F. Dodds, ed. *The Way Forward Beyond Agenda 21* (pp. 3–14). London: Earthscan.

Liniger, H., and Weingartner, R. 1998. Mountains and freshwater supply. *Unasylva* 49(195): 39–46.

Llambí, L. D., Smith, J. K., Pereira, N., Pereira, A. C., Valero, F., Monasterio, M., and Dávila, M. V. 2005. Participatory planning for biodiversity conservation in the high Tropical Andes: Are farmers interested? *Mountain Research and Development* 25: 200–205.

Lockwood, M. 2010. Good governance for terrestrial protected areas: A framework, principles and performance outcomes. *Journal of Environmental Management* 91: 754–766.

Lynch, O., and Maggio, G. 2000. *Mountain Laws and People: Moving towards Sustainable Development and Recognition of Community-based Property Rights*. Washington, DC: Center for Environmental Law.

Macchiavelli, A. 2009. Alpine tourism: Development contradictions and conditions for innovation. *Journal of Alpine Research* 97(1): 99–112.

MacKinnon, K. 2004. Challenges and opportunities for mountain protected areas: A perspective from the World Bank. In D. Harmon and G. L. Worboys, eds., *Managing Mountain Protected Areas: Challenges and Responses for the 21st Century* (pp. 26–34). Colledara: Andromeda.

Mallari, A. A., and Enote, J. E. 1996. Maintaining control: culture and tourism in the Pueblo of Zuni, New Mexico. In M. F. Price, ed., *People and Tourism in Fragile Environments* (pp. 19–31). Chichester, UK: Wiley.

Maselli, D. 2012. Promoting sustainable mountain development at the global level. *Mountain Research and Development* 32(S1): 64–70

Maselli, D., Kohler, T., Ariza Nino, C., and Greenwood, G. 2011. *Draft Synthesis Report on Progress and Perspectives in Sustainable Mountain Development: From Rio 1992 to Rio 2012 and Beyond*. Prepared for the Lucerne World Mountain Conference, 10–12 October 2011. Revised version, 1 November 2011.

Mauch, C., and Reynard, E. 2002. *The Evolution of the National Water Regime in Switzerland*. Chavanne-pres-Renens: Institut de Hautes Etudes en Administration Publique.

McNeely, J. 2009. Payments for ecosystem services: An international perspective. In K. N. Ninan, ed., *Conserving and Valuing Ecosystem Services and Biodiversity: Economic, Institutional and Social Challenges* (pp. 135–150). London: Earthscan.

McNeill, J. R. 1992. *The Mountains of the Mediterranean World: An Environmental History*. Cambridge, UK: Cambridge University Press.

Mehari, A., Schulz, B., and Depeweg, H. 2004. If and how expectations can be met? An evaluation of the modernized sptate irrigation systems in Eritrea. In *Food Production and Water: Social and Economic Issues of Irrigation and Drainage*. Proceedings of the 55th International Executive Council of the International Commission on Irrigation and Drainage (ICID) and Interregional Conference, 4–11 September 2004, Moscow.

Merz, J., Dangol, P. M., Dhakal, M. P., Dongol, B. S., Nakarmi, G., and Weingartner, R. 2006. Road construction impacts on stream suspended sediment loads in a nested catchment system in Nepal. *Land Degradation and Development* 17: 343–351.

Messerli, B. 2012. Global change and the world's mountains. *Mountain Research and Development* 32(S1): 55–63.

Messerli, B., and Ives, J. D., eds. 1997. *Mountains of the World: A Global Priority*. Carnforth, UK: Parthenon.

Micklin, P. 2010. The past, present, and future Aral Sea. *Lakes & Reservoirs: Research and Management* 15: 193–213.

Mittermeier, R. A., Gil, P. R., Hoffmann, M., Pilgrim, J., Brooks, T., Mittermeier, C. G., Lamoureux, J., and Fonseca, G. A. B. 2005. *Hotspots Revisited*. Chicago: University of Chicago Press.

Montgomery. L. 2002. ICTs in mountains. <http://www.mtnforum.org/emaildiscuss/discuss02/031102277.htm>.

Monz, C. 2000. Recreation resource assessment and monitoring techniques for mountain regions. In P. M. Godde, M. F. Price, and F. M. Zimmermann, eds., *Tourism and Development in Mountain Regions* (pp. 47–68). New York: CABI.

Moss, L. A. G., ed. 2006. *The Amenity Migrants: Seeking and Sustaining Mountains and Their Cultures*. New York: CABI.

Mountain Agenda. 1997. *Mountains of the World: Challenges for the 21st Century*. Bern: Mountain Agenda.

Mountain Agenda. 1998. *Mountains of the World: Water Towers for the 21st Century*. Bern: Mountain Agenda.

Mountain Agenda. 1999. *Mountains of the World: Tourism and Sustainable Mountain Development*. Bern: Mountain Agenda.

Mountain Agenda. 2001. *Mountains of the World: Mountains, Energy and Transport*. Bern: Mountain Agenda.

Mountain Agenda. 2002. *Mountains of the World: Sustainable Development in Mountain Areas: The Need for Adequate Policies and Instruments*. Bern: Mountain Agenda.

Muhweezi, A. B., Sikoyo, G. M., and Chemonges, M. 2007. Introducing a transboundary ecosystem management approach in the Mount Elgon region: The need for strengthened institutional collaboration. *Mountain Research and Development* 27: 215–219.

Mujica, E., and Rueda, J. L., eds. 1996. *El Desarollo Sostenible de Montañas en América Latina*. Lima: CONDESAN/CIP.

Nepal, S. K. 2002. Mountain ecotourism and sustainable development: Ecology, economics, and ethics. *Mountain Research and Development* 22: 104–109.

Nepal, S. K., Kohler, T., and Banzhaf, B. R. 2002. *Great Himalaya: Tourism and the Dynamics of Change in Nepal*. Bern: Swiss Foundation for Alpine Research.

Netting, R. M. 1981. *Balancing on an Alp: Ecological Change and Continuity in a Swiss Mountain Community*. Cambridge, UK: Cambridge University Press.

Neuman, F., Keenan, L., Sherchan, U., Sander, K., Huberman, D., and Karky, B. 2010. Payments for environmental services (PES): An overview of the options and challenges for mountain systems and people. *Mountain Forum Bulletin* 10(1): 5–9.

Nikšić, M., and Gašparović, S. 2010. Geographic and traffic aspects of possibilities for implementing ropeway systems in passenger transport. *Promet* 22(5): 389–398.

Nöthiger, C., and Elsasser, H. 2004. Natural hazards and tourism: New findings on the European Alps. *Mountain Research and Development* 24: 24–27.

Odermatt, S. 2004. Evaluation of mountain case studies by means of sustainability variables. *Mountain Research and Development* 24: 336–341.

Olimova, S., and Olimov, M. 2007. Labor migration from mountainous areas in the Central Asian region: Good or evil? *Mountain Research and Development* 27: 104–108.

Organisation for Economic Cooperation and Development (OECD). 1993. *OECD Core Set of Indicators for Environmental Performance Reviews*. OECD Environment Monographs 83. Paris: OECD.

Ostrom, E. 2009. The contribution of community institutions to environmental problem-solving. In A. Breton, G. Brosio, S. Dalmazzone, and G. Garrone, eds. *Governing the Environment* (pp. 87–112). Cheltenham, UK: Edward Elgar.

Ozenda, P. and Borel, J. L. 2003. The alpine vegetation of the Alps. In L. Nagy, G. Grabherr, C. Koerner, and D. B. A. Thompson, eds., *Alpine Biodiversity in Europe* (pp. 53–64). Berlin: Springer.

Padiukova, P., and Padiukova, O. 2007. Challenges for women's life in the mountain regions of the Kyrgyz Republic. Conference paper, Women of the Mountains Conference, 8–9 March 2007 at Utah Valley State College, Orem, Utah.

Pagiola, S. 2008. Payments for environmental services in Costa Rica. *Ecological Economics* 65: 712–724.

Papola, T. S. 2001. *Poverty in Mountain Areas of HKH Region: Some Basic Issues in Measurement, Diagnosis, and Alleviation*. Kathmandu: International Centre for International Mountain Development.

Parvez, S., and Rasmussen, S. F. 2004. Sustaining mountain economies: Sustainable livelihoods and poverty alleviation. In M. F. Price, L. Jansky and A. A. Iatsenia, eds., *Key Issues for Mountain Areas* (pp. 86–110). Tokyo: United Nations University Press.

Peine, J. D. 2004. Citizen-scientists conduct environmental monitoring in mountain protected areas. In D. Harmon and G. L. Worboys, eds., *Managing Mountain Protected Areas: Challenges and Responses for the 21st Century* (pp. 246–247). Colledara: Andromeda.

Permanent Secretariat of the Alpine Convention. 2011. *Sustainable Rural Development and Innovation*. Innsbruck: Alpine Convention.

Pezzey, J. 1989. Economic analysis of sustainable growth and sustainable development. Environmental Working Paper 15. Washington, DC: World Bank.

Pilgrim, S., Samson C., and Pretty, J. 2010. Ecocultural revitalization: Repleneshing community connections to the land. In S. Pilgrim and J. Pretty, eds., *Nature and Culture: Rebuilding Lost Connections* (pp. 235–256). London: Earthscan.

Porras, I. 2010. *Fair and Green? Social Impacts of Payments for Environmental Services in Costa Rica*. London: International Institute for Environment and Development.

Pratt, D. J. 2004. Democratic and decentralized institutions for sustainability in mountains. In M. F. Price, L. Jansky, and A. A. Iatsenia, eds., *Key Issues for Mountain Areas* (pp. 149–168). Tokyo: United Nations University Press.

Price, M. F. 1992. Patterns of the development of tourism in mountain environments. *GeoJournal* 27: 87–96.

Price, M. F. 1998. Mountains: Globally important ecosystems. *Unasylva* 49(195): 3–12.

Price, M. F. 1999. *Chapter 13 in Action: A Task Manager's Report*. Rome: Food and Agriculture Organization of the United Nations.

Price, M. F. 2003. Sustainable mountain development in Europe. In A. Mather and J. Bryden, eds., *Regional Sustainable Development Review: Europe, Encyclopedia of Life-Support Systems*. Oxford, UK: EOLSS Publishers. <http://www.eolss.net>.

Price, M. F., and Butt, N., eds. 2000. *Forests in Sustainable Mountain Development: A State-of-Knowledge Report for 2000*. Wallingford, UK: CABI International.

Price, M. F., and Hofer, T. 2005. The International Year of Mountains, 2002: Progress and prospects. In D. B. A. Thompson, M. F. Price, and C. A. Galbraith, eds., *The Mountains of Northern Europe: Conservation, Management, People and Nature* (pp. 11–22). Edinburgh: The Stationery Office.

Price, M. F., Jansky, L., and Iatsenia, A. A., eds. 2004. *Key Issues for Mountain Areas*. Tokyo: United Nations University Press.

Price, M. F., and Kim, E. G. 1999. Priorities for sustainable mountain development in Europe. *International Journal of Sustainable Development and World Ecology* 6, 203–219.

Price, M. F., Lysenko, I., and Gloersen, E. 2004. La delimitation des montagnes européennes [Delineating Europe's mountains]. *Revue de Geographie Alpine* 92(2): 61–86.

Price, M. F., and Messerli, B. 2002. Fostering sustainable mountain development: From Rio to the International Year of Mountains, and beyond. *Unasylva* 53(208): 6–17.

Price, M. F., Moss, L. A. G., and Williams, P. W. 1997. Tourism and amenity migration. In B. Messerli, and J. D. Ives, eds., *Mountains of the World: A Global Priority* (pp. 249–280). Carnforth, UK: Parthenon.

Pringle, I., Bajracharya, U., and Bajracharya, A. 2004. Innovating multimedia to increase accessibility in the Hills of Nepal. *Mountain Research and Development* 24: 292–297.

Quillacq, P., and Onida, M., eds. 2011. *Environmental Protection and Mountains: Is Environmental Law Adapted to the Challenges Faced by Mountain Areas?* Innsbruck: Permanent Secretariat of the Alpine Convention.

Rafa, M. 2004. Managing mountain areas in Catalonia (northeastern Spain) beyond protected areas: The role of the

private sector and NGOs. In D. Harmon and
G. L. Worboys, eds., *Managing Mountain Protected Areas: Challenges and Responses for the 21st Century* (pp. 250–254). Colledara: Andromeda.

Ramcharan, R. 2009. Why an economic core: Domestic transport costs. *Journal of Economic Geography* 9: 559–581.

Rieder, P., and Wyder, J. 1997. Economic and political framework for sustainability of mountain areas. In B. Messerli and J. D. Ives, eds., *Mountains of the World: A Global Priority* (pp. 85–102). Carnforth, UK: Parthenon.

Rijal, K., ed. 1999. *Energy Use in Mountain Areas: Trends and Patterns in China, India, Nepal and Pakistan.* Kathmandu: International Centre for Integrated Mountain Development.

Rodríguez-Rodríguez, D., Bomhard, B., Butchart, S. H. M., and Foster, M. N. 2011. Progress towards international targets for protected area coverage in mountains: A multi-scale assessment. *Biological Conservation* 144, doi:10.1016/j.biocon.2011.08.023.

Roe, E. M. 1996. Sustainable development and cultural theory. *International Journal of Sustainable Development and World Ecology* 3: 1–14.

Rogers, P. P., Kazi, F. J., and Boyd, J. A. 2008. *An Introduction to Sustainable Development.* London: Earthscan.

Romerio, F. 2008. Hydroelectric resources between state and market in the Alpine countries. In E. Wiegandt, ed., *Mountains: Sources of Water, Sources of Knowledge* (pp. 83–102). Dordrecht: Springer.

Royal Swedish Academy of Sciences. 2002. *The Abisko Document: Research for Mountain Area Development.* Ambio Special Report. Stockholm: Royal Swedish Academy of Sciences.

Rudaz, G. 2011. The causes of mountains: The politics of promoting a global agenda. *Global Environmental Politics* 11(4): 43–65.

Saffery, A. 2000. Mongolia's tourism development race: Case study from the Gobi Gurvusaikhan National Park. In P. M. Godde, M. F. Price, and F. M. Zimmermann eds., *Tourism and Development in Mountain Regions* (pp. 255–274). New York: CABI.

Schaffner, U., and Schaffner, R. 2001. Access road construction in Ethiopia and Yemen. In Mountain Agenda, *Mountains of the World: Mountains, Energy and Transport.* Bern: Mountain Agenda.

Schickhoff, U. 2001. The Karakorum Highway: Accelerating social and environmental change in a formerly secluded high mountain region. In Mountain Agenda, *Mountains of the World: Mountains, Energy and Transport.* Bern: Mountain Agenda.

Schofield, V, 2010. *Kashmir in Conflict: India, Pakistan and the Unending War.* London: IB Tauris.

Sène, E. H., and McGuire, D. 1997. Sustainable mountain development: Chapter 13 in action. In B. Messerli and J. D. Ives, eds., *Mountains of the World: A Global Priority* (pp. 447–453). Carnforth, UK: Parthenon.

Serquet, G., and Rebetez, M. 2011. Relationship between tourism demand in the Swiss Alps and hot summer air temperatures associated with climate change. *Climatic Change,* doi:10.1007/s10584-010-0012-6.

Shanty, F. G., and Mishra, P. P., eds. 2007. *Organized Crime: From Trafficking to Terrorism*, Vol. 1. Santa Barbara, CA: ABC-Clio.

Shivakoti, G. P., and Ostrom, E., eds. 2001. *Improving Irrigation Governance and Management in Nepal.* Oakland, CA: ICS Press.

Singh, R. B., Mal S., and Kala, C. P. 2009. Community responses to mountain tourism: A case in Bhyundar Valley, Indian Himalaya. *Journal of Mountain Science* 6: 394–404.

Soriano, C. R. R. 2007. Exploring the ICT and rural poverty reduction link: Community telecenters and rural livelihoods in Wu'an, China. *Electronic Journal of Information Systems in Developing Countries* 32: 1–15.

Speziale, K. L., Lambertucci, S. A., and Olsson, O. 2008. Disturbance from roads negatively affects Andean condor habitat use. *Biological Conservation* 141: 1765–1772.

Starkey, P. 2001. Animal power: Appropriate transport in mountain areas. In Mountain Agenda, *Mountains of the World: Mountains, Energy and Transport.* Bern: Mountain Agenda.

Starr, S. F. 2004. Conflict and peace in mountain societies. In M. F. Price, L. Jansky, and A. A. Iatsenia, eds., *Key Issues for Mountain Areas* (pp. 169–180). Tokyo: United Nations University Press.

Steinberg, M., and Taylor, M. 2007. Marginalizing a vulnerable cultural and environmental landscape: Opium poppy production in highland Guatemala. *Mountain Research and Development* 27: 318–321.

Stepp, J. R., Castaneda, H., and Cervone, S. 2005. Mountains and biocultural diversity. *Mountain Research and Development* 25: 223–227.

Stevens, S. F. 1993. *Claiming the High Ground: Sherpas, Subsistence and Environmental Change in the Highest Himalaya.* Berkeley: University of California Press.

Stevens, S. F., ed. 1997. *Conservation through Cultural Survival: Indigenous Peoples and Protected Areas.* Washington, DC: Island Press.

Stolton, S., and Dudley, N., eds. 2010. *Arguments for Protected Areas.* London: Earthscan.

Suau, C. 2010. Fog collection and sustainable architecture in Atacama coast. Proceedings of 5th International Conference on Fog, Fog Collection and Dew, 25–30 July 2010, Münster, Germany.

Tefera, B., and Sterk, G. 2008. Hydropower-induced land use change in Fincha'a watershed, western Ethiopia: Analysis and impacts. *Mountain Research and Development* 28:72–80.

Thongmak, S., and Hulse, D. 1993. The winds of change: Karen people in harmony with World Heritage. In E. Kemf, ed., *The Law of the Mother: Protecting Indigenous People in Protected areas Areas* (pp. 162–168). San Francisco: Sierra Club.

UN General Assembly. 2011. *Sustainable Mountain Development.* Report of the Secretary General.

UN General Assembly. 2012. *The Future We Want.* Resolution 66/288.

UNODC. 2010. *World Drug Report 2010.* Vienna: United Nations Office on Drugs and Crime.

UN World Tourism Organisation (UNWTO). 2011. *UNWTO Tourism Highlights 2011 Edition.* Madrid: UNWTO.

Valaoras, G. 2000. Conservation and development in Greek mountain areas. In P. M. Godde, M. F. Price, and F. M. Zimmermann, eds., *Tourism and Development in Mountain Regions* (pp. 69–83). New York: CABI.

Valaoras, G., Pistolas, K., and Sotiropoulou, H. Y. 2002. Ecotourism revives rural communities: The case of the Dadia Forest Reserve, Evros, Greece. *Mountain Research and Development* 22: 123–127.

Van de Hamsvoort, P. C. M., and Latacz-Lohmann, U. 1998. Sustainability: A review of the debate and an extension. *International Journal of Sustainable Development and World Ecology* 5: 99–110.

Van de Kerk, G., and Manuel, A. R. 2008. A comprehensive index for a sustainable society: The SSI—The Sustainable Society Index. *Ecological Economics* 66: 228–242.

Viviroli, D., Dürr, H. H., Messerli, B., Meybeck, M., and Weingartner, R. 2007. Mountains of the world, water towers for humanity: Typology, mapping and global significance. *Water Resources Research* 43(W07447), doi:10.1029/2006WR005653.

Viviroli, D., and Weingartner, R. 2008. "Water towers": A global view of the hydrological importance of mountains. *Mountains: Sources of Water, Sources of Knowledge. Advances in Global Change Research* 31: 15–20.

Viviroli, D., Weingartner, R., and Messerli, B. 2003. Assessing the hydrological significance of the world's mountains. *Mountain Research and Development* 23: 32–40.

von Dach, S. 2002. Integrated mountain development: A question of gender mainstreaming. *Mountain Research and Development* 22: 236–239.

Wallner, A., and Wiesmann, U. 2009. Critical issues in managing protected areas by multi-stakeholder participation: Analysis of a process in the Swiss Alps. *eco. mont* 1(1): 45–50.

Walsh, S., ed. 2002. *High Priorities: GEF's Contribution to Preserving and Sustaining Mountain Ecosystems*. Washington, DC: Global Environmental Facility.

Weingartner, R. 1998. The Alps: The water tower of Europe. In Mountain Agenda. *Mountains of the World: Water Towers for the 21st Century*. Bern: Mountain Agenda.

Weinthal, E., and Vengosh, A. 2010. Water and conflict: Moving from the global to the local. In R. Parker and M. Sommer, eds., *Routledge Handbook in Global Public Health* (pp. 265–272). New York: Routledge.

Wiesmann, U. 1998. *Sustainable Regional Development in Rural Africa: Conceptual Framework and Case Studies from Kenya*. African Studies 14. Bern: Geographica Bernensia.

Wiesmann, U. 1999. Striking a balance in community-based mass tourism. In Mountain Agenda, *Mountains of the World: Tourism and Sustainable Mountain Development*. Bern: Mountain Agenda.

Wiesmann, U., and Hurni, H., ed. 2011. *Research for Sustainable Development: Foundations, Experiences, and Perspectives*. Perspectives of the Swiss National Centre of Competence in Research (NCCR) Research North-South. University of Bern, Vol. 6. Bern: Geographica Bernensia.

Wolf, A. T. 2002. *Atlas of International Freshwater Agreements*. Nairobi: United Nations Environment Programme.

Worboys, G. L., Francis, W. L., and Lockwood, M., eds. 2010. *Connectivity Conservation Management: A Global Guide*. London: Earthscan.

World Bank. 2010. *Migration and Remittances Fact Book 2010*. Washington, DC: World Bank.

World Commission on Dams. 2010. *Dams and Development: A New Framework for Decision Making*. Report of the World Commission on Dams. London: Earthscan.

World Commission on Environment and Development (WCED). 1987. *Our Common Future*. Oxford, UK: Oxford University Press.

Wymann von Dach, S., Ott, C., Kläy, A., and Stillhardt, B. 2006. Will international pursuit of the Millennium Development Goals alleviate poverty in mountains? *Mountain Research and Development* 26: 4–8.

Yang, M., Hens, L., Ou, X., and De Wulf, R. 2009. Tourism: An alternative to development? *Mountain Research and Development* 29: 75–81

Yonghui, Y., Baiping, Z., Xiaoding, M., and Peng, M. 2006. Large-scale hydroelectric projects and mountain development on the upper Yangtze River. *Mountain Research and Development* 26: 109–114.

Zhou, L., and Liu, K. 2008. Community tourism as practiced in the mountainous Qiang region of Sichuan Province, China: A case study in Zhenghe Village. *Journal of Mountain Science* 5: 140–156.

INDEX

normal lapse rate, 51
North America, geologic provinces, 13*map*
North American Plate, 19
northern hemisphere conifer forests, 187–88
northern hemisphere mountain forests, 187–88
northers, 73
North Ossetia-Alania, natural resources, 354
Nothofagus: El-Niño-Southern Oscillation, 197
    forestline, New Zealand, 200
    predominance, New Zealand, 190
    timberline, southern hemisphere, 194
nuées ardentes, 28
nunatak, 143, 228

O
obsequent fault-line scarp, 32, 33*fig.*
oceans, origin, 21
ocean-to-continent convergence, 20*fig.*
ocean-to-ocean convergence, 20*fig.*
ogive, 112
Okanagan Mountain Park fire, 295
old snow, 88
Olympic marmot *(M. olympus)*, 239, 240*fig.*
Olympic Mountains: deflection, 45, 46
    forced ascent, 60
    funneling, 46*fig.*
    precipitation, 46
    snowline, 94
    treeline, 196*fig.*
Olympus Mons, Mars, 27
Omei Shan, 260
ophiolites, 17–18
ophiolite suites, 18, 24
opium poppy *(Papaver somniferum)*, 311, 311*fig.*
Oregon Coast Range, 169
origins of mountains, 11–39
    Aleutian-type island arc, 21–22
    Andean-type mountain belt, 24
    collisional mountain ranges, 24
    conveyor-belt system, 18–19
    dome mountains, 37
    erosional mountains, 24–26
    faulted mountains, 29–32
    folded mountains, 33–37
    plate tectonics, 16–21
    residual mountains, 26–27
    sea-floor spreading, 16–18
    theories of mountain origin, 15–16
    topographic inversion, 37
    volcanic mountains, 27–29
orogenesis, 12, 35
orogenic belts, 21
orogeny, 127
orographic effect, 46, 58, 59, 349
orographic snowline, 93
orography, 25, 26*fig.*
oroshi, 73
Ostrem, Gunnar, 112
Our Common Future, 334
out-migration, 340
outwash, 114
overdeepening, 119
overharvesting, 327
overthurst fault, 31, 35
overturned folding, 35
Owens Valley, California, 350*fig.*
oxidation, 135
oxisols, 177
oxygen deficit: mountain people, 276–78
    mountain wildlife, 231, 246
Ozarks, Missouri, 37

P
Pachamama (Mother Earth), sacrifice to, 274*fig.*
Pacific Decadal Oscillation (PDO), 197
Pacific Plate, 19
Pacific Ring of Fire, 22, 290
Pacific salmon, 77
Paleozoic age, 12*map*
Paleozoic passive margin, 23*fig.*
Palouse region, Washington state, 119
Pálsson, Sveinn, 110
Pamir, Tajikistan: irrigation from glacial rivers, 349*fig.*
    mixed farming, 320–21
    walnut orchards, 326
Pamir Mountain landslide, 148
Pangerango (Java), 4
Panzaleo (Ecuador), 255
Paracutín ash and cinder cone (Mexico), 27
Pardee, J. T., 120
partnerships for conservation and development, 347–48, 357–58
pashmina and other wool shawls, 289
passive-margin plateaus, 21
passive margins, 21
pastoralism. *See* mountain pastoralism
pastoral nomads, 314–17
paternoster lakes, 119
patterned ground, 140–42
payments for ecosystem services (PES), 357
Peattie, Roderick, 2
Penck, Albrecht and Walther, 129
people in the mountains, 267–300
    acclimatization, 277
    adrêt/ubac slopes, 280
    amenity migration, 273
    atmospheric pressure, 276–78
    barrier effect, 276
    biohazards, 294–96
    change and its impact, 296–97
    characteristics, 6
    circular migration, 272
    commercial flower/plant business, 284
    cultural minorities, 268–69, 296
    developing regions, 268
    drought, 293
    earthquakes, 289–90
    floods, 293–94
    food plants, 284
    forestry, 287
    geological conditions, 275–76
    glaciers, 281–82
    guerilla groups/revolutionaries, 276
    health effects, 276–77
    hill stations, 274
    hunting and gathering, 286
    hypsographic demography, 268
    infectious disease, 294
    integration into larger world, 296
    isolation, 276, 296
    livelihood systems, 285–89
    meteorological hazards, 293
    mining, 286–87
    number of residents, 268, 269
    oxygen deficit, 276–78
    permanent residents, 271–72
    pilgrimage, 274
    plant hunting, 284
    political integration, 296
    population density, 269
    precipitation, 279
    resilience of mountain people, 296
    risk of disaster or accident, 289–96

    seasonal economic migration, 273
    second-home ownership, 274
    semipermanent residents, 272–75
    silviculture, 287
    slope instability, 291–92
    snow and ice, 281
    snow avalanches, 292–93
    sports/tourism, 285
    streams, rivers, and lakes, 282–83
    sunlight, 279–80
    temperature, 278
    time share units, 274
    topography, 276
    trade and artisan livelihoods, 288–89
    transient people, 275
    urbanization, 270–71
    vegetation, 283–84
    volcanoes, 290–91
    wildfires, 295
    wildlife, 284–85
    wind, 280–81
    winter resorts, 278–79. *See also* agricultural settlement and land use
    attitudes toward mountains
perennials, 204, 208
perennial stream, 153
periglacial loess-steppe, 231
periglacial processes, 130
periglacial system, 132
PERMACLIM, 137
permafrost, 135–38
    climate change, 137–38
    continuous *vs.* discontinuous, 135, 136
    defined, 135
    modeling, 136–37
permafrost distribution modeling, 136–37
PERMAKART, 136
PERMAMAP, 136
PERMAMOD, 137
permanent residents, 271–72
permanent snow, 93
PERMEBAL, 137
perpetual spring, 55
Peruvian Andes landslide, 148
Peruvian herders, 312–13
PES, 357
Peter III of Aragon, 257
Petrarch (poet), 258
Phala nomads, 316
Phyllocladus spp., 190
physical weathering, 132–33
Pic du Midi, France, wind speed, 67
piedmont glacier, 106
pika *(Ochotona princeps)*, 235, 235*fig.*
"pile of pancakes" (multistoried lenticular clouds), 62
pilgrimage, 274
pine forests *(Pinus hartwegii)*, 190
Pinus cembra, 187
PlaNet Finance, 358
plantation agriculture, 311
plant hunting, 284
plateau, 3
plate boundary, 21
plate tectonics, 16–21
"Pleistocene grotesque giants," 226
plucking, 113
plunging anticline, 34*fig.*
plunging synclinal valley, 34*fig.*
plunging syncline, 34*fig.*
pocket gopher *(Thomomys* spp.): alpine tundra, 170
    burrowing, 235–36
    winter activities, 226